Youssef Elouerkhaoui

Credit Correlation

Theory and Practice

palgrave
macmillan

Youssef Elouerkhaoui
London, UK

Applied Quantitative Finance
ISBN 978-3-319-60972-0 ISBN 978-3-319-60973-7 (eBook)
DOI 10.1007/978-3-319-60973-7

Library of Congress Control Number: 2017947732

Cover credit: © Aeriform/Alamy Stock Foto
Designer credit: Fatima Jamadar

Printed on acid-free paper

This Palgrave Macmillan imprint is published by Springer Nature
The registered company is Springer International Publishing AG
The registered company address is: Gewerbestrasse 11, 6330 Cham, Switzerland

To my parents.
To my wife and my daughter Marwa

Preface

A quiet revival of credit correlation modelling has been taking place steadily in the background for the last few years. No, you don't need to be alarmed yet: we are not going back to pricing the exotic credit derivative products of yesteryears. But a more profound and meaningful shift in derivatives markets has been unfolding in the aftermath of the credit crisis. The products are now much simpler, transparent and very commoditized, but the management of the derivatives business as a whole is much more complex and involves many more subtleties that were not accounted for before. The pricing now transcends the simplistic risk-free valuation assumptions and involves the juxtaposition of (counterparty) credit risk, funding, liquidity and regulatory capital effects on top of the risk-free value. We do not price a trade in isolation, but it is part of a large portfolio of instruments whose valuation depends on the additional pricing adjustments that make up the all-in-price of the trade.

Nowhere is this more true than in credit derivatives' portfolios. The key risks that one has to deal with are the default correlation risk and the likelihood of credit spread jumps when one or multiple counterparties default. Even for a simple portfolio of vanilla, credit default swaps and credit indices computing the credit valuation adjustment when the counterparty defaults necessitate the modelling of default correlation effects between the single-name default events in this problem. The default correlation assumptions translate directly into wrong-way risk and gap risk effects that one needs to account for and manage accordingly. Similarly, the modelling of systemic risk losses that one would incur when trading with a CCP is similar to modelling a waterfall structure of a cash CDO where the joint defaults of the various clearing members, on the one hand, and the loss cushion provided by the guarantee fund, on the other hand, define the loss variable and the subordination level of this equivalent senior CDO tranche. This is yet another example where default correlation needs to be modelled properly.

With these applications in mind, the objective of this book is primarily to provide the mathematical tools and numerical implementation techniques that are needed to tackle rigourously the credit correlation challenges that we face in the post-crisis XVA modelling era.

Objectives, Audience and Structure

This book is a combination of a lot of the work that I did over the last 15 years on credit correlation modelling and a review of some of the most important mathematical tools and various contributions in the field.

I gave it the title: "Credit Correlation: From FTDs to XVAs—Theory and Practice". As we are now dealing with more XVA modelling issues (such as CVA, FVA, KVA, CCP and SIMM), many of the mathematical tools and techniques that we have to use time and time again stem from the general "Credit Correlation" modelling toolbox. With that in mind, the book is aimed at both students and researchers interested in credit modelling in general, but also XVA modellers who need solid theoretical foundations in credit correlation (and practical numerical implementation methods) that they would then use to address the various XVA (and credit) challenges that they face. By XVA here, I mean not only the micro-trading desk (standard) derivatives valuation adjustments (CVA, DVA and FVA) but also, more generally, macro-modelling issues when dealing with CCPs, systemic risk and regulatory capital.

Who is the intended audience for this book?
This is a book by a practitioner for practitioners, but it is also aimed at Ph.D. and MSc students in financial engineering, and researchers in mathematical finance.

It is based on solid mathematical foundations needed for credit modelling problems, especially point processes and marked point processes (MPPs), filtration enlargements, and intensities with respect to a given filtration. But its emphasis is geared towards applications to real-life problems, covering numerical implementation issues, calibration approaches and pricing and risk-related topics. The material is at an advanced level and assumes that the reader is familiar with the basics of credit modelling covered, for example, in the books by Schönbucher (2003) and O'Kane (2008).

This is not a collection of papers on credit correlation or a general survey of previously published material, but a coherent book on the topic that builds the theory from the ground up and uses it to solve very specific valuation problems. One of the aims of the book is to provide the reader with the theoretical and practical tools to tackle the challenges encountered when modelling CVA for credit portfolios, including general and specific wrong-way risks and gap risk inherently present in the default-enlarged filtrations used in credit markets.

How does it differ from other books in the field?
This book can be viewed as the next stage of the credit correlation evolution—in the post-crisis era. This would be one level of complexity after the standard textbooks by Schönbucher and O'Kane. The logical sequence, both in terms of coverage and advanced topics, would be: Schönbucher (2003) $\rightarrow$ O'Kane (2008) $\rightarrow$ Elouerkhaoui (2017).

What will you learn from this book?
The book by Dominic O'Kane is an excellent reference on standard single-name modelling, Gaussian copula and base correlation for pricing CDOs. The present book builds on that and expands the focus (for the post-crisis era) to cover the (credit) mathematical tools needed for XVA modelling.

Some of the key concepts covered include:

- Filtration enlargements, intensity processes and the generalized Dellacherie formula;
- Poisson shock models a la Marshall-Olkin—both the bottom-up version and the top-down version;
- A special focus on numerical implementation techniques;
- Advanced topics, including: Stochastic Recovery Modelling, Hedging in Incomplete Credit Markets (CR01 vs JTD), and Credit CVA.

I have also included up-to-date material on correlation skew rescaling, expected tranche loss interpolation, relative entropy minimization, CDO-squared and Stein (zero-bias) approximation.

How is this book organized?
Broadly speaking, the book follows a standard organization structure that would mimic the syllabus for a (graduate) course on credit correlation. It is organized in four parts. Part I gives the mathematical tools needed for credit correlation modelling. Part II reviews credit correlation modelling approaches and numerical implementation issues. Part III addresses some advanced credit

correlation topics, including dynamic credit modelling, stochastic recovery and hedging. Part IV gives a prelude to the next credit modelling challenges and describes CVA modelling for credit portfolios.

I start in Chap. 1 by giving the motivation for credit correlation models, namely: (stochastic) spread correlation is not default (event) correlation! This is well known in credit markets, but many XVA modellers do not seem to appreciate the distinction between the two. I give a (chronological) timeline of credit correlation modelling, from the early days of credit with the works of Lando and Duffie-Singleton to dynamic portfolio credit modelling (SPA, GPL, Schonbucher). It is quite fascinating to reflect on how far we have come (over a decade) in our understanding of (subtle) mathematics involved in credit models—many of these subtleties are only limited to credit modelling problems: filtration enlargements, intensity processes, copulas, top-down approaches, etc… I finish the book with a chapter on the next challenge: "CVA modelling". This is just to give a taste of what is coming up next. All the mathematical tools, credit models and concepts that have been exposed over the various chapters in the book culminate in a "deep" understanding of credit correlation, which is then leveraged to address the new XVA challenges of today and tomorrow.

Description of Contents by Chapter

We give below a detailed description of the contents by chapter.

Chapter 1: Introduction and Context. In this introduction, we start by presenting the main (portfolio) credit derivative contracts that we are interested in. To make precise what is meant by default correlation, in the context of credit portfolio modelling, we take a little detour and construct a toy model based on "correlated intensities". Our goal is to use this toy example to motivate the need for proper credit correlation modelling by showing that, ultimately, it only generates some second-order effects, which are not directly related to (proper) joint default events dependence. To fully appreciate the breadth and depth of the topic, we also give a brief timeline of default correlation modelling over the last two decades, which highlights the various pieces in the overall credit correlation puzzle that we will put together over the next few chapters.

Part I: Theoretical Tools

Chapter 2: Mathematical Fundamentals. In this chapter, we present the essential mathematical tools needed in the modelling of portfolio credit derivative products. This includes doubly stochastic Poisson processes, also known as Cox processes; point processes and their intensities, on some given filtration; and copula functions.

Chapter 3: Expectations in the Enlarged Filtration. In this chapter, we derive a formula of the conditional expectation with respect to the enlarged filtration. This is a generalization of the Dellacherie formula. We shall use this key result to compute the expectations that we encounter in the conditional jump diffusion framework. In particular, the conditional survival probability can be computed with our formula. We apply this result in Chap. 4 where conditional survival probability calculations, on the enlarged filtration, are carried out in details.

Chapter 4: Copulas and Conditional Jump Diffusions. Enlarging the economic state-variables' filtration by observing the default process of all available credits has some profound implications on the dynamics of intensities. Indeed, the sudden default of one credit triggers jumps in the spreads of all the other obligors. This is what we refer to as the "Conditional Jump Diffusion" effect. The aim of this chapter is to give a comprehensive and self-contained presentation of the CJD framework. In particular, we derive the default times' density function in the "looping" defaults model, and we study the equivalence between the copula approach and the conditional jumps framework. This is a key result that we will use, in practice, to calibrate non-observable default correlation parameters.

Part II: Correlation Models: Practical Implementation
Chapter 5: Correlation Demystified: A General Overview. This chapter gives a broad overview of default correlation modelling in the context of pricing and risk managing a correlation trading book. We cover both theoretical and practical market aspects, as well as numerical performance issues.

Chapter 6: Correlation Skew: A Black-Scholes Approach. In this chapter, we view the valuation of CDO tranches as an option pricing problem. The pay-off of a CDO tranche is a call-spread on the loss variable. By specifying the distribution of the loss variable at each time horizon, one would be able to value tranches. The standard way of defining this distribution is the base correlation approach. Here, we use a Black-Scholes analogy, and we define an implied volatility for each tranche. Then, given a Black volatility surface, we parameterize the loss distribution with a stochastic CEV model. We show that this parametric form gives a very good fit to the market tranche quotes. In addition, we give an application of the correlation skew Black approach to risk management and hedging.

Chapter 7: An Introduction to the Marshall-Olkin Copula. In this chapter, we study the "Marshall-Olkin" copula model in the context of credit risk modelling. This framework was traditionally used in reliability theory to model the failure rate in multi-component systems. The failure of each component is assumed to be contingent on some independent Poisson shocks. Our aim is to show that MO is a viable alternative to the Gaussian copula. This is done in three steps: (1) we introduce the MO model as the natural extension of a univariate Poisson process, (2) we discuss parametrization and calibration issues, and (3) we compare it with the standard Gaussian copula. We also show that the MO model can be used to reproduce the observed correlation skew in the CDO market. More recently, there has been renewed interest in the MO copula in the context of model risk management (see Morini 2011) and systemic risk modelling (see Gatarek and Jablecki 2016).

Chapter 8: Numerical Tools: Basket Expansions. In the next few chapters, we study some efficient numerical methods for the valuation of large basket credit derivatives. While the approaches are presented in the Marshall-Olkin copula model, most of the numerical techniques are generic and could be used with other copulas as well. The methods presented span a large spectrum of applied mathematics: Fourier transforms, changes of probability measure, numerical stable schemes, high-dimensional Sobol integration and recursive convolution algorithms.

Chapter 9: Static Replication. In principle, the direct (pricing) method requires, for each time step, 2^{n+1} values, corresponding to the set of all possible default combinations; as the size of the underlying basket increases, the number of default configurations explodes exponentially. This significant limitation restricts the applicability of the method to baskets under 10 or 11 credits. As an alternative, we develop a different approach, which is based on a static replication idea.

In this chapter, we describe how this static FTD replication is done: first, we show the relationship between kth-to-default and $(k-1)$th-to-default swaps; then, we apply this recursion step-by-step until we arrive at the complete FTD expansion.

Chapter 10: The Homogeneous Transformation. In general, the number of sub-FTDs in the replication formula is a function of n, the size of the basket, and k, the order of the basket default swap. The most time-consuming step in the evaluation is the generation of the sub-FTDs, for all possible combinations. If we had a homogeneous basket, then, for a given subset size l, all the FTD instruments would have exactly the same value, and the pricing equation would simplify substantially. In particular, the number of sub-FTDs to compute would reduce to one evaluation per l-subset, hence a total of $N(k,n) = k$ FTD evaluations for the whole kth-to-default swap.

The first (natural) approximation that we consider is to transform the original non-homogeneous basket to a homogeneous one while preserving some properties of the aggregate default distribution. In the approach described here, for each default order, we use the corresponding percentile of the aggregate default distribution, and we require that this quantity remains invariant with respect to the homogeneous approximation. We shall see that this transformation is exact for an FTD swap, and that, for higher-order defaults, the approximation gives very good results.

Chapter 11: The Asymptotic Homogeneous Expansion. The transformation, described in the previous chapter, produces a homogeneous portfolio, which mimics some properties of the aggregate default distribution, and can be used for the purposes of basket default swap valuation. By using this homogeneous portfolio, the numerical burden that comes with the pricing of large baskets is eased, and the valuation algorithm is significantly faster. The kth-to-default survival probability $Q_n^{[k]} = \mathbb{P}\big(\tau^{[k]} > T\big)$ for the n-dimensional homogeneous portfolio can be computed recursively as:

$$Q_n^{[k]} = \binom{n}{k-1}\left(Q_{n-1}^{[k-1]} - Q_n^{[k-1]}\right) + Q_n^{[k-1]}.$$

Unfortunately, this simple-looking recursion is numerically unstable. As one moves up the recursion tree, the numerical round-off errors propagate rapidly, and the resulting prices are completely erroneous. To address this issue, we take a different route: rather than using the recursive approach, we study the asymptotic behaviour of the homogeneous portfolio. We show, in this chapter, that the solution $Q_n^{[k]}$ admits an asymptotic series expansion, and we explain how to compute each term in the expansion.

Chapter 12: The Asymptotic Expansion. In this chapter, we relax the homogeneous portfolio assumption, and we derive an asymptotic series expansion of the kth-to-default Q-factor in the non-homogeneous case. We also show how to compute the conditional aggregate default distributions that appear in the expansion using the convolution recursion algorithm. The latter and other recursive methods have been traditionally used in actuarial mathematics to evaluate ruin probabilities and insurance premia.

Chapter 13: CDO-Squared: Correlation of Correlation. In this chapter, we analyse the "correlation of correlation" risk in the Marshall-Olkin copula framework. The valuation of higher-order correlation products such as "CDOs of CDOs" (also known as "CDO-Squared") is mainly driven by correlation of correlation effects. First, we extend the first-to-default replication method to baskets of basket products. Then, we develop an intuitive methodology for analysing this type of structures. The idea is to model each underlying basket security as a single-name process, and to derive its equivalent intensity process and its decomposition on the MO common market factors. This, in turn, defines the multivariate dependence between the underlying basket securities in the portfolio.

Chapter 14: Second Generation Models: From Flat to Correlation Skew. In this chapter, we review some popular correlation skew models. We give a brief description of each model and discuss the advantages and limitations of each modelling framework. This includes the stochastic correlation model, local correlation (and random factor loading), the Levy copula and the implied (hazard rate) copula.

The stochastic and local correlation models are so-called second generation models, which extend the Gaussian copula in an attempt to model tranche prices of the entire capital structure at a fixed time horizon. The Levy copula is also an extension of the Gaussian copula, which accounts for the correlation skew, but it also provides some dynamics for the expected loss process. And last but not least, the implied hazard rate copula is a nonparametric copula function, which constructs the implied distribution of the conditioning factor (and the associated conditional single-name probabilities) from the tranche prices directly.

Chapter 15: Third Generation Models: From Static to Dynamic Models. In this chapter, we review some of the most important dynamic credit models in the literature. We give a brief description of each model and discuss the advantages and limitations of each modelling framework. We also comment on the usefulness of each model for a given family of correlation products. The models discussed include: the Top-Down model (of Giesecke and Goldberg 2005), the Dynamic Generalized Poisson Loss model (of Brigo, Pallavicini, Torresetti 2007), the N+ Model (of Longstaff and Rajan 2008), the Markov Chain Portfolio Loss model (of Schönbucher 2005), and the SPA model (of Sidenius, Piterbarg, Andersen 2008).

Part III: Advanced Topics in Pricing and Risk Management

Chapter 16: Pricing Path-Dependent Credit Products. This chapter addresses the problem of pricing (soft) path-dependent portfolio credit derivatives whose pay-off depends on the loss variable at different time horizons. We review the general theory of copulas and Markov processes, and we establish the link between the copula approach and the Markov-Functional paradigm used in interest rates modelling. Equipped with these theoretical foundations, we show how one can construct a dynamic credit model, which matches the correlation skew at each tenor, by construction, and follows an exogenously specified choice of dynamics. Finally, we discuss the details of the numerical implementation, and we give some pricing examples in this framework.

Chapter 17: Hedging in Incomplete Markets. In this chapter, we present a methodology for hedging basket credit derivatives with single-name instruments. Because of the market incompleteness due to the residual correlation risk, perfect replication cannot be achieved. We allow for mean self-financing strategies and use a risk-minimization criterion to find the hedge. Managing credit risk is always a fine balance between hedging the jump-to-default exposure (JTD) or the credit spread exposure (CR01). Recently, this credit hedging dilemma (JTD vs CR01) is becoming very topical in the context of managing counterparty credit risk for large derivatives books.

Chapter 18: Min-Variance Hedging with Carry. In this chapter, we present the construction of the Min-Variance Hedging Delta operator used for basket products. Because of the market incompleteness –i.e. we cannot replicate a basket product with its underlying default swaps– min-variance hedging is the best thing that we can hope for. There will always be a residual correlation risk orthogonal to the sub-space of hedging instruments. We also present an extension of the standard MVH optimization to take into account the drift mismatch between the basket and the hedging portfolio. This defines the "Min-Variance Hedging Deltas with Carry".

Chapter 19: Correlation Calibration with Stochastic Recovery. In this chapter, we expand the base correlation framework by enriching it with Stochastic Recovery modelling as a way to address the model limitations observed in a distressed credit environment. We introduce the general class of conditional-functional recovery models, which specify the recovery rate as a function of the common conditioning factor of the Gaussian copula. Then, we review some of the most popular ones, such as: the Conditional Discrete model of Krekel (2008), the Conditional Gaussian of Andersen and Sidenius (2005) and the Conditional Mark-Down of Amraoui and Hitier (2008). We also look at stochastic recovery from an aggregate portfolio perspective and present a top-down specification of the problem. By establishing the equivalence between these two approaches, we show that the latter can provide a useful tool for analysing the structure of various stochastic recovery model assumptions.

Part IV: The Next Challenge

Chapter 20: New Frontiers in Credit Modelling: the CVA Challenge. In this chapter, we present a general framework for evaluating the (counterparty) credit valuation adjustment for CDO tranches. We shall see that given the "exotic" nature of the CVA derivative pay-off, we will have to leverage a variety of modelling techniques that have been developed over years for the correlation book; this includes default correlation modelling, credit index options' pricing, dynamic credit modelling and CDO-squared pricing.

References

L. Andersen, J. Sidenius, Extensions to the Gaussian Copula: Random recovery and random factor loadings. J. Credit Risk. **1**(1, Winter 2004/2005), 29–70 (2005)

S. Amraoui, S. Hitier, Optimal stochastic recovery for base correlation (Working Paper, BNP Paribas, 2008)

D. Brigo, A. Pallavicini, R. Torresetti, Calibration of CDO tranches with the dynamical generalized-Poisson Loss model. Risk (2007)

Y. Elouerkhaoui, *Credit Correlation: Theory and Practice* (Palgrave, London, 2017)

D. Gatarek, J. Jablecki, Modeling joint defaults in correlation-sensitive instruments. J. Credit Risk **12**(3) (2016)

K. Giesecke, L. Goldberg, A top down approach to multi-name credit (Working Paper, 2005)

M. Krekel, Pricing distressed CDOs with base correlation and stochastic recovery (Working Paper, Unicredit, 2008)

F. Longstaff, A. Rajan, An empirical analysis of the pricing of collateralized debt obligations. J. Finance. **63**(2), 529–563 (2008)

M. Morini, *Understanding and Managing Model Risk: A Practical Guide for Quants, Traders and Validators* (Wiley, Chichester, 2011)

D. O'Kane, *Modelling Single-Name and Multi-Name Credit Derivatives* (Wiley, Chichester, 2008)

P.J. Schönbucher, *Credit Derivatives Pricing Models: Models, Pricing and Implementation* (Wiley, Chichester, 2003)

P.J. Schönbucher, Portfolio losses and the term structure of loss transition rates: A new methodology for the pricing of portfolio credit derivatives (Working Paper, ETH Zurich, 2005)

J. Sidenius, V. Piterbarg, L. Andersen, A new framework for dynamic credit portfolio loss modeling. Int. J. Theor. Appl. Financ. **11**(2), 163–197 (2008)

Acknowledgements

There is a long list of people that I need to thank who have contributed directly or indirectly to making this book a reality.

First of all, from my years in academia, I would like to thank Monique Jeanblanc, Agnes Sulem, Rudiger Frey, Tomasz Bielecki, Nizar Touzi and Jean-Paul Laurent. They have contributed to a large extent in shaping my mathematical rigour and discipline when addressing the subtleties involved in credit default modelling.

I am grateful to my current and former colleagues at Citigroup: Mickey Bhatia, Vikram Prasad, Carey Lathrop, Sankar L, Arvind Rajan, Karim Berradi, Havard Sandvik, Giri Banigallapati, Sam Phillips, David Shelton, Martin Johansson and Alex Jackson. And I am also grateful to my former colleagues at UBS: Nick Jeffryes, Adam Johnson, Richard Whittle, Mike Connor, Dave Friedman, Neil Smout and Trevor Chilton. Many of the ideas developed in the book have been influenced by various discussions, projects and collaborations that we have had over the years.

I would like to acknowledge the contributions of my fellow researchers in the quant community with whom I have interacted over the years during conferences and seminars and who influenced many of the ideas developed in the book: Alex Lipton, Leif Andersen, Paul Glasserman, Dariusz Gatarek, Darrell Duffie, Damiano Brigo, Massimo Morini, Rama Cont, Luca Capriotti and Andrea Pallavicini.

A deep amount of gratitude goes to my parents and my family for their love and support over the years. And a big thank you to my wife and my daughter Marwa for putting up with me during the evenings and long weekends that I had to spend away from them while working on this book project. I can never thank you enough.

Contents

List of Figures

1

Introduction and Context

To set the context, we start this introduction with a presentation of the main (portfolio) credit derivative contracts that we are interested in. When we talk about portfolio credit derivative valuations, the first thing that we need to do is to generate a set of loss (or default) distributions, at different time horizons, from the single-name curves and some "correlation" assumptions. To make precise what is really meant by correlation, in the context of credit portfolio modelling, we take a little detour and construct a toy model based on "correlated intensities". Our goal here is to use this toy example to motivate the need for proper credit correlation modelling by showing that, ultimately, it only generates some noisy second-order effects, which are not really directly related to (proper) joint default events dependence. The conclusion from this exercise is: yes, default correlation models are needed. And it has taken a very long journey to get to where we are today in terms of credit correlation modelling expertise. To fully appreciate the breadth and depth of the topic, we shall give a brief timeline of default correlation modelling over the last two decades. It highlights the various pieces in the overall credit correlation puzzle that we will put together over the next few chapters. Enjoy the journey!

1.1 Synopsis of Credit Derivative Products

In this section, we give a brief description of the key credit derivative products traded in the market, namely: Credit Default Swaps, First-to-Default Swaps and Collateralized Debt Obligations.

Y. Elouerkhaoui, *Credit Correlation*, Applied Quantitative Finance,
DOI 10.1007/978-3-319-60973-7_1

1.1.1 Credit Default Swaps

A Credit Default Swap (CDS) is a derivative contract, similar to an insurance, which offers protection against the default of a given reference entity or a specific reference bond. The protection buyer pays a premium (or a fee) either upfront, at contract inception, or on a regular fixed schedule, to the protection seller until maturity of the contract or default of the reference entity, whichever happens first. In return, in the event of default, the protection seller pays the protection buyer a fixed amount equal to the notional of the swap minus the recovery value of the reference bond, thereby making the protection buyer whole.

Mathematically, the PV of a swap is given by the difference between the PV of the protection leg V_{premium}, and the PV of the premium leg $V_{\text{protection}}$.

Let us consider a credit default swap, maturing at time T, with payment schedule $\left(T_1, T_2, ..., T_{n_T} = T\right)$, daycount fractions (δ_i), and premium S. We denote by τ the default time of the reference entity; $D_T = \mathbf{1}_{\{\tau \leq T\}}$ is the default indicator; R is the recovery rate; and N is the trade notional. Then, we can write the PV of the CDS as the expectation of the trade discounted cash-flows:

$$V_{\text{premium}} = \mathbb{E}\left[N \sum_{i=1}^{n_T} p_{0,T_i} \left(1 - D_{T_i}\right) S \cdot \delta_i\right],$$

$$V_{\text{protection}} = \mathbb{E}\left[N \int_0^T p_{0,t} \left(1 - R\right) \cdot dD_t\right];$$

where $p_{0,T}$ is the risk-free discount factor maturing at time T.

Accrued in default. In practice, most credit default swaps trade with an accrued-in-default payment feature: if the default occurs during a coupon period, then the cash-flow payment that has been accrued up until the time of default time is included in the default payment. This gives the clean PV of the CDS contract.

Digital CDS. There are also Digital or Fixed Recovery CDS contracts where the payment at default is agreed beforehand as a fixed (digital) amount, which does not depend on the actual recovery value of the underlying reference entity. The combination of regular CDSs and fixed recovery CDSs can be used to bootstrap both the default probabilities and the recovery rates for a given reference entity.

Credit Linked Notes. Credit Default Swaps can be traded in a funded format as Credit Linked Notes, where the investor buys a defaultable Floating Rate Note (FRN), which pays a set of regular floating cash-flows, on the payment schedule dates—which are equal to a floating Libor rate plus a fixed spread-, and pays back the notional of the note at maturity conditional on survival of the reference entity. If a default event occurs, then all coupon payments stop, and a fraction of the notional payment, defined by the recovery rate of the defaulted entity, is paid back at the default termination date.

SNAC Convention. It is worth noting that after the CDS big-bang convention changes in 2010, most CDS do not trade at par anymore. Instead, a set of pre-defined regular coupons (100 bps for regular investment grade credit, and 500 bps for wider high yield credits) is applied to all CDS contract, and an upfront payment (to be exchanged between the two counterparties) is computed. This is referred to as the Standard North American Contract (SNAC) convention. CDS contracts are marked on an upfront basis, and a standard (ISDA) calculator is used to convert the upfronts to spreads and vice-versa (see, for example, Beumee et al. (2009) for more details).

1.1.2 First to Default Swaps

A First-to-Default Swap (FTD) or more generally an nth-to-Default Swap (NTD) is a credit default swap, which references a basket of underlying single-name credits. The credit event in this case is linked to the time of the first to default entity (or nth-to-default respectively). For an FTD, the protection buyer pays the premium on a regular schedule until maturity or the time of the first defaulting entity, i.e., if any single-name in the reference basket defaults, the swap terminates. On the protection leg side, at the first-to-default termination date, the counterparty makes a payment equal to the notional minus the recovery value of the reference single-name, which defaulted first. Similarly, for an kth-to-default swap, the credit event is defined as the (stopping) time when k underlying single-name credits in the basket have defaulted. The payment to be made by the protection seller, at the termination date, is equal to the notional minus the recovery value of the kth-defaulted entity, i.e., the one which defaulted last on the termination date.

Suppose we have a basket of n reference entities with default times $(\tau_1, ..., \tau_n)$, recovery rates $(R_1, ..., R_n)$, and notionals $(N_1, ..., N_n)$. We denote by $D_T^i = \mathbf{1}_{\{\tau_i \leq T\}}$, for $1 \leq i \leq n$, the default indicators of each obligor. We define the sequence of ordered default times $\left(\tau^{[1]}, \tau^{[2]}, ..., \tau^{[n]}\right)$, where $\tau^{[1]}$ is the time of the first-to-default event, $\tau^{[2]}$ is the time of the second-to-default and so forth:

$$\tau^{[1]} = \min_{1 \le i \le n} (\tau_i);$$
$$\tau^{[k]} = \min_{1 \le i \le n} \left(\tau_i : \tau_i > \tau^{[k-1]}\right), \text{ for } 2 \le k \le n.$$

The default indicators of the ordered default times are denoted by $D_T^{\left(\tau^{[k]}\right)} = \mathbf{1}_{\{\tau^{[k]} \le T\}}$, for $1 \le k \le n$.

Then, the PV of a first-to-default swap or more generally a kth -to-default swap is given by

$$V_{\text{premium}}^{[k]} = \left(\sum_{i=1}^{n} N_i\right) \mathbb{E}\left[\sum_{j=1}^{n_T} p_{0,T_j}\left(1 - D_{T_j}^{\left(\tau^{[k]}\right)}\right) S \cdot \delta_j\right];$$
$$V_{\text{protection}}^{[k]} = \sum_{i=1}^{n} \mathbb{E}\left[N_i \int_0^T p_{0,t}\left(1 - R_i\right) \cdot \left(1 - D_t^{\left(\tau^{[k]}\right)}\right) dD_t^i\right];$$

Essentially, this can be computed by generating the basket default distribution at different time horizons $T \ge 0$:

$$\mathbb{P}\left(D_T = k\right), \text{ for } 0 \le k \le n,$$

where the aggregate default counter is defined as

$$D_T = \sum_{i=1}^{n} D_T^i.$$

Thus, we can write the kth-to-default survival probabilities and the single-name kth-to-default densities as

$$\mathbb{E}\left[1 - D_T^{\left(\tau^{[k]}\right)}\right] = \mathbb{P}\left(D_T < k\right);$$
$$\mathbb{E}\left[\left(1 - D_T^{\left(\tau^{[k]}\right)}\right) dD_T^i\right] = \mathbb{E}\left[\mathbf{1}_{\{D_T - D_T^i = k-1\}}\left(1 - D_T^i\right) dD_T^i\right];$$

Those are plugged back into the formulas of the premium leg and protection leg PVs.

1.1.3 Collateralized Debt Obligations

Collateralized Debt Obligations (CDOs) are portfolio credit derivatives whose payoff is specified by the aggregate portfolio loss variable as opposed to the number of defaults (as with FTDs and NTDs). A CDO tranche is defined by an attachment point and a detachment point, which determine the range of losses that will impact the derivative contract payments. On the protection leg of the swap, the protection seller pays all the portfolio losses that fall within the attachment-detachment points' interval. The payoff is a call spread on the loss variable, where the first and second strikes are given by the attachment and detachment points. In other words, the protection seller is not liable for any losses below the attachment point or above the detachment point; but contractually, he must pay any incremental losses above the attachment point as long as they do not exceed the losses cap set by the detachment point. On the other hand, the protection buyer needs to pay a regular premium stream on the amortizing tranche notional, which decreases in line with the losses incurred on the tranche.

We talk about Synthetic CDOs when the underlying portfolio references a set of credit default swaps; sometimes they are also referred to as CSOs (Collateralized Synthetic Obligations). On the hand, when the portfolio contains a set of cash instruments (such as bonds or loans), we talk about Cash CDOs, which we refer to, more specifically, as CBOs (Collateralized Bond Obligations) when the underlyings are bonds, or CLOs (Collateralized Loan Obligations) when the underlyings are loans. In the case of Cash CDOs, the payoff can be much more complex as it originates from a waterfall structure, where the CDO payments are defined across the whole capital structure concurrently, with different priority rules between the tranches and a variety of OC/IC triggers, which could impact the cash-flows that are made to the various investors.

The aggregate loss variable for the portfolio $\Pi = (\tau_1, \ldots, \tau_n)$, normalized by the total portfolio notional, at time horizon T, is defined as

$$L_T = \frac{1}{\sum_{i=1}^{n} N_i} \left[\sum_{i=1}^{n} N_i \left(1 - R_i\right) D_T^i \right];$$

the corresponding normalized loss variable $M_T^{(\alpha,\beta)}$, for a tranche with attachment and detachment points α and β respectively, is given by the call-spread payoff on the portfolio loss variable:

$$
\begin{aligned}
M_T^{(\alpha,\beta)} &= \frac{1}{\beta-\alpha} \min\left(\max\left(L_T-\alpha\right), \beta-\alpha\right) \\
&= \frac{1}{\beta-\alpha}\left[(L_T-\beta)^+ - (L_T-\alpha)^+\right].
\end{aligned}
$$

The PV of the CDO tranche, which pays a premium S on the dates (T_j) is given by the difference between the premium PV, $V_{\text{premium}}^{[\alpha,\beta]}$, and the loss protection PV, $V_{\text{protection}}^{[\alpha,\beta]}$.

For the premium leg, the tranche notional amortizes with the realized losses, which gives the payoff

$$
V_{\text{premium}}^{[\alpha,\beta]} = \mathbb{E}\left[\sum_{j=1}^{n_T} p_{0,T_j}\left(1 - M_{T_j}^{(\alpha,\beta)}\right) S \cdot \delta_j\right];
$$

For the protection leg, we make a payment every time a loss is incurred on the tranche. Since, the portfolio loss process L_t and the tranche loss process $M_t^{(\alpha,\beta)}$ are pure jump processes. The CDO loss protection cash-flows correspond exactly to the increments of the tranche process $M_t^{(\alpha,\beta)}$, i.e., there is a payment when the process $M_t^{(\alpha,\beta)}$ jumps, which happens at every default time. Thus, the payoff of the protection leg is defined by the Stieljes integral

$$
V_{\text{protection}}^{[\alpha,\beta]} = \mathbb{E}\left[\int_{]0,T]} p_{0,t}\, dM_t^{(\alpha,\beta)}\right].
$$

Using the integration by parts formula and Fubini's theorem to interchange the order of integration, we can re-write the loss protection integral as

$$
V_{\text{protection}}^{[\alpha,\beta]} = p_{0,T}\mathbb{E}\left[M_T^{(\alpha,\beta)}\right] - \int_0^T \frac{\partial p_{0,t}}{\partial t}\mathbb{E}\left[M_t^{(\alpha,\beta)}\right] dt,
$$

where we have assumed independence between interest rates and defaults. Therefore, the pricing of CDOs boils down to generating an expected tranche loss curve at different time horizons $T \geq 0$

$$
\mathbb{E}\left[M_T^{(\alpha,\beta)}\right], \quad \text{for all } T \geq 0.
$$

If we know the density function of the portfolio loss variable, $\mathbb{P}(L_T \in dx)$, for all $T \geq 0$, the tranche loss expectation is given by $\mathbb{E}\left[M_T^{(\alpha,\beta)}\right] = \int_\alpha^\beta x\mathbb{P}(L_T \in dx)$.

We shall see in next few chapters how to specify the default correlation dependence between the individual single-name default indicators, and the type of numerical methods to use to generate the basket default distributions efficiently; this, in turn, gives the expected tranche loss curve (for CDOs), or the probability of the kth-to-default event (for NTDs), which feed into the pricing formula.

1.2 Motivation for Credit Correlation Models

To motivate the need for "proper" credit correlation models, we make a little detour and review the "correlated intensities" approach discussed by Duffie (1998) in the context of pricing First-to-Default baskets.

We have a basket of n default times $\Pi = (\tau_1, ..., \tau_n)$, with (stochastic) intensity processes $(\lambda_1, ..., \lambda_n)$. Without going into too many technical details at this stage, we say that λ is an intensity process for a stopping time τ if the compensated default process defined as

$$D_t - \int_0^t (1 - D_s)\,\lambda_s ds, \quad \text{for } t \geq 0,$$

is a martingale. Intuitively, it is the instantaneous default probability conditional on survival up to time t. Indeed, we can write

$$\mathbb{P}(\tau \in [t, t + dt) \,|\mathcal{G}_t) = \mathbb{E}\left[\lambda_t \,|\mathcal{G}_t\right] dt, \quad \text{for } t \geq \tau.$$

This can be thought of as a generalization of the intensity rate for a Poisson process.

We shall make two important assumptions and use two important results from Duffie (1998).

Assumption 1 We assume that we do not have instantaneous joint defaults, i.e., for all $i \neq j$, we have $\mathbb{P}\left(\tau_i = \tau_j\right) = 0$.

Lemma 1 (First-to-Default Intensity) *Under this assumption, we can show that the intensity process of the first-to-default time,* $\tau^{[1]} = \min\limits_{1\leq i\leq n} (\tau_i)$, *is simply given by the sum of the individual intensities.*

Proof We denote by M_t^i the compensated single-name default process:

$$M_t^i \triangleq D_t^i - \int_0^t \left(1 - D_s^i\right) \lambda_s^i ds, \text{ for } t \geq 0;$$

we define the equivalent process M_t for the first-to-default indicator $D_t^{(\tau^{[1]})} = \mathbf{1}_{\{\tau^{[1]} \leq t\}}$

$$M_t \triangleq D_t^{(\tau^{[1]})} - \int_0^t \left(1 - D_s^{(\tau^{[1]})}\right) \lambda_s ds, \text{ for } t \geq 0,$$

where λ is the sum of the single-name intensities, $\lambda_t = \sum_{i=1}^n \lambda_t^i$, and we want to prove that it is also a martingale. To this end, it suffices to observe that

$$D_t^{(\tau^{[1]})} = 1 - \prod_{i=1}^n \left(1 - D_t^i\right),$$

and to differentiate. Since we have assumed no-instantaneous joint defaults, we can write the differential as

$$\begin{aligned} dD_t^{(\tau^{[1]})} &= \sum_{i=1}^n dD_t^i \left[\prod_{j \neq i} \left(1 - D_t^j\right)\right] \\ &= \sum_{i=1}^n dM_t^i \left[\prod_{j \neq i} \left(1 - D_t^j\right)\right] + \sum_{i=1}^n \left[\prod_{j=1}^n \left(1 - D_t^j\right)\right] \lambda_t^i dt, \end{aligned}$$

which gives

$$dD_t^{(\tau^{[1]})} - \left(1 - D_t^{(\tau^{[1]})}\right) \left[\sum_{i=1}^n \lambda_t^i\right] dt = \sum_{i=1}^n dM_t^i \left[\prod_{j \neq i} \left(1 - D_t^j\right)\right] = dM_t;$$

the process $M_t = \int_0^t \sum_{i=1}^n dM_s^i \left[\prod_{j \neq i} \left(1 - D_s^j\right)\right]$ is also martingale (as all the products $\prod_{j \neq i} \left(1 - D_s^j\right)$ in the integrand are bounded by 1). ■

The second assumption that we need is the no-jump at default assumption.

Assumption 2 We assume that the first-to-default (conditional) survival probability does not jump at the time of default.

More formally, we define the jump process for a semi-martingale Y as $\Delta Y_t = Y_t - \lim_{s \uparrow t} Y_s$; and we have the following proposition.

Proposition 2 (Conditional Survival Probability). *Suppose we have a stopping time τ with a bounded intensity process λ. Fix a time horizon T, and let*

$$Y_t = \mathbb{E}\left[\exp\left(-\int_t^T \lambda_s ds\right) |\mathcal{G}_t\right], \ \textit{for all } t \leq T.$$

If we assume that the jump variable ΔY_τ is equal to zero almost surely, then the conditional survival probability is given by

$$\mathbb{P}\left(\tau > T\,|\mathcal{G}_t\right) = Y_t, \ \textit{for } t < \tau,$$

almost surely.

Proof Define the martingale Z as

$$Z_t = \mathbb{E}\left[\exp\left(-\int_0^T \lambda_s ds\right) |\mathcal{G}_t\right], \ \text{for all } t \leq T.$$

Expressing the process Y in terms of the martingale Z, and applying Ito's lemma, we get

$$dY_t = d\left[\exp\left(\int_0^t \lambda_s ds\right) Z_t\right] = \lambda_t Y_t dt + \exp\left(\int_0^t \lambda_s ds\right) dZ_t.$$

On the other hand, since λ is an intensity process for the stopping time τ, we have, by definition,

$$dD_t = (1 - D_t)\,\lambda_t^i dt + dM_t,$$

where M is a martingale. A second application of Ito's lemma (for jump processes) to the process U_t defined as the product $U_t = (1 - D_t)\,Y_t$ yields

$$dU_t = -Y_{t^-} dD_t + (1 - D_{t^-})\,dY_t - \Delta Y_t \Delta D_t.$$

Invoking the no-jump at default assumption, $\Delta Y_\tau = 0$, we know that the cross (jump) term is null, $\Delta Y_t \Delta D_t = 0$. Substituting the expressions of dY_t

and dD_t, the drift terms cancel out, and we are left with

$$dU_t = -Y_{t^-} dM_t + (1 - D_{t^-}) \exp\left(\int_0^t \lambda_s ds \right) dZ_t.$$

Hence, U is also a martingale with terminal condition $U_T = 1 - D_T$. Thus, we can write, on the set $\{t < \tau\}$,

$$Y_t = U_t = \mathbb{E}\left[U_T \,|\mathcal{G}_t \right] = \mathbb{E}\left[1 - D_T \,|\mathcal{G}_t \right] = \mathbb{P}\left(\tau > T \,|\mathcal{G}_t \right).$$

■

Now, we apply the result to the first-to-default (stopping) time $\tau^{[1]}$. Under assumptions 1 and 2, the FTD conditional survival probability is simply given by

$$\mathbb{P}\left(\tau^{[1]} > T \,|\mathcal{G}_t \right) = \mathbb{E}\left[\exp\left(-\int_t^T \left[\sum_{i=1}^n \lambda_s^i \right] ds \right) |\mathcal{G}_t \right], \quad \text{for all } \tau^{[1]} < t \leq T.$$

To include an interdependence (correlation) effect between the single-name events, we assume that the individual intensities λ^i follow some correlated diffusion dynamics. But it is important to note, at this junction, that the "correlations" built in this manner are "intensity-correlations" and not "default-events correlations" per se. And there is a big difference between the two concepts.

From the expression of the FTD survival probabilities, it can be seen that the effect of intensity correlations here is, in many ways, similar to a convexity (or a quanto) adjustment of some sort. Actually, numerically, it does imply a slight level of default correlation, but if we look closer, we realize that it is nothing more than an illusion. The effect itself is negligible—unless we have extremely high spreads or very long maturities, it would barely be noticeable.

As we mentioned earlier, default correlations and intensity correlations are two different things. In the former, we are looking at the actual default events themselves and estimating the likelihood of joint defaults. In the latter, we are considering the regular (observable) diffusion of "credit spreads" over time. Making the assumption that: (a) joint defaults cannot occur simultaneously, and (b) that the basket intensities (defined on the enlarged filtration $\mathcal{G}$) do not jump when one of the names defaults, would, by construction, kill any form of (genuine) default correlation. The "convexity adjustment" noise generated from the correlated diffusions is clearly not the answer.

We shall illustrate this point by choosing Gaussian HJM dynamics for the single-name intensities and estimating the (model) implied default correlations. We only use Gaussian dynamics for this illustration because they give simple closed form expressions that we can conveniently work with. But the Gaussian HJM dynamics suffer from the usual limitations: explosive process at long maturities and negative rates for high volatilities

We denote by $h^i_{t,T}$ the forward intensity for issuer i; the spot (instantaneous) intensity is given by $\lambda^i_t = h^i_{t,t}$. Thus, we can write the single-name survival probabilities $Q^i_{t,T}$ as: for $\tau_i < t \le T$,

$$Q^i_{t,T} = \mathbb{P}(\tau_i > T \,|\mathcal{G}_t) = \mathbb{E}\left[\exp\left(-\int_t^T \lambda^i_s ds\right)|\mathcal{G}_t\right] = \exp\left(-\int_t^T h^i_{t,s} ds\right).$$

Under the spot measure, we assume Gaussian HJM dynamics

$$dh^i_{t,T} = (\sigma_i(t,T) \cdot \Sigma_i(t,T))\, dt + \sigma_i(t,T) \cdot dW^i_t,$$

where $\Sigma_i(t,T) = \int_t^T \sigma_i(t,s)$ is the cumulative volatility function, W is an n-dimensional Brownian motion, and ρ^λ_{ij} are the single-name intensity correlations $\rho^\lambda_{ij} dt = \left\langle dW^i_t, dW^j_t \right\rangle$.

Note that we have excluded any potential default-induced jumps from the dynamics. This is because of the no-jump at default assumption that we have made; although all these quantities are defined with respect to the large filtration $\mathbb{G}$, for the purposes of this illustration exercise, we have assumed that jumps at default effects are excluded.

Proposition 3 (FTD Survival Probability). *Under the assumptions of this section, the expression of the FTD survival probability is given by:*

$$Q^{(\tau^{[1]})}_{0,T} = \left[\prod_{i=1}^n Q^i_{0,T}\right] \exp\left(\sum_{i<j} \int_0^T \rho^\lambda_{ij} \Sigma_i(s,T)\, \Sigma_j(t,T)\, ds\right).$$

Proof Under assumptions 1 and 2, the FTD survival probability, at time 0, is

$$Q^{(\tau^{[1]})}_{0,T} = \mathbb{E}\left[\exp\left(-\int_t^T \left[\sum_{i=1}^n \lambda^i_s\right] ds\right)\right].$$

The expression of the single-name intensities λ^i are given by integrating the SDE above, which we sum up to get the FTD intensity

$$\sum_{i=1}^{n} \lambda_t^i = \sum_{i=1}^{n} h_{0,t}^i + \sum_{i=1}^{n} \int_0^t (\sigma_i(s,t) \cdot \Sigma_i(s,t))\, ds + \int_0^t \sigma_i(s,t) \cdot dW_s^i.$$

The FTD survival probability then becomes

$$Q_{0,T}^{(\tau^{[1]})} = \left[\prod_{i=1}^{n} Q_{0,T}^i\right] \exp\left(-\int_0^T \left(\sum_{i=1}^{n} \int_0^t (\sigma_i(s,t) \cdot \Sigma_i(s,t))\, ds\right) dt\right)$$
$$\times \mathbb{E}\left[\exp\left(-\int_0^T \int_0^t \left(\sum_{i=1}^{n} \sigma_i(s,t)\right) \cdot dW_s^i dt\right)\right].$$

Using Fubini to exchange the order of integration, and the Gaussian expectation $\mathbb{E}\left[e^X\right] = e^{\mu + \frac{v}{2}}$, where μ and v are the mean and variance of the normal variable X, yields the result in the proposition. ■

As a corollary, we obtain the formula for the (implied) pairwise default correlations in this set-up.

Corollary 1 *The pairwise default correlations ρ_{ij}^D, over a time horizon T, are given by:*

$$\rho_{ij}^D = \frac{covar\left(D_T^i, D_T^j\right)}{\sqrt{var\left(D_T^i\right)}\sqrt{var\left(D_T^j\right)}} = \frac{Q_{0,T}^i Q_{0,T}^j \left(\exp\left(C_{ij}\right) - 1\right)}{\sqrt{Q_{0,T}^i\left(1 - Q_{0,T}^i\right) Q_{0,T}^j\left(1 - Q_{0,T}^j\right)}},$$

where the term C_{ij} is defined as

$$C_{ij} = \int_0^T \rho_{ij}^\lambda \Sigma_i(s,T)\, \Sigma_i(s,T)\, ds$$

With flat normal volatilities, $\Sigma_i(s,T) = \sigma_i(T-s)$, the cross-correlation term C_{ij} is given by

$$C_{ij} = \rho_{ij}^\lambda \sigma_i \sigma_j \frac{T^3}{3}.$$

Assuming "log-normal" (equivalent) volatilities of 50%, we set our normal spot volatilities to

$$\sigma_i = 0.5 \times h^i_{0,T}.$$

Now, we set the pairwise intensity correlations to their maximum values $\rho^{\lambda}_{ij} = 1$, and we analyze the levels of default correlations obtained as we vary spreads and maturities.

	1Y (%)	5Y (%)	10Y (%)	15Y (%)
10 bps	0.008	0.208	0.829	1.861
50 bps	0.042	1.029	4.068	9.060
100 bps	0.083	2.033	7.957	17.626
200 bps	0.165	3.970	15.309	34.034
500 bps	0.406	9.289	35.705	91.321
1000 bps	0.793	16.923	75.714	449.537

The first observation to be made is that, some numbers in this table, with high spreads and long maturities, do not look remotely similar to a correlation number. The reason is very simple: we have used a Gaussian diffusion of the intensities therefore there is a non-negligible probability of having (unrealistic) negative intensities. For illustration purposes, we have used quite extreme intensity correlations and volatility numbers. But under these extreme scenarios, when spreads and maturities are high, the FTD survival probability factors $Q^{(\tau^{[1]})}_{0,T}$, are not (valid) probabilities anymore (since their values would exceed 1).

If we analyze more reasonable levels, at 5 years, for example, we can see that, even under these extreme choices of intensity correlations and volatilities, the default correlation, for the 5-year tenor, can only get as high as 10% for an intensity level of 500 bps. Again, this is just an artefact of the convexity adjustment (noise) and not a real default correlation effect.

1.3 A Timeline of Credit Correlation Modelling

In this section, we give a brief historical perspective on the modelling of credit default correlation.

Pre-1998—Structural Models vs Reduced Form Models. Credit modelling was still at its infancy; most of the focus was on single-name default models (e.g., Jarrow and Turnbull (1995), Jarrow et al. (1997)) as opposed to correlation modelling. Although the first CDO structure (BISTRO—Broad

Index Secured Trust Offering) was issued by JP Morgan in 1997, the focus during this period was firmly on modelling the individual firm's default events in isolation. After some back and forth between structural and reduced-form approaches, reduced form models have emerged as the likely winner, and the next phase in modelling credit baskets on aggregate could start based on solid theoretical grounds in single-name modelling.

Lando (1998)—**Cox Process Approach**. David Lando, at the time, was a Ph.D. student (under Robert Jarrow) at Cornell University. His Ph.D. dissertation (Lando 1994) was about credit default risk modelling. In the third chapter of his dissertation (which was published under the title "On Cox Processes and Credit Risky Securities" in the Review of Derivatives Research), using a Cox process construction of the default time, he clarified many of the main building blocks needed to price any (single-name) credit derivative product. At the time, there was still a lot of competition between the structural approach and the reduced form approach. With supporters and opponents of each approach arguing the merits of their favourite choice, Lando (1998) has definitely helped make the (intensity) reduced-form popular with both practitioners and academics alike. The simplicity and the rigour of his approach was very appealing; and the resulting formulas were very elegant and easy to understand.

Duffie (1998)—**First-to-Default Baskets**. Darrell Duffie (Dean Witter Distinguished Professor of Finance at Stanford University) wrote a series of important papers (with his co-author, Kenneth Singleton) on basket credit modelling. This was many years before anyone else even appreciated the need for a rigorous treatment of default events' correlation. They laid out the ground work for modelling FTD baskets by defining rigorously the intensity process of a general stopping time, and they constructed the multivariate dependence with a multiple common shocks model. They also covered many subtle implementation details either using simulation algorithms or analytical approximations. Much later, this was studied and extended further by many authors, in the context of Marshall-Olkin copula modelling, but Duffie's contributions were undoubtedly groundbreaking at the time.

Li (2000)—**Gaussian Copula**. David Li, a statistician by training from the University of Waterloo specializing in actuarial mathematics, was the first to introduce the copula concept, formally, to the world of credit derivatives pricing. Up to that point, a number of Merton structural models had been used, where the Gaussian copula was used implicitly; but the reduced-form single-name credit curve constructions were not in line with the overall approach. Li, being very familiar with copulas used in insurance contracts to model joint mortality rates, was quick to spot the resemblance between an insurance policy and a first-to-default swap and resorted to copulas to define the joint default dependence in a mathematically sound framework. The influence of actuarial mathematics could be noticed in his original paper even in the non-standard notations he had used to denote conditional probabilities.

JP Morgan (2004)—Base Correlation. The credit strategists at JP Morgan (2004) wrote a research note on default correlation where they publicized the base correlation concept. Prior to that, many dealers were still using a set of separate "compound" correlations for each tranche, which allowed then to mark the individual tranches correctly; but there was no mathematical consistency between the tranches when bundled together (especially, for full-capital structure CDOs). Base correlation addressed the main shortcomings of compound correlations: by decomposing each tranche, in the capital structure, as the difference between two equity tranches and then marking the equity correlations, the base correlation curve offered a way out. It is always consistent; and therefore can be used to interpolate or extrapolate the pricing for other tranches and other portfolios. Base correlation has helped standardize the pricing in the market, and has contributed to the increased transparency and liquidity in index tranches both in the US and Europe.

Bear Stearns (2004)—(Tranche) Skew Rescaling. In order to rescale the base correlation skew curve to other bespoke portfolios which are different from the index, the industry needed a standard methodology to achieve this transformation. That is when the various so-called rescaling methodologies started to appear. Dealers had used different recipes to do the conversion. As with the base correlation approach, in order to accelerate the market convergence to a standard (or a set of standard) rescaling methods, the strategists at Bear Sterns published their research piece on Tranche Loss rescaling. From then on, the pace of convergence towards the standardization of bespoke skew rescaling methods started to pick up, and tranche loss rescaling emerged as one of winners, which made it onto the short list of standard approaches used by the market.

Societe Generale (2006)—(Portfolio) Skew Rescaling. The other standard method that has also proven to be quite popular with dealers (despite having a number of well-documented shortcomings) is Portfolio Loss rescaling. To shed some light on the possible rescaling methods and to make a rational (and somewhat scientific) comparison between the various methods, the strategist at Societe Generale published an important research note, which has helped categorize the various approaches with their pros and cons, and has probably contributed in establishing Tranche Loss and Portfolio Loss rescaling as the most popular ones, which offer the range of skews to be used for different portfolios.

Vasicek (1997)—**Large Homogenous Portfolio Approximation**. A part from the modelling of default correlation and the focus on calibrating market skews, there has been a parallel stream of work on numerical implementation of credit correlation pricing models. Given the large portfolios that we have to deal with, when pricing CDOs, the need for quick and efficient methods to compute prices and risks was critical. To kick off the trend, Oldrich Vasicek published an important paper on large homogeneous portfolio approximations. The idea was simple, yet very powerful. When we have a portfolio of identical single-names with similar default probabilities, by conditioning on the common factor in the Gaussian copula model, the individual single-name default probabilities become independent; once aggregated with identical notional contributions $1/n$, the (average) loss variable constructed in that way, by virtue of the law of large numbers, converges to the single-name default probability as n goes to infinity. Then, integrating over the Gaussian density of the common factor gives the LHP approximation of Vasicek (1997).

FFT and Recursion (2003). For the exact computation of the conditional loss distribution, in the early days of credit modelling, the Monte-Carlo simulation approach was the only available method at the time. But very quickly, most people realized that this would not be a scalable solution given the size of the underlying CDO portfolios and the growth in CDO market volumes. First, emerged the Fast Fourier Transform (FFT) method as an alternative where we could compute the loss distribution in only n^2 operations. This was made popular by Jean-Paul Laurent (a professor at Universite de Lyon) who, at time, was a consultant at BNP Paribas where he collaborated on various credit research projects with Jon Gregory. The popularity of the FFT method did not last for long as another even more powerful method—which was used successfully in actuarial mathematics—appeared shortly thereafter: the convolution recursion method. A number of papers made it popular very rapidly (this includes: Andersen et al. (2003), Hull and White (2004)), and most dealers switched their legacy pricers from Monte Carlo or FFT to Recursion.

Normal Proxy Method—Shelton (2004). Although the recursion method was quick and very efficient, we could still improve the performance even further by using adequate approximations. One such approximation, which proved very successful, in practice, is the Normal Proxy Method. This was made popular by David Shelton when he published a Citigroup Strategy Note describing the approach, where he applied it to both CDOs and CDO-Squared structures. In his paper, Shelton benchmarked the quality of the approximation and its performance benefits for pricing CDOs, but also, and more importantly, for computing hundreds of Greeks at the same time. Here we wish to approximate the loss variable, conditional on the common factor, for any bespoke portfolio, but we are not in a homogenous set-up anymore. The law of large numbers is not applicable, but we can use the Central Limit Theorem instead. As the number of names becomes large, the (conditional) loss variable converges to a normal distribution whose mean and variance can be computed very cheaply with trivial analytical formulas.

Stein Approximation—El Karoui et al. (2008). Obviously, the normal proxy method, although very popular and has proved very successful in practice, it suffers however from a few limitations where the simple approximation is not as accurate in specific edge cases. The main one is thin equity tranches where we are close to the zero-loss limit. This is not surprising, the loss distribution is bounded between 0 and 1 by construction; the normal distribution is not. In particular, the probability of having negative losses (although small) is not equal to zero, which clearly would have an impact on the edge cases close to zero. To address this limitation, Professor Nicole El Karoui (at Ecole Polytechnique), with Ying Jiao and David Kurtz, came up with a very useful improvement to the order-0 normal proxy using the zero-bias transformation method of Stein.

Second Generation Models—(2005–2006). Base correlation worked well in practice but many researchers were not completed satisfied with it. Although it can price index tranches correctly by construction, many of the interpolations and extrapolations that are then needed to be used for other portfolios, other tranches or other products (such as CDO^2) are not always easy to manage. Inspired, mainly, by similar models used in equity markets such as stochastic volatility, local volatility or jump diffusions, there were a few alternative proposals, which could offer a consistent pricing framework to be used, potentially, for all products and all portfolios, and hence reduce the need for ad-hoc interpolations. This includes: stochastic correlation in Burtschell et al. (2007), local correlation in Turc et al. (2005) or random factor loading in Andersen and Sidenius (2005), Levy copula in Baxter (2006), and implied (hazard rate) copula in Hull and White (2006). However, as theoretically appealing as they may appear, they did not really take off because they still struggled with calibrations to the index base correlation across tranches and tenors. Furthermore, the Greeks produced were not always intuitive as they implicitly assumed some built-in dynamics implied by the model, which could be quite different from the market deltas.

Marshall-Olkin Copula—Lindskog and McNeil (2003), **Elouerkhaoui** (2006). We cannot talk about second generation models and not mention the Marshall-Olkin copula. Most the second generation models that we described earlier are not, in essence, credit models. They are models for equity markets that have been adapted for credit. The Marshall-Olkin copula is different.

It is, fundamentally, a credit model, which tries to explain the portfolio loss distribution from the ground up by using a natural decomposition of the single-name defaults into common market factors and idiosyncratic shocks. This is very similar to the multiple Poisson shock models used in reliability theory to study the stability of a dynamic system. This type of models was considered very early on in credit modelling by Duffie (1998), and Duffie and Singleton (1999b), but was used and studied more extensively by Lindskog and McNeil (2003) and Elouerkhaoui (2006). In particular, it has been used, over the years, for pricing first-to-default baskets in a self-consistent manner that preserves all arbitrage-free bounds across different sub-portfolios, which is an important constraint that one needs to satisfy when one manages a large book of FTD swaps. It was also found naturally suited for modelling CDO loss distributions because of its inherent multi-modal distribution. Typically, with three common factors, the MO model would create default clustering effects around different parts of the capital structure, which naturally translates into prices of equity, mezzanine and senior tranches.

Arbitrage-Free Correlation Surface—(2005–2007). As people started to work on dynamic credit models where we want to diffuse the base-correlation surface over time, it was important to make sure that the starting point is arbitrage-free. Base correlation, with all its interpolation assumptions, does not always guarantee that the correlation surface, at the end, would not have arbitrage. Various researchers looked at this question carefully and made a few potential proposals. We mention here just two of the most important ones: the Expected Tranche Loss approach by Torresetti et al. (2007), and the Maximum Entropy (or Minimum Relative Entropy) of Vacca (2005). The first one is useful to ensure arbitrage-free conditions, across maturities, are verified, and the second one eliminates arbitrage in the strike dimension.

Third Generation Models—(2005–2007). Before the credit crisis, we have seen the emergence of new portfolio credit products whose payoff depends on the dynamics of the loss distribution over time. Obviously, none of the static credit models developed to date (including second generation models) was suitable to handle that new challenge. That is when a number of very rich dynamic credit models started to emerge. One of the main highlights is the Top-Down approach of Giesecke and Goldberg (2005). They turned the whole credit modelling problem upside down—literally. Rather than starting with the single-name defaults, assume a joint relationship between them, and then generate the aggregate loss distribution, instead they take the portfolio loss variable as the main modelling object that they work with. The allocation back to single-names (if needed) is done with Random-Thinning. Examples of the most important top-down models include: Dynamic GPL (Brigo et al.

2007), N+ model (Longstaff and Rajan 2008), Markov Chain Portfolio Loss (Schönbucher 2005), and the SPA model (2005).

Stochastic Recovery—(2008). In September 2006, Alex Lipton and Andrew Rennie organized a conference at Merril Lynch's headquarters in London, where they invited a line-up of credit modelling experts including both practitioners and academics to present their latest thinking on portfolio credit modelling, in general, and dynamic credit models in particular. The title of the event was: "Credit Correlation: Life After Copulas". The conference proceedings were published in a book by the same title. But shortly thereafter, the credit crisis started unfolding in 2007 and as the market dynamics changed, the focus moved away swiftly from dynamic credit models, which, all of a sudden, were not a priority anymore as less and less credit exotics were traded and the business shifted to simpler and more commoditized products. But we started to face a different problem: the base correlation, which until that point had performed very well, started to struggle with index tranche calibration. Given the rapid increase in systemic default risk, the super senior tranches, which did not have any value before, started to trade at levels around 10–20 basis points. The market was pricing the systemic default tail risk and the loss distribution started to its shift probability mass to the right of the distribution. The reason for the calibration failures was obvious. With a deterministic recovery value, the support of the loss distribution is, by construction, constrained to fit within the region defined by the loss-given-default upper bound. In other words, if the recovery value is 40%, then the loss variable cannot exceed its deterministic LGD, which, in this example, is 0.6. Hence, the need to make the recovery rate stochastic in order to expand the support of the loss distribution to the full [0, 1] interval. By doing so, the calibration of the base correlation curve would be able to distribute the probability mass across the whole interval in order to satisfy the constraints imposed at both ends of the distribution by the equity and super senior tranches. A frantic rush to fix the base correlation calibration failures has produced some very interesting stochastic recovery modelling frameworks that have been implemented by the industry. A few worth mentioning are: Conditional Mark-Down (Amraoui and Hitier 2008), Conditional Discrete (Krekel 2008) and Conditional Gaussian (Andersen and Sidenius 2005).

CVA modelling—(2009–Present). Last but not least, after the credit crisis in 2007, the investment banks' business model has changed dramatically. The products are simpler, but the risk management of the whole book is much more complex. There are no more forward-starting CDOs, TARNs, or Multi-Maturity CDO-Squareds to price, but new mathematical problems have emerged. Chief among them is Credit Valuation Adjustments (CVA).

When we price a bilateral OTC derivative contract, the standard approach is to assume that both counterparties will honour their contractual obligations and would not default during the lifetime of the trade. After the crisis, this assumption does not hold anymore: a few dealers have defaulted (Lehman Brothers, Bear Sterns) and many more were very close to defaulting, hence the necessity of computing credit valuation adjustments to reflect the price of the counterparty default premium in derivatives valuation. When we compute the CVA on a credit derivative position (or set of positions), for example, Credit Default Swaps or CDO tranches, we are back in a credit correlation modelling framework where we need to account for the correlation between the counterparty default, on the one hand, and the default of the reference entities in the CDS (or CDO) netting set portfolio, on the other hand. All the default correlation modelling machinery that was developed for credit derivatives can be re-used to tackle the new CVA modelling challenge. A few papers that have addressed CVA modelling for credit products include: Brigo and Capponi (2009), Crepey et al. (2010), Assefa et al. (2010), Lipton and Sepp (2009), Lipton and Savescu (2013).

Complex exotic credit derivative products are dead, but structured credit valuation adjustments are alive and kicking—they offer, by far, some of the most complex credit modelling challenges that we have faced to date, and the credit correlation tools that we have acquired over the years will come in very handy when addressing them.

The (exotic credit) king is dead! Long live the (credit valuation adjustment) king!

References

S. Amraoui, S. Hitier, Optimal stochastic recovery for base correlation (Working Paper, BNP Paribas, 2008)

L. Andersen, J. Sidenius, S. Basu, All your hedges in one basket. Risk, 67–72 (2003)

L. Andersen, J. Sidenius, Extensions to the Gaussian Copula: Random recovery and random factor loadings. J. Credit Risk **1**(1), Winter 2004/05, 29–70 (2005)

S. Assefa, T. Bielecki, S. Crepey, M. Jeanblanc, CVA computation for counterparty risk assessment in credit portfolios, in *Credit Risk Frontiers*, ed. by T. Bielecki, D. Brigo, F. Patras (Wiley, 2011), pp. 397–436

M. Baxter, Levy process dynamic modelling of single-name credits and CDO tranches (Working Paper, Nomura International plc, 2006)

J. Beumee, D. Brigo, D. Schiemert, G. Stoyle, *Charting a Course Through the CDS Big Bang* (Fitch Solutions, Quantitative Research, 2009)

D. Brigo, A. Capponi, Bilateral counterparty risk valuation with stochastic dynamical models and applications to credit default swaps (Working Paper, Fitch Solutions, 2009)

D. Brigo, K. Chourdakis, Counterparty risk for credit default swaps: impact of spread volatility and default correlation. Int. J. Theor. Appl. Financ. **12**, 1007–1026 (2009)

D. Brigo, A. Pallavicini, R. Torresetti, Calibration of CDO tranches with the dynamical generalized-poisson loss model. Risk (2007)

X. Burtschell, J. Gregory, J.P. Laurent, Beyond the Gaussian Copula: stochastic and local correlation. J. Credit Risk **3**(1), 31–62 (2007)

S. Crepey, M. Jeanblanc, B. Zargari, Counterparty risk on CDS in a markov chain Copula model with joint defaults, in *Recent Advances in Financial Engineering*, ed. by M. Kijima, C. Hara, Y. Muromachi, K. Tanaka (World Scientific, 2010), pp. 91–126

D. Duffie, First-to-default valuation (Working Paper, Graduate School of Business, Stanford University, 1998)

D. Duffie, K. Singleton, Modeling term structures of defaultable bonds. Rev. Financ. Stud. **12**(4), 687–720 (1999a)

D. Duffie, K. Singleton, Simulating correlated defaults (Working Paper, Graduate School of Business, Stanford University, 1999b)

N. El Karoui, Y. Jiao, D. Kurtz, Gauss and Poisson approximation: Applications to CDOs tranche pricing. J. Comput. Financ. **12**(2), 31–58 (2008)

Y. Elouerkhaoui, Etude des problèmes de corrélation and d'incomplétude dans les marchés de crédit, Ph.D. Thesis, Université Paris-Dauphine, 2006

K. Giesecke, L. Goldberg, A top down approach to multi-name credit (Working Paper, 2005)

J. Hull, A. White, Valuation of a CDO and *n*th-to-default CDS without Monte Carlo simulation. J. Deriv. **2**, 8–23 (2004)

J. Hull, A. White, Valuing credit derivatives using an implied copula approach. J. Deriv. **14**(2), 8–28 (2006)

R. Jarrow, D. Lando, S. Turnbull, A Markov model for the term structure of credit risk spreads. Rev. Financ. Stud. **10**(2), 481–523 (1997)

R. Jarrow, S. Turnbull, Pricing derivatives on financial securities subject to credit risk. J. Financ. **50**(1), 53–85 (1995)

M. Krekel, Pricing distressed CDOs with base correlation and stochastic recovery (Working Paper, Unicredit, 2008)

Lando, Three essays on contingent claims pricing, Ph.D. thesis, Cornell University, 1994

Lando, On cox processes and credit risky securities. Rev. Deriv. Res. **2**(2–3), 99–120

J.-P. Laurent, J. Gregory, Basket default swaps. CDOs and factor Copulas. J. Risk **7**(4), 103–122 (2005)

D.X. Li, On default correlation: A copula function approach. J. Fixed Income **9**, 43–54 (2000)

F. Lindskog, A.J. McNeil, Common poisson shock models: applications to insurance and credit risk modelling. Astin Bull. **33**, 209–238 (2003)
A. Lipton, A. Rennie, *Credit correlation: life after Copulas* (World Scientific, Singapore, 2008)
A. Lipton, A. Sepp, Credit value adjustment for credit default swaps via the structural default model. J. Credit Risk **5**, 123–146 (2009)
A. Lipton, I. Savescu, CDSs, CVA and DVA—a Structural Approach. Risk, 60–65 (2013)
F. Longstaff, A. Rajan, An empirical analysis of the pricing of collateralized debt obligations. J. Financ. **63**(2), 529–563 (2008)
L. McGinty, E. Beinstein, R. Ahluwalia, M. Watts, *Introducing Base Correlations* (Credit Derivatives Strategy, JP Morgan, 2004)
A. Reyfman, K. Ushakova, W. Kong, *How to value bespoke tranches consistently with standard ones* (Credit Derivatives Research, Bear Stearns, 2004)
P.J. Schönbucher, Portfolio losses and the term structure of loss transition rates: a new methodology for the pricing of portfolio credit derivatives (Working Paper, ETH Zurich, 2005)
D. Shelton, *Back To Normal* (Global Structured Credit Research, Citigroup, 2004)
J. Sidenius, V. Piterbarg, L. Andersen, A new framework for dynamic credit portfolio loss modeling. Int. J. Theor. Appl. Financ. **11**(2), 163–197 (2008)
R. Torresetti, D. Brigo, A. Pallavicini, Implied expected tranche loss surface from CDO data (Working Paper, 2007)
J. Turc, D. Benhamou, B. Hertzog, M. Teyssier, *Pricing Bespoke CDOs: Latest Developments* (Quantitative Strategy, Credit Research, Societe Generale, 2006)
J. Turc, P. Very, D. Benhamou, *Pricing CDOs with a Smile* (Quantitative Strategy, Credit Research, Societe Generale, 2005)
L. Vacca, Unbiased risk-neutral loss distributions. Risk, 97–101 (2005)
O. Vasicek, The loan loss distribution (Working Paper, KMV Corporation, 1997)

Part I

Theoretical Tools

2

Mathematical Fundamentals

In this chapter, we present the essential mathematical tools needed in the modelling of portfolio credit derivative products. This includes: doubly-stochastic Poisson processes, also known as Cox processes; point processes and their intensities, on some given filtration; and copula functions.

2.1 Credit Pricing Building Blocks

The key building blocks needed for pricing single-name credit products are: expectations of risky cash-flows, at fixed time horizons, conditional on survival; and expectations of recovery payments at the time of default. Those were derived by Lando (1998) in a doubly-stochastic Poisson process (also known as a Cox process) framework.

2.1.1 Cox Process

An inhomogeneous Poisson process N, with non-negative intensity function $h(.)$ is defined as a process with independent increments such that

$$\mathbb{P}(N_t - N_s = k) = \frac{\left(\int_s^t h(u)\,du\right)^k}{k!} \exp\left(-\int_s^t h(u)\,du\right), \text{ for } k = 0, 1, \ldots$$

Y. Elouerkhaoui, *Credit Correlation*, Applied Quantitative Finance,
DOI 10.1007/978-3-319-60973-7_2

Define τ as the first jump time of the Poisson process N. The probability of survival after time T, which is equivalent to $N_T = 0$, is

$$\mathbb{P}(\tau > T) = \mathbb{P}(N_t = 0) = \exp\left(-\int_0^T h(u)\,du\right).$$

This gives an operational recipe for simulating the first jump time of a Poisson process: let E_1 be a unit exponential random variable, then τ can be determined by setting

$$\tau = \inf\left\{t : \int_0^t h(u)\,du \geq E_1\right\}.$$

A Cox process is a generalized Poisson process where the intensity itself is stochastic: so, instead of having a time-dependent intensity function $h(.)$, the intensity is a stochastic process $h(., \omega)$. Conditional on each realization $\omega \in \Omega$, of the intensity process $h(., \omega)$, the counting process N is an inhomogeneous Poisson process with intensity $h(t, \omega)$.

Furthermore, we assume that the stochastic intensity can be written as the functional form

$$h(t, \omega) = \lambda(X_t),$$

where X is an $\mathbb{R}^d$-valued Ito process representing the background state variables in the economy, and $\lambda(.) : \mathbb{R}^d \to [0, \infty)$ is a non-negative, continuous function. The economic state variables would include: (risk-free) interest rates, equity prices, credit ratings and other macroeconomic variables, which would drive the likelihood of default; but they would exclude the actual default events of the obligor in question and other obligors in the economy. The stochastic intensity $\lambda(X_t)$ can be thought of as a (conditional) instantaneous default probability; in other words, conditional on the firm having survived up to time t, and for a given observed path of the state variables X up to time t, the probability of defaulting in the next instant, between t and $t + dt$, is equal to $\lambda(X_t)\,dt + o(dt)$.

More formally, if we work on a probability space $(\Omega, \mathcal{G}, \mathbb{P})$, where we have an an $\mathbb{R}^d$-valued Ito process X, and a unit exponential variable E_1, independent from X, we define the default time τ as

$$\tau = \inf\left\{t : \int_0^t \lambda(X_u)\,du \geq E_1\right\}.$$

This is the exact analogue to the Poisson construction algorithm, but here we are using a random intensity instead.

Having modeled the default time as the first jump time of a Cox process, we can now write the survival probability, conditional on X, as

$$\mathbb{P}\left(\tau > T \,\middle|\, (X_t)_{0\leq t\leq T}\right) = \exp\left(-\int_0^T \lambda\left(X_u\right) du\right),$$

which yields the expression of the survival probability by taking the expectation on both sides

$$\mathbb{P}\left(\tau > T\right) = \mathbb{E}\left[\exp\left(-\int_0^T \lambda\left(X_u\right) du\right)\right].$$

Next, we evaluate the basic three pricing building blocks that we need.

2.1.2 Three Building Blocks

First, we need to make precise the various filtrations that we work with. As with all credit modelling problems, we will have three filtrations: (1) the background filtration containing information about the state variables in the economy; (2) the default filtration, which tracks the obligor default events history; (3) the enlarged filtration, which combines both the economic state variables and the default events information; the three filtrations are defined as

$$\begin{aligned}\mathcal{F}_t &= \sigma\left\{X_s : 0 \leq s \leq t\right\};\\ \mathcal{H}_t &= \sigma\left\{\mathbf{1}_{\{\tau\leq t\}} : 0 \leq s \leq t\right\};\\ \mathcal{G}_t &= \mathcal{F}_t \vee \mathcal{H}_t.\end{aligned}$$

The conditional survival probability is given by the following lemma.

Lemma 5 (Conditional Survival Probability) *The survival probability conditional on (the whole path) $\mathcal{F}_\infty$ and (the default state) $\mathcal{H}_t$ is given by*

$$\mathbb{P}\left(\tau > T \,|\mathcal{F}_\infty \vee \mathcal{H}_t\right) = \mathbf{1}_{\{\tau>t\}} \exp\left(-\int_t^T \lambda_s ds\right).$$

Proof It suffices to observe that the conditional expectation $\mathbb{E}\left[\mathbf{1}_{\{\tau>t\}}\,|\mathcal{F}_\infty \vee \mathcal{H}_t\right]$ is 0 on the set $\{\tau \leq t\}$ and that $\{\tau > t\} \in \mathcal{H}_t$; then, using Bayes' rule, we obtain

$$\begin{aligned}\mathbb{P}(\tau > T\,|\mathcal{F}_\infty \vee \mathcal{H}_t) &= \mathbf{1}_{\{\tau>t\}}\mathbb{P}(\tau > T\,|\mathcal{F}_\infty \vee \mathcal{H}_t) \\ &= \mathbf{1}_{\{\tau>t\}}\frac{\mathbb{P}(\{\tau > T\}\cap\{\tau > t\}\,|\mathcal{F}_\infty)}{\mathbb{P}(\tau > t\,|\mathcal{F}_\infty)} \\ &= \mathbf{1}_{\{\tau>t\}}\frac{\mathbb{P}(\tau > T\,|\mathcal{F}_\infty)}{\mathbb{P}(\tau > t\,|\mathcal{F}_\infty)} \\ &= \mathbf{1}_{\{\tau>t\}}\frac{\exp\left(-\int_0^T \lambda_s ds\right)}{\exp\left(-\int_0^t \lambda_s ds\right)} = \mathbf{1}_{\{\tau>t\}}\exp\left(-\int_t^T \lambda_s ds\right).\end{aligned}$$

■

To price any credit risky (defaultable) contingent claim, we have to compute expectations of its discounted cash-flows, which can be one of three types:

1. Payment at Maturity $-X_T\mathbf{1}_{\{\tau>T\}}$: a cash-flow payment X_T, which is an $\mathcal{F}_T$-measurable variable, at a fixed time horizon T, if default has not occurred before time T;
2. Coupon Payments $-\mathbf{1}_{\{\tau>s\}}Y_s ds$: a stream of (continuous) payments specified by an $\mathcal{F}_t$-adapted process Y, which terminates when the default event happens;
3. Payment at Default $-Z_\tau$: A recovery rate payment, at the time of default τ, where Z is an $\mathcal{F}_t$-adapted process; the payment at default is the random variable $Z_\tau = Z_{\tau(\omega)}(\omega)$.

The key formulas are summarized in the next proposition.

Proposition 6 (Three Building Blocks) *We have the following conditional expectations for the three cash-flow types above.*

1. *Payment at Maturity:*

$$\mathbb{E}\left[e^{\left(-\int_t^T r_s ds\right)}X_T\mathbf{1}_{\{\tau>T\}}\,|\mathcal{G}_t\right] = \mathbf{1}_{\{\tau>t\}}\mathbb{E}\left[e^{\left(-\int_t^T (r_s+\lambda_s)ds\right)}X_T\,|\mathcal{F}_t\right].$$

2. *Coupon Payments:*

$$\mathbb{E}\left[\int_t^T e^{\left(-\int_t^s r_u du\right)}Y_s\mathbf{1}_{\{\tau>s\}}ds\,|\mathcal{G}_t\right] = \mathbf{1}_{\{\tau>t\}}\mathbb{E}\left[\int_t^T e^{\left(-\int_t^s (r_u+\lambda_u)du\right)}Y_s ds\,|\mathcal{F}_t\right].$$

3. *Payment at Default:*

$$\mathbb{E}\left[e^{\left(-\int_t^{\tau} r_u du\right)} Z_{\tau} \,|\mathcal{G}_t\right] = \mathbf{1}_{\{\tau>t\}}\mathbb{E}\left[\int_t^T \lambda_s e^{\left(-\int_t^s (r_u+\lambda_u)du\right)} Z_s ds \,|\mathcal{F}_t\right].$$

Note that the expectations for payments types 1 and 2, i.e., the payment at maturity and the coupon payments, are not dissimilar. In fact, the coupon payments integral is just a linear sum of multiple $\mathcal{F}_s$ -measurable payments conditional on survival after time s.

Proof First, we start with expectations conditional on survival at a fixed time horizon.

1. Using the law of iterated expectations and the conditional survival probability from the previous lemma, we have

$$\begin{aligned}\mathbb{E}\left[e^{\left(-\int_t^T r_s ds\right)} X_T \mathbf{1}_{\{\tau>T\}} \,|\mathcal{G}_t\right] &= \mathbb{E}\left[\mathbb{E}\left[e^{\left(-\int_t^T r_s ds\right)} X_T \mathbf{1}_{\{\tau>T\}} \,|\mathcal{F}_\infty \vee \mathcal{H}_t\right] |\mathcal{G}_t\right] \\ &= \mathbb{E}\left[e^{\left(-\int_t^T r_s ds\right)} X_T \mathbb{E}\left[\mathbf{1}_{\{\tau>T\}} \,|\mathcal{F}_\infty \vee \mathcal{H}_t\right] |\mathcal{G}_t\right] \\ &= \mathbf{1}_{\{\tau>t\}}\mathbb{E}\left[e^{\left(-\int_t^T (r_s+\lambda_s)ds\right)} X_T \,|\mathcal{G}_t\right]\end{aligned}$$

The last step that we need is to switch the conditioning on the enlarged filtration $\mathcal{G}_t$ to the conditioning on the background filtration $\mathcal{F}_t$. Notice the following sigma fields' inclusions

$$\mathcal{F}_t \subset \mathcal{F}_t \vee \mathcal{H}_t \subset \mathcal{F}_t \vee \sigma\left(E_1\right);$$

and recall from the Cox process construction the independence between the (threshold) exponential random variable E_1 and the sigma filed $\sigma\left(\widetilde{X_T}\right) \vee \mathcal{F}_t$, where $\widetilde{X_T} = e^{\left(-\int_t^T (r_s+\lambda_s)ds\right)} X_T$, so that we can write

$$\mathbb{E}\left[e^{\left(-\int_t^T (r_s+\lambda_s)ds\right)} X_T \,|\mathcal{F}_t \vee \sigma\left(E_1\right)\right] = \mathbb{E}\left[e^{\left(-\int_t^T (r_s+\lambda_s)ds\right)} X_T \,|\mathcal{F}_t\right],$$

which gives the final result.

2. The proof for expectations of coupon payment streams is exactly identical to the proof for cash-flow payments at a fixed time horizon T.

3. For recovery payments, we use the expression of the default time density conditional on $\mathcal{F}_\infty$: for all $s > t$,

$$\mathbb{P}(\tau \in ds \,|\tau > t, \mathcal{F}_\infty) = \frac{\partial}{\partial s}\mathbb{P}(\tau \leq s\,|\tau > t, \mathcal{F}_\infty) = \lambda_s \exp\left(-\int_t^s \lambda_u du\right).$$

Thus, we can write the recovery expectation in terms of the conditional default density as

$$\begin{aligned}\mathbb{E}\left[e^{\left(-\int_t^\tau r_u du\right)} Z_\tau \,|\mathcal{G}_t\right] &= \mathbb{E}\left[\mathbb{E}\left[e^{\left(-\int_t^\tau r_u du\right)} Z_\tau \,|\mathcal{F}_\infty \vee \mathcal{H}_t\right] |\mathcal{G}_t\right] \\ &= \mathbf{1}_{\{\tau>t\}}\mathbb{E}\left[\int_t^T \lambda_s e^{\left(-\int_t^s \lambda_u du\right)} \left[e^{\left(-\int_t^s r_u du\right)} Z_s\right] ds \,|\mathcal{G}_t\right] \\ &= \mathbf{1}_{\{\tau>t\}}\mathbb{E}\left[\int_t^T \lambda_s e^{\left(-\int_t^s (r_u+\lambda_u) du\right)} Z_s ds \,|\mathcal{F}_t\right];\end{aligned}$$

in the last line, we have replaced the conditioning on $\mathcal{G}_t$ with the conditioning on $\mathcal{F}_t$ following the same argument as before.

■

Next, we give a brief overview of the theory of point processes and the general definition of intensity processes with respect to a given filtration. This is a generalization of the results obtained in the Cox process framework. The choice of filtration (and its corresponding intensity process) is of critical importance when we work in a credit portfolio set-up with multiple default times. Single-name default intensities (on the enlarged filtration) can be distorted as (portfolio) default events occur, which creates some interesting default clustering patterns.

2.2 Point Processes, Filtrations and Intensities

We have seen in the previous section a definition of the default event (stopping) time based on a Cox process approach. The default time is constructed, from first principles, as the first jump time of Poisson process whose intensity is stochastic and driven by some economic state variables (Ito) process. It turns out that we can go one step further and model the default event, in a more general framework, as a stopping time with respect to a given filtration. But as soon as we do that, extra care and attention need to be given to the choice of (working) filtration and the exact definition of the intensity process (associated with this default time). Intuitively, when we have a general point process N_t, on some filtration $\mathcal{F}_t$, the heuristic definition of an ($\mathcal{F}_t$-) intensity is simply the conditional instantaneous jump probability:

$$\mathbb{P}(dN_t = 1 \,|\mathcal{F}_t) = \lambda_t dt + o(dt).$$

This concept has been formalized in a mathematically rigorous manner by Brémaud (1980). This is the definitive mathematical bible on point processes (and market point processes) that any credit modeler (keen on mathematical rigour) needs to refer it continuously.

Here we follow the pedagogical presentation in Brémaud (1980) and we summarize some of the key results, which we will need later in the sequel.

2.2.1 Counting Process

A point process on the half line $[0, \infty)$ can be represented in one of three ways: (a) as a sequence of non-negative random times; (b) as a discrete random measure; (c) or as the associated counting process. Here we shall use the point process terminology to refer to both the sequence of random times and their counting process interchangeably.

Definition 7 (*Point Process*) A point process on can be described by a sequence of non-negative random times, on some probability space $(\Omega, \mathcal{F}, \mathbb{P})$, such that

$$\begin{aligned} T_0 &= 0, \\ T_n &< \infty \Longrightarrow T_n < T_{n+1}. \end{aligned}$$

A realization of the point process is said to be non-explosive if $T_\infty = \lim_{n\uparrow\infty} T_n = +\infty$. For each realization T_n corresponds a counting function N, defined as

$$N_t = \begin{cases} n & \text{if } t \in [T_n, T_{n+1}), n \geq 0, \\ +\infty & \text{if } t \geq T_\infty. \end{cases}$$

The sequence T_n is called a point process; but sometimes the associated counting process N is also called a point process by abuse of notation. We say that it is non-explosive if $N_t < \infty$ a.s., for $t \geq 0$. And we say that the point process N_t is integrable if $\mathbb{E}[N_t] < \infty$, for $t \geq 0$.

We can also have a multivariate point process. The definition is given below.

Definition 8 (*Multivariate Point Process*) Let T_n be a point process, and let Z_n be a sequence of random variables in $\{1, 2, \ldots, k\}$. Define for all $1 \leq i \leq k$,

$$N_t(i) = \sum_{n\geq 1} \mathbf{1}\{T_n \leq t\} \mathbf{1}\{Z_n = i\}, \text{ for } t \geq 0.$$

The k-dimensional vector process $(N_t(1), \ldots, N_t(k))$ and the double sequence $(T_n, Z_n, n \geq 1)$ are called k-variate point processes. The processes $N_t(i)$ have no common jumps, i.e., for $i \neq j$, we have $\Delta N_t(i) \Delta N_t(j) = 0$ a.s., for $t \geq 0$.

2.2.2 Doubly Stochastic Poisson Process

Broadly speaking, a doubly stochastic Poisson process is generated with a two-step procedure: first, we simulate a full path of the intensity-driving background process X; then, given this realized path, we generate a Poisson process with intensity $\lambda_t = h(X_t)$. This is the Cox process construction that we have discussed in Sect. 2.1.1.

This definition can be generalized and extended formally to a larger class of intensity processes.

Definition 9 (*Doubly Stochastic Poisson Processes*) Let N_t be a point process adapted to some filtration $\mathcal{F}_t$, and let λ_t be a non-negative measurable process. Suppose that

$$\lambda_t \text{ is } \mathcal{F}_0\text{-measurable, for } t \geq 0,$$

and that

$$\int_0^t \lambda_s ds < \infty \text{ a.s., for } t \geq 0.$$

If we have, for all $0 \leq s \leq t$ and all $u \in \mathbb{R}$,

$$\mathbb{E}\left[e^{iu(N_t - N_s)} | \mathcal{F}_s\right] = \exp\left(\left(e^{iu} - 1\right) \int_s^t \lambda_v dv\right),$$

then the process N_t is called a $(\mathbb{P}, \mathcal{F}_t)$-doubly stochastic Poisson process with (stochastic) $\mathcal{F}_t$-intensity.

If the intensity λ_t is deterministic (and, therefore, is a time-dependent function $\lambda(t)$) then N_t is called an $(\mathbb{P}, \mathcal{F}_t)$-Poisson process.

If the filtration is restricted to its natural filtration $\mathcal{F}_t = \mathcal{F}_t^N$, then we say that N_t is a Poisson process with intensity $\lambda(t)$.

If the intensity is equal to one, $\lambda(t) = 1$, then N_t is a standard Poisson process.

Remark 10 From the definition above, it follows that:

- the Poisson process increments $N_t - N_s$ are independent from $\mathcal{F}_s$ conditionally on (the path) $\mathcal{F}_0$. Indeed, since λ_t is $\mathcal{F}_0$-measurable, we can write

$$\mathbb{E}\left[e^{iu(N_t-N_s)}\,|\mathcal{F}_s \vee \mathcal{F}_0\right] = \mathbb{E}\left[\exp\left(\left(e^{iu}-1\right)\int_s^t \lambda_v dv\right)|\mathcal{F}_0\right] = \mathbb{E}\left[e^{iu(N_t-N_s)}\,|\mathcal{F}_0\right];$$

- for all $0 \le s \le t$, the probability distribution of the increment $N_t - N_s$ conditional on $\mathcal{F}_s$, is given by

$$\mathbb{P}\left(N_t - N_s = k\,|\mathcal{F}_s\right) = \frac{\left(\int_s^t \lambda_v dv\right)^k}{k!} e^{-\int_s^t \lambda_v dv}, \text{ for } k \ge 0.$$

An alternative definition of a doubly stochastic process is based on a result due to Watanabe (1964): it offers a more general characterization, which can be extended to define the (stochastic) intensity for any general point process (which is not necessarily Poisson or doubly stochastic Poisson).

2.2.3 Watanabe's Characterization

Let N_t be a doubly stochastic Poisson process with an $\mathcal{F}_t$ -intensity λ_t. Using the $\mathcal{F}_s$-conditional probability distribution of the increment $N_t - N_s$, we can write

$$\mathbb{E}\left[N_t - N_s\,|\mathcal{F}_s\right] = \mathbb{E}\left[\int_s^t \lambda_u du\,|\mathcal{F}_s\right].$$

Suppose that the cumulative intensity is integrable, i.e., $\mathbb{E}\left[\int_0^t \lambda_u du\right] < \infty$, for all $t \ge 0$, then from the equation above the process N_t is also integrable $\mathbb{E}\left[N_t\right] < \infty$; hence, the process M_t defined as

$$M_t = N_t - \int_0^t \lambda_u du$$

is an $\mathcal{F}_t$-martingale. Furthermore, for all non-negative $\mathcal{F}_t$-predictable processes C_t, we have

$$\mathbb{E}\left[\int_0^\infty C_s dN_s\right] = \mathbb{E}\left[\int_0^\infty C_s \lambda_s ds\right].$$

This characterization can be used as a definition for the intensity of a doubly stochastic Poisson process.

Theorem 11 (Characterization of Doubly Stochastic Poisson Processes) *Let N_t be a point process adapted to some filtration $\mathcal{F}_t$, and let λ_t be a non-negative measurable process such that: for all $t \geq 0$, λ_t is -measurable, and $\int_0^t \lambda_s ds < \infty$ a.s.*

If the equality

$$\mathbb{E}\left[\int_0^\infty C_s dN_s\right] = \mathbb{E}\left[\int_0^\infty C_s \lambda_s ds\right]$$

holds for all non-negative $\mathcal{F}_t$-predictable processes C_t, then N_t is a doubly stochastic process with $\mathcal{F}_t$-intensity λ_t.

Watanabe in 1964 came up with the first important characterization property, which links point processes and martingales. His characterization result relates to Poisson processes.

Theorem 12 (Watanabe 1964) *Let N_t be a point process adapted to some filtration $\mathcal{F}_t$, and let $\lambda(t)$ be a locally integrable non-negative measurable function. Suppose that $M_t = N_t - \int_0^t \lambda(s) ds$ is an $\mathcal{F}_t$-martingale. Then, N_t is an $\mathcal{F}_t$-Poisson process with intensity $\lambda(t)$, i.e., for all $0 \leq s \leq t$, the increment $N_t - N_s$ is a Poisson random variable with parameter $\int_s^t \lambda(u) du$, which is independent from $\mathcal{F}_s$.*

2.2.4 Stochastic Intensity

In the general case, to define the $\mathcal{F}_t$-intensity for any point process (adapted to the filtration $\mathcal{F}_t$), we can use the previous doubly stochastic Poisson process characterization theorem.

Definition 13 (*Stochastic Intensity*) Let N_t be a point process adapted to some filtration $\mathcal{F}_t$, and let λ_t be a non-negative $\mathcal{F}_t$-progressive process such that $\int_0^t \lambda_s ds < \infty$ a.s, for all $t \geq 0$. If for all non-negative $\mathcal{F}_t$-predictable processes C_t, the equality

$$\mathbb{E}\left[\int_0^\infty C_s dN_s\right] = \mathbb{E}\left[\int_0^\infty C_s \lambda_s ds\right]$$

holds, then we say that the process N_t admits a $(\mathbb{P}, \mathcal{F}_t)$-intensity (or $\mathcal{F}_t$-intensity) λ_t.

Using this definition, we have the following integration theorem.

Theorem 14 (Integration Theorem) *If N_t admits an $\mathcal{F}_t$-intensity λ_t (where $\int_0^t \lambda_s ds < \infty$ a.s, for all $t \geq 0$), then N_t is non-explosive and*

- $M_t = N_t - \int_0^t \lambda_s ds$ *is an $\mathcal{F}_t$-local martingale;*
- *if X_t is an $\mathcal{F}_t$-predictable process such that $\mathbb{E}\left[\int_0^t |X_s| \lambda_s ds\right] < \infty, t \geq 0$, then $\int_0^t X_s dM_s$ is an $\mathcal{F}_t$-martingale;*
- *if X_t is an $\mathcal{F}_t$-predictable process such that $\int_0^t |X_s| \lambda_s ds < \infty, t \geq 0$, then $\int_0^t X_s dM_s$ is an $\mathcal{F}_t$-local martingale.*

The next martingale characterization theorem is the main result that we shall use to define the intensity for default (stopping) times.

Theorem 15 (Martingale Characterization of Intensity) *Let N_t be a non-explosive point process adapted to the filtration $\mathcal{F}_t$. Suppose that for some non-negative $\mathcal{F}_t$-progressive process λ_t and for all $n \geq 1$,*

$$M_{t \wedge T_n} = N_{t \wedge T_n} - \int_0^{t \wedge T_n} \lambda_s ds, \text{ is an } (\mathbb{P}, \mathcal{F}_t) \text{-martingale.}$$

Then, λ_t is the $\mathcal{F}_t$-intensity of point process N_t.

One can observe that, using this intensity martingale characterization property, the following equality holds:

$$\mathbb{E}\left[N_{t \wedge T_n} - N_{s \wedge T_n} \,|\mathcal{F}_s\right] = \mathbb{E}\left[\int_{s \wedge T_n}^{t \wedge T_n} \lambda_u du \,|\mathcal{F}_s\right],$$

which, by letting $n \uparrow \infty$, becomes

$$\mathbb{E}\left[N_t - N_s \,|\mathcal{F}_s\right] = \mathbb{E}\left[\int_s^t \lambda_u du \,|\mathcal{F}_s\right].$$

This is reminiscent of one of the classical definitions of intensity. In particular, if we assume that λ_t is right-continuous and bounded, then by applying successively the Lebesgue averaging theorem and the Lebesgue dominated convergence theorem, we can see that the $\mathcal{F}_t$-conditional probability of instantaneous jumps (in the point process) is equal to the $\mathcal{F}_t$-intensity

$$\lim_{t \downarrow s} \frac{1}{t-s} \mathbb{E}\left[N_t - N_s \left| \mathcal{F}_s \right. \right] = \lambda_s, \text{ a.s.}$$

2.2.5 Predictable Intensities

So far, we have given a characterization of an $\mathcal{F}_t$-intensity process for a point process N_t, but we have not said anything about its uniqueness. In general, the $\mathcal{F}_t$-intensity, as defined previously, is not unique. But we can always find a predictable version of the intensity, which is made unique by the predictability constraint.

The formal results regarding uniqueness and existence of predictable versions are given in the next two theorems.

Theorem 16 (Uniqueness of Predictable Intensities) *Let N_t be a point process adapted to the filtration $\mathcal{F}_t$. Let λ_t and $\widetilde{\lambda}_t$ be two $\mathcal{F}_t$-intensities of the point process N_t, which are $\mathcal{F}_t$-predictable, then*

$$\lambda_t(\omega) = \widetilde{\lambda}_t(\omega), \quad \mathbb{P}(d\omega)\, dN_t(\omega) \text{ -a.e.}$$

In particular, for $n \geq 1$, we have

$$\begin{aligned} \lambda_{T_n} &= \widetilde{\lambda}_{T_n}, & \text{on } \{T_n < \infty\}, \\ \lambda_t(\omega) &= \widetilde{\lambda}_t(\omega), & \lambda_t(\omega)\, dt \text{ and } \widetilde{\lambda}_t(\omega)\, dt\text{-a.e.} \end{aligned}$$

Theorem 17 (Existence of Predictable Versions of Intensities) *Let N_t be a point process with an $\mathcal{F}_t$-intensity λ_t. Then, one can find an $\mathcal{F}_t$-intensity $\widetilde{\lambda}_t$ that it predictable.*

Now when we talk about the $\mathcal{F}_t$-intensity of the point process N_t (as opposed to an $\mathcal{F}_t$-intensity), we are referring to the (unique) predictable version.

2.2.6 Change of Filtration

A very important result, which forms the foundation of everything that one does when working on enlarged (credit) filtrations –generated by the default events of a credit portfolio– is the change of filtration theorem. As one switches between the enlarged portfolio filtration and the individual single-name filtrations (or other sub-basket filtrations), one needs to pay special attention to the intensities that are used as they have a fundamental impact on all the conditional expectation calculations that one performs.

We state the change of filtration theorem next.

Theorem 18 (Change of Filtration for Intensities) *Let N_t be a point process with the $\mathcal{F}_t$-intensity λ_t. Let $\mathcal{G}_t$ be a sub-filtration of N_t smaller than $\mathcal{F}_t$, i.e.,*

$$\mathcal{F}_t^N \subset \mathcal{G}_t \subset \mathcal{F}_t, \quad t \geq 0.$$

Then, the process N_t admits a $\mathcal{G}_t$-intensity μ_t defined by

$$\mu_t(\omega) = \left(\frac{\lambda_u d\mathbb{P}du}{d\mathbb{P}du}\right)(t, \omega), \quad on\, \mathcal{P}(\mathcal{G}_t).$$

Loosely speaking, this can be re-stated as: if N_t is a point process with the $\mathcal{F}_t$-intensity λ_t, and if $\mathcal{G}_t$ is a sub-filtration of N_t, which is smaller than $\mathcal{F}_t$, then $\mu_t = \mathbb{E}\left[\lambda_t | \mathcal{G}_t\right]$ is the $\mathcal{G}_t$-intensity of N_t.

Now, looking more closely at a multivariate point process, we can describe the (discrete) conditional probability density of the embedded mark process $(Z_n, n \geq 0)$ in terms of the point process intensities. In practice, this is a useful property that usually comes in handy when we wish to implement efficient Monte-Carlo simulation algorithms for multi-name baskets.

Theorem 19 *Let $(T_n, Z_n, n \geq 0)$ be an m-variate point process, and let $N_t(i)$, for $1 \leq i \leq m$, be its associated counting processes. Let $\mathcal{F}_t$ be a filtration of the form*

$$\mathcal{F}_t = \mathcal{F}_0 \vee \left(\bigvee_{i=1}^{m} \mathcal{F}_t^{N(i)}\right),$$

where is $\mathcal{F}_t^{N(i)}$ the filtration of the process $N_t(i)$. Suppose that, for each $1 \leq i \leq m$, $N_t(i)$ admits the $\mathcal{F}_t$-intensity $\lambda_t(i)$. Then, for all $n \geq 1$,

$$\frac{\lambda_{T_n}(i)}{\lambda_{T_n}} = \mathbb{P}\left(Z_n = i \middle| \mathcal{F}_{T_n^-}\right), \; on\; \{T_n < \infty\},$$

where $\lambda_t = \sum_{i=1}^{m} \lambda_t(i)$ is the -intensity of the process $N_t = \sum_{i=1}^{m} N_t(i)$.

Broadly speaking, the ratio $\frac{\lambda_t(i)}{\sum_{i=1}^{m} \lambda_t(i)}$ is the probability of having a jump of type i, at time t, conditional on $\mathcal{F}_{t^-}$, and knowing that we have a jump in one of the m point processes $N_t(j)$ at time t. We could write it heuristically as

$$\mathbb{P}(dN_t(i) = 1 \,|\mathcal{F}_{t^-}, dN_t = 1) = \frac{\mathbb{P}(dN_t(i) = 1, dN_t = 1 \,|\mathcal{F}_{t^-})}{\mathbb{P}(dN_t = 1 \,|\mathcal{F}_{t^-})}$$
$$= \frac{\mathbb{P}(dN_t(i) = 1 \,|\mathcal{F}_{t^-})}{\mathbb{P}(dN_t = 1 \,|\mathcal{F}_{t^-})}$$
$$= \frac{\lambda_t(i)}{\lambda_t}.$$

2.2.7 Random Time Change

In the same way that any continuous local martingale can be represented as a (continuous) time-changed Brownian motion, there is a similar property for point processes, which can be re-casted as time-changed Poisson processes.

The basic result is given in the next theorem.

Theorem 20 (Time-Changed Poisson Process) *Let N_t be a point process with the $\mathcal{F}_t$-intensity λ_t and the $\mathcal{G}_t$ -intensity $\widetilde{\lambda}_t$, where $\mathcal{F}_t$ and $\mathcal{G}_t$ are filtrations of N_t such that*

$$\mathcal{F}_t^N \subset \mathcal{G}_t \subset \mathcal{F}_t.$$

Suppose that $N_\infty = \infty$, a.s. Define for each t, the $\mathcal{G}_t$-stopping time $\theta(t)$ as

$$\int_0^{\theta(t)} \widetilde{\lambda}_s ds = t.$$

Then, the point process $\widetilde{N}_t$ defined by the time change $\theta(t)$,

$$\widetilde{N}_t = N_{\theta(t)},$$

is a standard Poisson process (i.e., with $\mathcal{G}_t$-intensity 1)

Having defined, mathematically, the intensity for a general point process, on a given filtration–and appreciated the subtleties around the choice of filtration-, to proceed further, we need to provide an "operational" tool to construct these quantities and relate them to each other in some way. This is achieved by using the concept of copula functions, which we describe next.

2.3 Copulas

Default correlation has been, for a long time, a very ambiguous concept – shrouded in mystery and often misunderstood, or at least misinterpreted in one way or another. This state of fuzziness is probably due to the fact that our minds are trained to think in terms of Gaussian distributions. A multivariate Gaussian distribution is completely determined by its pair-wise correlations (and variances or its covariance matrix). This fact is very specific to normal distributions. When we are talking about "correlating default events" we are trying to specify the multivariate distribution of a set of Bernoulli variables which cannot be achieved by looking solely at the pairwise correlations; a more general tool is needed.

Consider two random variables (T_1, T_2), with a bivariate distribution $F(t_1, t_2)$, and marginals $F_1(t_1)$, $F_2(t_2)$; then, we have the following properties:

$$F(t_1, +\infty) = F_1(t_1), F(+\infty, t_2) = F_2(t_2);$$
$$F(t_1, -\infty) = F(-\infty, t_2) = F(-\infty, -\infty) = 0;$$
$$F(+\infty, +\infty) = 1.$$

Furthermore, the measure of the rectangle $[x_1, x_2] \times [y_1, y_2]$ is positive and given by:

$$\mathbb{P}(x_1 \leq T_1 \leq x_2, \; y_1 \leq T_2 \leq y_2) = F(x_2, y_2) - F(x_2, y_1) - F(x_1, y_2) + F(x_1, y_1) \geq 0.$$

When T_1 and T_2 are independent, the bivariate distribution is simply the product of the (univariate) marginals:

$$F(t_1, t_2) = F_1(t_1) \times F_2(t_2).$$

The problem of determining a bivariate distribution from its marginals has an infinite number of solutions. In particular, we have an upper and lower bound that are solutions to this problem. Fréchet (1957) has shown that the following condition holds:

$$\max(F_1(t_1) + F_2(t_2) - 1, 0) \leq F(t_1, t_2) \leq \min(F_1(t_1), F_2(t_2)).$$

This family of solutions can be parametrized elegantly by using the formalism of Copula functions. We present here a short summary of the main results regarding copulas (we refer to Embrechts et al. (2003) for more details).

2.3.1 Sklar's Theorem

Basically, a copula function is a function that links a set of univariate marginal distributions to a complete multivariate distribution.

The formal mathematical definition is given below (see Nelsen 1999).

Definition 21 (*Copula*) An n-dimensional copula is any function $C : [0, 1]^n \to [0, 1]$ with the following properties

- C is grounded, i.e., $C(u_1, \ldots, u_n) = 0$ for all $(u_1, \ldots, u_n) \in [0, 1]^n$ such that $u_k = 0$ for at least one k;
- C is n-increasing, i.e., the C-volume of all n-boxes whose vertices lie in $[0, 1]^n$ is positive:

$$\sum_{i_1=1}^{2} \cdots \sum_{i_n=1}^{2} (-1)^{i_1+\cdots+i_n} C\left(u_1^{i_1}, \ldots, u_n^{i_n}\right) \geq 0,$$

  for all $(u_1^1, \ldots, u_n^1)$ and $(u_1^2, \ldots, u_n^2)$ in $[0, 1]^n$ with $u_k^1 \leq u_k^2$, $1 \leq k \leq n$;
- C has margins C_k, which satisfy $C_k(u_k) = C(1, \ldots, 1, u_k, 1 \ldots, 1) = u_k$ for all u_k in $[0, 1]$.

This definition ensures that C is a multivariate uniform distribution.

For our purposes, we shall use the following (equivalent) operational definition.

Definition 22 (*Copula Function*) Let $U_1, U_2, \ldots, U_n$ be a set of n uniform random variables. Then, the joint distribution function

$$C(u_1, u_2, \ldots, u_n) = \mathbb{P}(U_1 \leq u_1, U_2 \leq u_2, \ldots, U_n \leq u_n)$$

is called a Copula function.

The link between copulas and the construction of multivariate distributions is given by Sklar's theorem.

For a set of n random variables $X_1, X_2, \ldots, X_n$ with univariate distributions $F_i(x_i) = \mathbb{P}(X_i \leq x_i)$, we can define their multivariate distribution

$F(x_1, x_2, \ldots, x_n)$ by a choice of Copula function as follows:

$$F(x_1, x_2, \ldots, x_n) \triangleq C(F_1(x_1), F_2(x_2), \ldots, F_n(x_n)).$$

To see that, it suffices to write:

$$\begin{aligned} C(F_1(x_1), F_2(x_2), \ldots, F_n(x_n)) &= \mathbb{P}(U_1 \leq F_1(x_1), U_2 \leq F_2(x_2), \ldots, U_n \leq F_n(x_n)) \\ &= \mathbb{P}\left(F_1^{-1}(U_1) \leq x_1, F_2^{-1}(U_2) \leq x_2, \ldots, F_n^{-1}(U_n) \leq x_n\right) \\ &= \mathbb{P}(X_1 \leq x_1, X_2 \leq x_2, \ldots, X_n \leq x_n) \\ &= F(x_1, x_2, \ldots, x_n). \end{aligned}$$

The converse result is also true. Sklar (1959) has shown that any multivariate distribution can be expressed as a Copula function.

Theorem 23 (Sklar's theorem) *Let F be an n-dimensional distribution function with margins $F_1, \ldots, F_n$. Then, there exists an n-copula C such that for all $(x_1, \ldots, x_n) \in \overline{\mathbb{R}}^n$,*

$$F(x_1, \ldots, x_n) = C(F_1(x_1), \ldots, F_n(x_n)).$$

If $F_1, \ldots, F_n$ are all continuous, then C is unique. Otherwise, C is uniquely determined on $Ran\ F_1 \times \cdots \times Ran\ F_n$. Conversely, if C is an n-copula and $F_1, \ldots, F_n$ are distribution functions, then the function F defined above is an n-dimensional distribution function with margins $F_1, \ldots, F_n$.

The mixed kth-order partial derivatives of a copula function C, $\frac{\partial^k C(\mathbf{u})}{\partial u_1 \ldots \partial u_k}$, exist for almost all $\mathbf{u}$ in $[0, 1]^n$; moreover, the partial derivatives are always bounded between 0 and 1,

$$0 \leq \frac{\partial^k C(\mathbf{u})}{\partial u_1 \ldots \partial u_k} \leq 1.$$

Now, every copula function C can be decomposed into its absolutely continuous part and its singular part:

$$C(u_1, \ldots, u_n) = A_C(u_1, \ldots, u_n) + S_C(u_1, \ldots, u_n),$$

where

$$\begin{aligned} A_C(u_1, \ldots, u_n) &= \int_0^{u_1} \cdots \int_0^{u_n} \frac{\partial^n C(s_1, \ldots, s_n)}{\partial s_1 \ldots \partial s_n} ds_1 \ldots ds_n, \\ S_C(u_1, \ldots, u_n) &= C(u_1, \ldots, u_n) - A_C(u_1, \ldots, u_n). \end{aligned}$$

If $C = A_C$ on $[0, 1]^n$, then C is said to be absolutely continuous; in this case, it will have a density distribution $\frac{\partial^n C(u_1,...,u_n)}{\partial u_1...\partial u_n}$.

If $C = S_C$ on $[0, 1]^n$, then C is said to be singular, and will have zero density, $\frac{\partial^n C(u_1,...,u_n)}{\partial u_1...\partial u_n} = 0$, almost everywhere in $[0, 1]^n$.

The Marshall-Olkin copula, which we will study later, is an important example of a copula function with a regular continuous part and a singular part on the diagonals.

Frechet-Hoeffding Bounds. Define the functions M^n, Π^n and W^n, on $[0, 1]^n$, as

$$\begin{aligned}
M^n(\mathbf{u}) &= \min(u_1, \ldots, u_n), \\
\Pi^n(\mathbf{u}) &= u_1 \ldots u_n, \\
W^n(\mathbf{u}) &= \max(u_1 + \cdots + u_n - n + 1, 0).
\end{aligned}$$

Note that the functions M^n and Π^n are n-copula functions, for all $n \geq 2$; the function W^n, on the other hand, is not a copula function for any $n \geq 3$.

The upper and lower bounds for a copula function are given by the Frechet-Hoeffding bounds inequality (Fréchet 1957).

Theorem 24 (Frechet-Hoeffding Bounds) *Let C be an n-copula function, then for every $\mathbf{u}$ in $[0, 1]^n$, we have*

$$W^n(\mathbf{u}) \leq C(\mathbf{u}) \leq M^n(\mathbf{u}).$$

The whole question now is: what is the best choice of copula function for our purposes.

The traditional copula used in the market explicitly or implicitly is the Gaussian Copula. It has the advantage of being easy to simulate and its correlation parameters happen to have a nice interpretation in the Firm Asset Value approach. That is effectively what is used, for example, in the CreditMetrics model (see Gupton et al. 1997). Other approaches are also possible: we can assume alternatively that individual obligor defaults are driven by the so-called "Shock models", where common market factors trigger the joint defaults of multiple credits simultaneously. The copula function that originates from this model is known as the Marshall-Olkin copula function.

2.3.2 Dependence Concepts

As we have seen previously, linear correlation (also known as Pearson's correlation) is not sufficient to quantify the dependence between two random variables. The tool that one should use to specify a well-posed dependence structure is the copula function, and as such, any metric, which better captures the random variables distributional dependence features, ought to be based on the copula function itself. Some of the most popular copula-based (dependence) metrics include: Kendall's Tau, Spearman's rho and the (upper) tail dependence coefficient. We review each one in turn.

Linear Correlation. The linear correlation (or Pearson's correlation) coefficient is defined as follows.

Definition 25 (*Linear Correlation*) Let X and Y be two random variables with non-zero finite variances, then the linear correlation coefficient $\rho(X, Y)$ is

$$\rho(X, Y) = \frac{\mathrm{Cov}(X, Y)}{\sqrt{\mathrm{Var}(X)}\sqrt{\mathrm{Var}(Y)}},$$

where $\mathrm{Cov}(X, Y) = \mathbb{E}[X \cdot Y] - \mathbb{E}[X]\mathbb{E}[Y]$ is the covariance of X and Y; $\mathrm{Var}(X)$ and $\mathrm{Var}(Y)$ are the variances of X and Y.

Pearson's correlation is a measure of linear dependence. In particular, if we have perfect linear dependence between the random variables X and Y, i.e., when $Y = aX + b$, for some fixed coefficients $a \neq 0$ and b, then the linear correlation is exactly equal to one: $|\rho(X, Y)| = 1$; and the converse result is also true.

Linear correlation is a natural dependence measure for elliptical distributions (such as the multivariate normal or the multivariate t-distribution). For all other distributions, the linear correlation coefficient can be very misleading. Even for elliptical distributions, it only really makes sense for the Gaussian distribution; for t-distributions, where we have heavier tails, the behaviour in the tail is parametrized differently and cannot be captured through the simplistic linear correlation coefficient.

Concordance. Let (X, Y) and $\left(\widetilde{X}, \widetilde{Y}\right)$ be two pairs of random variables with identical marginal distributions.

The probability of concordance between (X, Y) and $\left(\widetilde{X}, \widetilde{Y}\right)$ is given by

$$\mathbb{P}\left(\left(X - \widetilde{X}\right)\left(Y - \widetilde{Y}\right) > 0\right);$$

similarly, the probability of discordance is

$$\mathbb{P}\left(\left(X - \widetilde{X}\right)\left(Y - \widetilde{Y}\right) < 0\right).$$

The difference between the probabilities of concordance and discordance can be expressed in terms of the copula functions (see Nelsen 1999).

Theorem 26 (Difference of Concordance and Discordance Probabilities) *Let (X, Y) and $(\widetilde{X}, \widetilde{Y})$ be two pairs of continuous random variables, with common marginal distributions (F, G), and bivariate joint distributions H and $\widetilde{H}$ respectively. Their bivariate copula functions are C and $\widetilde{C}$ respectively: $H(x, y) = C(F(x), G(y))$, $\widetilde{H}(x, y) = \widetilde{C}(F(x), G(y))$. Let Q denote the difference between the probabilities of concordance and discordance*

$$Q = \mathbb{P}\left(\left(X - \widetilde{X}\right)\left(Y - \widetilde{Y}\right) > 0\right) - \mathbb{P}\left(\left(X - \widetilde{X}\right)\left(Y - \widetilde{Y}\right) < 0\right),$$

then, we have

$$Q = Q\left(C, \widetilde{C}\right) = 4 \int\int_{[0,1]^2} \widetilde{C}(u, v)\, dC(u, v) - 1.$$

This result will be used next to derive the expressions of the Kendall tau and Spearman rho coefficients.

Kendall's Tau and Spearman's Rho. The most important copula-based concordance measures, to be used with distributions other than the normal one, are: Kendall's tau and Spearman's rho.

Definition 27 (*Kendall's Tau*) For two random variables X and Y, Kendall's tau is defined as

$$\tau(X, Y) = \mathbb{P}\left(\left(X - \widetilde{X}\right)\left(Y - \widetilde{Y}\right) > 0\right) - \mathbb{P}\left(\left(X - \widetilde{X}\right)\left(Y - \widetilde{Y}\right) < 0\right),$$

where $(\widetilde{X}, \widetilde{Y})$ is an independent copy of the pair (X, Y).

This can be written in terms of the copula function.

Theorem 28 *Let X and Y be two continuous random variables with copula C, then their Kendall's tau coefficient is given by*

$$\tau(X, Y) = Q(C, C) = 4 \int\int_{[0,1]^2} C(u, v)\, dC(u, v) - 1.$$

Within the same class of concordance measures, we have the Spearman's rho coefficient, which is defined as follows.

Definition 29 (*Spearman's Rho*) For two random variables X and Y, Spearman's rho is defined as

$$\rho_S(X, Y) = 3\mathbb{P}\left(\left(X - \widetilde{X}\right)\left(Y - Y'\right) > 0\right) - \mathbb{P}\left(\left(X - \widetilde{X}\right)\left(Y - Y'\right) < 0\right),$$

where $(\widetilde{X}, \widetilde{Y})$ and (X', Y') are independent copies of the pair (X, Y).

Expressed in terms of the copula function, we have the following result.

Theorem 30 *Let X and Y be two continuous random variables with copula C, then their Spearman's rho coefficient is given by*

$$\rho_S(X, Y) = Q(C, \Pi) = 12 \int\int_{[0,1]^2} uv dC(u, v) - 3 = 12 \int\int_{[0,1]^2} C(u, v)\, du dv - 3.$$

If X and Y have marginal distributions F and G, we can use their uniform variates $U = F(X)$ and $V = G(Y)$, and can re-write the Spearman rho measure as

$$\begin{aligned}\rho_S(X, Y) &= 12 \int\int_{[0,1]^2} uv dC(u, v) - 3 = 12\mathbb{E}[UV] - 3 \\ &= \frac{\mathbb{E}[UV] - \frac{1}{4}}{\frac{1}{12}} = \frac{\mathrm{Cov}(X, Y)}{\sqrt{\mathrm{Var}(X)}\sqrt{\mathrm{Var}(Y)}} \\ &= \rho(F(X), G(Y)).\end{aligned}$$

Tail Dependence. Another important concept, especially for heavy-tailed distributions, is the tail dependence coefficient: it quantifies the amount of joint dependence in the tail of the distribution.

Definition 31 (*Tail Dependence*) Let X and Y be two continuous random variables with marginal distributions F and G. The upper tail dependence coefficient is defined as

$$\lim_{u \uparrow 1} \mathbb{P}\left(Y > G^{-1}(u) \middle| X > F^{-1}(u)\right) = \lambda_U,$$

if the limit $\lambda_U \in [0, 1]$ exists.

If $\lambda_U > 0$, we say that X and Y are asymptotically dependent in the upper tail.

If $\lambda_U = 0$, we say that X and Y are asymptotically independent in the upper tail.

We can re-write the upper tail conditional probability $\mathbb{P}\left(Y > G^{-1}(u) \middle| X > F^{-1}(u)\right)$ as

$$\frac{1 - \mathbb{P}\left(X \le F^{-1}(u)\right) - \mathbb{P}\left(Y \le G^{-1}(u)\right) + \mathbb{P}\left(X \le F^{-1}(u), Y \le G^{-1}(u)\right)}{1 - \mathbb{P}\left(X \le F^{-1}(u)\right)},$$

which gives an alternative definition in terms of the copula function.

Definition 32 (*Copula Tail Dependence*) The upper tail dependence λ_U for a bivariate copula function C is defined as

$$\lim_{u \uparrow 1} \frac{1 - 2u + C(u, u)}{1 - u} = \lambda_U,$$

if the limit λ_U exists.

Similarly, we can also define a lower tail dependence measure in a symmetric way

$$\lim_{u \downarrow 0} \frac{C(u, u)}{u} = \lambda_L.$$

2.3.3 Elliptical Copulas

Elliptical distributions are typically the most commonly used multivariate distribution functions. They enjoy many of the multivariate normal distribution tractability features; and they can also be simulated very easily.

Elliptical copulas are the copula functions generated from elliptical distributions.

Definition 33 (*Elliptical Distributions*) Let $\mathbf{X}$ be an n-dimensional random variable; Fix a real vector $\mu \in \mathbb{R}^n$, a positive definite, symmetric matrix $n \times n$ matrix Σ, and a real function $\phi(.)$. We say that $\mathbf{X}$ has an elliptical distribution, $\mathbf{X} \sim E_n(\mu, \Sigma, \phi)$, with parameters μ, Σ, ϕ if the characteristic function of the vector $\mathbf{X} - \mu$ is a function of the quadratic form $\mathbf{t}^T \Sigma \mathbf{t}$:

$$\varphi_{X-\mu}(\mathbf{t}) = \mathbb{E}\left[\exp\left(\mathbf{t}^T (\mathbf{X} - \mu)\right)\right] = \phi\left(\mathbf{t}^T \Sigma \mathbf{t}\right).$$

The most important elliptical copulas are the Gaussian copula and the t-copula.

Gaussian Copula. The copula of the n-variate normal distribution with Gaussian correlation matrix R is given by

$$C_R^G(\mathbf{u}) = \Phi_R^n\left(\Phi^{-1}(u_1), \ldots, \Phi^{-1}(u_n)\right),$$

where Φ_R^n denotes the n-dimensional multivariate standard normal distribution with correlation matrix R; Φ^{-1} denotes the inverse of the univariate standard normal distribution. In the bivariate case, we can write the Gaussian copula function as

$$C_R^G(u_1, u_2) = \int_{-\infty}^{\Phi^{-1}(u_1)} \int_{-\infty}^{\Phi^{-1}(u_2)} \frac{1}{2\pi\sqrt{1-R_{12}^2}} \exp\left(-\frac{x^2 - 2R_{12}xy + y^2}{2\left(1-R_{12}^2\right)}\right) dxdy.$$

Student t-copula. If the vector $\mathbf{X}$ can be represented as

$$\mathbf{X} \stackrel{d}{=} \mu + \frac{\sqrt{\nu}}{\sqrt{S}}\mathbf{Z},$$

where $\mu \in \mathbb{R}^n$, and the random variables $S \sim \chi_\nu^2$ and $Z \sim \mathcal{N}_n(\mathbf{0}, \Sigma)$ are independent, then $\mathbf{X}$ has an n-dimensional multivariate t_ν-distribution, with mean μ (for $\nu > 1$) and covariance matrix $\frac{\nu}{\nu-2}\Sigma$ (for $\nu > 2$).

The (Student) t-copula is then defined as

$$C_{\nu,R}^t(\mathbf{u}) = t_{\nu,R}^n\left(t_\nu^{-1}(u_1), \ldots, t_\nu^{-1}(u_n)\right),$$

where $R_{ij} = \frac{\Sigma_{ij}}{\sqrt{\Sigma_{ii}\Sigma_{jj}}}$, is $t_{\nu,R}^n$ is the n-dimensional multivariate t-distribution with parameters (ν, R), and is the inverse of the univariate t_ν-distribution with ν degrees of freedom. In the bivariate case, the expression of the copula function is

$$C_{\nu,R}^t(u_1, u_2) = \int_{-\infty}^{t_\nu^{-1}(u_1)} \int_{-\infty}^{t_\nu^{-1}(u_2)} \frac{1}{2\pi\sqrt{1-R_{12}^2}} \left(1 + \frac{x^2 - 2R_{12}xy + y^2}{\nu\left(1-R_{12}^2\right)}\right)^{-\frac{\nu+2}{2}} dxdy.$$

2.3.4 Archimedean Copulas

An interesting class of copula functions used in finance insurance applications to model asymmetric losses and gains is the Archimedean copula. Unlike their elliptical counterparts, which are derived from a given family of multivariate distributions, Archimedean copulas are instead constructed, by hand, from a set of generator functions, thereby creating a rich variety of dependence structures. In addition to the asymmetric properties offered by these copulas, they also enjoy easy-to-implement closed-form expressions, which is very useful in numerical applications.

We start with the definition of a pseudo-inverse function needed to construct the Archimedean copula.

Definition 34 (*Pseudo-inverse*) Let $\varphi : [0, 1] \to [0, \infty]$ be a continuous, strictly decreasing function, with $\varphi(1) = 0$. The pseudo-inverse of φ is the function $\varphi^{[-1]} : [0, \infty] \to [0, 1]$ defined as

$$\varphi^{[-1]}(x) = \begin{cases} \varphi^{-1}(x), \text{ for } 0 \leq x \leq \varphi(0), \\ 0, \text{ if } \varphi(0) < x \leq \infty. \end{cases}$$

We can now give a general definition of bivariate Archimedean copulas (see Nelsen 1999).

Definition 35 (*Bivariate Archimedean Copula*) Let $\varphi : [0, 1] \to [0, \infty]$ be a continuous, strictly decreasing function, with $\varphi(1) = 0$, and let $\varphi^{[-1]}$ be its pseudo-inverse. Define the bivariate function $C : [0, 1]^2 \to [0, 1]$ by

$$C(u, v) = \varphi^{[-1]}(\varphi(u) + \varphi(v)).$$

Then, the function C is a copula if and only if is φ convex. Copulas of this form are called Archimedean copulas; and φ is the generator of the copula. If $\varphi(0) = \infty$, we say that φ is a strict generator and C is a strict Archimedean copula.

We give a few popular examples.

Example 36 (Gumbel Copula). Let $\varphi(x) = (-\ln x)^\theta$, where $\theta \geq 1$. Its copula function $C_\theta(u, v)$ is called a Gumbel copula:

$$C_\theta(u, v) = \exp\left(-\left((-\ln u)^\theta + (-\ln v)^\theta\right)^{\frac{1}{\theta}}\right).$$

Example 37 (Clayton Copula). Let $\varphi(x) = \frac{x^{-\theta}-1}{\theta}$, where $\theta \in [-1, \infty)$, $\theta \neq 0$. Its copula function $C_\theta(u, v)$ is called a Clayton copula:

$$C_\theta(u, v) = \max\left(\left(u^{-\theta} + v^{-\theta} - 1\right)^{-\frac{1}{\theta}}, 0\right).$$

Example 38 (Frank Copula). Let $\varphi(x) = -\ln \frac{e^{-\theta x}-1}{e^{-\theta}-1}$, where $\theta \in \mathbb{R}$, $\theta \neq 0$. Its copula function $C_\theta(u, v)$ is called a Frank copula:

$$C_\theta(u, v) = -\frac{1}{\theta} \ln\left(1 + \frac{\left(e^{-\theta u} - 1\right)\left(e^{-\theta v} - 1\right)}{e^{-\theta} - 1}\right).$$

To generalize to an n-dimensional Archimedean copula, we can construct, by extension, the function C^n as

$$C^n(\mathbf{u}) = \varphi^{[-1]}(\varphi(u_1) + \cdots + \varphi(u_n)),$$

but we also need to show that it is indeed a copula function under some conditions.

First, we define what it is meant by a completely monotone function: we say that a function $g(x)$ is completely monotone on the interval I, if it has derivatives of all orders, which alternate in sign, i.e., it satisfies

$$(-1)^k \frac{d^k}{dx^k} g(x) \geq 0,$$

for all $k \geq 0$, and all x in the interior of the interval I.

We can now state the following theorem from Kimberling (1974), which gives necessary and sufficient conditions for the function C^n to be a copula.

Theorem 39 (Kimberling 1974) *Let $\varphi : [0,1] \to [0,\infty]$ be a continuous, strictly decreasing function, with $\varphi(1) = \infty$ and $\varphi(1) = 0$, and let $\varphi^{[-1]}$ be its pseudo-inverse. The n-dimensional function $C^n : [0,1]^n \to [0,1]$ defined by*

$$C^n(\mathbf{u}) = \varphi^{[-1]}(\varphi(u_1) + \cdots + \varphi(u_n)),$$

is a n-copula, for all $n \geq 2$, if and only if φ^{-1} is completely monotone on $[0,\infty)$.

Some n-dimensional examples are given below.

Example 40 (Gumbel Copula). Let $\varphi_i(t) = (-\ln t)^{\theta_i}$, with $\theta_i \geq 1$, for $1 \leq i \leq n$, be the generators of the Gumbel copula. The n-dimensional extension of the Gumbel family of copula functions is an n-copula if $\theta_1 \leq \cdots \leq \theta_n$.

Example 41 The Archimedean copula family defined with generators $\varphi_i(t) = (t^{-1} - 1)^{\theta_i}$, for $\theta_i \geq 1$, is indeed an n-copula if $\theta_1 \leq \cdots \leq \theta_n$.

2.3.5 Marshall-Olkin Copulas

Next, we discuss the general class of Marshall-Olkin copula functions. We start with the bivariate case, then generalize to the n-dimensional case.

Suppose we have a system with two components subject to some independent shocks, which can trigger the failure of one of the components separately or both components at the same time. Thus, we have three independent Poisson processes with intensities $(\lambda_1, \lambda_2, \lambda_{12})$ respectively. Their corresponding first jump times are $(\theta_1, \theta_2, \theta_{12})$: they are independent exponentially-distributed variables with parameters $(\lambda_1, \lambda_2, \lambda_{12})$ respectively. We denote by τ_1 and τ_2 the failure time of the two components.

The joint survival probability function of the two components, $\overline{H}(T_1, T_2)$, is given by

$$\begin{aligned}\overline{H}(T_1, T_2) &= \mathbb{P}(\tau_1 > T_1, \tau_2 > T_2) = \mathbb{P}(\theta_1 > T_1)\,\mathbb{P}(\theta_2 > T_2)\,\mathbb{P}(\theta_{12} > \max(T_1, T_2)) \\ &= \exp(-\lambda_1 T_1)\exp(-\lambda_2 T_2)\exp(-\lambda_{12}\max(T_1, T_2)).\end{aligned}$$

Similarly, the univariate survival probabilities, $\overline{F_1}(T_1)$ and $\overline{F_2}(T_2)$, are

$$\begin{aligned}\overline{F_1}(T_1) &= \mathbb{P}(\tau_1 > T_1) = \mathbb{P}(\theta_1 > T_1)\,\mathbb{P}(\theta_{12} > \max(T_1, T_2)) = \exp(-\lambda_1 T_1 - \lambda_{12}\max(T_1, T_2)), \\ \overline{F_2}(T_2) &= \mathbb{P}(\tau_2 > T_2) = \mathbb{P}(\theta_2 > T_2)\,\mathbb{P}(\theta_{12} > \max(T_1, T_2)) = \exp(-\lambda_2 T_2 - \lambda_{12}\max(T_1, T_2)).\end{aligned}$$

Define the ratios $\alpha_1 = \frac{\lambda_{12}}{\lambda_1+\lambda_{12}}$ and $\alpha_2 = \frac{\lambda_{12}}{\lambda_2+\lambda_{12}}$, and substitute the univariate (survival) marginals into the bivariate (survival) distribution function

$$\overline{H}(T_1, T_2) = \overline{F_1}(T_1)\,\overline{F_2}(T_2)\min\left(\left(\overline{F_1}(T_1)\right)^{-\alpha_1}, \left(\overline{F_2}(T_2)\right)^{-\alpha_2}\right).$$

This leads to the family of Marshall-Olkin copula functions

$$C_{\alpha_1,\alpha_2}(u_1, u_2) = \min\left(u_1^{1-\alpha_1}u_2, u_1 u_2^{1-\alpha_2}\right).$$

This copula function has both an absolutely continuous part

$$A_{C_{\alpha_1,\alpha_2}} = \frac{\partial^2}{\partial u_1 \partial u_2} C_{\alpha_1,\alpha_2}(u_1, u_2) = \begin{cases} u_1^{-\alpha_1}, \text{ if } u_1^{\alpha_1} > u_2^{\alpha_2}, \\ u_2^{-\alpha_2}, \text{ if } u_1^{\alpha_1} > u_2^{\alpha_2}; \end{cases}$$

and a singularity on the region defined by the curve $\left\{u_1^{\alpha_1} = u_2^{\alpha_2}\right\}$, where we have a simultaneous failure of both components at some time θ_{12}.

Kendall's tau, Spearman's rho and upper tail dependence, for this copula function, can be computed easily and are given by

$$\begin{aligned}\tau\left(C_{\alpha_1,\alpha_2}\right) &= \frac{\alpha_1\alpha_2}{\alpha_1 + \alpha_2 - \alpha_1\alpha_2}, \\ \rho_S\left(C_{\alpha_1,\alpha_2}\right) &= \frac{3\alpha_1\alpha_2}{2\alpha_1 + 2\alpha_2 - \alpha_1\alpha_2}, \\ \lambda_U &= \min(\alpha_1, \alpha_2).\end{aligned}$$

For the n-dimensional generalization, we have n components, and $2^n - 1$ common shocks, which can trigger the failure of one or more components in the system. We denote by Π_n the set of all non-empty subset of $\{1, \ldots, n\}$. We have $2^n - 1$ independent Poisson shocks N^{π}, $\pi \in \Pi_n$, with intensities λ_{π}, which can trigger the failure of the components in the subset π only. Their first jump times are θ_{π}. The failure time of each individual component is then given by: $\tau_i = \min\{\theta_{\pi} : i \in \pi\}$.

The (τ_i, τ_j)-bivariate marginal of the Marshall-Olkin copula is also a Marshall-Olkin copula with parameters

$$\alpha_i = \frac{\sum_{\pi: i\in\pi, j\in\pi} \lambda_{\pi}}{\sum_{\pi: i\in\pi} \lambda_{\pi}} \text{ and } \alpha_j = \frac{\sum_{\pi: i\in\pi, j\in\pi} \lambda_{\pi}}{\sum_{\pi: j\in\pi} \lambda_{\pi}};$$

the Kendall tau and Spearman rho coefficients are given by

$$\tau\left(C_{\alpha_i,\alpha_j}\right) = \frac{\alpha_i\alpha_j}{\alpha_i + \alpha_j - \alpha_i\alpha_j} \text{ and } \rho_S\left(C_{\alpha_i,\alpha_j}\right) = \frac{3\alpha_i\alpha_j}{2\alpha_i + 2\alpha_j - \alpha_i\alpha_j}.$$

The n-dimensional Marshall-Olkin copula function provides a very rich joint dependence structure, with enough flexibility to capture the granular joint probabilities of every combination of sub-defaults; but with 2^n combinations to deal with, the problem explodes rapidly as the number of names grows. We shall see later in Chap. 7 that we can build a parsimonious parametrization of the model, which gives the desired default clustering properties that we need while maintaining numerical tractability.

References

R. Barlow, F. Proschan, *Statistical Theory of Reliability and Life Testing* (Silver Spring, Maryland, 1981)

P. Brémaud, *Point Processes and Queues: Martingale Dynamics* (Springer, New York, 1980)

P. Embrechts, F. Lindskog, A. McNeil, Modelling dependence with copulas and applications to risk management, in *Handbook of Heavy Tailed Distributions in Finance*, ed. by S.T. Rachev (Amsterdam, Elsevier/North-Holland, 2003)

M. Fréchet, Les tableaux de corrélation dont les marges sont données. Annales de l'Université de Lyon, Sciences Math ématiques et Astronomie **20**, 13–31 (1957)

G.M. Gupton, C.C. Finger, M. Bhatia, *CreditMetrics-Technical Document* (Morgan Guaranty Trust Co, New York, 1997)

H. Joe, *Multivariate Models and Dependence Concepts* (Chapman & Hall, London, 1997)

N. Johnson, S. Kotz, *Distributions in Statistics: Continuous Multivariate Distributions* (Wiley, New York, 1972)

C. Kimberling, A Probabilistic Interpretation of Complete Monotonicity. Aequationes Mathematicae **10**, 152–164 (1974)

D. Lando, On cox processes and credit risky securities. Rev. Deriv. Res. **2**(2/3), 99–120 (1998)

F. Lindskog, A. McNeil, Common poisson shock models: applications to insurance and credit risk modelling. ASTIN Bullet. **33**(2), 209–238 (2003)

A. Marshall, I. Olkin, A multivariate exponential distribution. J. Am. Stat. Assoc. **62**, 30–44 (1967)

R. Nelsen, *An Introduction to Copulas* (Springer, New York, 1999)

A. Sklar, Fonctions de répartition à *n* dimensions et leurs marges. Publications de l'Institut de Statistique de l'Universit é de Paris **8**, 229–231 (1959)

S. Watanabe, Additive functionals of Markov processes and Levy systems. Jpn. J. Math. **34**, 53–79 (1964)

3

Expectations in the Enlarged Filtration

In this chapter, we derive a formula of the conditional expectation with respect to the enlarged filtration. This is a generalization of the Dellacherie formula. We shall use this key result to compute the expectations that we encounter in the conditional jump diffusion framework. In particular, the conditional survival probability can be computed with our formula. We apply this result in Chap. 4 where conditional survival probability calculations, on the enlarged filtration, are carried out in details.

3.1 The Dellacherie Formula

We work on probability space $(\Omega, \mathcal{G}, P)$ where we have a set of default times $(\tau_1, \ldots, \tau_n)$, and an $\mathbb{R}^d$-valued Itô process $(X_t)_{t \geq 0}$, describing the evolution of the state-variables in the economy. $\{\mathcal{F}_t\}$ is the background filtration generated by X. And $\left\{\mathcal{H}_t^i\right\}$ are the single-name default filtrations $\mathcal{H}_t^i \triangleq \sigma\left(\mathbf{1}_{\{\tau_i \leq s\}} : 0 \leq s \leq t\right)$. The enlarged filtration $\{\mathcal{G}_t\}$ contains both the individual default filtrations and the background diffusion: $\mathcal{G}_t = \mathcal{F}_t \vee \bigvee_{i=1}^n \mathcal{H}_t^i$.

We start with the following result established in Dellacherie (1970).

Lemma 42 *Let Y be a $\mathcal{G}$-measurable random variable. Then, we have*

$$\mathbb{E}\left[Y \left| \mathcal{F}_t \vee \mathcal{H}_t^i\right.\right] = \mathbf{1}_{\{\tau_i \leq t\}} \mathbb{E}\left[Y \left| \mathcal{F}_t \vee \mathcal{H}_\infty^i\right.\right] + \mathbf{1}_{\{\tau_i > t\}} \frac{\mathbb{E}\left[Y \times \mathbf{1}_{\{\tau_i > t\}} \left| \mathcal{F}_t\right.\right]}{\mathbb{E}\left[\mathbf{1}_{\{\tau_i > t\}} \left| \mathcal{F}_t\right.\right]}.$$

Y. Elouerkhaoui, *Credit Correlation*, Applied Quantitative Finance,
DOI 10.1007/978-3-319-60973-7_3

In order to compute the conditional expectation with respect to the filtration $\{\mathcal{F}_t \vee \mathcal{H}_t^i\}$, one needs to consider the two possible default states. On each set, representing whether a default has occurred before t or not, the conditional distribution is different. In general, for n default times $(\tau_1, ..., \tau_n)$, we have a set of 2^n default states; one has to compute the conditional expectation with respect to the enlarged filtration $\{\mathcal{G}_t\} = \{\mathcal{F}_t \vee \bigvee_{i=1}^n \mathcal{H}_t^i\}$ on each default scenario. To be more precise, we introduce the following notations.

Notation. At time t, each default state is represented by $\pi \in \Pi_\mathbf{n}$, where $\Pi_\mathbf{n}$ is the set of all subsets of $\{1, ..., n\}$. $\pi = \emptyset$ means that there has been no defaults before t; $\pi = \{1, ..., n\}$ means that every obligor has already defaulted. If $\pi = \{j_1, ..., j_k\}$ for some indexes $j_m \in \{1, ..., n\}$, then these indicate the obligors that have defaulted. To the subset π , we associate the indicator $D_t^{(\pi)}$, which is equal to 1 if we are in the default state (π) or 0 otherwise:

$$D_t^{(\pi)} \triangleq \left[\prod_{j\in\pi} \left(D_t^j\right)\right] \times \left[\prod_{j\notin\pi} \left(1 - D_t^j\right)\right]. \tag{3.1}$$

We also define the filtration $\left\{\mathcal{G}_t^{(\pi)}\right\}$ as:

$$\mathcal{G}_t^{(\pi)} \triangleq \mathcal{F}_t \vee \left[\bigvee_{j\in\pi} \mathcal{H}_\infty^j\right] = \mathcal{F}_t \vee \left[\bigvee_{j\in\pi} \sigma\left(\tau_j\right)\right]. \tag{3.2}$$

For instance, if $n = 2$, then we have 4 possible default states: $\Pi_\mathbf{2} = \{\emptyset, \{1\}, \{2\}, \{1, 2\}\}$; the default state indicators are

$$D_t^{(\emptyset)} \triangleq \mathbf{1}_{\{\tau_1>t\}}\mathbf{1}_{\{\tau_2>t\}};\ D_t^{(\{1\})} \triangleq \mathbf{1}_{\{\tau_1\le t\}}\mathbf{1}_{\{\tau_2>t\}};$$
$$D_t^{(\{2\})} \triangleq \mathbf{1}_{\{\tau_1>t\}}\mathbf{1}_{\{\tau_2\le t\}};\ D_t^{(\{1,2\})} \triangleq \mathbf{1}_{\{\tau_1\le t\}}\mathbf{1}_{\{\tau_2\le t\}}.$$

The filtrations $\left\{\mathcal{G}_t^{(\emptyset)}\right\}\left\{\mathcal{G}_t^{(\{1\})}\right\}\left\{\mathcal{G}_t^{(\{2\})}\right\}\left\{\mathcal{G}_t^{(\{1,2\})}\right\}$ follow immediately from (3.2).

3.2 Generalized Dellacherie Formula

Next, we state our conditional expectation result.

Proposition 43 (Conditional expectation w.r.t. the enlarged filtration) *Let Y be a $\mathcal{G}$-measurable random variable. Then, we have*

$$\mathbb{E}\left[Y\left|\mathcal{G}_t\right.\right] = \sum_{\pi\in\Pi_{\mathbf{n}}} D_t^{(\pi)} \frac{\mathbb{E}\left[Y \times \prod_{j\notin\pi}\left(1-D_t^j\right)\left|\mathcal{G}_t^{(\pi)}\right.\right]}{\mathbb{E}\left[\prod_{j\notin\pi}\left(1-D_t^j\right)\left|\mathcal{G}_t^{(\pi)}\right.\right]}.$$

The proof is based on a repeated use of Lemma 42.

Proof We shall proceed by induction. By Lemma 42, the property is satisfied for $n=1$.

Assume that the formula is true for n and let us prove that it holds for $n+1$.

Define for $n\geq 1$, the filtration $\left\{\mathcal{F}_t^{(n)}\right\} \triangleq \{\mathcal{F}_t \vee \bigvee_{i=1}^n \mathcal{H}_t^i\}$.

Applying Lemma 42 to the filtration $\left\{\mathcal{F}_t^{(n+1)}\right\} = \left\{\mathcal{F}_t^{(n)} \vee \mathcal{H}_t^{n+1}\right\}$, we get

$$\mathbb{E}\left[Y\left|\mathcal{F}_t^{(n+1)}\right.\right] = D_t^{n+1}\mathbb{E}\left[Y\left|\mathcal{F}_t^{(n)} \vee \mathcal{H}_\infty^{n+1}\right.\right] + \left(1-D_t^{n+1}\right)\frac{\mathbb{E}\left[Y\times\left(1-D_t^{n+1}\right)\left|\mathcal{F}_t^{(n)}\right.\right]}{\mathbb{E}\left[\left(1-D_t^{n+1}\right)\left|\mathcal{F}_t^{(n)}\right.\right]}. \tag{3.3}$$

Observing that $\left\{\mathcal{F}_t^{(n)} \vee \mathcal{H}_\infty^{n+1}\right\} = \{[\mathcal{F}_t \vee \mathcal{H}_\infty^{n+1}] \vee [\bigvee_{i=1}^n \mathcal{H}_t^i]\}$, the first term of equation (3.3) can be expanded by applying the induction relationship:

$$\begin{aligned}
&D_t^{n+1}\mathbb{E}\left[Y\left|\mathcal{F}_t^{(n)} \vee \mathcal{H}_\infty^{n+1}\right.\right] \\
&= D_t^{n+1}\mathbb{E}\left[Y\left|\left[\mathcal{F}_t \vee \mathcal{H}_\infty^{n+1}\right] \vee \left[\bigvee_{i=1}^{n} \mathcal{H}_t^i\right]\right.\right] \\
&= D_t^{n+1}\sum_{\pi_n\in\Pi_{\mathbf{n}}} D_t^{(\pi_n)} \frac{\mathbb{E}\left[Y\times\prod_{j\notin\pi_n}\left(1-D_t^j\right)\left|\left[\mathcal{F}_t\vee\mathcal{H}_\infty^{n+1}\right]\vee\left[\bigvee_{j\in\pi_n}\mathcal{H}_\infty^j\right]\right.\right]}{\mathbb{E}\left[\prod_{j\notin\pi_n}\left(1-D_t^j\right)\left|\left[\mathcal{F}_t\vee\mathcal{H}_\infty^{n+1}\right]\vee\left[\bigvee_{j\in\pi_n}\mathcal{H}_\infty^j\right]\right.\right]}.
\end{aligned} \tag{3.4}$$

Define a partition of $\Pi_{\mathbf{n+1}}$: $\Pi_{\mathbf{n+1}} = \Pi^+_{\mathbf{n+1}} \cup \Pi^-_{\mathbf{n+1}}$, and $\Pi^+_{\mathbf{n+1}} \cap \Pi^-_{\mathbf{n+1}} = \emptyset$, where $\Pi^+_{\mathbf{n+1}}$ is the set of all subsets containing $(n+1)$, and $\Pi^-_{\mathbf{n+1}}$ is the set of all subsets not containing $(n+1)$. They are constructed as:

$$\Pi^+_{\mathbf{n+1}} = \{\pi_n \cup \{n+1\} : \pi_n \in \Pi_{\mathbf{n}}\},$$
$$\Pi^-_{\mathbf{n+1}} = \{\pi_n \cup \emptyset : \pi_n \in \Pi_{\mathbf{n}}\}.$$

Equation (3.4), then, becomes

$$\ldots = \sum_{\pi_{n+1} \in \Pi^+_{\mathbf{n}}} D_t^{(\pi_{n+1})} \frac{\mathbb{E}\left[Y \times \prod_{j \notin \pi_{n+1}} \left(1 - D_t^j\right) \middle| \mathcal{F}_t \vee \left[\bigvee_{j \in \pi_{n+1}} \mathcal{H}^j_\infty\right]\right]}{\mathbb{E}\left[\prod_{j \notin \pi_{n+1}} \left(1 - D_t^j\right) \middle| \mathcal{F}_t \vee \left[\bigvee_{j \in \pi_{n+1}} \mathcal{H}^j_\infty\right]\right]}. \tag{3.5}$$

The second term in Eq. (3.3),

$$\left(1 - D_t^{n+1}\right) \frac{\mathbb{E}\left[Y \times \left(1 - D_t^{n+1}\right) \middle| \mathcal{F}_t^{(n)}\right]}{\mathbb{E}\left[\left(1 - D_t^{n+1}\right) \middle| \mathcal{F}_t^{(n)}\right]},$$

can also be computed by the induction relationship. To expand the numerator, we apply it to the variable $\left[Y \times \left(1 - D_t^{n+1}\right)\right]$:

$$\mathbb{E}\left[Y \times \left(1 - D_t^{n+1}\right) \middle| \mathcal{F}_t^{(n)}\right] = \sum_{\pi_n \in \Pi_{\mathbf{n}}} D_t^{(\pi_n)} \frac{\mathbb{E}\left[\left[Y \times \left(1 - D_t^{n+1}\right)\right] \times \prod_{j \notin \pi_n} \left(1 - D_t^j\right) \middle| \mathcal{G}_t^{(\pi_n)}\right]}{\mathbb{E}\left[\prod_{j \notin \pi_n} \left(1 - D_t^j\right) \middle| \mathcal{G}_t^{(\pi_n)}\right]},$$

and to expand the denominator, we apply it to the variable $\left(1 - D_t^{n+1}\right)$:

$$\mathbb{E}\left[\left(1 - D_t^{n+1}\right) \middle| \mathcal{F}_t^{(n)}\right] = \sum_{\pi'_n \in \Pi_{\mathbf{n}}} D_t^{(\pi'_n)} \frac{\mathbb{E}\left[\left[\left(1 - D_t^{n+1}\right)\right] \times \prod_{j \notin \pi'_n} \left(1 - D_t^j\right) \middle| \mathcal{G}_t^{(\pi'_n)}\right]}{\mathbb{E}\left[\prod_{j \notin \pi'_n} \left(1 - D_t^j\right) \middle| \mathcal{G}_t^{(\pi'_n)}\right]}.$$

The ratio is obtained by observing that $D_t^{(\pi_n)} D_t^{(\pi_n')} = 0$ if $\pi_n \neq \pi_n'$, so that we are left in the denominator with one term, which corresponds to $\pi_n = \pi_n'$.

$$\frac{\mathbb{E}\left[Y \times \left(1 - D_t^{n+1}\right) \middle| \mathcal{F}_t^{(n)}\right]}{\mathbb{E}\left[\left(1 - D_t^{n+1}\right) \middle| \mathcal{F}_t^{(n)}\right]}$$

$$= \sum_{\pi_n \in \Pi_\mathbf{n}} D_t^{(\pi_n)} \frac{\mathbb{E}\left[\left[Y \times \left(1 - D_t^{n+1}\right)\right] \times \prod_{j \notin \pi_n} \left(1 - D_t^j\right) \middle| \mathcal{G}_t^{(\pi_n)}\right]}{\mathbb{E}\left[\prod_{j \notin \pi_n} \left(1 - D_t^j\right) \middle| \mathcal{G}_t^{(\pi_n)}\right] \times \left[D_t^{(\pi_n)} \frac{\mathbb{E}\left[\left[\left(1 - D_t^{n+1}\right)\right] \times \prod_{j \notin \pi_n} \left(1 - D_t^j\right) \middle| \mathcal{G}_t^{(\pi_n)}\right]}{\mathbb{E}\left[\prod_{j \notin \pi_n} \left(1 - D_t^j\right) \middle| \mathcal{G}_t^{(\pi_n)}\right]}\right]}$$

$$= \sum_{\pi_n \in \Pi_\mathbf{n}} D_t^{(\pi_n)} \frac{\mathbb{E}\left[\left[Y \times \left(1 - D_t^{n+1}\right)\right] \times \prod_{j \notin \pi_n} \left(1 - D_t^j\right) \middle| \mathcal{G}_t^{(\pi_n)}\right]}{\mathbb{E}\left[\left[\left(1 - D_t^{n+1}\right)\right] \times \prod_{j \notin \pi_n} \left(1 - D_t^j\right) \middle| \mathcal{G}_t^{(\pi_n)}\right]}. \tag{3.6}$$

Using the $\Pi_{\mathbf{n+1}}$ corresponding default states, Eq. (3.6) becomes

$$\left(1 - D_t^{n+1}\right) \frac{\mathbb{E}\left[Y \times \left(1 - D_t^{n+1}\right) \middle| \mathcal{F}_t^{(n)}\right]}{\mathbb{E}\left[\left(1 - D_t^{n+1}\right) \middle| \mathcal{F}_t^{(n)}\right]} = \sum_{\pi_{n+1} \in \Pi_\mathbf{n}^-} D_t^{(\pi_{n+1})} \frac{\mathbb{E}\left[[Y] \times \prod_{j \notin \pi_{n+1}} \left(1 - D_t^j\right) \middle| \mathcal{G}_t^{(\pi_{n+1})}\right]}{\mathbb{E}\left[\prod_{j \notin \pi_{n+1}} \left(1 - D_t^j\right) \middle| \mathcal{G}_t^{(\pi_{n+1})}\right]}. \tag{3.7}$$

Combining (3.5) and (3.7) ends the proof. ■

References

C. Dellacherie, Un exemple de la théorie générale des processus, Séminaires de Probabilités IV. Lecture Notes in Math. 124 (Springer, Berlin, 1970), pp. 60–70

C. Dellacherie, *Capacités et processus stochastiques* (Springer, Berlin, 1972)

C. Dellacherie, P.A. Meyer, *Probabilités et potentiel* (Hermann, Paris, 1975)

C. Dellacherie, P.A. Meyer, A propos du travail de Yor sur les grossissements des tribus, Séminaire de Probabilités XII. Lecture Notes in Math. 649 (Springer, Berlin, 1978) pp. 69–78

4

Copulas and Conditional Jump Diffusions

Enlarging the economic state-variables' filtration by observing the default process of all available credits has some profound implications on the dynamics of intensities. Indeed, the sudden default of one credit triggers jumps in the spreads of all the other obligors. This is what we refer to as the "Conditional Jump Diffusion" effect. The aim of this chapter is to give a comprehensive and self-contained presentation of the CJD framework. In particular, we derive the default times' density function in the "looping" defaults model, and we study the equivalence between the copula approach and the conditional jumps framework. This is a key result that we will use, in practice, to calibrate non-observable default correlation parameters.

4.1 Introduction

The intensity of each default time studied on its natural filtration has a different dynamic from the one observed on the enlarged filtration where the default information of the other obligors is taken into account. This fundamental observation was made in the papers by Jeanblanc and Rutkowski (2000a, b) and Elliott et al. (2000). Enlarging the working filtration or, in financial terms, increasing the flow of information available to investors, profoundly alters the intensity dynamics. A sudden default in one credit can create a shock wave in the market, which translates into revising its estimate of the default likelihood of the other obligors; this, in turn, triggers a jump in their credit spreads. This phenomenon will be referred to henceforth as the "Conditional Jump Diffusion" effect.

Y. Elouerkhaoui, *Credit Correlation*, Applied Quantitative Finance,
DOI 10.1007/978-3-319-60973-7_4

The first example of conditional jump diffusion dynamics was given in Kusuoka (1999). This example is also studied in Jeanblanc and Rutkowski (2000b). The infectious defaults model of Davis and Lo (2001b) assumes a CJD dynamic, where, upon default, the intensities jump proportionally to their pre-default levels, then revert back to the pre-default state. Deriving survival probabilities in the general CJD framework is far from being a trivial task. Jarrow and Yu (2001) have studied a simplified version of the model where closed-form formulas can be derived. They assume that the CJD coupling is unidirectional. In other words, the universe of credits is split into two subsets: a default event in the first subgroup triggers jumps in the spreads of the other obligors; the default of an obligor in the second set, on the other hand, has no impact on the spreads of the first.

Copulas and CJD dynamics are two sides of the same coin. A choice of copula implies a specification of a CJD dynamic, and vice-versa. Copulas and CJDs are equivalent in that sense. This was first studied in the paper by Schönbucher and Schubert (2001). This seminal work bridges the gap between the mathematician's approach to default correlation, where it is merely about a choice of copula function, and the trader's approach where default correlation translates into a windfall P&L in the event of default.

The aim of this chapter is to review and complement the literature on the modelling of default correlation in the so-called Conditional Jump Diffusion framework by giving a comprehensive and self-contained presentation of this approach. First, we present a general formula of the default times' multivariate distribution in the "looping" defaults model. Then, using the generalized Dellacherie formula, we consider the converse proposition and establish the equivalence between the copula approach and CJD dynamics along the lines of Schönbucher and Schubert (2001).

4.2 The Model

We work in an economy represented by a probability space $(\Omega, \mathcal{G}, P)$ and a time horizon $T^* \in (0, \infty)$, on which is given a d-dimensional Brownian motion W. We assume that the probability space $(\Omega, \mathcal{G}, P)$ is rich enough to support a set of n non-negative random variables $(\tau_1, ..., \tau_n)$ representing the default times of the obligors in the economy. Further, we assume, for convenience, that $\mathbb{P}(\tau_i = 0) = 0$ and $\mathbb{P}(\tau_i > t) > 0$ for any $t \in \mathbb{R}_+$.

We also introduce an $\mathbb{R}^d$-valued Itô process $(X_t)_{t \geq 0}$, describing the evolution of the state-variables in the economy, which solves the following SDE

$$dX_t = \alpha(X_t)\, dt + \beta(X_t)\, dW_t,$$

for some Lipschitz functions $\alpha_k : \mathbb{R}^d \to \mathbb{R}$ and $\beta_{kj} : \mathbb{R}^d \to \mathbb{R}^d$, $1 \le k \le d$, $1 \le j \le d$.

Filtrations. We denote by $\{\mathcal{F}_t\}_{0 \le t \le T^*}$ the filtration generated by X and augmented with the P -null sets of $\mathcal{G}$:

$$\mathcal{F}_t = \sigma\left(X_s : 0 \le s \le t\right) \vee \mathcal{N}.$$

We introduce, for each obligor i, the right-continuous process $D_t^i \triangleq \mathbf{1}_{\{\tau_i \le t\}}$ indicating whether the firm has defaulted or not. We denote by $\{\mathcal{H}_t^i\}$ the filtration generated by this process

$$\mathcal{H}_t^i \triangleq \mathcal{F}_t^{D^i} = \sigma\left(D_s^i : 0 \le s \le t\right).$$

We define the following filtrations:

1. The collective filtration of the economic state variables and the default processes
$$\mathcal{G}_t \triangleq \mathcal{F}_t \vee \left[\bigvee_{i=1}^{n} \mathcal{H}_t^i\right].$$

2. For each obligor i, the filtration generated by the state variables and the default processes of all other firms other than i
$$\mathcal{G}_t^{-i} \triangleq \mathcal{F}_t \vee \mathcal{H}_t^{-i},$$
$$\mathcal{H}_t^{-i} \triangleq \left[\bigvee_{j \ne i} \mathcal{H}_t^j\right] = \mathcal{H}_t^1 \vee \ldots \vee \mathcal{H}_t^{i-1} \vee \mathcal{H}_t^{i+1} \vee \ldots \vee \mathcal{H}_t^n.$$

3. The firm specific information generated by the default filtration of i and the state variables' filtration
$$\mathcal{G}_t^i \triangleq \mathcal{F}_t \vee \mathcal{H}_t^i.$$

We assume that the filtration $\{\mathcal{F}_t\}$ has the martingale invariance property with respect to the filtration $\{\mathcal{G}_t\}$, i.e., that hypothesis $(\mathcal{H})$ holds.

Hypothesis $(\mathcal{H})$. Every $\{\mathcal{F}_t\}$-square-integrable martingale is a $\{\mathcal{G}_t\}$-square-integrable martingale.

This implies, in particular, that the $\{\mathcal{F}_t\}$ -Brownian motion is a Brownian motion in the enlarged filtration $\{\mathcal{G}_t\}$.

Assumption 1. We assume that the probability of instantaneous joint defaults is equal to zero, i.e., $\mathbb{P}\left(\tau_i = \tau_j\right) = 0$, for $i \neq j$.

Intensities. We shall use the following definition of an intensity process for a stopping time τ with respect to a given filtration $\{\mathcal{I}_t\}$. We refer to Brémaud (1980) for details.

Definition 44 (*Intensity process*) Let τ be an $\{\mathcal{I}_t\}$ -stopping time and let λ be a non-negative $\{\mathcal{I}_t\}$-predictable process such that, for all $t \geq 0$, $\int_0^t \lambda_s ds < \infty$ almost surely. We say that λ is an $\{\mathcal{I}_t\}$-intensity of the stopping time τ if

$$M_t \triangleq D_t - \int_0^{t\wedge\tau} \lambda_s ds$$

is an $\{\mathcal{I}_t\}$-martingale. The process M_t is called the compensated point process.

Remark 45 *From Definition* 1, *the intensity is not uniquely defined after the occurrence of the default time. Indeed, if* λ_t *is an intensity process, then* $\lambda_t^1 = \lambda_t \mathbf{1}_{\{t\leq\tau\}}$ *and* $\lambda_t^2 = \lambda_t^1 + \theta \mathbf{1}_{\{\tau>t\}}$, *for all* $\theta \geq 0$, *are also intensity processes for* τ.

For each obligor i, we define two types of intensities: the first one with respect to the firm-specific filtration $\left\{\mathcal{G}_t^i\right\}$ and the second one with respect to the enlarged filtration $\{\mathcal{G}_t\}$.

Firm-Specific-Information Setting. We assume that τ_i has an intensity $\widetilde{h}^i$ with respect to the filtration $\left\{\mathcal{G}_t^i\right\}$. We know that there exists an $\{\mathcal{F}_t\}$-adapted process h^i such that $\mathbf{1}_{\{\tau_i>t\}}\widetilde{h}_t^i = \mathbf{1}_{\{\tau_i>t\}}h_t^i$ (see, for instance, Jeanblanc and Rutkowski (2000b) for details). The process h^i is the $\{\mathcal{F}_t\}$-adapted version of the $\left\{\mathcal{G}_t^i\right\}$ -intensity, i.e.,

$$h^i \text{ is } \{\mathcal{F}_t\}\text{-adapted, and } D_t^i - \int_0^{t\wedge\tau_i} h_s^i ds \text{ is a } \left\{\mathcal{G}_t^i\right\}\text{-martingale.}$$

Enlarged-Information Setting. We assume that τ_i has an intensity $\widetilde{\lambda}^i$ with respect to the enlarged filtration $\{\mathcal{G}_t\}$. There exists a $\left\{\mathcal{G}_t^{-i}\right\}$-adapted process λ^i such that $\mathbf{1}_{\{\tau_i>t\}}\widetilde{\lambda}_t^i = \mathbf{1}_{\{\tau_i>t\}}\lambda_t^i$. The process λ^i is the $\left\{\mathcal{G}_t^{-i}\right\}$-adapted version of the $\{\mathcal{G}_t\}$-intensity, i.e.,

$$\lambda^i \text{ is } \left\{\mathcal{G}_t^{-i}\right\}\text{-adapted, and } D_t^i - \int_0^{t\wedge\tau_i} \lambda_s^i ds \text{ is a } \{\mathcal{G}_t\}\text{-martingale.}$$

Expectations. If we work in the firm-specific filtration $\{\mathcal{G}_t^i\}$, it is well-known that the conditional survival probabilities can be computed as

$$\mathbb{P}\left(\tau_i > T \left| \mathcal{G}_t^i \right.\right) = \mathbf{1}_{\{\tau_i > t\}} \mathbb{E}\left[\exp\left(-\int_t^T h_s^i ds\right) | \mathcal{F}_t\right], \text{ for } T \geq t. \quad (4.1)$$

However, when we consider the general case where we work on the enlarged filtration $\{\mathcal{G}_t\}$, the conditional survival probability cannot be computed in a straightforward way. To address this issue, we shall use the change of measure technique introduced by Kusuoka (1999): under the new measure the "circular" nature of this type of "looping" default models is broken and calculations can be easily carried out.

Assumption 2. For practical applications, we make the assumption that the intensities are Markov-functionals of the background process X and the default indicators $\mathbf{D} \triangleq \left(D^1, ..., D^n\right)$. We assume that

$$\lambda_t^i = \lambda^i\left(X_t, D_t^1, ..., D_t^{i-1}, 0, D_t^{i+1}, ..., D_t^n\right),$$

for some bounded continuous functions $\lambda^i(.,.) : \mathbb{R}^d \times \{0, 1\}^n \to \mathbb{R}_+$, which are $\mathcal{C}^2$ in the first argument. This is similar to the Markovian setting considered in Frey and Backhaus (2004).

4.3 Interacting Itô and Point Processes

In this section, we consider a general model of "looping" defaults where the default point processes impact intensities, which, in turn, drive the default processes. This type of "circular" dependence between Itô and point processes has been studied in Becherer and Schweizer (2005). They have addressed, in particular, the question of existence and uniqueness of the solution.

Construction via a Change of Measure. To construct this non-standard dependence structure, we use a change of measure method, which extends the argument in Kusuoka (1999). The basic idea is to start with some probability space $\left(\Omega, \mathcal{G}', \left\{\mathcal{G}_t'\right\}, P'\right)$ on which we are given a Brownian motion W and a set of well defined independent default times with constant $\left(P', \left\{\mathcal{G}_t'\right\}\right)$-intensities equal to 1. Assume $\mathcal{G}_0'$ is trivial, $\mathcal{G}_{T^*}' = \mathcal{G}'$, and $\left\{\mathcal{G}_t'\right\}$ satisfies the usual conditions. Then, define the probability measure P as

$$\frac{dP}{dP'} = \mathcal{E}\left(\sum_{i=1}^{n}\int\left(\lambda_t^i - 1\right)\left(dD_t^i - \left(1 - D_t^i\right)dt\right)\right)_{T^*}. \tag{4.2}$$

Using Girsanov theorem, we can see that this measure change is such that λ_t^i is the $\left(P, \left\{\mathcal{G}_t'\right\}\right)$-intensity of τ_i. Again, by Girsanov, W is a local $\left(P, \left\{\mathcal{G}_t'\right\}\right)$-martingale; its quadratic covariance process $\langle W\rangle$ is the same under P and P', hence it is a $\left(P, \left\{\mathcal{G}_t'\right\}\right)$-Brownian motion. Define $(\Omega, \mathcal{G}, \{\mathcal{G}_t\}, P)$ as the P -completion of $\left(\Omega, \mathcal{G}', \left\{\mathcal{G}_t'\right\}, P'\right)$, one can verify that $\{\mathcal{G}_t\}$ satisfies the usual conditions under P. Therefore, W is a $(P, \{\mathcal{G}_t\})$ -Brownian motion.

We shall use the change of measure construction to derive the default times' density function in Proposition 46.

Non-Standard SDEs. Applying Itô's lemma to the intensity process $\lambda_t^i = \lambda^i(X_t, \mathbf{D}_t)$, one finds that the looping defaults model in a Markovian setting can be described as:

$(\lambda, \mathbf{D})$ is the solution of the following system of SDEs

$$d\lambda_t^i = \alpha^i(X_t, \mathbf{D}_t)\,dt + \beta^i(X_t, \mathbf{D}_t)\,dW_t + \sum_{\substack{j=1\\ j\neq i}}^{n}\Delta^{ij}(X_t, \mathbf{D}_t)\,dM_t^j, \tag{4.3}$$

the functions $\alpha^i(.,.) : \mathbb{R}^d \times \{0,1\}^n \to \mathbb{R}$, for $i = 1, ..., n$; $\beta^{il}(.,.) : \mathbb{R}^d \times \{0,1\}^n \to \mathbb{R}$, for $i = 1, ..., n$, $l = 1, ..., d$; $\Delta^{ij}(.,.) : \mathbb{R}^d \times \{0,1\}^n \to \mathbb{R}$ for $i = 1, ..., n$, $j = 1, ..., n$, are given by

$$\alpha^i(\mathbf{x}, \mathbf{y}) = \mathcal{L}^X\lambda^i(\mathbf{x}, \mathbf{y}) + \mathcal{L}^{\mathbf{D}}_{[\mathbf{x}]}\lambda^i(\mathbf{x}, \mathbf{y}),$$

$$\beta^{il}(\mathbf{x}, \mathbf{y}) = \sum_{k=1}^{d}\frac{\partial\lambda^i(\mathbf{x}, \mathbf{y})}{\partial x_k}\beta^{kl}(\mathbf{x}),$$

$$\Delta^{ij}(\mathbf{x}, \mathbf{y}) = \lambda^i\left(\mathbf{x}, \mathbf{y}^{-j}\right) - \lambda^i(\mathbf{x}, \mathbf{y}),$$

where $\mathcal{L}^X$ is the infinitesimal generator of the $\mathbb{R}^d$-valued diffusion process X_t, and $\mathcal{L}^{\mathbf{D}}_{[\mathbf{x}(\omega)]}$ denotes the infinitesimal generator of the Markovian process $\mathbf{D}$, for a given path of the background process $X_t(\omega) = \mathbf{x}(\omega)$.

This is not a standard SDE: the coefficients defining the intensities depend on the default state, and the default state vector depends in turn on the intensities. An example of this class of models is studied next.

The Jarrow and Yu Model[1]. The looping defaults model of Jarrow and Yu follows the SDE

$$d\lambda_t^i = \sum_{\substack{j=1 \\ j\neq i}}^{n} \Delta^{ij} dD_t^j, \tag{4.4}$$

for some constant jumps $\Delta^{ij} \in \mathbb{R}$. This is a particular case of Eq. (4.3), where

$$\begin{aligned}
\alpha^i(\mathbf{x}, \mathbf{y}) &= \sum_{i=1}^{n} (1 - \mathbf{y}(i))\, \lambda^i(\mathbf{x}, \mathbf{y})\, \Delta^{ij};\\
\beta^{il}(\mathbf{x}, \mathbf{y}) &= 0;\\
\Delta^{ij}(\mathbf{x}, \mathbf{y}) &= \Delta^{ij}\,, \Delta^{ii} = 0.
\end{aligned}$$

The intensity process λ^i is, then, given by

$$\lambda_t^i = \lambda_0^i + \sum_{\substack{j=1 \\ j\neq i}}^{n} \Delta^{ij} D_t^j. \tag{4.5}$$

Our first result is an analytical expression of the default times' multivariate density.

Proposition 46 (Default Times' Multivariate Density) *Let* $(t_1, t_2, \ldots, t_n) \in \mathbb{R}_+^n$*, and suppose that*

$$t_{\pi(1)} \leq t_{\pi(2)} \leq \ldots \leq t_{\pi(n)},$$

where the mapping $\pi(.) : \{1, \ldots, n\} \to \{1, \ldots, n\}$ *is a monotonic permutation of* $(t_1, t_2, \ldots, t_n)$*. Then, the default times' multivariate density is given by*

$$f(t_1, \ldots, t_n) = \prod_{i=1}^{n} \left[\lambda_0^{\pi(i)} + \sum_{j=1}^{i-1} \Delta^{\pi(i)\pi(j)} \right] \exp\left(-\sum_{j=1}^{i} \left[\lambda_0^{\pi(i)} + \sum_{k=1}^{j-1} \Delta^{\pi(i)\pi(k)} \right] \left(t_{\pi(j)} - t_{\pi(j-1)} \right) \right), \tag{4.6}$$

with the convention $t_{\pi(0)} = 0$.

Example. For $n = 3$, suppose $t_2 \leq t_1 \leq t_3$, then

$$
\begin{aligned}
f(t_1, t_2, t_3) = \lambda_0^2 \left(\lambda_0^1 + \Delta^{12}\right) \left(\lambda_0^3 + \Delta^{32} + \Delta^{31}\right) \\
\times \exp\left(-\lambda_0^2 t_2\right) \\
\times \exp\left(-\left(\lambda_0^1 t_2 + \left(\lambda_0^1 + \Delta^{12}\right)(t_1 - t_2)\right)\right) \\
\times \exp\left(-\left(\lambda_0^3 t_2 + \left(\lambda_0^3 + \Delta^{32}\right)(t_1 - t_2) + \left(\lambda_0^3 + \Delta^{32} + \Delta^{31}\right)(t_3 - t_1)\right)\right).
\end{aligned}
$$

Proof Extending the argument of Kusuoka (1999), we use a change of measure technique. We assume that we have some probability measure P' under which the default times $(\tau_1, ..., \tau_n)$ are independent random variables exponentially distributed with parameter 1. So that the P'-density of $(\tau_1, ..., \tau_n)$ is

$$f'(t_1, ..., t_n) \triangleq \mathbb{P}'(\tau_1 \in dt_1, ..., \tau_n \in dt_n) = \exp(-(t_1 + ... + t_n)).$$

Define the probability measure P (as in Eq. (4.2))

$$\frac{dP}{dP'} = L_{T^*}, \; P'\text{-a.s.},$$

where L satisfies, for $t \in [0, T^*]$,

$$L_t = 1 + \sum_{i=1}^{n} \int_{]0,t]} L_{s-} \phi_s^i \left(dD_s^i - \left(1 - D_s^i\right) ds\right). \tag{4.7}$$

ϕ^i is given by

$$\phi_t^i \triangleq \lambda_t^i - 1 = \lambda_0^i + \sum_{\substack{j=1 \\ j \neq i}}^{n} \Delta^{ij} D_t^j - 1. \tag{4.8}$$

By Girsanov's theorem, the intensity of the default time τ_i under P' is equal to $\left(1 + \phi_t^i\right) \times 1 = \lambda_t^i$.

The Doléans-Dade martingale L can also be written as the product

$$L_t = \prod_{i=1}^{n} L_t^i, \text{ for } t \in [0, T^*],$$

where L^i, $1 \leq i \leq n$, is defined as

$$L_t^i = 1 + \int_{]0,t]} L_{s-}^i \phi_s^i \left(dD_s^i - \left(1 - D_s^i\right) ds\right). \tag{4.9}$$

The solution of the SDE (4.9) is

$$\begin{aligned} L_t^i &= \exp\left(-\int_0^{\tau_i \wedge t} \left(\lambda_s^i - 1\right) ds\right) \left[\mathbf{1}_{\{\tau_i > t\}} + \mathbf{1}_{\{\tau_i \le t\}} \lambda_{\tau_i}^i\right] \\ &= \exp\left((\tau_i \wedge t) - \Lambda_{\tau_i \wedge t}^i\right) \left[\mathbf{1}_{\{\tau_i > t\}} + \mathbf{1}_{\{\tau_i \le t\}} \lambda_{\tau_i}^i\right], \end{aligned}$$

where Λ^i denotes the hazard rate process $\Lambda_t^i \triangleq \int_0^t \lambda_s^i ds$.

To compute the density function at $(t_1, ..., t_n)$, we proceed as follows.

We fix an arbitrary positive number $\alpha > 0$, and we compute, for $(\epsilon_1, ..., \epsilon_n) \in [-\alpha, \alpha]^n$, the expression of

$$\mathbb{P}\left(\tau_1 \in (t_1 - \epsilon_1, t_1], ..., \tau_n \in (t_n - \epsilon_n, t_n]\right).$$

To this end, we choose a time horizon T^* such that $T^* > \max_{1 \le i \le n} (t_i) + \alpha$ (e.g., $T^* = 1 + \max_{1 \le i \le n} (t_i) + \alpha$), and we use the change of probability measure

$$\frac{dP}{dP'} = L_{T^*} = \prod_{i=1}^{n} L_{T^*}^i, \ P'\text{-a.s.}$$

Thus, we have

$$\begin{aligned} &\mathbb{P}\left(\tau_1 \in (t_1 - \epsilon_1, t_1], ..., \tau_n \in (t_n - \epsilon_n, t_n]\right) \\ &= \mathbb{E}^{P'}\left[\mathbf{1}\left\{\tau_1 \in (t_1 - \epsilon_1, t_1], ..., \tau_n \in (t_n - \epsilon_n, t_n]\right\} \times L_{T^*}\right] \\ &= \int_{u_1 \in (t_1 - \epsilon_1, t_1]} \cdots \int_{u_n \in (t_n - \epsilon_n, t_n]} \exp\left(-\left(u_1 + ... + u_n\right)\right) \\ &\quad \times \left[\prod_{i=1}^{n} \exp\left(u_i - \Lambda_{u_i}^i\right)\left[\lambda_{u_i}^i\right]\right] du_1 ... du_n \\ &= \int_{u_1 \in (t_1 - \epsilon_1, t_1]} \cdots \int_{u_n \in (t_n - \epsilon_n, t_n]} \left[\prod_{i=1}^{n} \exp\left(-\Lambda_{u_i}^i\right)\left[\lambda_{u_i}^i\right]\right] du_1 ... du_n. \end{aligned} \tag{4.10}$$

The first equality follows from the change of measure, the second equality uses the fact that on the set $\{(\tau_1, ..., \tau_n) \in (t_1 - \epsilon_1, t_1] \times ... \times (t_n - \epsilon_n, t_n]\}$, we have

$$L_{T^*}^i = \exp\left(\tau_i - \Lambda_{\tau_i}^i\right)\left[\lambda_{\tau_i}^i\right],$$

since $\tau_i \le t_i + \alpha < T^*$.

Consider the monotonic permutation of the default times $\pi : \tau_{\pi(1)} \leq \tau_{\pi(2)} \leq ... \leq \tau_{\pi(n)}$, and let us derive explicitly the expressions of the intensity and hazard rate for $\{\tau_{\pi(1)} \leq ... \leq \tau_{\pi(n)}\}$.

By (4.5), the intensity $\lambda^{\pi(i)}_{\tau_{\pi(i)}}$ at time $t = \tau_{\pi(i)}$ is simply

$$\lambda^{\pi(i)}_{\tau_{\pi(i)}} = \lambda^{\pi(i)}_0 + \sum_{j=1}^{i-1} \Delta^{\pi(i)\pi(j)}. \tag{4.11}$$

The hazard rate is computed by (4.5) and using the fact that $\lambda^{\pi(i)}$ is piece-wise constant on the intervals $[\tau_{\pi(j-1)}, \tau_{\pi(j)})$, for $1 \leq j \leq i$:

$$\begin{aligned}\Lambda^{\pi(i)}_{\tau_{\pi(i)}} &= \sum_{j=1}^{i} \Lambda^{\pi(i)}_{\tau_{\pi(j)}} - \Lambda^{\pi(i)}_{\tau_{\pi(j-1)}} \\ &= \sum_{j=1}^{i} \lambda^{\pi(i)}_{\tau_{\pi(j-1)}} \left(\tau_{\pi(j)} - \tau_{\pi(j-1)}\right) \\ &= \sum_{j=1}^{i} \left[\lambda^{\pi(i)}_0 + \sum_{k=1}^{j-1} \Delta^{\pi(i)\pi(k)}\right] \left(\tau_{\pi(j)} - \tau_{\pi(j-1)}\right). \end{aligned} \tag{4.12}$$

By introducing the permutation π, (4.10) becomes

$$\begin{aligned}&\mathbb{P}\left(\tau_1 \in (t_1 - \epsilon_1, t_1], ..., \tau_n \in (t_n - \epsilon_n, t_n]\right) \\ &= \int_{u_1 \in (t_1 - \epsilon_1, t_1]} \cdots \int_{u_n \in (t_n - \epsilon_n, t_n]} \left[\prod_{i=1}^{n} \exp\left(-\Lambda^{\pi(i)}_{u_{\pi(i)}}\right)\left[\lambda^{\pi(i)}_{u_{\pi(i)}}\right]\right] du_1 ... du_n.\end{aligned}$$

Taking the limit $\epsilon_i \to 0$, for $1 \leq i \leq n$, we arrive at the expression of the density under the P-measure

$$f(t_1, ..., t_n) = \left[\prod_{i=1}^{n} \exp\left(-\Lambda^{\pi(i)}_{t_{\pi(i)}}\right)\left[\lambda^{\pi(i)}_{t_{\pi(i)}}\right]\right].$$

Substituting the intensity and the hazard rate by their expressions (4.11) and (4.12) ends the proof.■

Marked Point Process Representation. An alternative way of constructing the looping defaults' model is to use the total hazard rate construction developed by Norros (1986) and Shaked and Shanthikumar (1987). This approach was used in Yu (2004).

Assumption 1 excludes simultaneous defaults, we can therefore define the sequence of strictly ordered default times $(T_0, T_1, ..., T_n) : T_0 = 0 < T_1 < ... < T_n$, as well as the identity of the defaulted obligor $(Z_0, Z_1, ..., Z_n)$:

$$T_0 = 0,\ Z_0 = 0;$$
$$T_k = \min\{\tau_i : 1 \le i \le n,\ \tau_i > T_{k-1}\};$$
$$Z_k = i \text{ if } T_k = \tau_i.$$

When Assumption 1 is not satisfied, at each default time, multiple defaults can occur, which means that the size of the mark space is 2^n. This is the case, for example, in the Marshall-Olkin copula. The marked point process (T_n, Z_n) is called the failure process associated to $(\tau_1, ..., \tau_n)$ (see Norros (1986)). For each $1 \le i \le n$, the point process D_t^i is given by

$$\tau_i = \min\{T_k : Z_k = i\};$$
$$D_t^i = \sum_{k=1}^{n} \mathbf{1}_{\{T_k \le t\}} \mathbf{1}_{\{Z_k = i\}}.$$

The internal history of the process $\left(D_t^1, ..., D_t^n\right)$, $\mathcal{G}_t = \bigvee_{i=1}^n \mathcal{F}_t^{D^i}$, satisfies the following properties (see Brémaud (1980) Chap III T2):

$$\mathcal{G}_{T_n} = \sigma\left(T_0, Z_0, ..., T_n, Z_n\right);$$
$$\mathcal{G}_{T_n^-} = \sigma\left(T_0, Z_0, ..., T_{n-1}, Z_{n-1}, T_n\right);$$
$$\mathcal{G}_{T_n} = \mathcal{G}_{T_n^-} \vee \sigma\left(Z_n\right).$$

The compensators Λ^i w.r.t. the internal history can be written in regenerative form (Brémaud 1980 Chap III T7)

$$\Lambda_t^i = \sum_{k=1}^{n} \mathbf{1}_{\{T_{k-1} \le t < T_k\}} A_t^{(k)}\left(i; T_1, Z_1, ..., T_{k-1}, Z_{k-1}\right), \tag{4.13}$$

where each $A_\cdot^{(k)}(i; ...)$ is a deterministic function of its arguments. The stochastic compensator Λ^i is a piece-wise deterministic function evaluated at default times. To clarify the notations, we underline that $A^{(k)}_{\cdot}(i; ...)$ are deterministic functions

$$A_t^{(k)}(i; ...) \; : \; ([0, +\infty) \times \{1, ..., n\})^{k-1} \rightarrow \mathbb{R}^+$$
$$(T_1, Z_1, ..., T_{k-1}, Z_{k-1}) \rightarrow A_t^{(k)}(i; T_1, Z_1, ..., T_{k-1}, Z_{k-1}),$$

but the compensator Λ^i is stochastic

$$\Lambda_t^i(\omega) = \sum_{k=1}^{n} \mathbf{1}_{\{T_{k-1}(\omega) \leq t < T_k(\omega)\}} A_t^{(k)}(i; T_1(\omega), Z_1(\omega), ..., T_{k-1}(\omega), Z_{k-1}(\omega)).$$

It is well known that if: (a) $\mathbb{P}(0 < \tau_i < \infty) = 1$, (b) $\mathbb{P}\left(\tau_i = \tau_j\right) = 0$, for $i \neq j$, (c) the compensators Λ^i are continuous, then the random variables $\Lambda^i_\infty = \Lambda^i_{\tau_i}$, for $1 \leq i \leq n$, are independent and exponentially distributed with parameter 1 (see Norros (1986) Th 2.1). Hence, the default times $(\tau_1, ..., \tau_n)$ are mapped to some independent exponential variables $(\theta_1, ..., \theta_n)$. Norros establishes that this mapping is bijective and thus provides a way of constructing the default times in terms of independent exponentials.

The mapping $\psi : \mathbb{R}^n_+ \rightarrow \overline{\mathbb{R}}^n_+ : \psi(\tau_1, ..., \tau_n) = (\theta_1, ..., \theta_n)$ is called the total hazard function of $(\tau_1, ..., \tau_n)$. It is almost inverse ψ^*, in the sense that $\psi^*(\psi(\tau_1, ..., \tau_n)) = (\tau_1, ..., \tau_n)$ a.s., is obtained by the following algorithm.

Define $A_\theta^{(k),*}(i; ...) = \inf\left\{t : A_t^{(k)}(i; ...) > \theta\right\}$. For a given $(\theta_1, ..., \theta_n) \in \mathbb{R}^n_+$, and proceed as follows:

$$\begin{cases} T_1 = \min\left\{A_{\theta_i}^{(1),*}(i; \emptyset) : i \in \{1, ..., n\}\right\}; \\ Z_1 = \min\left\{i : A_{\theta_i}^{(1),*}(i; \emptyset) = T_1, i \in \{1, ..., n\}\right\}; \\ \rightarrow \tau_{Z_1} = T_1 \end{cases}$$

$$\begin{cases} T_{k+1} = \min\left\{A_{\theta_i}^{(k+1),*}\left(i; (T_l, Z_l)_{1 \leq l \leq k}\right) : i \in \{1, ..., n\} \setminus \{Z_1, ..., Z_k\}\right\}; \\ Z_{k+1} = \min\left\{i : A_{\theta_i}^{(k+1),*}\left(i; (T_l, Z_l)_{1 \leq l \leq k}\right) = T_{k+1}, i \in \{1, ..., n\} \setminus \{Z_1, ..., Z_k\}\right\} \\ \rightarrow \tau_{Z_{k+1}} = T_{k+1}; \end{cases}$$

$$\begin{cases} T_n = A_{\theta_n}^{(n),*}\left(Z_n; (T_l, Z_l)_{1 \leq l \leq k-1}\right); \\ \rightarrow \tau_{Z_n} = T_n; \end{cases}$$

See Algorithm 2.4 in Norros (1986) for further details.

This provides another way of constructing the looping defaults model. Indeed, one can establish the link between the intensity (4.5) and the functions

$A^{(k)}(i; ...)$ of the regenerative form (4.13). Observe first that the compensator Λ_t^i is given by

$$\Lambda_t^i = \int_0^t \left(1 - D_s^i\right) \lambda_s^i ds,$$

and let us write it down explicitly for an example where obligor i is the third default, i.e. $T_3 = \tau_i$, $Z_3 = i$.

$0 \le t < T_1 : \Lambda_t^i = \lambda_0^i t = A_t^{(1)}(i)$

$T_1 \le t < T_2 : Z_1 \neq i : \Lambda_t^i = \lambda_0^i T_1 + \left(\lambda_0^i + \Delta^{iZ_1}\right)(t - T_1) = A_t^{(2)}(i; T_1, Z_1)$

$T_2 \le t < T_3 : Z_2 \neq i : \Lambda_t^i = \lambda_0^i T_1 + \left(\lambda_0^i + \Delta^{iZ_1}\right)(T_2 - T_1) + \left(\lambda_0^i + \Delta^{iZ_1} + \Delta^{iZ_2}\right)(t - T_2) = A_t^{(3)}(i; T_1, Z_1, T_2, Z_2)$

$T_3 \le t < T_4 : Z_3 = i : \Lambda_t^i = \lambda_0^i T_1 + \left(\lambda_0^i + \Delta^{iZ_1}\right)(T_2 - T_1) + \left(\lambda_0^i + \Delta^{iZ_1} + \Delta^{iZ_2}\right)(T_3 - T_2) = A_t^{(4)}(i; T_1, Z_1, T_2, Z_2, T_3, Z_3)$

$T_4 \le t : \Lambda_t^i = \Lambda_{T_3}^i$

In general, the looping defaults model (4.5) can be expressed in regenerative form (4.13) as follows:

$$A_t^{(k)}(i; T_1, Z_1, ..., T_{k-1}, Z_{k-1}) = \sum_{j=1}^{k-1} \left[\lambda_0^i + \sum_{l=1}^{j-1} \Delta^{iZ_l} \right] (T_j - T_{j-1}) \prod_{l=1}^{j-1} \mathbf{1}_{\{Z_l \neq i\}}$$
$$+ \left[\lambda_0^i + \sum_{l=1}^{k-1} \Delta^{iZ_l} \right] (t - T_{k-1}) \prod_{l=1}^{k-1} \mathbf{1}_{\{Z_l \neq i\}}.$$

The marked point process representation (T_n, Z_n) offers an alternative proof of Proposition 46. One could write, for example,

$$\begin{aligned}
&\mathbb{P}(\tau_1 \in dt_1, ..., \tau_n \in dt_n) \\
&= \mathbb{P}\left(\tau_{\pi(1)} \in dt_{\pi(1)}, ..., \tau_{\pi(n)} \in dt_{\pi(n)}\right) \\
&= \mathbb{P}\left(T_1 \in dt_{\pi(1)}, Z_1 = \pi(1), ..., T_n \in dt_{\pi(n)}, Z_n = \pi(n)\right) \\
&= \prod_{k=1}^{n} \mathbb{P}\left(T_k \in dt_{\pi(k)}, Z_k = \pi(k) \,\middle|\, T_1 = t_{\pi(1)}, Z_1 = \pi(1), ..., \right. \\
&\qquad \left. T_{k-1} = t_{\pi(k-1)}, Z_{k-1} = \pi(k-1)\right).
\end{aligned}$$

The double sequence $(T_k, Z_k)_{1 \le k \le n}$ defines a marked point process with counting measure

$$\mu(\omega, dt \times dz) : (\Omega, \mathcal{G}) \to ((0,\infty) \times E, (0,\infty) \otimes \mathcal{E}),$$

$$\int_0^t \int_E H(\omega, s, z)\, \mu(\omega, dt \times dz) = \sum_{k=1}^{n} H(\omega, T_k(\omega), Z_k(\omega))\, \mathbf{1}_{\{T_k(\omega) \le t\}},$$

and predictable compensator

$$\nu(\omega, dt \times dz) = \Phi_t(\omega, dz)\, \lambda_t^{\mu} dt,$$

where λ_t^{μ} is a non-negative $\{\mathcal{G}_t\}$-predictable process and $\Phi_t(\omega, dz)$ is a probability transition kernel from $(\Omega \times [0,\infty), \mathcal{G} \otimes \mathcal{B}_+)$ into $(E, \mathcal{E})$. The pair $\left(\lambda_t^{\mu}, \Phi_t(dz)\right)$ is called the $(P, \mathcal{G}_t)$-local characteristics of $\mu(dt \times dz)$. Here, we have

$$\lambda_t^{\mu} = \sum_{i=1}^{n} \left(1 - D_t^i\right) \lambda_t^i;$$

$$\Phi_t(i) = \frac{\left(1 - D_t^i\right) \lambda_t^i}{\lambda_t^{\mu}}, \text{ for } i \in \{1, ..., n\},$$

with the convention $\Phi_t(.) = 0$ if $\lambda_t^{\mu} = 0$. The density of (T_k, Z_k), conditional on $\mathcal{G}_{T_{k-1}}$, is given by (see Brémaud (1980) Chap VIII T7)

$$g^{(k)}(\omega, t, z) : \left(\Omega \times [0,\infty), \mathcal{G}_{T_{k-1}} \otimes \mathcal{B}_+\right) \to (E, \mathcal{E})$$
$$g^{(k)}(\omega, t, z) \triangleq \mathbb{P}\left(T_k \in dt, Z_k = i \,\middle|\, \mathcal{G}_{T_{k-1}}\right)(\omega)$$
$$= \Phi_t(\omega, i)\, \lambda_t^{\mu}(\omega) \exp\left(-\int_{T_{k-1}(\omega)}^{t} \lambda_t^{\mu}(\omega)\, ds\right) dt.$$

Reverting back to the default times' multivariate density and using the expression of the conditional density $g^{(k)}(\omega, t, z)$, we can express each term in the product as

$$\mathbb{P}\left(T_k \in dt_{\pi(k)}, Z_k = \pi(k) \,\middle|\, \mathcal{G}_{T_{k-1}}\right)$$
$$= \left(1 - D_{t_{\pi(k)}^-}^{\pi(k)}\right) \lambda_{t_{\pi(k)}}^{\pi(k)} \exp\left(-\int_{T_{k-1}(\omega)}^{t_{\pi(k)}} \lambda_s^{\mu} ds\right) dt_{\pi(k)}.$$

Conditioning on the path $\left\{T_1 = t_{\pi(1)}, Z_1 = \pi(1), ..., T_{k-1} = t_{\pi(k-1)}, Z_{k-1} = \pi(k-1)\right\}$, and using the fact that the intensities are piece-wise constant between jumps, we get

$$\begin{aligned} &\mathbb{P}\left(T_k \in dt_{\pi(k)}, Z_k = \pi(k) \,\middle|\, T_1 = t_{\pi(1)}, Z_1 = \pi(1), ..., \right. \\ &\left. T_{k-1} = t_{\pi(k-1)}, Z_{k-1} = \pi(k-1)\right) \\ &= \lambda_{t_{\pi(k)}}^{\pi(k)} \exp\left(-\lambda_{t_{\pi(k-1)}}^{\mu}\left(t_{\pi(k)} - t_{\pi(k-1)}\right)\right) dt_{\pi(k)}. \end{aligned} \tag{4.14}$$

The intensity $\lambda_{t_{\pi(k)}}^{\pi(k)}$ at time $t = t_{\pi(k)}$ is given by

$$\lambda_{t_{\pi(k)}}^{\pi(k)} = \lambda_0^{\pi(k)} + \sum_{j=1}^{i-1} \Delta^{\pi(i)\pi(j)}, \tag{4.15}$$

and the intensity of the failure process $\lambda_{t_{\pi(k-1)}}^{\mu}$ is given by the sum of the remaining intensities associated with the obligors that have survived after $t_{\pi(k-1)}$

$$\lambda_{t_{\pi(k-1)}}^{\mu} = \sum_{i=1}^{n}\left(1 - D_{t_{\pi(k-1)}}^{i}\right)\lambda_{t_{\pi(k-1)}}^{i} = \sum_{i=1}^{n}\left(1 - D_{t_{\pi(k-1)}}^{\pi(i)}\right)\lambda_{t_{\pi(k-1)}}^{\pi(i)} = \sum_{i=k}^{n}\lambda_{t_{\pi(k-1)}}^{\pi(i)}; \tag{4.16}$$

Putting (4.14), (4.15) and (4.16) together gives the result after some basic algebra:

$$\begin{aligned} f(t_1, ..., t_n) &= \prod_{k=1}^{n} \lambda_{t_{\pi(k)}}^{\pi(k)} \exp\left(-\lambda_{t_{\pi(k-1)}}^{\mu}\left(t_{\pi(k)} - t_{\pi(k-1)}\right)\right) \\ &= \left[\prod_{k=1}^{n} \lambda_{t_{\pi(k)}}^{\pi(k)}\right] \exp\left(-\sum_{k=1}^{n}\sum_{i=k}^{n} \lambda_{t_{\pi(k-1)}}^{\pi(i)}\left(t_{\pi(k)} - t_{\pi(k-1)}\right)\right) \\ &= \left[\prod_{k=1}^{n} \lambda_{t_{\pi(k)}}^{\pi(k)}\right] \exp\left(-\sum_{i=1}^{n}\sum_{k=1}^{i} \lambda_{t_{\pi(k-1)}}^{\pi(i)}\left(t_{\pi(k)} - t_{\pi(k-1)}\right)\right) \\ &= \left[\prod_{k=1}^{n} \lambda_{t_{\pi(k)}}^{\pi(k)}\right] \exp\left(-\sum_{i=1}^{n} \Lambda_{t_{\pi(i)}}^{\pi(i)}\right) \\ &= \left[\prod_{i=1}^{n} \left[\lambda_{t_{\pi(i)}}^{\pi(i)}\right] \exp\left(-\Lambda_{t_{\pi(i)}}^{\pi(i)}\right)\right]; \end{aligned}$$

we have inverted the summation order in the third equality, introduced the cumulative hazard rate in the fourth equality and replaced the index k by i in the last line.

4.4 The Copula Approach

In this section, we consider the problem of computing the conditional survival probability, $\mathbb{P}(\tau_i > T \mid \mathcal{G}_t)$, in a copula framework. Using the generalized Dellacherie formula, we derive an analytical result depending on the default state that we are in. This approach was first studied in Schönbucher and Schubert (2001). However, the set-up here is more general since we consider a "time-dependent" conditional copula to model the default times' multivariate dependence. The theory of conditional copulas is presented next.

The Theory of Conditional Copulas. Patton (2001) has extended the existing theory of (unconditional) copulas to the conditional case by introducing the so-called "conditional copula". This tool can be used in the modelling of time-varying conditional dependence. Here, we follow Patton's approach by giving the formal definitions of a conditional multivariate distribution and a conditional copula, and Sklar's theorem for conditional distributions.

Let $\mathcal{A}$ be some arbitrary sub-σ-algebra.

Definition 47 (*Conditional multivariate distribution*) An n-dimensional conditional distribution function is a function $H(|\mathcal{A}) : \overline{\mathbb{R}}^n \times \Omega \to [0, 1]$ with the following properties:

- $H(|\mathcal{A})$ is grounded, i.e., for almost every $\omega \in \Omega$, $H(x_1, ..., x_n |\mathcal{A})(\omega) = 0$ for all $(x_1, ..., x_n) \in \overline{\mathbb{R}}^n$ such that $x_k = -\infty$ for at least one k;
- $H(|\mathcal{A})$ is n-increasing, i.e., for almost every $\omega \in \Omega$, the H-volume of all n-boxes in $\overline{\mathbb{R}}^n$ is positive:

$$V_H\left(\left[x_1^1, x_1^2\right] \times ... \times \left[x_n^1, x_n^2\right]\right) = \sum_{i_1=1}^{2} ... \sum_{i_n=1}^{2} (-1)^{i_1+....+i_n} H\left(x_1^{i_1}, ..., x_n^{i_n} |\mathcal{A}\right)(\omega) \geq 0,$$

  for all $\left(x_1^1, ..., x_n^1\right)$ and $\left(x_1^2, ..., x_n^2\right)$ in $\overline{\mathbb{R}}^n$ with $x_k^1 \leq x_k^2$, $1 \leq k \leq n$;
- $H(\infty, ..., \infty |\mathcal{A})(\omega) = 1$, for almost every $\omega \in \Omega$.
- $H(x_1, ..., x_n |\mathcal{A})(.) : \Omega \to [0, 1]$ is a measurable function on $(\Omega, \mathcal{A})$, for each $(x_1, ..., x_n) \in \overline{\mathbb{R}}^n$.

The margins of an n-dimensional conditional distribution function are conditional distribution functions.

Definition 48 (*Conditional copula*) An n-dimensional conditional copula is a function $C(|\mathcal{A}) : [0,1]^n \times \Omega \to [0,1]$ with the following properties:

- $C(|\mathcal{A})$ is grounded, i.e., for almost every $\omega \in \Omega$, $C(u_1, ..., u_n |\mathcal{A})(\omega) = 0$ for all $(u_1, ..., u_n) \in [0,1]^n$ such that $u_k = 0$ for at least one k;
- $C(|\mathcal{A})$ is n-increasing, i.e., for almost every $\omega \in \Omega$, the C-volume of all n-boxes in $[0,1]^n$ is positive:

$$\sum_{i_1=1}^{2} \cdots \sum_{i_n=1}^{2} (-1)^{i_1+....+i_n} C\left(u_1^{i_1}, ..., u_n^{i_n} |\mathcal{A}\right)(\omega) \geq 0,$$

  for all $\left(u_1^1, ..., u_n^1\right)$ and $\left(u_1^2, ..., u_n^2\right)$ in $\overline{\mathbb{R}}^n$ with $u_k^1 \leq u_k^2$, $1 \leq k \leq n$;
- $C(|\mathcal{A})$ has (conditional) margins $C_k(|\mathcal{A})$, which satisfy, for almost every $\omega \in \Omega$, $C_k(u_k |\mathcal{A})(\omega) = C(1, ..., 1, u_k, 1..., 1 |\mathcal{A})(\omega) = u_k$ for all u_k in $[0,1]$.
- $C(u_1, ..., u_n |\mathcal{A})(.) : \Omega \to [0,1]$ is a measurable function on $(\Omega, \mathcal{A})$, for each $(u_1, ..., u_n) \in [0,1]^n$.

Alternatively, a conditional copula can be defined as a conditional distribution function with uniformly distributed conditional margins.

Theorem 49 (Sklar's theorem for conditional distributions) *Let $H(|\mathcal{A})$ be an n-dimensional conditional distribution function with conditional margins $F_1(|\mathcal{A}), ..., F_n(|\mathcal{A})$. Then, there exists a conditional n-copula $C(|\mathcal{A}) : [0,1]^n \times \Omega \to [0,1]$ such that for almost every $\omega \in \Omega$, and for all $(x_1, ..., x_n) \in \overline{\mathbb{R}}^n$,*

$$H(x_1, ..., x_n |\mathcal{A})(\omega) = C(F_1(x_1 |\mathcal{A})(\omega), ..., F_n(x_n |\mathcal{A})(\omega) |\mathcal{A})(\omega).$$

If for almost every $\omega \in \Omega$, the functions $x_i \to F_i(x_i |\mathcal{A})(\omega)$ are all continuous, then $C(|\mathcal{A})(\omega)$ is unique. Otherwise, $C(|\mathcal{A})(\omega)$ is uniquely determined on the product of the values taken by $x_i \to F_i(x_i |\mathcal{A})(\omega)$, $i = 1, ..., n$. Conversely, if $C(|\mathcal{A})$ is a conditional n-copula and $F_1(|\mathcal{A}), ..., F_n(|\mathcal{A})$ are conditional distribution functions, then the function $H(|\mathcal{A})$ defined above is a conditional n-dimensional distribution function with conditional margins $F_1(|\mathcal{A}), ..., F_n(|\mathcal{A})$.

As a corollary to Theorem 49, we can construct the conditional copula from any conditional multivariate distribution.

Definition 50 (*Generalized-inverse*). The generalized-inverse of a univariate distribution function F is defined as

$$F^{-1}(u) = \inf\{x : F(x) \geq u\}, \text{ for all } u \in [0,1].$$

Corollary 51 *Let $H(|\mathcal{A})$ be an n-dimensional conditional distribution with continuous conditional margins $F_1(|\mathcal{A}), ..., F_n(|\mathcal{A})$. Then, there exists a unique conditional copula $C(|\mathcal{A}) : [0,1]^n \times \Omega \to [0,1]$ such that*

$$C(u_1, ..., u_n |\mathcal{A})(\omega) = H\left(F_1^{-1}(u_1|\mathcal{A})(\omega), ..., F_n^{-1}(u_n|\mathcal{A})(\omega) |\mathcal{A}\right)(\omega),$$

for almost every $\omega \in \Omega$, and for all $(u_1, ..., u_n) \in [0,1]^n$.

Given a set of conditional marginal distributions and a conditional copula, we can construct a joint conditional distribution, and from any given joint conditional distribution, we can extract the conditional margins and the conditional copula.

Next, we make the following assumptions.

Model. As before, we denote by $\tau_1, ..., \tau_n$ a set of non-negative variables on a probability space $(\Omega, \mathcal{G}, P)$, such that $\mathbb{P}(\tau_i = 0) = 0$ and $\mathbb{P}(\tau_i > t) > 0$ for any $t \in R_+$. We set $D_t^i = \mathbf{1}_{\{\tau_i \leq t\}}$ and we denote by $\{\mathcal{H}_t^i\}$ the associated filtration: $\mathcal{H}_t^i = \sigma(D_s^i : s \leq t)$. We are also given an It ô process X and its filtration $\{\mathcal{F}_t\}$ augmented with the P-null sets of $\mathcal{G}$, and $\mathcal{F}_0$ is trivial. The investor filtration, in this model, is $\{\mathcal{G}_t\}$: $\mathcal{G}_t = \mathcal{F}_t \vee \bigvee_{i=1}^n \mathcal{H}_t^i$. The firm-specific information is denoted by $\{\mathcal{G}_t^i\}$: $\mathcal{G}_t^i = \mathcal{F}_t \vee \mathcal{H}_t^i$.

On the firm-specific filtration $\{\mathcal{G}_t^i\}$, the default time τ_i has an $\{\mathcal{F}_t\}$-adapted intensity h^i, i.e., $D_t^i - \int_0^{t\wedge\tau_i} h_s^i ds$ is a $\{\mathcal{G}_t^i\}$-martingale. So that

$$\mathbb{P}\left(\tau_i > T \middle| \mathcal{G}_t^i\right) = \left(1 - D_t^i\right) \mathbb{E}\left[\exp\left(-\int_t^T h_s^i ds\right) |\mathcal{F}_t\right].$$

On the enlarged filtration $\{\mathcal{G}_t\}$, τ_i has a $\left\{\mathcal{G}_t^{-i}\right\}$-adapted intensity λ^i, i.e., $D_t^i - \int_0^{t\wedge\tau_i} \lambda_s^i ds$ is a $\{\mathcal{G}_t\}$-martingale.

Our goal is to establish a formula of the conditional expectation

$$\mathbb{P}(\tau_i > T |\mathcal{G}_t).$$

Furthermore, we assume that the multivariate dependence structure is described by the process $\left(\overline{C}_t^{\tau}\right)_{t\geq 0}$, where

$$\overline{C}_t^{\tau} : [0,1]^n \times \Omega \times [0,\infty) \to [0,1]$$

is the conditional (survival) copula, $\overline{C}_t^{\tau}(.) \triangleq \overline{C}^{\tau}(.\,|\mathcal{F}_t)$, i.e., for almost every $\omega \in \Omega$, and for all $(t_1, ..., t_n) \in [0,\infty)^n$,

$$\mathbb{P}(\tau_1 > t_1, ..., \tau_n > t_n\,|\mathcal{F}_t)(\omega) = \overline{C}_t^{\tau}(\mathbb{P}(\tau_1 > t_1\,|\mathcal{F}_t)(\omega), ..., \mathbb{P}(\tau_n > t_n\,|\mathcal{F}_t)(\omega))(\omega).$$

One can formally construct the process $\left(\overline{C}_t^{\tau}\right)_{t\geq 0}$ as follows.

Denote by $G_t(x_1, ..., x_n) = \mathbb{P}(\tau_1 \leq x_1, ..., \tau_n \leq x_n\,|\mathcal{F}_t)$ the conditional multivariate distribution function, and by $G_t^i(x) = \mathbb{P}(\tau_i \leq x\,|\mathcal{F}_t)$ the conditional margins. Define the generalized inverse of $G_t^i(.)$, $I_t^i(u) = \inf\left\{x : G_t^i(x) \geq u\right\}$; for almost every $\omega \in \Omega$, the functions $x \to \mathbb{P}(\tau_i \leq x\,|\mathcal{F}_t)(\omega)$ are continuous. The conditional copula C_t^{τ} is then given by

$$C_t^{\tau}(u_1, ..., u_n)(\omega) = G_t\left(I_t^1(u_1)(\omega), ..., I_t^n(u_n)(\omega)\right)(\omega),$$

for almost every $\omega \in \Omega$, and for all $(u_1, ..., u_n) \in [0,1]^n$. The conditional survival copula, which links the marginal survival functions, is given by

$$\overline{C}_t^{\tau}(u_1, ..., u_n)(\omega) = \sum_{i_1=1}^{2} ... \sum_{i_n=1}^{2} (-1)^{i_1+...+i_n} C_t^{\tau}\left(v_1^{i_1}, ..., v_n^{i_n}\right)(\omega),$$

where $v_i^1 = 1 - u_i$ and $v_i^2 = 1$, and $(u_1, ..., u_n) \in [0,1]^n$.

Assumption 3. We assume that, for each $t \geq 0$, and for almost every $\omega \in \Omega$, the function

$$\begin{aligned} \overline{C}_t^{\tau}(.,\omega) &: [0,1]^n \to [0,1] \\ (u_1, ..., u_n) &\to \overline{C}_t^{\tau}(u_1, ..., u_n)(\omega) \end{aligned}$$

is absolutely continuous.

Comparison with the Threshold Copula. In practice, one can construct the copula process by considering a Cox-process approach as in Lando (1998). In this subsection, we compare the conditional time-dependent copula with the threshold copula used in Schönbucher and Schubert (2001). Let $\left(h_t^i\right)_{t\geq 0}$ be an $\{\mathcal{F}_t\}$-adapted nonnegative càdlàg process. Set

$$\tau_i \triangleq \inf\left\{t : \exp\left(-\int_0^t h_s^i ds\right) \leq U_i\right\}, \tag{4.17}$$

where U_i is a random variable uniformly distributed on [0, 1] and independent of $\mathcal{F}_\infty$. One can check that h^i is the $\{\mathcal{F}_t\}$-adapted intensity of τ_i with respect to the firm-specific filtration $\left\{\mathcal{G}_t^i\right\}$. Assume that the distribution of the n-dimensional vector $(U_1, ..., U_n)$ is defined by the (static) survival copula function $\overline{C}^U : [0, 1]^n \rightarrow [0, 1]$,

$$\mathbb{P}\left(U_1 > u_1, ..., U_n > u_n\right) = \overline{C}^U\left(u_1, ..., u_n\right).$$

$\overline{C}^U(.)$ is referred to as the default thresholds (survival) copula. Following a remark in Jouanin et al. (2001), one can relate this copula to the default times' conditional survival probability: for almost every $\omega \in \Omega$, we have

$$\begin{aligned}
&\mathbb{P}\left(\tau_1 > t_1, ..., \tau_n > t_n \left|\mathcal{F}_t\right.\right)(\omega) \\
&= \mathbb{E}\left[\mathbb{P}\left(\tau_1 > t_1, ..., \tau_n > t_n \left|\mathcal{F}_\infty\right.\right)(\omega)\left|\mathcal{F}_t\right.\right](\omega) \\
&= \mathbb{E}\left[\mathbb{P}\left(U_1 > \exp\left(-\int_0^{t_1} h_s^1(\omega)\, ds\right), ..., U_n > \exp\left(-\int_0^{t_n} h_s^n(\omega)\, ds\right)\left|\mathcal{F}_\infty\right.\right)(\omega)\left|\mathcal{F}_t\right.\right](\omega) \\
&= \mathbb{E}\left[\mathbb{P}\left(U_1 > \exp\left(-\int_0^{t_1} h_s^1(\omega)\, ds\right), ..., U_n > \exp\left(-\int_0^{t_n} h_s^n(\omega)\, ds\right)\right)\left|\mathcal{F}_t\right.\right](\omega) \\
&= \mathbb{E}\left[\overline{C}^U\left(\exp\left(-\int_0^{t_1} h_s^1(\omega)\, ds\right), ..., \exp\left(-\int_0^{t_n} h_s^n(\omega)\, ds\right)\right)\left|\mathcal{F}_t\right.\right](\omega);
\end{aligned}$$

the first equality follows from the law of iterated expectation; the second equality is from the construction of the default times (4.17); the third equality is due to the independence of the threshold variables $(U_1, ..., U_n)$ from $\mathcal{F}_\infty$ and the fact that the $\int_0^{t_i} h_s^i ds$ are $\mathcal{F}_\infty$ -measurable; and the fourth equality follows from the definition of the threshold copula.

The default times' conditional copula process $\left(\overline{C}_t^\tau\right)_{t\geq 0}$ can then be constructed via

$$\overline{C}_t^\tau\left(\mathbb{E}\left[e^{-\int_0^{t_1} h_s^1 ds}\left|\mathcal{F}_t\right.\right], ..., \mathbb{E}\left[e^{-\int_0^{t_n} h_s^n ds}\left|\mathcal{F}_t\right.\right]\right) = \mathbb{E}\left[\overline{C}^U\left(e^{-\int_0^{t_1} h_s^1 ds}, ..., e^{-\int_0^{t_n} h_s^n ds}\right)\left|\mathcal{F}_t\right.\right]. \tag{4.18}$$

This provides a practical way of generating the conditional copula process that was formally introduced in the previous section. By choosing an absolutely continuous threshold copula, one can ensure that Assumption 3 is verified.

Forward Intensity. A convenient way of parameterizing the conditional survival probabilities $\mathbb{P}\left(\tau_i > T \,\middle|\, \mathcal{F}_t \vee \mathcal{H}_t^i\right)$ and $\mathbb{P}\left(\tau_i > T \,|\mathcal{G}_t\right)$ is to introduce the forward intensity processes $\left(h_{t,T}^i\right)_{t \geq 0}$ and $\left(\lambda_{t,T}^i\right)_{t \geq 0}$.

Definition 52 (*Forward Intensity*). Assume that $\mathbb{P}\left(\tau_i > 0 \,|\mathcal{F}_t\right) = 1$, and $\mathbb{P}\left(\tau_i > T \,|\mathcal{F}_t\right) > 0$, for all $T \in \mathbb{R}^+$.

Let $F_t^i\left(T\right) \triangleq \mathbb{P}\left(\tau_i > T \,|\mathcal{F}_t\right)$, for all $T \in \mathbb{R}^+$, denote the conditional survival probability. We assume that $F_t^i\left(T\right)$ is a.s. continuously differentiable in the T-variable. Then, the $\{\mathcal{F}_t\}$-adapted forward intensity at T is defined as

$$h_{t,T}^i = -\frac{\partial \log F_t^i\left(T\right)}{\partial T}, \text{ for all } T \in \mathbb{R}^+. \tag{4.19}$$

Integrating (4.19), we obtain

$$\begin{aligned}\mathbb{P}\left(\tau_i > T \,|\mathcal{F}_t\right) &= \mathbb{P}\left(\tau_i > 0 \,|\mathcal{F}_t\right) \exp\left(-\int_0^T h_{t,s}^i ds\right)\\ &= \exp\left(-\int_0^T h_{t,s}^i ds\right), \text{ for } T \in \mathbb{R}^+.\end{aligned}$$

Thus, we can write the conditional survival probability w.r.t. $\mathcal{F}_t \vee \mathcal{H}_t^i$ as

$$\begin{aligned}\mathbb{P}\left(\tau_i > T \,\middle|\, \mathcal{F}_t \vee \mathcal{H}_t^i\right) &= \left(1 - D_t^i\right) \frac{\mathbb{P}\left(\tau_i > T \,|\mathcal{F}_t\right)}{\mathbb{P}\left(\tau_i > t \,|\mathcal{F}_t\right)}\\ &= \left(1 - D_t^i\right) \exp\left(-\int_t^T h_{t,s}^i ds\right), \text{ for } T \in \mathbb{R}^+.\end{aligned}$$

Similarly, we define the $\left\{\mathcal{G}_t^{-i}\right\}$-adapted forward intensity.

Definition 53 (*Forward Intensity*). Assume that $\mathbb{P}\left(\tau_i > 0 \,\middle|\, \mathcal{G}_t^{-i}\right) = 1$, and $\mathbb{P}\left(\tau_i > T \,\middle|\, \mathcal{G}_t^{-i}\right) > 0$, for all $T \in \mathbb{R}^+$.

Set $H_t^i\left(T\right) \triangleq \mathbb{P}\left(\tau_i > T \,\middle|\, \mathcal{G}_t^{-i}\right)$, for all $T \in \mathbb{R}^+$, and define the $\left\{\mathcal{G}_t^{-i}\right\}$-adapted forward intensity as:

$$\lambda_{t,T}^i \triangleq -\frac{\partial \log H_t^i\left(T\right)}{\partial T}, \text{ for all } T \in \mathbb{R}^+. \tag{4.20}$$

Then, we have

$$\mathbb{P}\left(\tau_i > T \left|\mathcal{G}_t^{-i}\right.\right) = \exp\left(-\int_0^T \lambda_{t,s}^i ds\right), \text{ for } T \in \mathbb{R}^+.$$

$$\mathbb{P}\left(\tau_i > T \left|\mathcal{G}_t\right.\right) = \left(1 - D_t^i\right) \exp\left(-\int_t^T \lambda_{t,s}^i ds\right), \text{ for } T \in \mathbb{R}^+.$$

At this point, the forward intensity process is merely a parameterization of the conditional survival probability. Next, we establish the link between this parameterization and the standard "spot" intensities.

Link Between Forward and Spot Intensity. Conditional survival probabilities and intensities are linked via the following result due to Aven (1985).

Proposition 54 (Aven 1985). *Let* $\left(\Omega, \mathcal{G}, \left\{\widetilde{\mathcal{G}}_t\right\}_{t\in[0,T^*]}, P\right)$ *be a filtered probability space and* $D_t = \mathbf{1}_{\{\tau \le t\}}$ *with* τ *a* $\left\{\widetilde{\mathcal{G}}_t\right\}$*-stopping time. Let* $\{\varepsilon_n\}_{n=1}^{\infty}$ *be a sequence, which decreases to zero and let* $Y_t^n, t \in \left[0, T^*\right]$ *be a measurable version of the process*

$$Y_t^n = \frac{1}{\varepsilon_n} \mathbb{E}\left[D_{t+\varepsilon_n} - D_t \left|\widetilde{\mathcal{G}}_t\right.\right].$$

Assume that there are non-negative and measurable processes $\widetilde{g}_t$ *and* y_t, $t \in \left[0, T^*\right]$ *such that:*

1. *for each* t,

$$\lim_{n\to\infty} Y_t^n = \widetilde{g}_t \text{ a.s.;}$$

2. *for each* t*, there exists for almost all* $\omega \in \Omega$*, an* $n_0 = n_0(t, \omega)$ *such that*

$$\left|Y_s^n(\omega) - \widetilde{g}_s(\omega)\right| \le y_s(\omega), \forall s \le t, n \ge n_0;$$

where

$$\int_0^t y_s ds < \infty, \text{ a.s.}, t \in \left[0, T^*\right];$$

then $\widetilde{M}_t = D_t - \int_0^t \widetilde{g}_s ds$ *is a* $\left\{\widetilde{\mathcal{G}}_t\right\}$*-martingale, and* $\int_0^t \widetilde{g}_s ds$ *is the compensator of* D_t.

The relationship between intensity and conditional survival probability follows directly from this result (see, for instance, Schönbucher and Schubert (2001)).

Lemma 55 *Let* $\widetilde{H}_t(T)$ *denote the conditional survival probability*

$$\widetilde{H}_t(T) = \mathbb{P}\left(\tau > T \left| \widetilde{\mathcal{G}}_t\right.\right).$$

Assume that $\widetilde{H}_t(T)$ *is differentiable from the right with respect to* T *at* $T = t$, *and that the assumptions of Proposition 54 are satisfied. Then, the intensity process of* D_t *is given by*

$$\widetilde{g}_t = -\frac{\partial \widetilde{H}_t(T)}{\partial T}\,|_{T=t}\,.$$

Remark 56 *Note that* $\widetilde{g}_t = 0$ *after* τ *and that* $\widetilde{g}_t$ *is* $\left\{\widetilde{\mathcal{G}}_t\right\}$*-adapted.*

Using our forward intensity parameterization (and assuming that the technical assumption of Proposition 54 are satisfied), we can recover the $\{\mathcal{F}_t\}$-adapted intensity process of τ_i. Set $\widetilde{F}^i_t(T) \triangleq \mathbb{P}\left(\tau_i > T \left|\mathcal{F}_t \vee \mathcal{H}^i_t\right.\right)$, and $F^i_t(T) \triangleq \mathbb{P}(\tau_i > T\,|\mathcal{F}_t)$. The process $\widetilde{h}^i_t$, which is $\left\{\mathcal{F}_t \vee \mathcal{H}^i_t\right\}$-adapted, and the $\{\mathcal{F}_t\}$-adapted version h^i_t are related by

$$\widetilde{h}^i_t = \mathbf{1}_{\{\tau_i > t\}} h^i_t, \text{ for } t \geq 0.$$

The expression of $\widetilde{h}^i_t$ can be obtained by Lemma 55:

$$\begin{aligned}
\widetilde{h}^i_t &= -\frac{\partial \widetilde{F}^i_t(T)}{\partial T}\,|_{T=t} \\
&= -\frac{\partial}{\partial T}\left[\mathbf{1}_{\{\tau_i > t\}} \frac{F^i_t(T)}{F^i_t(t)}\right]|_{T=t} \\
&= -\frac{\partial}{\partial T}\left[\mathbf{1}_{\{\tau_i > t\}} \exp\left(-\int_t^T h^i_{t,s} ds\right)\right]|_{T=t} \\
&= \mathbf{1}_{\{\tau_i > t\}} h^i_{t,t}.
\end{aligned}$$

Thus, we have the following relationship between the $\{\mathcal{F}_t\}$-adapted spot and forward intensities: for all $t < \tau_i$,

$$h^i_t = h^i_{t,t}. \tag{4.21}$$

Similarly, we can recover the $\left\{\mathcal{G}^{-i}_t\right\}$ -adapted intensity: for all $t < \tau_i$,

$$\lambda^i_t = \lambda^i_{t,t}. \tag{4.22}$$

Equations (4.22) and (4.21) can be viewed as the equivalent of the well-known relationship between a forward rate and a short rate in interest rate modelling.

Computing the Conditional Survival Probability. Using our generalized Dellacherie formula, the conditional survival probability $H_t^i(T) = \mathbb{P}(\tau_i > T \,|\mathcal{G}_t)$ can be obtained as

$$H_t^i(T) = \sum_{\pi \in \Pi_{\mathbf{n}}} D_t^{(\pi)} H_t^{i,(\pi)}(T), \tag{4.23}$$

where

$$H_t^{i,(\pi)}(T) \triangleq \frac{\mathbb{E}\left[\left(1 - D_T^i\right) \times \prod_{j \notin \pi} \left(1 - D_t^j\right) \middle| \mathcal{G}_t^{(\pi)}\right]}{\mathbb{E}\left[\prod_{j \notin \pi} \left(1 - D_t^j\right) \middle| \mathcal{G}_t^{(\pi)}\right]},$$

$$D_t^{(\pi)} \triangleq \left[\prod_{j \in \pi} \left(D_t^j\right)\right] \times \left[\prod_{j \notin \pi} \left(1 - D_t^j\right)\right],$$

$$\mathcal{G}_t^{(\pi)} \triangleq \mathcal{F}_t \vee \left[\bigvee_{j \in \pi} \mathcal{H}_\infty^j\right] = \mathcal{F}_t \vee \left[\bigvee_{j \in \pi} \sigma\left(\tau_j\right)\right].$$

It is clear that all the terms in the sum (4.23) such that $i \in \pi$ are equal to zero. For all the remaining ones, the conditional distribution of default depends on the default state π observed at time t, and will vary from one state to the next. We derive the formula of the conditional expectation in the pre-default state: $\pi = \emptyset$. Then, we do the same when one or more defaults have occurred before $t : |\pi| > 0$.

Pre-default. We have the following result.

Proposition 57 (Pre-default). *If no default has occurred before time t, then the $\{\mathcal{G}_t\}$-forward intensity is given by*

$$\lambda_{t,T}^{i,(\emptyset)} = h_{t,T}^i \exp\left(-\int_0^T h_{t,s}^i ds\right) \frac{\frac{\partial}{\partial x_i} \overline{C}_t^\tau\left(e^{-\int_0^t h_{t,s}^1 ds}, \ldots, e^{-\int_0^T h_{t,s}^i ds}, \ldots, e^{-\int_0^t h_{t,s}^n ds}\right)}{\overline{C}_t^\tau\left(e^{-\int_0^t h_{t,s}^1 ds}, \ldots, e^{-\int_0^T h_{t,s}^i ds}, \ldots, e^{-\int_0^t h_{t,s}^n ds}\right)}. \tag{4.24}$$

Proof On the set $\prod_{j=1}^{n}\left(\mathbf{1}_{\{\tau_j>t\}}\right)$, the conditional survival probability is given by

$$H_t^{i,(\emptyset)}(T) = \frac{\mathbb{E}\left[\left(1-D_T^i\right)\times\prod_{j=1}^{n}\left(1-D_t^j\right)|\mathcal{F}_t\right]}{\mathbb{E}\left[\prod_{j=1}^{n}\left(1-D_t^j\right)|\mathcal{F}_t\right]}.$$

The numerator can be computed as

$$\begin{aligned}
&\mathbb{E}\left[\left(1-D_T^i\right)\times\prod_{j=1}^{n}\left(1-D_t^j\right)|\mathcal{F}_t\right]\\
&=\overline{C}_t^\tau\left(\mathbb{E}\left[\left(1-D_t^1\right)|\mathcal{F}_t\right],\ldots,\mathbb{E}\left[\left(1-D_T^i\right)|\mathcal{F}_t\right],\ldots,\mathbb{E}\left[\left(1-D_t^n\right)|\mathcal{F}_t\right]\right)\\
&=\overline{C}_t^\tau\left(e^{-\int_0^t h_{t,s}^1 ds},\ldots,e^{-\int_0^T h_{t,s}^i ds},\ldots,e^{-\int_0^t h_{t,s}^n ds}\right);
\end{aligned}$$

the first equality is from the definition of the survival copula process, and the second equality uses the definition of the $\{\mathcal{F}_t\}$-forward intensities. The denominator is computed similarly. Differentiating with respect to T,

$$\lambda_{t,T}^{i,(\emptyset)} \triangleq -\frac{1}{H_t^{i,(\emptyset)}(T)}\frac{\partial H_t^{i,(\emptyset)}(T)}{\partial T},$$

we obtain the result of the proposition. ■

Post-default. If one or more defaults have occurred, the conditional distribution of τ_i changes, which results in a different expression of the $\{\mathcal{G}_t\}$-forward intensity.

Proposition 58 (Post-default). *If k obligors indexed by $\pi = \{j_1, \ldots, j_k\}$ have defaulted before time t, and their default times are $\{t_{j_1}, \ldots, t_{j_k}\}$ respectively. Then, the $\{\mathcal{G}_t\}$-forward intensity is given by*

$$\lambda_{t,T}^{i,(\pi)} = h_{t,T}^i \exp\left(-\int_0^T h_{t,s}^i ds\right)\frac{\frac{\partial}{\partial x_i}\frac{\partial^k}{\partial x_{j_1}\ldots\partial x_{j_k}}\overline{C}_t^\tau\left(e^{-\int_0^{\Theta_1} h_{t,s}^1 ds},\ldots,e^{-\int_0^{\Theta_n} h_{t,s}^n ds}\right)}{\frac{\partial^k}{\partial x_{j_1}\ldots\partial x_{j_k}}\overline{C}_t^\tau\left(e^{-\int_0^{\Theta_1} h_{t,s}^1 ds},\ldots,e^{-\int_0^{\Theta_n} h_{t,s}^n ds}\right)},\tag{4.25}$$

where

$$\Theta_j = t_j, \textit{for } j \in \pi = \{j_1, ..., j_k\}\,;$$
$$\Theta_i = T, \textit{for } j = i;$$
$$\Theta_j = t, \textit{ otherwise.}$$

Proof On the set $\left[\prod_{m=1}^{k} \mathbf{1}_{\{\tau_{jm} \in dt_{jm}\}}\right]\left[\prod_{j \notin \{j_1,...,j_k\}} \mathbf{1}_{\{\tau_j > t\}}\right]$, the conditional survival probability is

$$H_t^{i,(\pi)}(T) = \frac{\mathbb{E}\left[\left(1 - D_T^i\right) \times \prod_{j \notin \pi}\left(1 - D_t^j\right) \times \prod_{m=1}^{k} \mathbf{1}_{\{\tau_{jm} \in dt_{jm}\}} | \mathcal{F}_t\right]}{\mathbb{E}\left[\prod_{j \notin \pi}\left(1 - D_t^j\right) \times \prod_{m=1}^{k} \mathbf{1}_{\{\tau_{jm} \in dt_{jm}\}} | \mathcal{F}_t\right]}.$$

Using the conditional survival copula process and the $\{\mathcal{F}_t\}$-forward intensities, we have

$$\mathbb{E}\left[\left(1 - D_T^i\right) \times \prod_{j \notin \pi}\left(1 - D_t^j\right) \times \prod_{m=1}^{k} \mathbf{1}_{\{\tau_{jm} > t_{jm}\}} | \mathcal{F}_t\right]$$
$$= \overline{C}_t^{\tau}\left(e^{-\int_0^{\Theta_1} h_{t,s}^1 ds}, ..., e^{-\int_0^{\Theta_n} h_{t,s}^n ds}\right),$$

where

$$\Theta_j = t_j, \text{ for } j \in \pi = \{j_1, ..., j_k\}\,; \Theta_i = T; \text{ and } \Theta_j = t, \text{ otherwise.}$$

Hence, the k-density is given by

$$\mathbb{E}\left[\left(1 - D_T^i\right) \times \prod_{j \notin \pi}\left(1 - D_t^j\right) \times \prod_{m=1}^{k} \mathbf{1}_{\{\tau_{jm} \in dt_{jm}\}} | \mathcal{F}_t\right]$$
$$= \frac{\partial^k}{\partial \Theta_{j_1} ... \partial \Theta_{j_k}} \overline{C}_t^{\tau}\left(e^{-\int_0^{\Theta_1} h_{t,s}^1 ds}, ..., e^{-\int_0^{\Theta_n} h_{t,s}^n ds}\right)$$
$$= \left[\prod_{m=1}^{k} h_{t,T}^{j_m} \exp\left(-\int_0^{t_{jm}} h_{t,s}^{j_m} ds\right)\right] \frac{\partial^k}{\partial x_{j_1} ... \partial x_{j_k}} \overline{C}_t^{\tau}\left(e^{-\int_0^{\Theta_1} h_{t,s}^1 ds}, ..., e^{-\int_0^{\Theta_n} h_{t,s}^n ds}\right).$$

The denominator is computed similarly

$$\mathbb{E}\left[\left(1-D_t^i\right)\times\prod_{j\notin\pi}\left(1-D_t^j\right)\times\prod_{m=1}^{k}\mathbf{1}_{\{\tau_{jm}\in dt_{jm}\}}|\mathcal{F}_t\right]$$
$$=\left[\prod_{m=1}^{k}h_{t,T}^{jm}\exp\left(-\int_0^{t_{jm}}h_{t,s}^{jm}ds\right)\right]\frac{\partial^k}{\partial x_{j_1}...\partial x_{j_k}}\overline{C}_t^{\tau}\left(e^{-\int_0^{\theta_1}h_{t,s}^1ds},...,e^{-\int_0^{\theta_n}h_{t,s}^nds}\right),$$

where

$$\theta_j=t_j,\text{ for }j\in\pi=\{j_1,...,j_k\};\,\theta_i=t;\text{ and }\theta_j=t,\text{ otherwise.}$$

Differentiating the conditional survival probability $H_t^{i,(\pi)}(T)$,

$$H_t^{i,(\pi)}(T)=\frac{\frac{\partial^k}{\partial x_{j_1}...\partial x_{j_k}}\overline{C}_t^{\tau}\left(e^{-\int_0^{\Theta_1}h_{t,s}^1ds},...,e^{-\int_0^{\Theta_n}h_{t,s}^nds}\right)}{\frac{\partial^k}{\partial x_{j_1}...\partial x_{j_k}}\overline{C}_t^{\tau}\left(e^{-\int_0^{\theta_1}h_{t,s}^1ds},...,e^{-\int_0^{\theta_n}h_{t,s}^nds}\right)},$$

with respect to T, we get the final result:

$$\lambda_{t,T}^{i,(\pi)}\triangleq-\frac{1}{H_t^{i,(\pi)}(T)}\frac{\partial H_t^{i,(\pi)}(T)}{\partial T}.$$

■

The formulas of Schönbucher and Schubert are obtained in our framework by setting $T=t$, using the relationship between spot and forward intensity, namely $\lambda_t^i=\lambda_{t,t}^i$, $h_t^i=h_{t,t}^i$, and observing that

$$\overline{C}_t^{\tau}\left(e^{-\int_0^{\Theta_1}h_{t,s}^1ds},...,e^{-\int_0^{\Theta_n}h_{t,s}^nds}\right)=\overline{C}^U\left(e^{-\int_0^{\Theta_1}h_s^1ds},...,e^{-\int_0^{\Theta_n}h_s^nds}\right),\tag{4.26}$$

since for $\Theta_j\leq t$, all the $\int_0^{\Theta_j}h_s^jds$ are $\{\mathcal{F}_t\}$-measurable and Eq. (4.18) simplifies to (4.26).

Regenerative Form of the Intensity. Equations (4.24) and (4.25) offer another derivation of the regenerative form of the intensity. Indeed, using the notations of Sect. 4.3 where the marked point process representation was introduced, one can write:

$$\lambda_t^i = \sum_{k=1}^{n} \mathbf{1}_{\{T_{k-1} \le t < T_k\}} a_t^{(k)} \left(i; T_1, Z_1, ..., T_{k-1}, Z_{k-1}\right), \tag{4.27}$$

where

$$a_t^{(k)} \left(i; T_1, Z_1, ..., T_{k-1}, Z_{k-1}\right) = h_t^i \exp\left(-\int_0^t h_s^i ds\right) \frac{\overline{C}^U_{x_i x_{Z_1} ... x_{Z_{k-1}}}}{\overline{C}^U_{x_{Z_1} ... x_{Z_{k-1}}}}, \tag{4.28}$$

and the lighter notations $\overline{C}^U_{x_i x_{Z_1} ... x_{Z_{k-1}}}$ and $\overline{C}^U_{x_{Z_1} ... x_{Z_{k-1}}}$ are short for

$$\begin{aligned}
\overline{C}^U_{x_i x_{Z_1} ... x_{Z_{k-1}}} &= \frac{\partial}{\partial x_i} \frac{\partial^{k-1}}{\partial x_{Z_1} ... \partial x_{Z_{k-1}}} \overline{C}^U \left(e^{-\int_0^{\Theta_1} h_s^1 ds}, ..., e^{-\int_0^{\Theta_n} h_s^n ds}\right), \\
\overline{C}^U_{x_{Z_1} ... x_{Z_{k-1}}} &= \frac{\partial^{k-1}}{\partial x_{Z_1} ... \partial x_{Z_{k-1}}} \overline{C}^U \left(e^{-\int_0^{\Theta_1} h_s^1 ds}, ..., e^{-\int_0^{\Theta_n} h_s^n ds}\right), \\
\Theta_{Z_j} &= T_j, \text{ for } 1 \le j \le k-1; \ \Theta_i = t; \text{ and } \Theta_j = t \text{ otherwise.}
\end{aligned}$$

Dynamics. Assuming that the $\{\mathcal{F}_t\}$ -adapted intensities $\left(h_t^1, ..., h_t^n\right)_{t \ge 0}$ follow a diffusion, we can derive the dynamics of $\left(\lambda_t^i\right)_{t \ge 0}$ by applying Itô's lemma to the regenerative form (4.27). At each default time T_k, for $1 \le k \le n$ (and given that obligor i has not defaulted), the jump of λ_t^i is given by

$$\begin{aligned}
\Delta \lambda_{T_k}^i = \lambda_{T_k}^i - \lambda_{T_k^-}^i &= a_{T_k}^{(k+1)} (i) - \lim_{t \uparrow T_k} a_t^{(k)} (i) \\
&= h_{T_k}^i e^{-\int_0^{T_k} h_s^i ds} \left[\frac{\overline{C}^U_{x_i x_{Z_1} ... x_{Z_k}}}{\overline{C}^U_{x_{Z_1} ... x_{Z_k}}} - \frac{\overline{C}^U_{x_i x_{Z_1} ... x_{Z_{k-1}}}}{\overline{C}^U_{x_{Z_1} ... x_{Z_{k-1}}}} \right].
\end{aligned} \tag{4.29}$$

Between jump times, $\left(\lambda_t^i\right)_{t \ge 0}$ follows a diffusion. For $T_{k-1} \le t < T_k$, it coincides with $a_t^{(k)} (i)$ whose dynamics can be obtained by Itô's lemma:

$$\begin{aligned}
\frac{da_t^{(k)} (i)}{a_t^{(k)} (i)} &= \left[\left(\frac{dh_t^i}{h_t^i} - h_t^i dt \right) - \sum_{j \notin \{Z_1, ..., Z_{k-1}\}} \left(\frac{\overline{C}^U_{x_i x_j x_{Z_1} ... x_{Z_{k-1}}}}{\overline{C}^U_{x_i x_{Z_1} ... x_{Z_{k-1}}}} - \frac{\overline{C}^U_{x_j x_{Z_1} ... x_{Z_{k-1}}}}{\overline{C}^U_{x_{Z_1} ... x_{Z_{k-1}}}} \right) h_t^j e^{-\int_0^t h_s^j ds} \right] \\
&= \left[\left(\frac{dh_t^i}{h_t^i} - h_t^i dt \right) - \sum_{j \notin \{Z_1, ..., Z_{k-1}\}} \left(\frac{\overline{C}^U_{x_i x_j x_{Z_1} ... x_{Z_{k-1}}} \overline{C}^U_{x_{Z_1} ... x_{Z_{k-1}}}}{\overline{C}^U_{x_i x_{Z_1} ... x_{Z_{k-1}}} \overline{C}^U_{x_j x_{Z_1} ... x_{Z_{k-1}}}} - 1 \right) a_t^{(k)} (j) \, dt \right].
\end{aligned}$$

Note that the index $j \notin \{Z_1, ..., Z_{k-1}\}$ covers all the non-defaulted obligors including i. Thus, the dynamics of $\left(\lambda_t^i\right)_{t \geq 0}$ are given by:

for $t < \tau_i$,

$$\frac{d\lambda_t^i}{\lambda_{t-}^i} = \left[\left(\frac{dh_t^i}{h_t^i} - h_t^i dt\right) - \left(\frac{\overline{C}^U_{x_i x_i x_{Z_1} \dots x_{Z_{k-1}}} \overline{C}^U_{x_{Z_1} \dots x_{Z_{k-1}}}}{\overline{C}^U_{x_i x_{Z_1} \dots x_{Z_{k-1}}} \overline{C}^U_{x_i x_{Z_1} \dots x_{Z_{k-1}}}} - 1\right) \lambda_{t-}^i dt\right]$$
$$+ \left[\sum_{\substack{j \notin \{Z_1, \dots, Z_{k-1}\} \\ j \neq i}} \left(\frac{\overline{C}^U_{x_i x_j x_{Z_1} \dots x_{Z_{k-1}}} \overline{C}^U_{x_{Z_1} \dots x_{Z_{k-1}}}}{\overline{C}^U_{x_i x_{Z_1} \dots x_{Z_{k-1}}} \overline{C}^U_{x_j x_{Z_1} \dots x_{Z_{k-1}}}} - 1\right) \left(dD_t^j - \lambda_{t-}^j dt\right)\right]. \tag{4.30}$$

Defining a version of the intensity λ_t^i, which does not jump after τ_i,

$$\lambda_{\tau_i}^i = \lambda_{\tau_i^-}^i,$$

we can express (4.30) using the marked point process $\mu\,(dt \times dz)$:

$$\frac{d\lambda_t^i}{\lambda_{t-}^i} = \frac{dh_t^i}{h_t^i} + \left[\left(1 - \frac{\overline{C}^U_{x_i x_i x_{Z_1} \dots x_{Z_{k-1}}} \overline{C}^U_{x_{Z_1} \dots x_{Z_{k-1}}}}{\overline{C}^U_{x_i x_{Z_1} \dots x_{Z_{k-1}}} \overline{C}^U_{x_i x_{Z_1} \dots x_{Z_{k-1}}}}\right) \lambda_t^i - h_t^i\right] dt$$
$$+ \int_E \Delta\,(i, \omega, t, z)\,(\mu\,(dt \times dz) - \nu\,(dt \times dz))\,, \tag{4.31}$$

where the jump size $\Delta\,(i, \omega, t, z)$ is given by

$$\Delta\,(i, \omega, t, z) = \begin{cases} \dfrac{\overline{C}^U_{x_i x_z x_{Z_1(\omega)} \dots x_{Z_{k-1}(\omega)}} \overline{C}^U_{x_{Z_1(\omega)} \dots x_{Z_{k-1}(\omega)}}}{\overline{C}^U_{x_i x_{Z_1(\omega)} \dots x_{Z_{k-1}(\omega)}} \overline{C}^U_{x_z x_{Z_1(\omega)} \dots x_{Z_{k-1}(\omega)}}} - 1, \text{ if } z \neq i, \\ 0, \text{ if } z = i. \end{cases} \tag{4.32}$$

The intensity λ^i has a diffusion part, which is defined by the dynamics of the $\{\mathcal{F}_t\}$-adapted intensity h^i , and jumps upon the default of the other obligors. This is an equivalent representation of the copula model as another member of the class of "looping" default models considered in Sect. 3. Note that, if the tresholds $(U_1, ..., U_2)$ are independent, i.e.,

$$C\,(u_1, ..., u_n) = \prod_{i=1}^{n} u_i, \text{ for all } u_i \in [0, 1]\,,$$

then,

$$\frac{\overline{C}^U_{x_i x_z x_{z_1(\omega)} \ldots x_{z_{k-1}(\omega)}} \overline{C}^U_{x_{z_1(\omega)} \ldots x_{z_{k-1}(\omega)}}}{\overline{C}^U_{x_i x_{z_1(\omega)} \ldots x_{z_{k-1}(\omega)}} \overline{C}^U_{x_z x_{z_1(\omega)} \ldots x_{z_{k-1}(\omega)}}} = 1, \text{ for } z \neq i,$$

$$\frac{\overline{C}^U_{x_i x_i x_{z_1(\omega)} \ldots x_{z_{k-1}(\omega)}} \overline{C}^U_{x_{z_1(\omega)} \ldots x_{z_{k-1}(\omega)}}}{\overline{C}^U_{x_i x_{z_1(\omega)} \ldots x_{z_{k-1}(\omega)}} \overline{C}^U_{x_i x_{z_1(\omega)} \ldots x_{z_{k-1}(\omega)}}} = 0, \text{ for } z = i,$$

and Eq. (4.31) degenerates to

$$\frac{d\lambda_t^i}{\lambda_{t-}^i} = \frac{dh_t^i}{h_t^i} + \left[\lambda_t^i - h_t^i\right] dt.$$

This is consistent with the fact that, when the default tresholds are independent, the intensities λ^i and h^i coincide.

4.5 Numerical Examples

The goal of this section is to illustrate the relationship between CJD dynamics and the copula approach. We start with the Jarrow and Yu model, for $n = 2$, and we look at the default correlation implied by a given conditional jump size. Then, we consider the inverse problem where we study the default contingent jump implied by a Gaussian copula.

Case Study: n = 2. We consider the example of 2 default times with interacting intensities:

$$\lambda_t^1 = \lambda_0^1 + \Delta^{12} \mathbf{1}_{\{\tau_2 \leq t\}},$$
$$\lambda_t^2 = \lambda_0^2 + \Delta^{21} \mathbf{1}_{\{\tau_1 \leq t\}},$$

with $\Delta^{12} > \lambda_0^1$ and $\Delta^{22} > \lambda_0^2$. And let α_{12} and α_{21} denote the proportional jump ratios: $\Delta^{12} = \alpha_{12}\lambda_0^1$, $\Delta^{21} = \alpha_{21}\lambda_0^2$.

Figure 4.1 shows how the pair-wise default correlation varies with the jump size for different time horizons. When the jump size is zero, the two default times are independent and their default correlation is zero. On the other hand, when the jump size goes to infinity, the default correlation goes to its maximum value. In this example, the two intensities are identical, and the highest achievable default correlation is 1. Intuitively, an infinite intensity corresponds to the

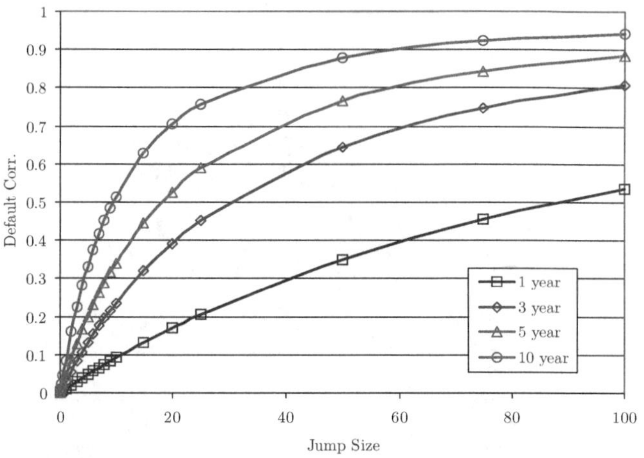

Fig. 4.1 Default correlation as a function of the jump size ratio $\alpha = \alpha_{12} = \alpha_{21}$, for different time horizons T. The intensities are $\lambda_0^1 = \lambda_0^2 = 100$ bps

default state. In other words, an infinite jump size implies that the default of one obligor triggers the default of the other.

The copula function implied by these CJD dynamics is depicted in Fig. 4.2. For convenience, we have plotted the copula in Gaussian coordinates, i.e., we use the transformation: $(\theta_1, \theta_2) \rightarrow \left(\Phi^{-1}\left(\mathbb{P}\left(\tau_1 \leq \theta_1\right)\right), \Phi^{-1}\left(\mathbb{P}\left(\tau_2 \leq \theta_2\right)\right)\right)$.

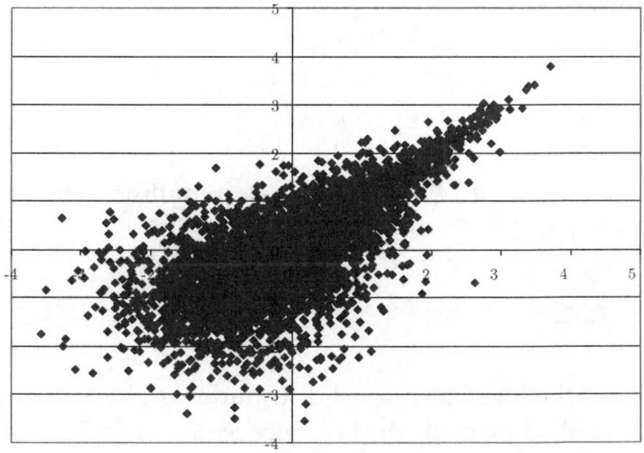

Fig. 4.2 Conditional jump diffusion copula, with a jump ratio of $\alpha_{12} = \alpha_{21} = 3.58$, which corresponds to a 5 year-default correlation of 0.15 for $\lambda_0^1 = \lambda_0^2 = 100$ bps

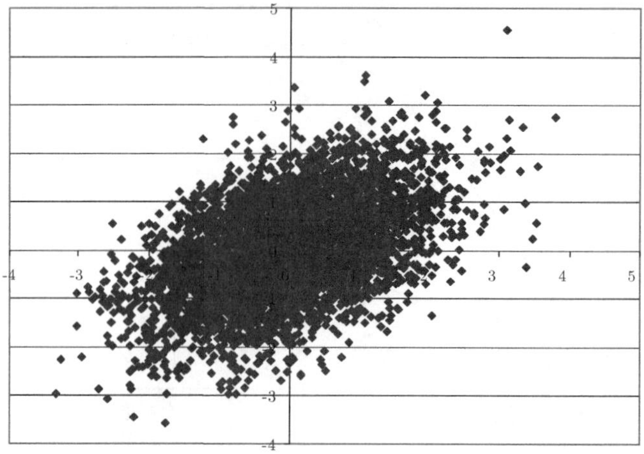

Fig. 4.3 Gaussian copula with an asset correlation of 0.497, which corresponds to a 5 year-default correlation of 0.15 for $\lambda_0^1 = \lambda_0^2 = 100$ bps

In order to do a comparison with the standard Gaussian copula (depicted in Fig. 4.3), we calibrate the parameters of both copulas such that the default correlation is the same. In this example, the 5-year default correlation is 15%. Since the default correlation is the same, the "slope" of the copula functions is preserved, but the behaviour in the tail is very different.

Gaussian copula. Now, we investigate the inverse problem. We consider two default times (τ_1, τ_2) with marginals

$$\mathbb{P}(\tau_1 > T) = e^{-h_1 T},$$
$$\mathbb{P}(\tau_2 > T) = e^{-h_2 T},$$

for some fixed $h_1 \in \mathbb{R}_+$, $h_2 \in \mathbb{R}_+$, and whose joint distribution is defined via a Gaussian copula, i.e.,

$$\mathbb{P}(\tau_1 \le T_1, \tau_2 \le T_2) = \Phi_2\left(\Phi^{-1}(\mathbb{P}(\tau_1 \le T_1)), \Phi^{-1}(\mathbb{P}(\tau_2 \le T_2))\right),$$

where $\Phi_2(.,.)$ is the bivariate normal distribution and $\Phi^{-1}(.)$ is the inverse normal function. And we study the link between the default contingent jump size and correlation.

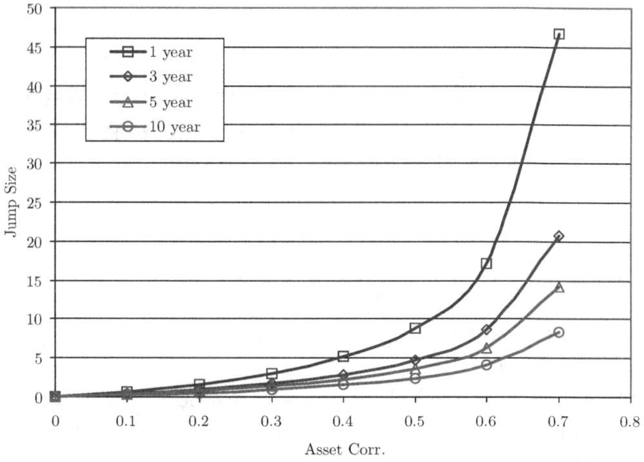

Fig. 4.4 Jump size ratio implied by a Gaussian copula as a function of asset correlation. The intensities are $h_1 = h_2 = 100$ bps

Figure 4.4 shows how the jump size ratio varies with asset correlation. As correlation increases, the jump size ratio increases as well. For a correlation of 1, the jump size goes to infinity. In other words, the default of one obligor triggers the default of the other one and vice-versa.

4.6 Conclusion

The objective of this chapter was to introduce the conditional jump diffusion framework and its relationship with default correlations and copulas. This was done in two steps. First, we have studied conditional jump diffusion dynamics and their implied default times' multivariate dependence. Then, we have considered the inverse problem and examined the consequences of a choice of copula on the dynamics of intensities. Thus, establishing the equivalence between CJD and Copulas. This equivalence principle has a few useful applications. It can be used as a sanity check, to verify whether a specific calibrated copula function implies reasonable jumps. The analysis of implied jumps can be a powerful tool for finding aberrations in the calibration procedure. On the other hand, one can have a view on the jump in default and would rather build a default model, which reflects that view. This is very similar, in spirit, to calibrating a term structure interest rate model, then infer implied forward volatilities, or building a Brace-Gatarek-Musiela (BGM) style model, which is consistent with one's view of forward volatilities. Both approaches are useful and complement each other.

Note

1. In the Jarrow and Yu paper, a simplified version of the general model is considered. They assume that the coupling matrix $\left[\Delta^{ij}\right]$ is upper-triangular in order to break the circular dependence.

References

T. Aven, A Theorem for determining the compensator of a counting process. Scand. J. Stat. **12**(1), 69–72 (1985)

D. Becherer, M. Schweizer, Classical solutions to reaction diffusion systems for hedging problems with interacting itô and point processes. Ann. Appl. Probab. **15**(2), 1111–1144 (2005)

P. Brémaud, *Point Processes and Queues: Martingale Dynamics* (Springer, New York, 1980)

M. Davis, V. Lo, Infectious defaults. Quant. Financ. **1**, 305–308 (2001a)

M. Davis, V. Lo, Modelling default correlation in bond portfolios, in Mastering Risk, Volume 2: Applications, ed. by Carol Alexander (Financial Times, Prentice Hall, 2001b), pp. 141–151

R.J. Elliott, M. Jeanblanc, M. Yor, On models of default risk. Math. Financ. **10**(2), 179–195 (2000)

R. Frey, J. Backhaus, Portfolio credit risk models with interacting default intensities: a Markovian approach (Working Paper, Department of Mathematics, University of Leipzig, 2004)

R. Jarrow, F. Yu, Counterparty risk and the pricing of defaultable securities. J. Financ. **LVI**(5), 1765–1800 (2001)

M. Jeanblanc, M. Rutkowski, Modelling of default risk: an overview, in *Mathematical Finance: Theory and Practice*, ed. by J. Yong, R. Cont (Higher Education Press, Beijing, 2000a), pp. 171–269

M. Jeanblanc, M. Rutkowski, Modelling of default risk: mathematical tools (Working Paper, Université d'Evry and Warsaw University of Technology, 2000b)

J.F. Jouanin, G. Rapuch, G. Riboulet, T. Roncalli, Modelling dependence for credit derivatives with copulas (Working Paper, Groupe de Recherche Operationnelle, Credit Lyonnais, 2001)

S. Kusuoka, A remark on default risk models. Adv. Math. Econ. **1**, 69–82 (1999)

D. Lando, On Cox processes and credit risky securities. Rev. Deriv. Res. **2**(2/3), 99–120 (1998)

I. Norros, A compensator representation of multivariate life length distributions with applications. Scand. J. Stat. **13**, 99–112 (1986)

A. Patton, Modelling time-varying exchange rate dependence using the conditional Copula (Working Paper 2001–09, University of California, San Diego, 2001)

P.J. Schönbucher, D Schubert, Copula-dependent default risk in intensity models (Working Paper, Department of Statistics, Bonn University, 2001)
M. Shaked, G. Shanthikumar, The multivariate hazard construction. Stoch. Process. Appl. **24**, 241–258 (1987)
A. Sklar, Fonctions de répartition à n dimensions et leurs marges. Publications de l'Institut de Statistique de l'Universit é de Paris **8**, 229–231 (1959)
F. Yu, Correlated defaults and the valuation of defaultable securities (Working Paper, University of California, Irvine, 2004)

Part II

Correlation Models: Practical Implementation

5

Correlation Demystified: A General Overview

This chapter gives a broad overview of default correlation modelling in the context of pricing and risk managing a correlation trading book. We cover both theoretical and practical market aspects, as well as numerical performance issues.

5.1 Base Correlation

Over the years, we have seen increased liquidity in CDO tranche trading, which resulted in a well-established observable market for default correlation. Dealers are quoting a two-way market on a pre-specified set of tranches referenced to a given index portfolio. By inverting the Gaussian copula formula, one finds the implied level of correlation that would match the quoted tranche premiums. As with the Black-Scholes option model, there is not a single correlation number that would match all tranches, with various attachment points, at the same time. Supply and demand factors combined with the credit views and risk appetite of market participants would explain the observed discrepancy between correlations across the capital structure. The table below gives an example of market bid/offer premiums for the European iTraxx index tranches:

0–3%	23.3	24.3
3–6%	134	137
6–9%	44	47
9–12%	28	32.3
12–22%	14.2	15.5

Y. Elouerkhaoui, *Credit Correlation*, Applied Quantitative Finance,
DOI 10.1007/978-3-319-60973-7_5

The index level is 37 bps. The (0–3%) tranche is quoted in points upfront for a tranche paying 500 bps running.

Next, we explain some key concepts such as base correlation and compound correlation in a formal manner.

5.1.1 One-Factor Gaussian Copula

We give a formal definition of the one-factor Gaussian copula function.

Definition 59 (*One-factor Gaussian Copula*) The one-factor Gaussian copula with parameter $\rho \in [0, 1)$ is defined as

$$C\left(u_1, \ldots, u_n\right) \triangleq \int_{-\infty}^{+\infty} \left[\prod_{i=1}^{n} \Phi\left(\frac{\Phi^{-1}\left(u_i\right) - \sqrt{\rho} y}{\sqrt{1-\rho}}\right)\right] \phi\left(y\right) dy,$$

where $\Phi\left(.\right)$, $\Phi^{-1}\left(.\right)$ and $\phi\left(.\right)$ are the standard normal distribution function, its inverse and its density function respectively.

This formal definition can be understood by considering a simplified firm value model as in Schönbucher (2001) for example. The default of obligor i is triggered when the asset value of the firm, denoted by V_i, is below a given threshold. V_i is assumed to follow a normal distribution. The relationship between default time and asset value is given by

$$\left\{\tau_i \leq T\right\} \Longleftrightarrow \left\{V_i \leq \Phi^{-1}\left(\mathbb{P}\left(\tau_i \leq T\right)\right)\right\}.$$

The asset values of different obligors are correlated. Their joint dependence is defined via a common factor Y, which follows a standard normal distribution, and idiosyncratic standard normal noises $\epsilon_1, \ldots, \epsilon_n$:

$$V_i \triangleq \sqrt{\rho} Y + \sqrt{1-\rho} \epsilon_i,$$

where Y and $\epsilon_1, \ldots, \epsilon_n$ are i.i.d. standard normally distributed. The linear correlation between the asset values of two obligors is ρ. This coefficient, which is used to parameterize the family of one-factor Gaussian copulas, is sometimes called an asset correlation. Conditional on a given value of the systemic factor Y, the asset values are independent; hence, the default times are independent as well. This is the set-up of a conditionally independent defaults model.

One can write down the default times' copula function by conditioning on Y and using the law of iterated expectations:

$$
\begin{aligned}
&\mathbb{P}(\tau_1 \leq T_1, \ldots, \tau_n \leq T_n) \\
&= \int_{-\infty}^{+\infty} \mathbb{P}(\tau_1 \leq T_1, \ldots, \tau_n \leq T_n \,|Y = y)\, \phi(y)\, dy \\
&= \int_{-\infty}^{+\infty} \mathbb{P}\left(V_1 \leq \Phi^{-1}(\mathbb{P}(\tau_1 \leq T_1)), \ldots, V_n \leq \Phi^{-1}(\mathbb{P}(\tau_n \leq T_n))\,|Y = y\right) \phi(y)\, dy \\
&= \int_{-\infty}^{+\infty} \mathbb{P}\left(\epsilon_1 \leq \frac{\Phi^{-1}(\mathbb{P}(\tau_1 \leq T_1)) - \sqrt{\rho} y}{\sqrt{1-\rho}}, \ldots, \epsilon_n \leq \frac{\Phi^{-1}(\mathbb{P}(\tau_n \leq T_n)) - \sqrt{\rho} y}{\sqrt{1-\rho}}\right) \phi(y)\, dy \\
&= \int_{-\infty}^{+\infty} \left[\prod_{i=1}^{n} \Phi\left(\frac{\Phi^{-1}(u_i) - \sqrt{\rho} y}{\sqrt{1-\rho}}\right)\right] \phi(y)\, dy.
\end{aligned}
$$

The one-factor Gaussian copula is the standard model used to quote CDO tranches in the market.

5.1.2 Pricing CDOs

Let us consider the pricing of a CDO tranche, which covers the losses of a given portfolio between two thresholds $0 \leq K_1 < K_2 \leq 1$.

Letting R_i denote the recovery rate of obligor i, we define the (normalized) portfolio loss process as

$$L_t \triangleq \frac{1}{n} \sum_{i=1}^{n} (1 - R_i)\, D_t^i.$$

The loss variable for the tranche (K_1, K_2) is defined as

$$M_t^{K_1,K_2} = \min\left(\max\left(L_t - K_1, 0\right), K_2 - K_1\right).$$

The portfolio loss process L_t, and the tranche loss process $M_t^{K_1,K_2}$ are pure jump processes. The CDO cash-flow payments correspond to the increments of $M_t^{K_1,K_2}$, i.e., there is a payment when the process $M_t^{K_1,K_2}$ jumps, which happens at every default time. The payoff of the protection leg of a CDO is therefore defined as the Stieljes integral

$$\text{protection\_leg} \triangleq \int_{]0,T]} p_{0,t}\, dM_t^{K_1,K_2},$$

where $p_{0,t} \triangleq \exp\left(-\int_0^t r_s ds\right)$ denote the risk-free discount factor maturing at time T. Letting $(T_0 = 0, T_1, \ldots, T_N)$ denote the cash-flow dates, $\delta_i \triangleq T_i - T_{i-1}$, the payment fractions and S the tranche premium, the payoff of the premium leg is defined as:

$$\text{premium\_leg} \triangleq S \times \sum_{i=1}^{n} p_{0,T_i} \left[(K_2 - K_1) - M_{T_i}^{K_1,K_2} \right] \delta_i.$$

The value of the CDO tranche is given by the expected value of the discounted payoff under the risk neutral measure.

Using the integration by part formula and Fubini's theorem to interchange the order of integration, we can re-write the protection integral as

$$\mathbb{E}\left[\int_{]0,T]} p_{0,t} dM_t^{K_1,K_2} \right] = p_{0,T} \mathbb{E}\left[M_T^{K_1,K_2} \right] - \int_0^T \frac{\partial p_{0,t}}{\partial t} \mathbb{E}\left[M_t^{K_1,K_2} \right] dt,$$

Similarly, to compute the value of the premium leg, we need to know the expected tranche losses at times T_i: $\mathbb{E}\left[M_{T_i}^{K_1,K_2} \right]$.

Thus, the pricing of CDO tranches boils down to computing the expected values of tranche payoff at different time horizons:

$$C_t(K_1, K_2) \triangleq \mathbb{E}\left[M_t^{K_1,K_2} \right], \text{ for } 0 \leq t \leq T. \tag{5.1}$$

For $t \geq 0$, if we know the density function $f_t(.)$ of the portfolio loss variable L_t:

$$f_t(x) \triangleq \mathbb{P}(L_t \in dx), \tag{5.2}$$

then, the expectation (5.1) can be written as

$$\mathbb{E}\left[M_t^{K_1,K_2} \right] = \int_{K_1}^{K_2} (x - K_1) f_t(x) dx + (K_2 - K_1)(1 - F_t(x)), \tag{5.3}$$

where $F_t(x) = \int_{-\infty}^{x} f_t(z) dz$ is the cumulative probability function of L_t. With a given copula, such as the one-factor copula, it is easy to compute the density function $f_t(.)$ using techniques such as the FFT (see Laurent and Gregory 2005), the convolution recursion (see Andersen et al. 2003) or the Normal Proxy Method (Shelton 2004).

5.1.3 Large Homogenous Portfolio

In order to study the qualitative behaviour of default correlation properties, we can use the Large Homogenous Portfolio approximation (LHP) to generate the portfolio loss distribution in a Gaussian copula model. This is a very neat closed form solution, which is quick and easy to implement, yet captures the key properties of the loss distribution and the impact of default correlation on pricing. Originally, it was introduced by Vasicek (1997) in a Firm Value model, but was then used more broadly in reduced form factor models (see, for example, Schönbucher (2001)).

We consider a homogeneous portfolio where the single name probabilities are identical; in particular, they have the same default probabilities, the same recoveries and the same correlation assumptions

$$\mathbb{P}(\tau_i \leq T) = p_T, \text{ for all } 1 \leq i \leq n,$$
$$R_i = R, \text{ for all } 1 \leq i \leq n.$$

We work in the standard Gaussian copula framework where the single-name defaults are correlated through a common factor Y,

$$V_i = \sqrt{\rho} Y + \sqrt{1-\rho}\,\epsilon_i,$$

where Y and $\epsilon_1, \ldots, \epsilon_n$ are i.i.d. standard normally distributed. The default of each name i is triggered if the asset value V_i exceeds the default threshold $\Phi^{-1}(p_T)$.

In the homogeneous set-up, the (normalized) loss variable L_T is given by the portfolio aggregate default fraction $X_T \triangleq \frac{1}{n}\sum_{i=1}^{n} D_T^i$, rescaled with the loss-given-default $(1-R)$,

$$L_T = \frac{(1-R)}{n}\sum_{i=1}^{n} D_T^i = (1-R)\cdot X_T.$$

Conditional on the common factor Y, the individual single-name defaults are independent and their conditional default probabilities are given by

$$p_T(y) = \mathbb{P}(\tau_i \leq T \,|Y = y) = \Phi\left(\frac{\Phi^{-1}(p_T) - y\sqrt{\rho}}{\sqrt{1-\rho}}\right).$$

From the law of large numbers, it follows that the fraction $\frac{1}{n}\sum_{i=1}^{n}\mathbf{1}_{\{\tau_i\leq T\}}$ converges to its average $p_T(y)$,

$$X_{T|Y=y} = \frac{1}{n}\sum_{i=1}^{n}\mathbf{1}_{\{\tau_i\leq T\}} \rightarrow \mathbb{E}\left[\mathbf{1}_{\{\tau_i\leq T\}}|Y=y\right] = p_T(y).$$

Thus, to compute the default rate distribution X_T, we condition on the common factor Y, we assume that the number of names goes to infinity $n \rightarrow +\infty$, and we approximate it as follows.

For all $0 \leq x \leq 1$,

$$\begin{aligned}\mathbb{P}(X_T \leq x) &= \int_{-\infty}^{+\infty}\mathbb{P}(X_T \leq x\,|Y=y)\,\varphi(y)\,dy\\ &\simeq \int_{-\infty}^{+\infty}\mathbb{P}(p_T(y) \leq x\,|Y=y)\,\varphi(y)\,dy\\ &= \int_{-\infty}^{+\infty}\mathbf{1}_{\{p_T(y)\leq x\}}\varphi(y)\,dy = \int_{p_T^{-1}(x)}^{+\infty}\varphi(y)\,dy = \Phi\left(-p_T^{-1}(x)\right),\end{aligned}$$

where we have substituted the average X_T with its limit $p_T(y)$, and used the fact that the function is $p_T(y)$ is monotonically decreasing.

Finally, replacing the inverse $p_T^{-1}(x)$ with its expression, we obtain

$$\mathbb{P}(X_T \leq x) = \Phi\left(\frac{\sqrt{1-\rho}\,\Phi^{-1}(x) - \Phi^{-1}(p_T)}{\sqrt{\rho}}\right).$$

Taking the first-derivative with respect to the default fraction x gives the corresponding density function

$$\mathbb{P}(X_T \in dx) = \sqrt{\frac{1-\rho}{\rho}}\exp\left(\frac{1}{2}\Phi^{-1}(x) - \frac{1}{2\rho}\left(\sqrt{1-\rho}\,\Phi^{-1}(x) - \Phi^{-1}(p_T)\right)^2\right).$$

Proposition 60 (Large Homogeneous Portfolio) *In the large homogeneous portfolio set-up (LHP), the single names are identical and correlated through a one-factor Gaussian copula model. When the number of names is large, $n \rightarrow +\infty$, the distribution of the portfolio default rate $X_T = \frac{1}{n}\sum_{i=1}^{n}\mathbf{1}_{\{\tau_i\leq T\}}$ is given by*

$$\mathbb{P}(X_T \leq x) = \Phi\left(\frac{\sqrt{1-\rho}\,\Phi^{-1}(x) - \Phi^{-1}(p_T)}{\sqrt{\rho}}\right).$$

This approximation works remarkably well in practice; it provides, in many cases, a very simple and intuitive formula, which can be used to study the qualitative behaviour of the loss distribution and the impact of default correlations.

5.1.4 FFT and Recursion

In the general case of non-homogeneous portfolios, we need to work out the portfolio loss distribution using efficient numerical methods. Conditioning on the common factor Y, the single-name default indicators, at a fixed time horizon T, $D_T^i = \mathbf{1}_{\{\tau_i \leq T\}}$, for $1 \leq i \leq n$, are independent. The problem that we need to solve is how to generate the (conditional) probability distribution of the loss variable $L_{T|Y=y} = \sum_{i=1}^{n} l_i \cdot D_{T|Y=y}^i$.

This can be done, in general, by a brute force method where we enumerate all the possible 2^n combinations of default scenarios, aggregate the losses in each scenario, and generate the loss distribution accordingly. Clearly, this is an explosive algorithm as a function of the portfolio size, and would not be manageable for more than a handful of names, say $n \leq 10$.

Fortunately, the curse of dimensionality, in this problem, can be addressed easily by Fast Fourier Transform (FFT) methods or Convolution Recursions –popular in the actuarial literature.

Fast Fourier Transform. In this subsection, we review briefly the Fourier transform inversion formula for a discrete random variable X taking values in $\{0, 1, \ldots, N-1\}$.

Let $\widetilde{p} = \{p_0, p_1, \ldots, p_{N-1}\}$ be its probability function, and $\varphi(s)$ its discrete Fourier Transform,

$$\varphi(s) = \sum_{i=0}^{N-1} e^{isk} p_k.$$

Evaluating the Fourier transform $\varphi(s)$ at the points $s_j = \frac{2\pi j}{N}$ for $j = 0, 1, \ldots, N-1$, gives the following system of equations:

$$\left\{ \varphi(s_j) = \sum_{i=0}^{n-1} e^{iks_j} p_k \right\}_{0 \leq j \leq n-1}.$$

By letting $\widetilde{\varphi} = \{\varphi(s_0), \varphi(s_1), \ldots, \varphi(s_{N-1})\}$ denote the vector of Fourier transform values, the system of equations can be written in matrix form as:

$$\widetilde{\varphi} = F\widetilde{p},$$

where F is the $N \times N$ matrix

$$F = \left(e^{is_j k}\right)_{\substack{0 \le j \le N-1 \\ 0 \le k \le N-1}}.$$

Since we have

$$\sum_{k=0}^{N-1} e^{is_j k} e^{-is_l k} = \begin{cases} \frac{e^{i(s_j - s_l)n} - 1}{e^{i(s_j - s_l)} - 1} = 0 \text{ for } j \neq l \\ N \qquad\qquad \text{for} j = l \end{cases},$$

then, the inverse matrix F^{-1} is given by:

$$F^{-1} = \left(\frac{1}{n} e^{-is_k j}\right)_{\substack{0 \le j \le N-1 \\ 0 \le k \le N-1}},$$

and the probability function $\widetilde{p}$ is recovered from the Fourier transform values $\varphi\left(s_j\right)$ as:

$$p_k = \frac{1}{N} \sum_{j=0}^{N-1} \varphi\left(s_j\right) e^{-is_j k}.$$

In our set-up, the Fourier transform of the loss variable conditional on Y, which is a sum of independent single-name Bernoulli variables, can be computed, trivially, as the product of the individual Fourier transforms

$$\begin{aligned} \mathcal{F}^{L_{T|Y=y}}(u) &= \mathbb{E}\left[\exp\left(iuL_{T|Y=y}\right)\right] = \mathbb{E}\left[\exp\left(iu \sum_{i=1}^{n} l_i D^i_{T|Y=y}\right)\right] \\ &= \prod_{i=1}^{n} \mathbb{E}\left[\exp\left(iu \cdot l_i D^i_{T|Y=y}\right)\right] = \prod_{i=1}^{n} \left(p^i_T(y) e^{iu \cdot l_i} + \left(1 - p^i_T(y)\right)\right). \end{aligned}$$

Using the FFT inversion algorithm, the loss distribution can then be recovered in n^2 operations, which is significantly faster than the 2^n brute-force method.

But we can do even better with the Convolution Recursion method.

Convolution Recursion. For a set of n independent discrete random variables $(Y_1, \ldots, Y_n)$, we want to derive the distribution of the sum

$$S_n \triangleq Y_1 + \cdots + Y_n.$$

The generating function of the random variable S_n,

$$\varphi_{S_n}(x) \triangleq \sum_{k=0}^{\infty} \mathbb{P}(S_n = k)\ x^k,$$

is given by the product of the generating functions of the variables Y_i (since they are all independent)

$$\varphi_S(x) = \prod_{i=1}^{n} \varphi_{Y_i}(x);$$

and its distribution $p_{S_n}(s) = \mathbb{P}(S_n = s)$ is given by the n-fold convolution of the distributions p_{Y_i}

$$p_{S_n}(s) = \bigotimes_{i=1}^{n} p_{Y_i}(s)\ . \tag{5.4}$$

This convolution product is computed by applying the following formula recursively

$$\begin{aligned} p_{S_{k+1}}(s) &= p_{S_k} \otimes p_{Y_{k+1}}(s) \\ &= \sum_{y=0}^{s} p_{Y_{k+1}}(y)\, p_{S_k}(s-y)\,, \text{ for } 1 \le k \le n-1. \end{aligned} \tag{5.5}$$

In our case, we need to generate the aggregate loss distribution $S_n = \sum_{i=1}^{n} l_i X_i$, where $(X_1, \ldots, X_n)$ are Bernoulli variables with parameters $\widetilde{p} = (p_1, \ldots, p_n)$.

The sum S_n will be computed recursively as follows:
for $0 \le k \le n-1$

$$p_{S_{k+1}}(s) = p_{k+1} p_{S_k}(s - l_{k+1}) + (1 - p_{k+1})\, p_{S_k}(s)\,,\ 0 \le s \le \sum_{i=1}^{k+1} l_i,$$

with the convention

$$p_{S_k}(-x) = 0, x \ge 0,$$

and the distribution of the empty sum S_0 is defined as

$$\begin{cases} p_{S_0}(0) = 1, \\ p_{S_0}(s) = 0 \text{ for } s > 0. \end{cases}$$

The performance of the recursion method is a linear function of the portfolio size n; this is even more efficient than the standard FFT method, which is a n^2 algorithm. That is the reason why recursion became the preferred numerical method for pricing CDOs, and was adopted rapidly as a market standard.

However, as the CDO market evolved into a more mature and established market, the size of CDO trading books expanded manifold and the performance of the pricing algorithms for large dealers with sizable positions became more critical, especially when computing risk measures, capital metrics (such as the Comprehensive Risk Measure), and CVA (Counterparty Valuation Adjustments).

In the quest for speed, there are a number of fast and accurate approximations that have emerged: the Normal Approximation, The Poisson Approximation and the Stein Method.

5.1.5 Normal, Poisson and Stein Approximations

In this section, we give a brief description of the most popular approximations that have been used in the pricing of CDOs, namely: the Normal Proxy method, the Poisson Proxy method and Stein's Approximation.

Normal Proxy Method. In the Normal Proxy method, to generate the portfolio loss distribution, we invoke the well-known Central Limit Theorem, which states that the sum of independent random variables converges to a Normal distribution as the number of variables goes to infinity. This was first introduced in the context of CDO (and CDO-Squared) modelling by Shelton (2004).

Conditional on the common factor Y, the loss variable is $L_{T|Y=y} = \sum_{i=1}^{n} l_i \cdot D^i_{T|Y=y}$ is defined as a sum of n independent single-name random variables; hence, for large n, it converges to a normal variable $N(\mu_Y, \sigma_Y)$ with (conditional) mean μ_Y and (conditional) volatility σ_Y,

$$L_{T|Y=y} = \sum_{i=1}^{n} l_i \cdot D^i_{T|Y=y} \underset{n\to\infty}{\to} N(\mu_y, \sigma_y),$$

where the mean and variance are given by

$$\mu_y = \mathbb{E}\left[L_T \,|Y = y\right] = \sum_{i=1}^{n} l_i \cdot p_T^i(y),$$

$$\sigma_y^2 = \mathbb{E}\left[L_T^2 \,|Y = y\right] - \mu_y^2 = \sum_{i=1}^{n} l_i^2 \cdot p_T^i(y)\left(1 - p_T^i(y)\right).$$

This property is made more precise by the Lyaponov condition.

Proposition 61 (Lyaponov Condition) *Consider a sequence of independent random variables* $(X_i)_{1\leq i}$, *with finite expected values* μ_i *and variances* σ_i^2. *Define the sum* $s_n^2 = \sum_{i=1}^{n}\sigma_i^2$. *If for some positive* $\delta > 0$, *the Lyaponov condition,*

$$\lim_{n\to\infty}\frac{1}{s_n^{2+\delta}}\sum_{i=1}^{n} E\left[|X_i - \mu_i|^{2+\delta}\right] = 0,$$

is satisfied, then the sum $\sum_{i=1}^{n}\frac{X_i-\mu_i}{s_n}$ *converges, in distribution, to a standard normal variable, as* n *goes to infinity*

$$\sum_{i=1}^{n}\frac{X_i - \mu_i}{s_n} \xrightarrow{d} N(0, 1).$$

Now, by virtue of the normal approximation, we can write the conditional expected tranche loss in closed form

$$\begin{aligned}\mathbb{E}^{(\mu_y,\sigma_y)}\left[M_T^{(K_1,K_2)} \,|Y = y\right] &= (\mu_y - K_1)\,\Phi\left(\frac{\mu_y - K_1}{\sigma_y}\right) + \sigma_y\varphi\left(\frac{\mu_y - K_1}{\sigma_y}\right)\\ &\quad - (\mu_y - K_2)\,\Phi\left(\frac{\mu_y - K_2}{\sigma_y}\right) + \sigma_y\varphi\left(\frac{\mu_y - K_2}{\sigma_y}\right);\end{aligned}$$

it is simply the normal Black formula for a call-spread payoff. Integrating with respect to the common factor gives the expectation of the tranche loss variable $M_T^{(K_1,K_2)}$,

$$\mathbb{E}\left[M_T^{(K_1,K_2)}\right] = \int_{-\infty}^{+\infty}\mathbb{E}\left[M_T^{(K_1,K_2)} \,|Y = y\right]\varphi(y)\,dy.$$

Obviously, since the normal distribution allows for negative values of the loss variable, this leads to higher errors for equity tranches (especially, below the 3% strike); the errors are even more magnified when we consider thin tranchelets in that region of the distribution.

That been said, in practice, this approximation was found to be extremely fast and surprisingly accurate in most use cases.

Poisson Proxy Method. As an alternative to the Normal Proxy method, we can use a Poisson approximation instead. Fundamentally, since the loss variable is generated from a Marked Point Process, it is very natural to approximate it with some Poisson distribution, which, in this case, would replicate its mean. The Poisson Proxy does not suffer from the negative losses limitation of the Normal distribution; however, it does not have an analogous strong theoretical justification provided by the Central Limit Theorem.

To apply the Poisson approximation to CDOs, we consider a portfolio with similar LGDs—Loss-Given-Default—per name, $l_1 = \cdots = l_n = \frac{1-R}{n}$, so that the loss variable L_T can be re-written in terms of the aggregate default counting process $D_T = \sum_{i=1}^n D_T^i$, $L_T = \frac{1-R}{n} D_T$. In this case, we want to compute call prices of the form

$$\mathbb{E}\left[(L_T - K)^+\right] = \mathbb{E}\left[\left(\frac{1-R}{n} D_T - K\right)^+\right] = \frac{1-R}{n}\mathbb{E}\left[\left(D_T - \widetilde{k}\right)^+\right],$$

where $\widetilde{k} \in \mathbb{N}^+$ is defined as the integer part of the rescaled strike $K \cdot \frac{n}{1-R}$, as $\widetilde{k} = \left[K \cdot \frac{n}{1-R}\right] + 1$.

Approximating the (conditional) portfolio default counter with a Poisson distribution, we write

$$\mathbb{P}\left(D_T = k \,|Y = y\right) = e^{-\lambda_y} \frac{\lambda_y^{k-1}}{(k-1)!}, \text{ for } k \geq 0,$$

where the portfolio Poisson intensity is defined as $\lambda_y = \sum_{i=1}^n p_T^i(y)$; thus, the conditional expected tranche loss is simply

$$\begin{aligned}
\mathbb{E}^{(\lambda_y)}\left[M_T^{(K_1,K_2)} \,|Y = y\right] &= \frac{1-R}{n} \sum_{k=0}^{\infty} \frac{\lambda_y^k}{k!} e^{-\lambda_y} \left[(k - \widetilde{k}_1)^+ - (k - \widetilde{k}_2)^+\right] \\
&= \sum_{k=0}^{\infty} \frac{\lambda_y^k}{k!} e^{-\lambda_y} \left[\left(k\frac{1-R}{n} - K_1\right)^+ - \left(k\frac{1-R}{n} - K_2\right)^+\right].
\end{aligned}$$

Typically, a Binomial distribution $B(n, p)$ can be reasonably approximated with a Poisson distribution when $np < 10$; in our set-up, this is equivalent to $\sum_{i=1}^n p_T^i(y) < 10$.

Stein Approximation. To improve the accuracy of the Normal and Poisson proxies presented above, El Karoui et al. (2008) derived first-order correction terms based Stein's method and the zero-bias transformation. Stein's method was introduced in Stein (1972) as a useful mathematical tool to study the Gaussian approximation of the Central Limit Theorem. This was then extended to the Poisson distribution by Chen (1975).

In the Gaussian case, we have the following theorem.

Theorem 62 (First-Order Gaussian Correction) *Let $X_1, \ldots, X_n$ be n independent zero-mean random variables, such that $\mathbb{E}\left[X_i^4\right]$, for $1 \leq i \leq n$, exist. Define the sum $S_n = \sum_{i=1}^n X_i$ and its variance $\sigma_{S_n}^2 = Var\left[S_n\right]$. For any function h such that $\left\| h'' \right\|$ exists, the normal approximation $\Phi_{\sigma_{S_n}}(h)$ of the expectation $\mathbb{E}\left[h\left(S_n\right)\right]$,*

$$\Phi_{\sigma_{S_n}}(h) = \frac{1}{\sigma_{S_n}\sqrt{2\pi}} \int_{-\infty}^{+\infty} h(x) \exp\left(-\frac{x^2}{2\sigma_{S_n}^2}\right) dx,$$

has a first-order correction term

$$C_h^{\Phi} = \frac{\mu^{(3)}}{2\sigma_{S_n}^4} \Phi_{\sigma_{S_n}}\left(\left(\frac{x^2}{3\sigma_{S_n}^2} - 1\right) x h(x)\right),$$

where $\mu^{(3)} = \sum_{i=1}^n \mathbb{E}\left[X_i^3\right]$; and the error of the corrected approximation is bounded.

Applying it to CDOs, we set the function h to a call payoff $h(x) = (x - K)^+$, and we compute the corrector term for each call

$$C_h^{\Phi} = \frac{\mu_{(3)}}{6\sigma_{S_n}^3} K \varphi(K);$$

substituting into the expression of the conditional expected tranche loss, we can write

$$\mathbb{E}\left[M_T^{(K_1,K_2)} | Y = y\right] \backsimeq \mathbb{E}^{(\mu_y,\sigma_y)}\left[M_T^{(K_1,K_2)} | Y = y\right] + \left[\frac{\mu_y^{(3)}}{6\sigma_y^3}\left(K_1 - \mu_y\right)\varphi\left(K_1 - \mu_y\right) - \frac{\mu_y^{(3)}}{6\sigma_y^3}\left(K_2 - \mu_y\right)\varphi\left(K_2 - \mu_y\right)\right],$$

where the third moment $\mu_y^{(3)}$ is given by

$$\mu_y^{(3)} = \sum_{i=1}^{n} l_i^3 p_T^i(y) \left(1 - p_T^i(y)\right) \left(1 - 2p_T^i(y)\right).$$

Similarly, for the Poisson approximation, we have the following theorem.

Theorem 63 (First-Order Poisson Correction) *Let $Y_1, \ldots, Y_n$ be n independent non-negative integer random variables, such that $\mathbb{E}\left[Y_i^3\right]$, for $1 \leq i \leq n$, exist. Define the sum $S_n = \sum_{i=1}^{n} X_i$, with mean $\lambda_{S_n} = \mathbb{E}[S_n]$ and variance $\sigma_{S_n}^2 = Var[S_n]$. For any bounded function h defined on $\mathbb{N}^+$, the Poisson approximation $\mathcal{P}_{\lambda_{S_n}}(h)$ of the expectation $\mathbb{E}[h(S_n)]$,*

$$\mathcal{P}_{\lambda_{S_n}}(h) = \sum_{k=0}^{n} \frac{\left(\lambda_{S_n}\right)^k}{k!} e^{-\lambda_{S_n}} h(k),$$

has a first-order correction term

$$C_h^{\mathcal{P}} = \frac{\sigma_{S_n}^2 - \lambda_{S_n}}{2} \mathcal{P}_{\lambda_{S_n}}\left(\Delta^2 h\right),$$

where the difference operator Δ is defined as $\Delta h(x) = h(x+1) - h(x)$; and the error of the corrected approximation is bounded.

When h is a call function $h(x) = (x-k)^+$, where k is a positive integer, the difference operator is given by $\Delta^2 h(x) = \mathbf{1}_{\{x=k-1\}}$, and the Poisson corrector term becomes

$$C_h^{\mathcal{P}} = \frac{\sigma_{S_n}^2 - \lambda_{S_n}}{2(k-1)!} e^{-\lambda_{S_n}} \lambda_{S_n}^{k-1}.$$

We apply the result to CDOs on portfolios with similar LGDs per name, $l_1 = \cdots = l_n = \frac{1-R}{n}$. The loss variable L_T can be written as a function of the default counter D_T as $L_T = \frac{1-R}{n} D_T$. Applying the Poisson correction to each the call payoff, we have

$$C_h^{\mathcal{P}} = \frac{1}{2} \sum_{i=1}^{n} \left(p_T^i(y)\right)^2 e^{-\lambda_y} \frac{\lambda_y^{k-1}}{(k-1)!}.$$

Plugging it back into the conditional expected tranche loss yields

$$\mathbb{E}\left[M_T^{(K_1,K_2)}|Y=y\right] \backsimeq \mathbb{E}^{(\lambda_y)}\left[M_T^{(K_1,K_2)}|Y=y\right]$$
$$-\frac{1-R}{2n}\left[\sum_{i=1}^{n}\left(p_T^i(y)\right)^2\right]e^{-\lambda_y}\left(\frac{\lambda_y^{\widetilde{k}_1-1}}{(\widetilde{k}_1-1)!}-\frac{\lambda_y^{\widetilde{k}_2-1}}{(\widetilde{k}_2-1)!}\right).$$

5.1.6 Compound Correlation

As mentioned earlier, the one-factor Gaussian copula has been used by dealers to quote the standardized CDO tranches traded in the market. Since the prices of various tranches are driven by supply and demand factors, a single correlation parameter is not sufficient to reproduce market prices. Inverting the pricing formula of the one-factor Gaussian copula, one would find the implied correlation, which matches the market price of each tranche. This implied correlation is referred to as "Compound Correlation".

Definition 64 (*Compound Correlation*) For a given CDO tranche with attachment points (K_1, K_2) and quoted premium S^{K_1,K_2}, let $G^{K_1,K_2}(S,\rho)$ denote the model price using the one-factor Gaussian copula with parameter ρ. We call compound correlation, the value of the parameter ρ such that

$$G^{K_1,K_2}\left(S^{K_1,K_2},\rho\right)=0. \tag{5.6}$$

If a compound correlation exists, i.e., if the relationship (5.6) is invertible, then this offers a way to compare different tranches on a relative value basis. Unfortunately, it turns out that the model price is not a monotonic function of compound correlation. Therefore, it is not guaranteed that we can always find a solution. Moreover, in some instances, we can find more than one value of correlation, which satisfies (5.6). This usually happens with Mezzanine tranches, which are not very correlation-sensitive. This behaviour is well documented (see McGinty et al. 2004) and has motivated the base correlation approach that we describe next.

Solving for compound correlations in the previous example, we get the following results.

0–3%	20.08%	18.57%
3–6%	5.92%	6.17%
6–9%	13.56%	14.19%
9–12%	20.82%	22.42%
12–22%	29.54%	30.43%

5.1.7 Base Correlation Curve

We can view each CDO tranche with attachment and detachment points (K_1, K_2) as the difference between two equity tranches: $(0, K_2)$ and $(0, K_1)$. This can be checked easily from the definition of the payoff:

$$M_T^{K_1,K_2} = M_T^{0,K_2} - M_T^{0,K_1}, \text{ for all } T \geq 0.$$

Therefore, to price any CDO tranche it suffices to have the whole continuum of equity tranches $(0, K)$, for $K \in [0, 1]$. Each one of these equity tranches can be valued with a different one-factor Gaussian copula correlation $\rho(0, K)$. The function $\rho(0, K) : [0, 1] \to [0, 1]$ is called the "Base Correlation" curve.

Definition 65 (*Base Correlation*) The base correlation curve is a function $\rho(0, K) : [0, 1] \to [0, 1]$, which parameterizes the prices of all equity tranches $(0, K)$. In other words, the price of the $(0, K)$-tranche is given by the one-factor Gaussian copula model with parameter $\rho(0, K)$.

Furthermore, the value of any tranche with attachment and detachment points (K_1, K_2) and quoted premium S^{K_1,K_2}, is given by

$$G^{0,K_2}\left(S^{K_1,K_2}, \rho(0, K_2)\right) - G^{0,K_1}\left(S^{K_1,K_2}, \rho(0, K_1)\right). \tag{5.7}$$

Using the standard tranches quoted in the market, one would proceed with a bootstrapping algorithm to find the base correlation curve, which reproduces the market quotes. The popularity of this method lies in the fact that the function

$$h(x) = G^{0,K}(S, x), \text{ for a given } K \in [0, 1] \text{ and } S \in \mathbb{R}_+,$$

is monotonic. Hence, we can always invert the relationship (5.7) for each attachment point.

Mathematically, base correlation is just another way of parameterizing the density function $f_T(.)$ of the portfolio loss L_T. Indeed, given a base correlation curve $(\rho(0, K))_{0 \leq K \leq 1}$, one can compute the value of all "equity tranches" $(C_T(0, K))_{0 \leq K \leq 1}$:

$$C_T(0, K) \triangleq \mathbb{E}\left[M_T^{0,K}\right].$$

Assuming that, for $T \geq 0$, the function $K \to C_T(0, K)$ is $\mathcal{C}^2$, we can recover the density function as:

$$f_T(K) = -\frac{\partial^2 C_T(0, K)}{\partial K^2}. \tag{5.8}$$

This follows directly from Eq. (5.3). This is similar to the Breeden and Litzenberger (1978) formula in options theory where the implied density of the forward stock price is obtained from the continuum of call prices at different strikes.

Solving for base correlations in the previous example, we get the following results.

0–3%	20.08%	18.57%
0–6%	29.60%	27.43%
0–9%	37.10%	34.12%
0–12%	42.54%	38.50%
0–22%	56.04%	49.28%

Now, given a suitable interpolation/extrapolation method of the base correlation curve, we can price consistently any CDO tranche on the same index portfolio. As long as the interpolation is at least twice-continuously differentiable (i.e., is a $\mathcal{C}^2$-function) and satisfies the loss distribution no-arbitrage conditions, the prices of tranches with different attachment and detachment points will be sensible and consistent with the (market) quoted standard index tranches.

The next question is: how should one mark the correlation skew curve for a "Bespoke" CDO tranche referencing a different portfolio?

The answer adopted by the industry is given by the Skew Rescaling Methodologies.

5.2 Skew Rescaling

Skew Rescaling is the process whereby we map a bespoke base correlation curve, in a consistent manner, to an market observed index base correlation curve, through a rescaling process that keeps some (appropriate) "riskiness" metric invariant between the two portfolios. Fundamentally, the base correlation parametrization represents an allocation of the different levels of risk across the capital structure; and as such, the underlying idea behind the rescaling method is that, by matching the same risk levels between the bespoke and benchmark index portfolios, we can translate the market-implied capital structure observed for the index into an equivalent capital structure allocation for the bespoke portfolio. This idea of quantifying the level of riskiness in tranches by comparing some well-defined risk metric is not new. In fact, in most rating agency models, the (risk) rating of a CDO tranche is assigned through a quantile of the portfolio loss distribution generated from a historical Monte-Carlo simulation.

More formally, we have the following definition.

We consider the universe of all credit names and we denote by Π_n the set of all subsets in our credit universe. A credit portfolio is defined as any subset π in Π_n.

We define a portfolio risk metric as the function $R_T^{\pi} : [0, 1] \to \mathbb{R}^+$, for all portfolios $\pi \in \Pi_n$, and all maturities $T \geq 0$.

Definition 66 (*Skew Rescaling*) The base correlation skew rescaling method defines the mapping between the benchmark index base correlation curve, $\rho^I(K, T)$, for all strikes $K \in [0, 1]$ and all maturities $T \geq 0$, and the equivalent bespoke base correlation $\rho^{\pi}(K, T)$, by matching the chosen portfolio risk metric function $R_T^{\pi}(.)$,

$$\rho^{\pi}(K, T) = \rho^I\left(\left(R_T^{\pi}\right)^{-1}\left(R_T^I(K)\right), T\right).$$

Remark 67 Depending on the choice of rescaling methodology, the mapping function $R_T^{\pi}(.)$ may not always be invertible. One criterion for a good rescaling method is its robustness with respect to the range of bespoke portfolio spreads. In particular, a rescaled base correlation curve has to exist first, but it also needs to be sensible. Some rescaled skews for very wide or very tight portfolios could fall outside the range of marked index strikes (either lower than the first strike or higher than the last strike), hence the associated rescaled correlation values may be non-sensical if they are given by an ad-hoc extrapolation method.

One of the first papers published on skew rescaling was Reyfman et al. Reyfman et al. (2004), which introduced the "Tranche Loss" rescaling method. Since then, there has been a number of rescaling methods used, at different points in time, by various dealers. For a good overview, we refer to the excellent research piece by Turc et al. (2006), which gives a useful comparison between the various rescaling methods used by different market participants. We give a brief summary of some of their findings.

There are three main skew rescaling methods that we shall discuss: Portfolio Loss Rescaling, Tranche Loss Rescaling and Loss Probability Rescaling.

5.2.1 Portfolio Loss Rescaling

In this method, we rescale the correlation strike for each portfolio with its expected portfolio loss (this is a better posed problem than rescaling with the portfolio average spread, which was historically a simple back-of-the-envelope calculation used to compare the risk in different portfolios):

$$R_T^{\pi}\left(K^{\pi}\right) = R_T^{I}\left(K^{I}\right) \rightarrow \frac{K^{\pi}}{\mathbb{E}\left[L_T^{\pi}\right]} = \frac{K^{I}}{\mathbb{E}\left[L_T^{I}\right]}.$$

This is also referred to sometimes as "Moneyness Rescaling". Using the Black-Scholes analogy—where we diffuse the (conditional) expected loss variable, the equivalent forward, at maturity T, is simply given by the expected loss at time 0: $F_{0,T}^{\pi} = \mathbb{E}\left[L_T^{\pi}\right]$; and as such, when $K = F_{0,T}^{\pi}$, the equity option payoff is At-The-Money, and the ratio $\frac{K^{\pi}}{\mathbb{E}[L_T^{\pi}]} = \frac{K^{\pi}}{F_{0,T}^{\pi}}$ represents the moneyness percentage of the equity option strike K with respect to the Expected Loss Forward $F_{0,T}^{\pi}$. Using the terminology from volatility skew modelling, this could be viewed as a "Sticky-Strike" parametrization of the correlation skew surface. The method itself is intuitive and easy to implement, but it has a few limitations:

1. it does not account for the portfolio dispersion. So, a portfolio with tight names and one very wide name close to defaulting, will be treated similarly to another homogeneous portfolio with the same duration weighted average spread (DWAS). This will distort the pricing for the junior part of the capital structure, especially equity tranches;
2. for very tight (or very wide) bespoke portfolios (compared to the benchmark index level), some of the rescaled bespoke strikes could over-shoot outside the range of quoted index strikes, (either higher than the last index strike or lower than the first index strike) and the corresponding correlations would be given by spurious extrapolated levels;
3. there is a discontinuity in P&L at the time of default. When one of the names (either in the bespoke or the index) defaults, it drops out of the portfolio and yields a jump in the expected loss ratio, which translates, in this case, to a jump in rescaled bespoke strike.

5.2.2 Probability Loss Rescaling

The skew rescaling, in this method, is based on matching the probability that portfolio losses do not exceed a given threshold, i.e., the probability that the equity tranche is not completely wiped out by the default losses:

$$R_T^{\pi}\left(K^{\pi}\right) = R_T^{I}\left(K^{I}\right) \rightarrow \mathbb{P}\left(L_T^{\pi} \leq K^{\pi}\right) = \mathbb{P}\left(L_T^{I} \leq K^{I}\right).$$

This rescaling is not as simple as multiplying two different ratios (as in Portfolio Loss Rescaling), but requires an inversion of the loss probability relationship, since computing the latter needs an input correlation, which, in turn, depends on the strike K^{π}. The main advantage of the method is that it takes dispersion into account and does not produce discontinuities at the time of default. On the flip side, in addition to having to carry out a numerical inversion to find the solution, for very wide portfolios (compared to the index), there is a range of small strikes that do not admit a solution. Even for the tight strikes where we have a solution, the correlation numbers would still be problematic as they would come mainly from the extrapolation below the first index strike.

Using the volatility skew analogy, this method could also be viewed as the equivalent of a "Sticky-Delta" parametrization for the correlation skew surface.

5.2.3 Tranche Loss Rescaling

In this method, instead of rescaling with the Portfolio Expected Loss, we use the Tranche Expected Loss:

$$R_T^{\pi}\left(K^{\pi}\right) = R_T^{I}\left(K^{I}\right) \rightarrow \frac{\mathbb{E}\left[L_T^{\pi} \cdot \mathbf{1}_{\{L_T^{\pi} \leq K^{\pi}\}}\right]}{\mathbb{E}\left[L_T^{\pi}\right]} = \frac{\mathbb{E}\left[L_T^{I} \cdot \mathbf{1}_{\{L_T^{I} \leq K^{I}\}}\right]}{\mathbb{E}\left[L_T^{I}\right]}.$$

The expected Tranche Loss ratio is a fairly standard metric used in credit correlation markets to quantify the allocation of the portfolio expected loss across the capital structure. It gives investors a good indicator of relative value opportunities across different tranches on the same index

In practice, the method works well across a large range of wide and tight portfolios. The expected tranche loss ratio is a monotonic function of strike, which varies between 0 and 1, by construction. Thus, we are guaranteed that a solution always exists (unlike Probability Loss Rescaling). Furthermore, the rescaled strikes tend, for most portfolios, to be within the range of quoted benchmark index strikes, and are less prone to extrapolation instability issues.

This method also takes dispersion into account but is not immune to the discontinuities at default that we have encountered with Portfolio Loss Rescaling.

In summary, we have the following comparison table of the pros and cons of each method.

	Portfolio loss	Probability loss	Tranche loss
Ease of implementation	Positive	Neutral	Neutral
Dispersion	Negative	Positive	Positive
Tight portfolios	Neutral	Negative	Positive
Wide portfolios	Neutral	Neutral	Positive
Discontinuity at default	Negative	Positive	Negative

5.2.4 Mapping, Blending and Interpolation

Mapping and Blending. The choice of the benchmark index skew that we need to use for each bespoke portfolio is very important. Whether we should use a European index (iTraxx), a US index (CDX), an Investment Grade (IG) index or a High Yield (HY) index depends on the closest index, which represents the salient features of the bespoke portfolio to be priced. Once the right index is chosen, we can then apply the skew rescaling methodology. For portfolios that have a mixture of names with different characteristics (e.g., 50% Europe, 50% US), dealers have used various "blended" skews that combine the two skew curves proportionally, which then get rescaled to the portfolio DWAS as usual, with the standard rescaling methods.

Once we have constructed the base correlation to use, two questions remain: (1) how should one interpolate across strikes for a fixed maturity; (2) what kind of interpolation should be used (for the same strike) across the time dimension. We address each question in turn.

Interpolation and Tranchlets. To interpolate across strikes, one could use a simple linear or cubic interpolation, or a slightly more sophisticated expected tranche loss (convexity-preserving) interpolation. The benefit of linear interpolations is simplicity. But they have a major drawback: they produce arbitrage opportunities due to the implied discontinuities in the loss density. Remember that the density function is given by the second derivative of the expected tranche loss curve with respect to strike. And as such, the base correlation curve needs to be at least C^2 to avoid the curve discontinuities at the attachment points.

One way to avoid that effect is to assume a Cubic Spline interpolation. This is clearly a better choice since we remove the obvious discontinuities: both the base correlation curve and its derivative with respect to strike are continuous. But even with a spline interpolation, there is the possibility of having negative densities for thin tranches. The additional constraint that we need is to maintain the convexity of the expected tranche loss curve. This can be achieved in one of two ways. Either we relax the cubic spline interpolation scheme, and use instead a second (or third) order polynomial with a judicious choice of constraints for the problem at hand. Or instead of interpolating the base correlation curve, interpolate the expected tranche loss curve directly with a convexity-preserving algorithm. Working with the expected tranche loss is a very natural modelling primitive to use, but the interpolation algorithm, while not overly complex, is still more involved than simple traditional cubic spline interpolations or variations thereof.

For extrapolations outside the market quoted strike range (typically, below the 3% strike), traders have used ad-hoc rules of thumb inline with general market expectations. These unobservable strikes are usually benchmarked against tranchelets (thin tranches of 1% width below the 3% strike), where we can get some market quotes occasionally, or against other (wider) bespoke tranches whose rescaled strikes fall within the 0–3% range and therefore are directly linked to where the extrapolated correlations are marked. There is no universally accepted method to use. This process is more an art than a science where a combination of market discovery and the appropriate level of reserves are used to account for illiquidity of the unobservable correlation marks.

The Time Dimension. For a fixed strike, the interpolation problem across the time dimension is similar is some way to an interest rate or a credit curve construction. Typically, we can have either a flat (or linear) zero-correlation curve, or a flat forward-correlation curve. What we mean by that is the following: when we consider, for example, the 5 year and 10 year correlations in isolation, to price all the cash-flows of the 5-year trade, we only use the flat 5-year (zero) correlation; similarly for a 10 year trade, all the cash-flows from today up to 10 years, use exactly the same flat 10-year zero correlation. And there is no consistency between the two. With flat forward-correlation, the process consists in splitting the cash-flows of the 10-year trade in 2 groups: all the cash-flows before year 5 are priced with the 5-year correlation, and all the cash-flows between 5 and 10 years are priced with the forward (5 into 10) correlation that ensures that we match the prices of the 10 year tranches. Clearly, if there is arbitrage in the market and those prices are not consistent, there may be instances when the forward (5 into 10) correlation is negative. This is similar to what happens with rates or credit curves.

5.3 CDO^2

CDO-Squared trades refer to CDO tranches where the underlying portfolio itself consists of a set of bespoke CDO tranches. The default correlation modeling, in this case, introduces one more layer of complexity. Not only do we need to use the correct (inner) rescaled correlation skew curve for each underlying (bespoke) tranche in order to recover its market price, but we also have to model the outer correlation between the different tranches as well. This needs to be done consistently to ensure that the effect of overlapping names in the underlying CDO portfolios is accounted for. Early attempts to price CDO^2 trades have used so-called "copula of copula" methods, where an exogenous CDO^2 copula is used to link the underlying sub-portfolio loss variables. These methods were doomed to fail from the start as they were missing one important factor in pricing: the amount of portfolio overlap between the underlying CDO sub-portfolios. Consistent pricing requires the construction, by hand, of a "Loss Copula", explicitly from the underlying single-name copula of the entire master portfolio, thus, ensuring, by definition, that any overlaps are accounted for. We refer, for example, to Li and Liang (2005) where such a construction is explained in detail.

Here we give a brief overview of the approach—in a standard Gaussian copula framework—and we emphasize the key building blocks of the method, namely, the Loss Skew Maps and the Loss Copula. We shall study the correlation profile in CDO-Squared structures, in more detail, in Chap. 13, where we use a Marshall-Olkin model to construct the outer CDO^2 Loss Copula. Similar ideas are leveraged again to construct the dynamic Markov-Functional model in Chap. 16, and to generate the CVA for CDOs in Chap. 20.

Our goal is to develop a consistent methodology for applying the base correlation skew approach to CDO^2 products, which satisfies three basic requirements:

1. we can re-price all the underlying CDOs correctly, i.e., if we degenerate the CDO^2 to one of its inner bespoke CDOs, we can recover the same prices given by the rescaled base correlation curve;
2. if we turn off the base correlation skew effect, by using a flat base correlation as input, we can recover the same CDO^2 prices as in the standard Gaussian copula model;
3. we apply the same base correlation skew curve used for the bespoke Master CDO portfolio to the CDO^2 structure, so that different "outer" correlations are used for different CDO^2 subordination levels, which reproduce the bespoke correlation curve of the Master portfolio.

We consider a CDO^2 with m underlying CDO tranches and a Master Portfolio of n names. The sub-portfolios $(\pi_j)_{1\leq j\leq m}$ are identified with the notional matrix $(N_{i,j})_{\substack{1\leq i\leq n\\1\leq j\leq m}}$ where $N_{i,j}$ represents the notional of name i in sub-portfolio j. The portfolio loss and tranche loss variables for each underlying CDO are defined as, for all $1 \leq j \leq m$,

$$L_t^j = \sum_{i=1}^{n} N_{i,j} (1 - R_i) D_t^i$$

$$M_t^{\left(K_1^j, K_2^j, j\right)} = \min\left(\max\left(L_t^j - K_1^j, 0\right), K_2^j - K_1^j\right),$$

where $\left(K_1^j, K_2^j\right)$ are the corresponding attachment and detachment points for CDO j. And the CDO^2 portfolio and tranche loss variables are defined as

$$L_t = \sum_{j=1}^{m} M_t^{\left(K_1^j, K_2^j, j\right)},$$

$$M_t^{K_1,K_2} = \min\left(\max\left(L_t - K_1, 0\right), K_2 - K_1\right),$$

with (K_1, K_2) are the attachment and detachment points for the CDO^2 tranche.

The CDO^2 payoff $M_t^{K_1,K_2}$ is a multivariate function of the underlying loss variables L_t^j,

$$M_t^{K_1,K_2} = f\left(L_t^1, L_t^2, \ldots, L_t^m\right).$$

To compute its expectation, we need the univariate marginal loss distributions, $\Phi_{L_t^j}(.)$, and the joint (loss) copula function linking the losses together $\mathbb{C}^L(x_1, \ldots, x_m)$,

$$\mathbb{E}\left[M_t^{K_1,K_2}\right] = \int_0^1 \cdots \int_0^1 f\left(\Phi_{L_t^1}^{-1}(x_1), \ldots, \Phi_{L_t^m}^{-1}(x_m)\right) \mathbb{C}^{[L]}(x_1, \ldots, x_m)\, dx_1 \ldots dx_m.$$

5.3.1 Loss Copula

As we have seen previously, for each loss variable L_t^j, the market-implied loss distribution $\Phi_{L_t^j}(.)$ is given by its base correlation curve, $\rho_j(0, K)$, which is obtained by taking the first-derivative of the continuum of equity prices $C_T^j(0, K)$, with respect to strike,

$$\Phi_{L_t^j}(x) = P\left(L_t^j \leq x\right) = 1 - \frac{\partial C_T^j(0, K)}{\partial K}|_{K=x} .$$

In order to ensure the correct (micro) structure of the copula function, which accounts for portfolio overlaps, we cannot just take any exogenously specified copula function; rather, we must construct it using a bottom-up approach: we start with a set of correlated default times $\left(\tau_1^*, \ldots, \tau_n^*\right)$, whose joint multivariate dependence is defined through a (default times) copula function $\mathbb{C}^{[\tau^*]}(x_1, \ldots, x_n)$—such as the standard Gaussian copula for example– then we construct and use its implied loss copula function $\mathbb{C}^{[L^*]}(x_1, \ldots, x_m)$.

We proceed in three steps.

1. First, we compute the (unskewed) marginal loss distributions $\Phi_{L_t^{j,*}}(x)$.
2. Then, we simulate the auxiliary default times $\left(\tau_1^*, \ldots, \tau_n^*\right)$ in order to compute the sub-portfolios (unskewed) loss variables $L_t^{j,*}$, which are then inverted with the corresponding marginals $\Phi_{L_t^{j,*}}(x)$; thus generating the uniform variates, $u_j^* = \Phi_{L_t^{j,*}}\left(L_t^{j,*}\right)$, whose joint dependence follows the copula function $\mathbb{C}^{[\tau^*]}(x_1, \ldots, x_n)$.
3. Finally, by applying the market-implied probability distribution $\Phi_{L_t^j}(.)$,

$$L_t^j = \Phi_{L_t^j}^{-1}\left(\Phi_{L_t^{j,*}}\left(L_t^{j,*}\right)\right),$$

we recover the correct market-implied marginals, while the joint density function is still defined by the auxiliary (default times') copula $\mathbb{C}^{[\tau^*]}(x_1, \ldots, x_n)$.

Thus, we have constructed a set of (sub-portfolio) loss variables $\left(L_t^j\right)_{1 \leq i \leq n}$, correlated through the bottom-up single-name copula function $\mathbb{C}^{[\tau^*]}(.)$, thereby taking into account the portfolio overlaps and the correct ordering of default events, while, at the same time, preserving the market-implied marginal loss distributions $\Phi_{L_t^j}(.)$.

5.3.2 Conditional Loss Copula

For the auxiliary default times' copula $\mathbb{C}^{[\tau^*]}(x_1, \ldots, x_n)$, if we use the standard Gaussian copula (or any other conditionally-independent factor copula), the pricing is done by conditioning on the common factor Y, so that the single-name default indicators D_t^i are independent, which is very convenient when computing the distributions of the sub-portfolio loss variables.

With the Normal Proxy method, for example, the loss variables $\left(L_t^{j,Y}\right)_{1 \leq j \leq m}$, conditional on Y, are approximated by a set of normal variables, whose conditional means $\mu_j^y = \mathbb{E}\left[L_t^j \,|Y = y\right]$ and conditional covariance matrix $\sigma_{j,k}^y = \mathbb{E}\left[L_t^j L_t^k \,|Y = y\right] - \mathbb{E}\left[L_t^j \,|Y\right] \mathbb{E}\left[L_t^k \,|Y = y\right]$ are known analytically and given by simple calculations.

Hence, the evaluation of the CDO^2 payoff boils down to computing the conditional expectation

$$\mathbb{E}\left[M_t^{K_1,K_2} \,|Y\right] = \int \ldots \int M_t^{K_1,K_2} \Phi_m^{\left(\mu_j^Y, \sigma_{j,k}^Y\right)} \left(L_t^{1,Y}, \ldots, L_t^{m,Y}\right) dL_t^{1,Y} \ldots dL_t^{m,Y},$$

as a multi-dimensional integral with a (Monte-Carlo) numerical integration method. And finally, we integrate, over the values of the central shock Y, to obtain the (CDO^2) expected tranche loss $\mathbb{E}\left[M_t^{K_1,K_2}\right]$.

As explained in the previous section, to price the CDO^2 consistently with the correlation skew of the underlying sub-portfolios, we need to use the same loss copula $\mathbb{C}^{[L^*]}(x_1, \ldots, x_n)$ (as in the no-skew case), but we have to apply the correct marginals through the skew maps. With the Normal Proxy method, we have one more level of complexity to address. All the calculations are done conditionally on Y: instead of working with the loss skewed and unskewed variables L_t^j and $L_t^{j,*}$, we operate on their conditional counterparts $L_t^{j,Y} = \mathbb{E}\left[L_t^j \,|Y\right]$ and $L_t^{j,*,Y} = \mathbb{E}\left[L_t^{j,*} \,|Y\right]$ respectively.

Indeed, conditional on the common shock Y, the loss variables $L_t^{j,*}$ have a set of well-known conditional marginal distributions $\Phi_{L_t^{j,*}}^Y(x) = \mathbb{P}\left(L_t^{j,*} \leq x \,|Y\right)$, and are linked with a "Conditional Loss Copula", $\mathbb{C}^{[L^*|Y]}(.)$,

$$\mathbb{C}^{[L^*|Y]}\left(\Phi_{L_t^{1,*}}^Y(x_1), \ldots, \Phi_{L_t^{m*}}^Y(x_m)\right) = \mathbb{P}\left(L_t^{1,*} \leq x_1, \ldots, L_t^{m,*} \leq x_m \,|Y\right).$$

As before, in the Normal Proxy method, the conditional loss variables' marginal distributions are, in fact, Gaussian, and the conditional loss copula is also a multivariate Gaussian copula. We also know the market-implied (loss) marginals, $\Phi_{L_t^j}(.)$, that we need to match, but we do not have immediately the conditional market-implied marginals $\Phi^Y_{L_t^j}(.)$.

To proceed, we need to derive them by solving the following inverse problem

$$\Phi_{L_t^j}(x) = \int_{-\infty}^{+\infty} \Phi^y_{L_t^j}(x) \frac{\exp\left(-\frac{y^2}{2}\right)}{\sqrt{2\pi}} dy.$$

To this end, we can apply the same loss copula approach. But now we consider the copula function of the extended set of $(m+1)$ variables $\left(L_t^{1,*}, \ldots L_t^{m,*}, Y\right)$. They have well-known marginals $\left(\Phi_{L_t^{1,*}}(x_1), \ldots, \Phi_{L_t^{m,*}}(x_m), \Phi_Y(x_{m+1})\right)$, and the idea is to use the $(m+1)$-copula function, which links them and to apply the new (target) skewed marginals $\left(\Phi_{L_t^1}(x_1), \ldots, \Phi_{L_t^m}(x_m), \Phi_Y(x_{m+1})\right)$, by re-mapping them appropriately.

The skewed and unskewed loss distributions are $\Phi_{L_t^j}(x) = \mathbb{P}\left(L_t^j \leq x\right)$ and $\Phi_{L_t^{j,*}}(x) = \mathbb{P}\left(L_t^{j,*} \leq x\right)$ respectively.

The copula which links the unskewed loss variable and the common factor can be written as

$$\mathbb{C}^{L_t^{j,*},Y}\left(\mathbb{P}\left(L_t^{j,*} \leq x\right), \mathbb{P}(Y \leq y)\right) = \mathbb{P}\left(L_t^{j,*} \leq x, Y \leq z\right) = \int_{-\infty}^{z} \Phi^y_{L_t^{j,*}}(x) \frac{\exp\left(-\frac{y^2}{2}\right)}{\sqrt{2\pi}} dy,$$

or expressed in terms of uniform variates

$$\mathbb{C}^{L_t^{j,*},Y}(u, v) = \int_{-\infty}^{\Phi_Y^{-1}(v)} \Phi^y_{L_t^{j,*}}\left(\Phi^{-1}_{L_t^{j,*}}(u)\right) \frac{\exp\left(-\frac{y^2}{2}\right)}{\sqrt{2\pi}} dy.$$

Plugging the marginal skewed distributions, $\Phi_{L_t^j}(x)$ and $\Phi_Y(x)$, in the bivariate copula function $\mathbb{C}^{L_t^j,Y}(u, v)$ gives the joint default probability that we need

$$\begin{aligned}\mathbb{P}\left(L_t^j \leq x, Y \leq z\right) &= \mathbb{C}^{L_t^j,Y}\left(\mathbb{P}\left(L_t^j \leq x\right), \mathbb{P}(Y \leq z)\right) = \mathbb{C}^{L_t^{j,*},Y}\left(\mathbb{P}\left(L_t^j \leq x\right), \mathbb{P}(Y \leq z)\right) \\ &= \mathbb{C}^{L_t^{j,*},Y}\left(\Phi_{L_t^j}(x), \Phi_Y(z)\right) = \int_{-\infty}^{z} \Phi^y_{L_t^{j,*}}\left(\Phi^{-1}_{L_t^{j,*}}\left(\Phi_{L_t^j}(x)\right)\right) \frac{\exp\left(-\frac{y^2}{2}\right)}{\sqrt{2\pi}} dy,\end{aligned}$$

which yields the conditional skewed density, $\Phi^{Y}_{L_t^j}(x) = \mathbb{P}\left(L_t^j \leq x \,|Y\right)$, by Bayes' rule

$$\Phi^{y}_{L_t^j}(x) = \mathbb{P}\left(L_t^j \leq x \,|Y = y\right) = \Phi^{y}_{L_t^{j,*}}\left(\Phi^{-1}_{L_t^{j,*}}\left(\Phi_{L_t^j}(x)\right)\right).$$

Proposition 68 (Conditional Skewed Loss Distribution) *The skewed loss distribution, conditional on the common factor Y, is given by a simple mapping using the inverse of the unskewed loss distribution and the market-implied skewed distribution*

$$\Phi^{y}_{L_t^j}(x) = \mathbb{P}\left(L_t^j \leq x \,|Y = y\right) = \Phi^{y}_{L_t^{j,*}}\left(\Phi^{-1}_{L_t^{j,*}}\left(\Phi_{L_t^j}(x)\right)\right).$$

Algorithm 69 *We can then summarize the implementation algorithm for pricing the CDO^2 payoff in the normal proxy method as follows.*

- *Conditional on $Y = y$, simulate m correlated Gaussian variables $\left(L_t^{1,*,y}, \ldots, L_t^{m,*,y}\right)$ with mean μ_j^y and covariance matrix $\sigma_{j,k}^y$.*
- *Set the conditional skewed loss variable as $L_t^{j,y} = \Phi^{-1}_{L_t^j}\left(\Phi_{L_t^{j,*}}\left(L_t^{j,*,y}\right)\right)$.*
- *Compute the expectation of the conditional CDO^2 payoff*

$$\mathbb{E}\left[M_t^{K_1,K_2} \,|Y = y\right] = \mathbb{E}\left[f\left(L_t^{1,y}, \ldots, L_t^{m,y}\right) |Y = y\right].$$

- *Integrate over all values of Y.*

5.3.3 Bespoke CDO^2 Skew

With the construction presented above, we have ensured that two of the main requirements for CDO^2 modelling are satisfied; namely that we reproduce the underlying CDO prices, and that we match the CDO^2 prices in the degenerated flat correlation case. The last requirement is consistency with the bespoke correlation skew curve of the Master CDO^2 portfolio. Remember that, qualitatively, the risk profile of a CDO^2 tranche is similar to that of a thin tranche on the aggregate Master portfolio. The two structures are not exactly equivalent but they behave similarly in most states of the world. There are a few combinations where the order of defaults can imply a different outcome for the two structures, but overall, their behaviour is fairly identical most of the time.

Typically, CDO2 and Bespoke CDOs are risk-managed together in the same book, therefore it is important to ensure consistent correlation skew marking between the two.

In this section, we explain how to achieve the bespoke skew consistency.

In fact, this is quite simple to do in this framework. Indeed, we have the marginals of the underlying loss variables $\Phi_{L_t^j}(x) = \mathbb{P}\left(L_t^j \leq x\right)$, and we have the loss copula $\mathbb{C}^{[L]}(.)$, which links them, parametrized with the outer correlation parameter, ρ^O, of the auxiliary default times copula $\mathbb{C}^{[\tau]}(.)$. Therefore, the loss distribution of the (global) master portfolio L_t^G, defined as the sum of the (inner) underlying loss variable,

$$L_t^G \triangleq \sum_{j=1}^{m} L_t^j,$$

can also be generated with the CDO^2 inputs ρ^O and $\Phi_{L_t^j}(.)$. Indeed, conditioning on the common factor Y , its Fourier transform is given by

$$\mathbb{E}\left[\exp\left(iuL_t^G\right)\right] = \int \mathbb{E}\left[\exp\left(iu\sum_{j=1}^{m} L_t^j\right) |Y = y\right] \varphi(y)\, dy,$$

which is computed, as before, with the same m-dimensional numerical integral

$$\mathbb{E}\left[\exp\left(iuL_t^G\right)|Y\right] = \int \ldots \int \mathbb{E}\left[\exp\left(iu\sum_{j=1}^{m} \Phi_{L_t^j}^{-1}\left(\Phi_{L_t^{j,*}}(x_j)\right)\right)\right] \Phi_m^{\left(\mu_j^Y,\sigma_{j,k}^Y\right)}(x_1,\ldots,x_m)\, dx_1 \ldots dx_m.$$

So, instead of using one flat outer correlation ρ^O to generate the loss copula, we can go one step further and, as in the CDO base correlation approach, define a CDO^2 base correlation curve $\rho^O(K)$, where each CDO^2 is priced as the difference between two equity CDO^2 tranches with different correlation levels for each attachment/detachment point. In order to map the CDO^2 subordination level to the corresponding bespoke master CDO strike, we use standard probability matching techniques, whereby for each CDO^2 tranche, we find the "equivalent" master CDO by matching the probabilities of exceeding different subordination levels, i.e., the probability of the tranche being completely wiped out.

5.3.4 Summary

In summary, we have presented, in this section, a consistent extension of the base correlation model for CDO^2 trades. This was achieved by: (a) using the appropriate functional form of the underlying loss variables' mapping, so that the underlying inner CDOs are re-priced correctly; (b) introducing a loss copula function constructed with a bottom-up approach to ensure that portfolio overlaps are accounted for; and (c) defining a base correlation curve for the CDO^2 outer correlation parameter that matches the bespoke base correlation curve for the equivalent master CDO portfolio.

5.4 Expected Tranche Loss Surface

As we have seen previously, once we have succeeded in calibrating a base correlation curve on the standard index tranches, the choice of interpolation and extrapolation methods, both in strike space and maturity space, will complete the construction of the whole surface, but it is very difficult to ensure that it is completely arbitrage-free. In fact, even with a set of genuinely arbitrage-free index tranches (generated, for example, with a well-posed copula model such as local correlation or stochastic correlation), it is not always possible to find a base correlation curve that reproduces those prices.

Now, we go back and revisit this question by exploring more general methods for constructing arbitrage-free base correlation surfaces. To do so, instead of working with the base correlation curve, we operate directly on the Expected Tranche Loss curve. Indeed, the expected tranche loss is the main quantity needed to compute CDO tranche prices. It plays the role of a survival probability curve for Credit Default Swaps or a discount factor curve for Interest Rate Swaps. Once we have calibrated our expected tranche loss curve, we can then convert it back to base correlation space if we wish to do so—this may not be always possible as base correlation is not capable of reproducing the entire set of all arbitrage-free prices. This is the approach adopted, for example, by Torresetti et al. (2007) and Walker (2006).

Having an arbitrage-free ETL surface, at time 0, takes on even more importance as we move to the study of dynamic portfolio loss models (such as Schönbucher (2005); Sidenius et al. (2008); Bennani (2005)), where the entire loss distribution cube is diffused through time. A pre-requisite is to have an arbitrage-free term structure of losses as of today, and to impose Heath-Jarrow-Morton type conditions on the diffusion dynamics to preserve the arbitrage-free property of the forward surfaces as well. We shall study some of the main dynamic portfolio loss models in Chap. 15.

5.4.1 The Problem

To formulate the mathematical problem that we have to solve, we write the prices of the benchmark index tranches that we want to match in terms of the Expected Tranche Loss curves.

We consider a set of index tranches, with maturities $0 < T_1 < T_2 < \cdots < T_n$, and strikes $0 < K_1 < K_2 < \cdots < K_{m-1} < K_m = 1$, and we set $K_0 = 0$. For example, for the iTraxx index, we have $\{T_i\}_{1 \leq i \leq n} = \{3 \text{ (years)}, 5 \text{ (years)}, 7 \text{ (years)}, 10 \text{ (years)}\}, \{K_j\}_{1 \leq j < m} = \{3\,(\%), 6\,(\%), 9\,(\%), 12\,(\%), 22\,(\%)\}$.

So, we have, in total, $(n \times (m-1))$-tranches to re-price by matching their break-even spreads $S_{i,j} = S\left(T_i; K_{j-1}, K_j\right)$ (or the upfront $U_{i,j} = U\left(T_i; K_{j-1}, K_j\right)$, in the case of equity tranches), for all $1 \leq i \leq n$, and $1 \leq j \leq m-1$.

The break-even spread is given by the ratio of the protection leg and the risky annuity:

$$S_{i,j} = S\left(T_i; K_{j-1}, K_j\right) = \frac{\text{ProtectionLeg}\left(T_i; K_{j-1}, K_j\right) - U_{i,j}}{\text{Annuity}\left(T_i; K_{j-1}, K_j\right)} = \frac{\Pi_{i,j}}{A_{i,j}}.$$

To simplify the presentation, we omit the exact financial details of the premium leg (payment frequency and accrued in default) and assume that the spread is paid daily (in practice, when the accrued-in-default is included, the continuous integral is very close to the correct value).

$$\Pi_{i,j} = \mathbb{E}\left[\int_{]0,T_i]} p_{0,t} dM_t^{K_{j-1},K_j}\right],$$

$$A_{i,j} = \mathbb{E}\left[\int_0^{T_i} p_{0,t}\left[1 - M_t^{K_{j-1},K_j}\right] dt\right].$$

In addition to the index tranches, we also need to match the spread of the CDS index as well, for all maturities,

$$S_{i,m} = S\left(T_i; K_0, K_m\right) = S\left(T_i; 0, 1\right).$$

The (normalized) Mezzanine expected tranche loss $M_t^{K_{j-1},K_j}$ can be written in terms of the Equity expected tranche losses at the respective attachment and detachment points as

$$M_t^{K_{j-1},K_j} = \frac{K_j}{K_j - K_{j-1}} M_t^{0,K_j} - \frac{K_{j-1}}{K_j - K_{j-1}} M_t^{0,K_{j-1}}.$$

We discretize the protection leg and risky annuity integrals and substitute the Mezz tranche loss with the difference of the two equities; we obtain

$$\Pi_{i,j} = \mathbb{E}\left[\sum_{k=1}^{t_k \le T_i} p_{0,\frac{t_k+t_{k-1}}{2}} \left(M_{t_k}^{K_{j-1},K_j} - M_{t_{k-1}}^{K_{j-1},K_j}\right)\right],$$

$$A_{i,j} = \mathbb{E}\left[\sum_{k=1}^{t_k \le T_i} p_{0,t_k} \left[1 - M_{t_k}^{K_{j-1},K_j}\right] (t_k - t_{k-1})\right].$$

The time interpolation can be done with a (classical) constant forward or linear zero-coupon assumption, and the expected tranche loss curves can then be bootstrapped, from tranche prices, like an interest rate discount factor curve.

5.4.2 Equity Tranche Forward Rate Curve

For each strike K, the (normalized) expected tranche loss curve can be viewed as a discount factor curve,

$$D_{0,T}^{(K)} \triangleq \mathbb{E}\left[M_T^{0,K}\right], \text{ for all } T \ge 0.$$

We can define the equivalent equity tranche loss forward rate as

$$f_{0,T}^{(K)} = -\frac{\partial \log D_{0,T}^{(K)}}{\partial T} \Longleftrightarrow D_{0,T}^{(K)} = \exp\left(-\int_0^T f_{0,s}^{(K)} ds\right),$$

and the equivalent equity tranche loss zero-coupon rate as

$$Z_{0,T}^{(K)} = -\frac{\log D_{0,T}^{(K)}}{T} \Longleftrightarrow D_{0,T}^{(K)} = \exp\left(-Z_{0,T}^{(K)} \cdot T\right).$$

Therefore, we could interpolate each $D_{0,T}^{(K)}$, for all $T \geq 0$, with standard methods similar to the ones used in interest rates. With a Constant Forward interpolation, the forward rate $f_{0,T}^{(K)}$ is assumed to be flat between node points; and with a Linear Zero-Coupon interpolation, the zero coupon rate $Z_{0,T}^{(K)}$ is interpolated linearly. Our preferred approach is constant forwards.

The bootstrapping algorithm proceeds then sequentially from the first strike to the last one, and from the shortest maturity to the longest one. We start with the first equity tranche $(0, K_1)$, and we bootstrap the curve $\left(D_{0,T_1}^{(K_1)}, D_{0,T_2}^{(K_1)}, \ldots, D_{0,T_n}^{(K_1)}\right)$ from the (tranche) swap rates $\left(S_{1,1}, S_{2,1}, \ldots, S_{n,1}\right)$; then equipped with the first curve $\left(D_{0,T_j}^{(K_1)}\right)_{1\leq j\leq m}$, and using the next set of (tranche) swap rates $\left(S_{j,2}\right)_{1\leq j\leq m}$ for the mezzanine tranche (K_1, K_2), we bootstrap the next curve $\left(D_{0,T_j}^{(K_2)}\right)_{1\leq j\leq m}$; and so forth until we obtain the entire surface of equity tranche loss discount factors $\left(D_{0,T_j}^{(K_i)}\right)_{1\leq j\leq m}$ for all $1 \leq i \leq n$.

Remark 70 For the 100% strike, $K_m = 1$, we use the swap rate of the index CDS, $S_{i,m} = S\left(T_i; 0, 1\right)$, and bootstrap the (index) discount factor curve $\left(D_{0,T_j}^{(K_m)}\right)_{1\leq j\leq m}$.

Remark 71 As with interest rates, we need to make sure that the curves are arbitrage-free in the time dimension by removing negative forward rates from the curve. Similarly, in the strike dimension, since we have bootstrapped the curves one-by-one for each strike, there is no guarantee that they will be arbitrage-free; we also need to ensure that they are non-decreasing in K to exclude any negative densities.

Other authors, such as Torresetti et al. (2007) or Walker (2006), have approached the problem slightly differently. Instead of using a bootstrapping approach, they solved for the ETL surface that minimizes the pricing error of the benchmark calibration instruments subject to the no-arbitrage constraints in strike and maturity. This is also quite common in interest rate modelling: there are many interest rate curve construction methods that focus on building "smooth" yield curves that match market prices in the least-square sense while ensuring no-arbitrage relationships are maintained.

5.5 Entropy Maximization

Given the tranche discount factor curves $\left(D_{0,T_j}^{(K_i)}\right)_{1\leq j\leq m}$ for all $1 \leq i \leq n$, now we need to build the rest of the loss distributions across the capital structures $D_{0,T_j}^{(K)}$ for all $K \in [0, 1]$. This can be done, for example, by Entropy Minimization methods.

Entropy minimization approaches have been used traditionally in finance to model equity volatility surfaces (see Buchen and Kelly (1996); Avellaneda (1997)); but there are also a number of interesting applications in the context of portfolio credit modelling (see Vacca (2005); Dempster et al. (2007); Li and Zheng (2009)).

5.5.1 Principle of Maximum Entropy

The principle of maximum entropy was first introduced by Jaynes (1957). Motivated by information theory concepts, he used the maximum-entropy estimate as a criterion for inferring a probability distribution based on partial information. In our case, the problem we need to solve is essentially similar to the volatility surface in Equities: given a finite set of option prices at different strikes, how can we extract a risk-neutral probability density that matches those prices? Clearly, this is a (mathematically) ill-posed problem since we typically have significantly more unknown variables than constraints. To proceed, we need to add an additional objective function to optimize.

Because entropy is, in fact, a measure of uncertainty in the system, finding the probability distribution that maximizes entropy is, in some sense, the most optimal distribution that exists given the information that we know. Any other distribution with lower entropy (therefore, lower uncertainty) has necessarily more implicit (or explicit) structure imposed on it, and as such is, undoubtedly, using an expanded information set larger than the one available to us. This is the main philosophy behind Jaynes' argument for choosing the maximum entropy criterion. In statistical terms, it can be viewed as a non-parametric estimation method of the probability distribution, which leads to the most unbiased distribution given the constraints of the problem.

5.5.2 Problem Formulation

We discretize the loss distribution on a grid of losses $D = \{l_0 = 0, l_1, l_2, \ldots, l_N = 1\}$, and we define the loss distribution, at each time horizon T, with a vector of probabilities $(p_T(0), p_T(1), \ldots, p_T(N))$:

$$p_T(i) \triangleq \mathbb{P}(L_T = l_i), \text{ for } 0 \le i \le N.$$

We have, in total, $N + 1$ (positive) variables to find, $p_T(i) \ge 0$, which satisfy the following constraints

$$\sum_{i=0}^{N} p_T(i) = 1,$$

$$\sum_{i=0}^{N} p_T(i) E^{(K_j)}(l_i) = D_{0,T}^{(K_j)}, \text{ for all } 1 \le j \le n.$$

where the function $E^{(K)}(x)$ defines the $(0, K)$-equity tranche payoff

$$E^{(K)}(x) = \min(x, K), \text{ for } x \in [0, 1].$$

In general, we have an undetermined system since the number of constraints is less than the number of unknowns. In order to reduce the search space (of possible probability distributions) to one unique solution, we resort to the maximum entropy criterion.

The entropy is defined as

$$S = -\sum_{i=0}^{N} p_T(i) \log(p_T(i)).$$

The (optimal) unbiased distribution with the least amount of (arbitrary) information above and beyond the known information from the imposed constraints is the one that maximizes the entropy. This is a standard optimization problem with linear constraints, which can be solved with the Lagrange multipliers method.

The Lagrangien $\mathcal{L}$ is defined as

$$\begin{aligned}\mathcal{L} = & -\sum_{i=0}^{N} p_T(i) \log(p_T(i)) \\ & - (\lambda_0 - 1)\left[\sum_{i=0}^{N} p_T(i) - 1\right] - \sum_{j=1}^{m} \lambda_j \left[\sum_{i=0}^{N} p_T(i) E^{(K_j)}(l_i) - D_{0,T}^{(K_j)}\right],\end{aligned}$$

where are $\lambda_0, \lambda_1, \ldots, \lambda_m$ the Lagrangien multipliers for each one of our $m+1$ constraints. Necessary conditions for existence of a maximum are given by the first-order derivatives

$$\frac{\partial \mathcal{L}}{\partial p_T(i)} = 0, \text{ for } 0 \leq i \leq N;$$

and the sufficient conditions for the existence of a solution are inferred from the second-order derivatives (Hessian) matrix, $\left[\frac{\partial^2 \mathcal{L}}{\partial p_T(i) \partial p_T(j)}\right]_{i,j}$, by ensuring it is negative definite. The optimal solution is given in terms of the Lagrange multipliers as

$$p_T(i) = \exp\left(-\lambda_0 - \lambda_1 E^{(K_1)}(l_i) - \cdots - \lambda_m E^{(K_m)}(l_i)\right), \text{ for } 0 \leq i \leq N.$$

Linear independence of the matrix gives a necessary and sufficient condition for the existence of a solution. It has been shown by Agmon et al. (1979) that this is verified if, for any vector $c(j) \neq 0$, $1 \leq j \leq m$, we have the following inequalities

$$\min_{0 \leq i \leq N}\left[\sum_{j=1}^{m} c(j) E^{(K_j)}(l_i)\right] < \sum_{j=1}^{m} c(j) D_{0,T}^{(K_j)} < \max_{0 \leq i \leq N}\left[\sum_{j=1}^{m} c(j) E^{(K_j)}(l_i)\right].$$

In our context, this set of inequalities is equivalent to ensuring that basic no-arbitrage relationships hold. In particular, the expected loss must always be bounded by the lowest and highest possible losses. Indeed, this can be seen by replacing $D_{0,T}^{(K_j)}$ with its definition and inverting the summation order:

$$\sum_{j=1}^{m} c(j) D_{0,T}^{(K_j)} = \sum_{j=1}^{m} c(j) \sum_{i=0}^{N} p_T(i) E^{(K_j)}(l_i) = \sum_{i=0}^{N} p_T(i) \sum_{j=1}^{m} c(j) E^{(K_j)}(l_i).$$

Substituting the solution in the $(m+1)$ constraints yields

$$\exp(\lambda_0) = \sum_{i=0}^{N} \exp\left(-\sum_{j=1}^{m} \lambda_j E^{(K_j)}(l_i)\right),$$

$$D_{0,T}^{(K_j)} \exp(\lambda_0) = \sum_{i=0}^{N} E^{(K_j)}(l_i) \exp\left(-\sum_{j=1}^{m} \lambda_j E^{(K_j)}(l_i)\right), \text{ for } 1 \leq j \leq m.$$

Replacing $\exp(\lambda_0)$ with its expression as a function of $(\lambda_1, \cdots, \lambda_m)$ gives the following system of non-linear equations

$$\sum_{i=0}^{N}\left(D_{0,T}^{(K_j)} - E^{(K_j)}(l_i)\right)\exp\left(-\sum_{j=1}^{m}\lambda_j E^{(K_j)}(l_i)\right) = 0, \text{ for } 1 \le j \le m,$$

which can be solved numerically. This problem can also be solved more elegantly by considering its dual problem instead.

5.5.3 Dual Problem

The dual (potential) function is defined as

$$\mathcal{L}^*(\lambda) = \log\left(\sum_{i=0}^{N}\exp\left(-\sum_{j=1}^{m}\lambda_j E^{(K_j)}(l_i)\right)\right) + \sum_{j=1}^{m}\lambda_j D_{0,T}^{(K_j)}.$$

The dual problem to maximum entropy consists in finding the minimum of the potential function $\mathcal{L}^*(\lambda)$ in order to determine the set of Lagrange multipliers λ:

$$\inf_{\lambda\in\mathbb{R}^m} \mathcal{L}^*(\lambda),$$

which is a much easier problem to solve since it is unconstrained and the dual function is always convex. This minimization can be solved numerically with any suitable gradient-based optimization method. Simplex-type direct search algorithms can also be used.

5.5.4 Regularization

For practical applications, given potential noise in the input data, it is often better to solve a "regularized" version of the problem. This is done by moving the constraints into the maximization objective function as a penalty term. The maximum entropy problem then becomes

$$\sup_{p_T\in L_1(D)} -\sum_{i=0}^{N} p_T(i)\log(p_T(i)) - \frac{1}{2\theta}\sum_{i=1}^{m}\omega_i^2\left[\sum_{i=0}^{N} p_T(i)\,E^{(K_j)}(l_i) - D_{0,T}^{(K_j)}\right]^2,$$

for some penalization constant $\theta > 0$, which controls the trade-offs between accurate replication of the constraints and smoothness of the distribution; and

$\omega_i > 0$ are the weights assigned to each constraint. To use the original solution to the equivalent dual problem, Decarreau et al. (1992) have reformulated this optimization as a new constrained optimization:

$$\sup_{p_T \in L_1(D),\, \epsilon \in \mathbb{R}^m} \quad -\sum_{i=0}^{N} p_T(i) \log\left(p_T(i)\right) - \frac{1}{2\theta} \left\|\epsilon\right\|^2,$$

subject to the linear constraints

$$\sum_{i=0}^{N} p_T(i)\, E^{(K_j)}(l_i) - D_{0,T}^{(K_j)} = \frac{\epsilon_i}{\omega_i}, \text{ for } 1 \le i \le m,$$

where $\|.\|$ is the Euclidean norm on $\mathbb{R}^m$. The dual function for this new problem is

$$\mathcal{L}^*(\lambda) = \log\left(\sum_{i=0}^{N} \exp\left(-\sum_{j=1}^{m} \lambda_j E^{(K_j)}(l_i)\right)\right) + \sum_{j=1}^{m} \lambda_j D_{0,T}^{(K_j)} + \frac{1}{2}\theta \left\|\widetilde{\lambda}\right\|^2,$$

where $\widetilde{\lambda}$ is the set of Lagrange multipliers rescaled by the corresponding weight $\widetilde{\lambda} = \left(\frac{\lambda_1}{\omega_1}, \ldots, \frac{\lambda_m}{\omega_m}\right)$. Thus, we can proceed and use the same numerical techniques to solve this dual problem.

5.5.5 Minimum Relative Entropy

By definition, the maximum entropy distribution is the one with the least structure, which satisfies the constraints. So, it is the closest to the (benchmark) uniform distribution, in the entropy sense, since, in absolute terms, the uniform distribution is the one with maximum entropy (and no additional constraints). If we know more information about the problem at hand, and we have an a-priori candidate distribution that we would like to mirror while satisfying the constraints, we can use the concept of Relative Entropy, which quantifies the (information) distance between two distributions. In that sense, minimizing the relative entropy gives the closest distribution compared to the benchmark one that satisfies the constraints of the problem. The minimum relative entropy between two distributions p and q is defined as:

$$L(q\,|p) = \sum_{i=0}^{N} q(i) \log\left(\frac{q(i)}{p(i)}\right).$$

This is typically used, in finance, when we have an empirical distribution estimated in the historical measure, which could be a good indicator of the general qualitative shape of the risk-neutral one. In the portfolio credit modelling context, this offers another way of doing Skew Rescaling between market-implied index distributions and bespoke portfolio distributions. In this case, the constraint we have to match is the new bespoke portfolio expected loss (relative to the index expected portfolio loss), and the rest of the distribution gets readjusted accordingly.

So, the minimum relative entropy skew rescaling would be done by minimizing the distance $L(q \mid p)$,

$$\sum_{i=0}^{N} q_T(i) \log\left(\frac{q_T(i)}{p_T(i)}\right),$$

subject to the constraints

$$\sum_{i=0}^{N} q_T(i) = 1,$$

$$\sum_{i=0}^{N} q_T(i) \cdot l_i = \mathbb{E}\left[L_T^{\pi}\right],$$

where $\mathbb{E}\left[L_T^{\pi}\right]$ is the expected loss of the bespoke portfolio π to be rescaled. The solution follows the same steps as before and can be expressed in terms of the Lagrange multipliers as

$$q_T(i) = p_T(i) \frac{e^{-\lambda \cdot l_i}}{\sum_{i=0}^{n} e^{-\lambda \cdot l_i}}, \text{ for } 0 \leq i \leq N,$$

which can be solved for by inverting the expected loss constraint

$$\sum_{i=0}^{N} p_T(i) \frac{e^{-\lambda \cdot l_i}}{\sum_{i=0}^{n} e^{-\lambda \cdot l_i}} \cdot l_i = \mathbb{E}\left[L_T^{\pi}\right].$$

If we have more market information on some parts of the capital structure (for example, the equity or the senior tranche) that we would like to account for in pricing, then we can impose more constraints on the optimization problem and solve it accordingly. There are also some interesting applications of the relative entropy method in (Cash) CLO space, where it could be used to price the whole set of CLO transactions relative to a (benchmark) CLO Discount Margin (DM) stack by minimizing the pricing distributions of (CDR, CPR, CRR) scenarios (see, for example, Li and Zheng (2009)).

5.6 Concluding Remarks

In this chapter, we gave a very general overview of default correlation modelling concepts and techniques. Starting with the market standard pricing approach, we have introduced the concept of a base correlation curve, which views a CDO tranche as the difference between two equity tranches, then marks each one of them with a different Gaussian correlation level. This approach has proven to be very popular with market practitioners because it is easy to implement, it is self-consistent and it enables one to price other non-standard tranches by interpolating (or extrapolating) the base correlation curve.

To mark the base correlation curve for other bespoke portfolios (different from the market traded index), market practice is to use the so-called skew rescaling methodology, which maps a tranche on the bespoke portfolio to an equivalent tranche on the index. Similarly for CDO^2 products, base correlation needs to be extended: this is done by introducing a carefully crafted outer-correlation copula function that links the loss variables of the underlying sub-portfolios. Although the various interpolation (and extrapolation) methods of the base correlation curve used by market participants perform well in practice, they do, however, require a lot of care and attention to ensure that the model-implied dynamics are still arbitrage free.

A useful tool to benchmark the base correlation model against a model-independent arbitrage-free correlation surface is the Expected Tranche Loss (ETL) surface. Without going through the base correlation parametrization, one can instead use the observable market tranche prices directly and extract the ETL-discount factors. The approach is similar to a bootstrapping algorithm used for interest rate yield curves. This is particularly useful when we want to check that the input market prices are arbitrage-free. Any potential calibration failures can then be explained from inconsistencies in market data as opposed to being a hidden artefact of the choice of base correlation parametrization. The ETL approach is helpful in getting consistency in the time dimension; the equivalent tool in the strike dimension is the Maximum Entropy Method. It provides a model-independent arbitrage-free calibration of the loss distribution, which is consistent with market prices and has minimum bias.

We have also discussed the various numerical implementation methods used for CDO pricing, ranging from Fast Fourier Transform (FFT) and Recursion methods to faster proxy methods, such as the Normal approximation, the Poisson approximation and Stein's method.

References

N. Agmon, Y. Alhassid, R. Levine, An algorithm for finding the distribution of maximal entropy. J. Comput. Phys. **30**, 250–259 (1979)

L. Andersen, J. Sidenius, S. Basu, All your hedges in one basket. Risk, 67–72 (2003)

A. Antonov, S. Mechkov, T. Misirpashaev, Analytical techniques for synthetic CDOs and credit default risk measures (Working Paper, Numerix, 2005)

M. Avellaneda, Minimum-entropy calibration of asset pricing models. Int. J. Theor. Appl. Financ. **1**(4), 447 (1998)

M. Avellaneda, C. Friedman, R. Holmes, D. Samperi, Calibrating volatility surfaces via relative entropy minimization. Appl. Math. Financ. **4**(1), 37–64 (1997)

N. Bennani, The forward loss model: a dynamic term structure approach for the pricing of portfolio credit derivatives (Working Paper, 2005)

D.T. Breeden, R.H. Litzenberger, Prices of state-contingent claims implicit in options prices. J. Bus. **51**(4)

P.W. Buchen, M. Kelly, The maximum entropy distribution of an asset inferred from option prices. J. Financ. Quantit. Anal. **31**(1), 143–159 (1996)

L.H.Y. Chen, Poisson approximations for dependent trials. Ann. Probab. **3**, 534–545 (1975)

A. Decarreau, D. Hilhorst, C. Lemarechal, J. Navaza, Dual methods in entropy maximization: application to some problems in crystallography. SIAM J. Optim. **2**(2), 173–197 (1992)

M.A.H. Dempster, E.A. Medova, S.W. Yang, Empirical Copulas for CDO tranche pricing using relative entropy. Int. J. Theor. Appl. Financ. **10**(4), 679–701 (2007)

N. El Karoui, Y. Jiao, D. Kurtz, Gauss and Poisson approximation: applications to CDOs tranche pricing. J. Comput. Financ. **12**(2), 31–58 (2008)

J. Hull, A. White, Valuation of a CDO and nth-to-default CDS without monte carlo simulation. J. Deriv. **2**, 8–23 (2004)

E.T. Jaynes, Information theory and statistical mechanics. Phys. Rev. **106**, 620–630 (1957)

J.-P. Laurent, J. Gregory, Basket default swaps. CDOs and factor Copulas. J. Risk **7**(4), 103–122 (2005)

D.X. Li, M. Liang, CDO squared pricing using a Gaussian mixture model with transformation of loss distribution (Working Paper, Barclays Capital, 2005)

Y. Li, Z. Zheng, A top-down model for cash CLO (Working Paper, 2009)

L. McGinty, E. Beinstein, R. Ahluwalia, M. Watts, *Introducing Base Correlations* (Credit Derivatives Strategy, JP Morgan, 2004)

A. Reyfman, K. Ushakova, W. Kong, *How to Value Bespoke Tranches Consistently with Standard Ones* (Credit Derivatives Research, Bear Stearns, 2004)

P.J. Schönbucher, Factor models for portfolio credit risk. J. Risk Financ. **3**(1), 45–56 (2001)

P.J. Schönbucher, Portfolio losses and the term structure of loss transition rates: a new methodology for the pricing of portfolio credit derivatives (Working Paper, ETH Zurich, 2005)

D. Shelton, *Back To Normal* (Global Structured Credit Research, Citigroup, 2004)

J. Sidenius, V. Piterbarg, L. Andersen, A new framework for dynamic credit portfolio loss modeling. Int. J. Theor. Appl. Financ. **11**(2), 163–197 (2008)

C. Stein, *A Bound For the Error In the Normal Approximation to the Distribution of a Sum of Dependent Random Variables* (University of California Press, Berkeley, Proc. Sixty Berkely Symp. Math. Statis. Probab., 1972), pp. 583–602

R. Torresetti, D. Brigo, A. Pallavicini, Implied expected tranche loss surface from CDO data (Working Paper, 2007)

J. Turc, D. Benhamou, B. Hertzog, M. Teyssier, *Pricing Bespoke CDOs: Latest Developments* (Quantitative Strategy, Credit Research, Societe Generale, 2006)

L. Vacca, Unbiased risk-neutral loss distributions. Risk, 97–101, 2005

O. Vasicek, The loan loss distribution (Working Paper, KMV Corporation, 1997)

M. Walker, CDO models—towards the next generation: incomplete markets and term structure (Working Paper, 2006)

6

Correlation Skew: A Black-Scholes Approach

In this chapter, we view the valuation of CDO tranches as an option pricing problem. The payoff of a CDO tranche is a call-spread on the loss variable. By specifying the distribution of the loss variable at each time horizon, one would be able to value tranches. The standard way of defining this distribution is the base correlation approach. Here, we use a Black-Scholes analogy and we define an implied volatility for each tranche. Then, given a Black volatility surface, we parameterize the loss distribution with a Stochastic CEV model. We show that this parametric form gives a very good fit to the market tranche quotes. In addition, we give an application of the correlation skew Black approach to risk management and hedging.

6.1 Introduction

The standard approach for pricing CDO tranches is the base correlation model. Using a Gaussian copula pricer with a base correlation curve, we can define implicitly the whole distribution of the loss variable. While the approach is simple and easy to implement, it suffers from a number of shortcomings. The interpolation in the strike dimension, the extrapolation below the first strike and above the last strike, and the interpolation in the maturity dimension can sometimes lead to unacceptable loss distributions and breach the no-arbitrage conditions. It is also not very easy to use for hedging purposes and running coherent stress tests.

Here, we take a different route and we view the pricing of CDOs as an option-pricing problem. Each CDO tranche is a call spread on the loss variable. Using Black-type dynamics, we can define an implied volatility for each tranche

Y. Elouerkhaoui, *Credit Correlation*, Applied Quantitative Finance,
DOI 10.1007/978-3-319-60973-7_6

attachment and detachment points. This would result in a complete Black volatility surface to which we could apply the standard machinery from option theory. For instance, we can parameterize the loss distribution using a Stochastic CEV model. For hedging and risk management purposes, we shall use the ATM vol and Vol of Vol parameters to describe the movements of the base correlation skew curve.

6.2 Building a Black-Scholes Model

We denote by L_T the loss variable, at time T, for a given portfolio

$$L_T \triangleq \frac{1}{n} \sum_{i=1}^{n} (1 - R_i)\, D_T^i,$$

where $D_T^i = \mathbf{1}_{\{\tau_i \leq T\}}$ are the default indicators, and R_i are the recovery rates. The payoff of a tranche, $M_T^{K_1,K_2}$, with attachment and detachment points K_1 and K_2 respectively, is given by the call-spread formula applied to the loss variable

$$M_T^{K_1,K_2} \triangleq (L_T - K_2)^+ - (L_T - K_1)^+ .$$

The value of the tranche today is the expectation of the discounted payoff

$$V = \mathbb{E}\left[p_{0,T} M_T^{K_1,K_2}\right] = \mathbb{E}\left[p_{0,T}\,(L_T - K_2)^+\right] - \mathbb{E}\left[p_{0,T}\,(L_T - K_1)^+\right],$$

where $p_{0,T}$ is the risk-free discount factor with maturity T. To introduce the Black-Scholes dynamics, we define the conditional loss variable $S_{t,T}$, at time t, as

$$S_{t,T} \triangleq \mathbb{E}\left[L_T \,|\mathcal{G}_t\right],$$

where $\{\mathcal{G}_t\}$ is the enlarged filtration containing both the background filtration and the default filtrations. At maturity $t = T$, the (terminal) conditional loss variable, $S_{T,T}$, is simply given by the loss variable at time T,

$$S_{T,T} = L_T.$$

And at inception $t = 0$, the initial value is given by the expected loss:

$$S_{0,T} = \frac{1}{n} \sum_{i=1}^{n} (1 - R_i) \left(1 - \mathbb{E}\left[\exp\left(-\int_0^T \lambda_s^i ds\right)\right]\right),$$

where λ^i are the individual single-name intensities. Note that, by construction, the process $\left(S_{t,T}\right)_{0\leq t\leq T}$ is a $\{\mathcal{G}_t\}$-martingale.

We assume that $S_{t,T}$ follows a Black-Scholes diffusion of the form

$$\frac{dS_{t,T}}{S_{t,T}} = \sigma_{BS} dW_t.$$

Then, for each maturity T, and each attachment point K, we use a different Black volatility $\sigma_{BS}(T, K)$ to value the option $\mathbb{E}\left[(L_T - K)^+\right]$.

Remark 72 $S_{t,T}$ is a $\mathcal{G}_t$-martingale; hence, by virtue of the martingale representation theorem, it can be written in general as the sum of diffusion term and default-induced jump terms

$$S_{t,T} = S_{0,T} + \int_0^t \Sigma_t \cdot dW_t + \sum_{i=1}^n \int_0^t \eta_t \left(dD_t^i - \lambda_t^i dt\right).$$

This means that, with our Black-Scholes parametrization assumption, the implied black volatilities will exhibit a very pronounced smile/skew effect due the jumps in the real dynamics. This is not dissimilar to what has been done traditionally in other asset classes (Equities, FX, Rates), where the dynamics of the spot process may include either jumps or stochastic volatility components; nonetheless, the Black parametrization is still used to generate the implied distribution at maturity by fitting an entire Black volatility surface $\sigma_{BS}(T, K)$.

6.3 Stochastic CEV Model

Given the implied Black volatility surface $\sigma_{BS}(T, K)$, we can use interpolations to get values of other parts of the capital structure. However, there is no guarantee that the implied distributions would be arbitrage free. Instead, following standard practice in options markets, we can parameterize the loss distribution in an arbitrage-free manner, by using, for example, a Stochastic CEV model (also known as the SABR model).

The SABR dynamics are given by

$$\begin{aligned} dS_{t,T} &= \alpha_t \left(S_{t,T}\right)^\beta dW_t^1, S_{0,T} = f \\ d\alpha_t &= \nu\alpha_t dW_t^2, \alpha_0 = \alpha \\ \left\langle dW_t^1, dW_t^2\right\rangle &= \rho dt. \end{aligned}$$

Using Hagan's formula (see Hagan et al. 2002), we can express the Black volatilities in terms of the Stochastic CEV model parameters

$$\sigma_{BS}(T,K)=\frac{\alpha}{(fK)^{\frac{1-\beta}{2}}\left\{1+\frac{(1-\beta)^2}{24}\log\left(\frac{f}{K}\right)^2+\frac{(1-\beta)^4}{1920}\log\left(\frac{f}{K}\right)^4+\ldots\right\}}\cdot\left(\frac{z}{\chi(z)}\right)$$
$$\left\{1+\left[\frac{(1-\beta)^2}{24}\frac{\alpha^2}{(fK)^{1-\beta}}+\frac{1}{4}\frac{\rho\beta\nu\alpha}{(fK)^{\frac{1-\beta}{2}}}+\frac{2-3\rho^2}{24}\nu^2\right]\right\}T+\ldots,$$

where

$$z=\frac{\nu}{\alpha}(fK)^{\frac{1-\beta}{2}}\log\left(\frac{f}{K}\right),$$
$$\chi(z)=\log\left\{\frac{\sqrt{1-2\rho z+z^2}+z-\rho}{1-\rho}\right\}.$$

6.4 Calibration Example

In this section, we give a numerical example to illustrate the method. We use the following base correlation skew curve. The portfolio duration-weighted average spreads (DWAS) at the 1-year, 3-year and 5-year tenors are: 8.5 bps, 23.2 bps and 38.2 bps.

Strike (%)	1 year (%)	3 year (%)	5 year (%)
0	19	6	0
3	31.6	17.9	11.6
6	40.2	27	23
9	45.3	33.5	31.7
12	49.4	38.5	38.7
22	57.7	52.1	55.7

First, we compute the expected tranche loss for each strike, which is then converted to a call price by subtracting the value of the forward (i.e., the expected portfolio loss). Indeed, we have

$$
\begin{aligned}
C(T,K) &= \mathbb{E}\left[(L_T - K)^+\right] = \int_K^{+\infty} (x-K)\,\mathbb{P}(L_T \in dx) \\
&= \mathbb{E}[L_T] - \left[\int_0^K x\mathbb{P}(L_T \in dx) + K\int_K^{+\infty} \mathbb{P}(L_T \in dx)\right] \\
&= \mathbb{E}[L_T] - \mathbb{E}[\min(L_T, K)].
\end{aligned}
$$

By using a Black-Scholes (log-normal) model, a Normal Black model and a CEV model (with exponent $\beta = 0.5$), we get the following corresponding volatilities.

The next table shows the (equivalent) call prices, the Implied Black-Scholes volatilities, the Normal Black volatilities and CEV volatilities, at different strikes, for the 5-year tenor, $T = 5$.

Strike (%)	$C(T,K)$ (%)	$\sigma_{BS}(T,K)$ (%)	$\sigma_{Normal}(T,K)$ (%)	$\sigma_{CEV}(T,K)$ (%)
3	0.3617	36	0.8592	5.6
6	0.2095	44.4	1.5418	8.2
9	0.1679	49.6	2.1845	10.3
12	0.1461	53.2	2.7902	12
22	0.1092	59.9	4.6382	16.2

The second table shows the same information for the 3-year tenor, $T = 3$.

Strike (%)	$C(T,K)$ (%)	$\sigma_{BS}(T,K)$ (%)	$\sigma_{Normal}(T,K)$ (%)	$\sigma_{CEV}(T,K)$ (%)
3	0.0657	61.4	0.9476	7.5
6	0.0372	68.9	1.6370	10.3
9	0.0280	73.5	2.2817	12.5
12	0.0225	76.4	2.8848	14.1
22	0.0164	83.6	4.8413	18.8

And the last table shows the information for the 1-year tenor, $T = 1$.

Strike (%)	$C(T,K)$ (%)	$\sigma_{BS}(T,K)$ (%)	$\sigma_{Normal}(T,K)$ (%)	$\sigma_{CEV}(T,K)$ (%)
3	0.0051	158.6	1.2419	13.3
6	0.0034	168.9	2.2108	17.9
9	0.0026	173.7	3.1069	21.1
12	0.0021	177.6	3.9772	23.9
22	0.0011	179.8	6.5758	30

Now, we can use a least-square optimization routine to best fit the parameters of the Stochastic CEV model. The results are given in the following table.

Expiry (T)	Beta (β)	Vol (α_0)	VolVol (ν)	Rho (ρ)
1	0.5	1.6364	2.1370	−0.69
3	0.5	0.9504	1.1695	−0.72
5	0.5	0.3993	0.7413	−0.30

In Figs. 6.1, 6.2 and 6.3, we show the quality of the StochCEV calibration. Using the parameters of the calibrated StochCEV, we compute the implied Black-Scholes volatilities and compare them with the input BS volatilities from the base correlation skew.

The calibration for the 3-year and 5-year tenors is excellent. At the 1-year tenor, the fit is not as accurate, but it is still very good. This is not very surprising; it is a well-known fact that stochastic volatility models work well for longer dated options, but do not have the same level of explanatory power for short-dated skews, which are more sensitive to jumps. The latter are better explained by jump diffusion models.

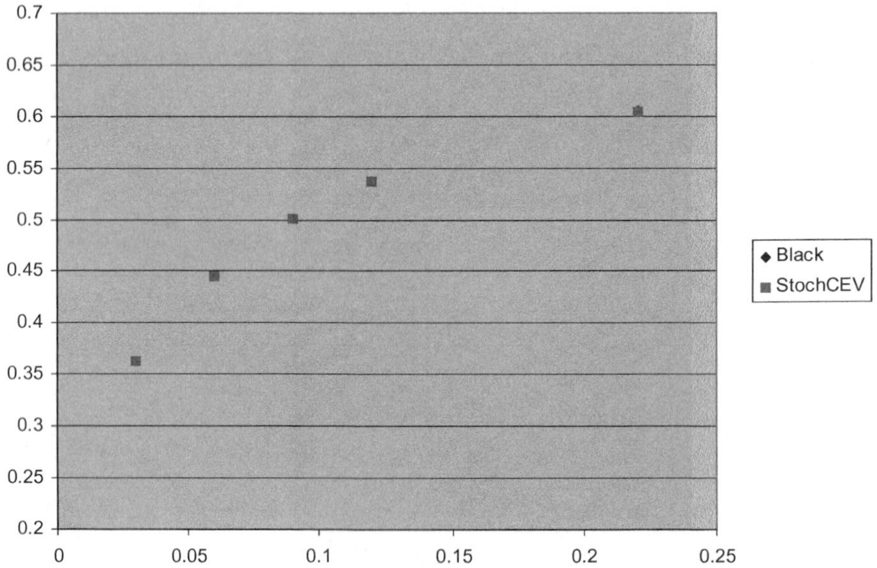

Fig. 6.1 Calibration of StochCEV model for the 5-year tenor. Comparison of the Black volatilities from the calibrated StochCEV model and the input Black volatilities

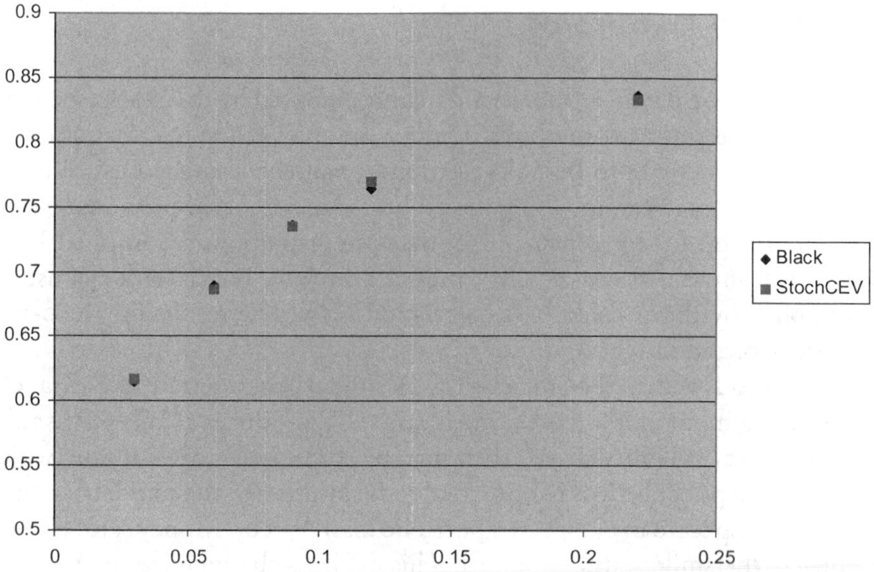

Fig. 6.2 Calibration of StochCEV model for the 3-year tenor. Comparison of the Black volatilities from the calibrated StochCEV model and the input Black volatilities

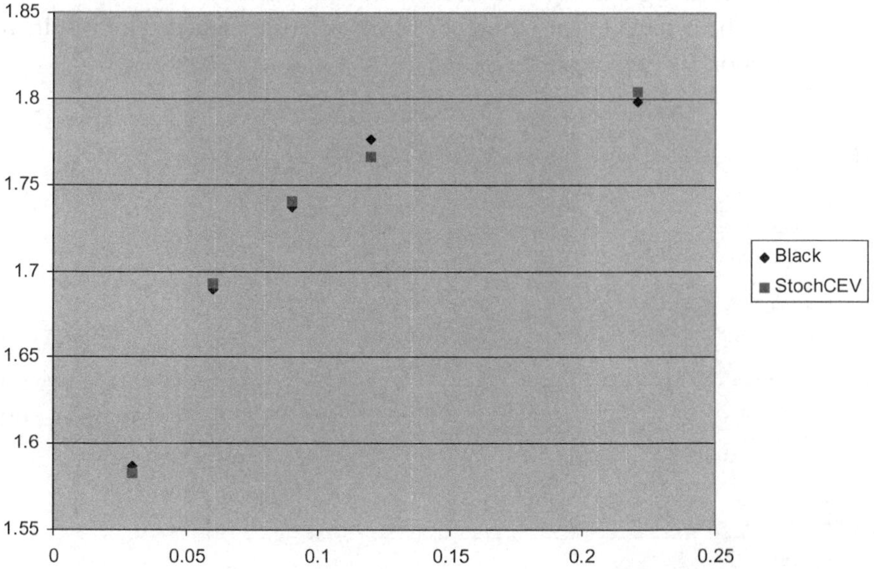

Fig. 6.3 Calibration of StochCEV model for the 1-year tenor. Comparison of the Black volatilities from the calibrated StochCEV model and the input Black volatilities

6.5 Skew Dynamics

In this section, we compare the skew dynamics implied by the Stochastic CEV model with the standard methods used by the market for rescaling the base correlation skew curves to bespoke portfolios, namely: Tranche Loss rescaling and Portfolio Loss rescaling. Using the calibrated model of the previous section, we bump all spreads by a factor of 2, then we recompute the implied Black volatilities from the Stochastic CEV model, which we benchmark against the ones implied by the rescaled base correlation skew. The results for the 5-year tenor are depicted in Fig. 6.4.

When spreads widen, the Stochastic CEV model behaves more like a Tranche Loss rescaling method. However, Stochastic CEV produces a “volatility smile” as opposed to a “volatility skew”; therefore, by shifting the forward rate (which is the expected portfolio loss in our set-up), the location of the standard tranche strikes with respect to the (expected portfolio loss) At-The-Money rate changes, and hence, the smile starts to appear. This explains the increase in the Black volatility at the 3%-strike point. The ATM rate in the base case is 1.9714%, and moves up to 3.8642% in this spread widening scenario.

Similarly, in the tightening scenario, we multiply the spreads by 50%, and compare the three rescaling methods in Black volatility space. The results for the 5-year tenor are depicted in Fig. 6.5.

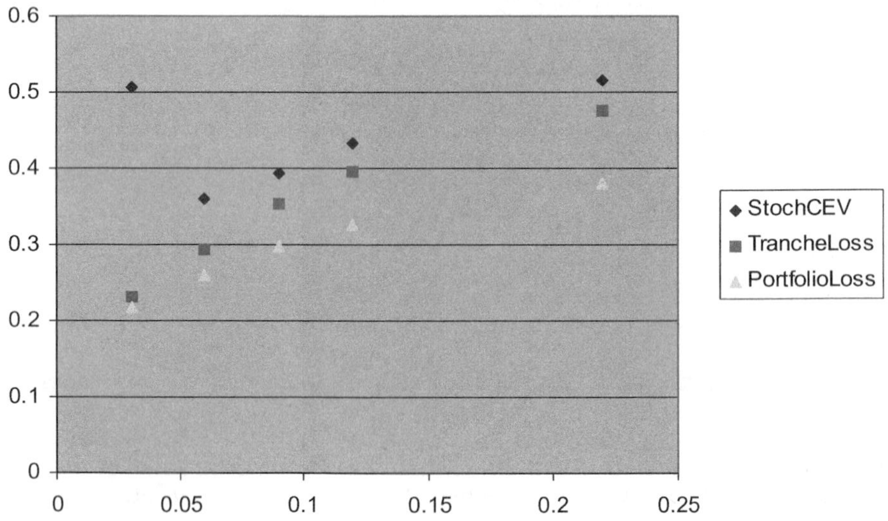

Fig. 6.4 Comparison of the rescaling properties of the StochCEV model with TrancheLoss and PortfolioLoss rescaling for a spread-widening scenario. All curves are multiplied by a factor of 2

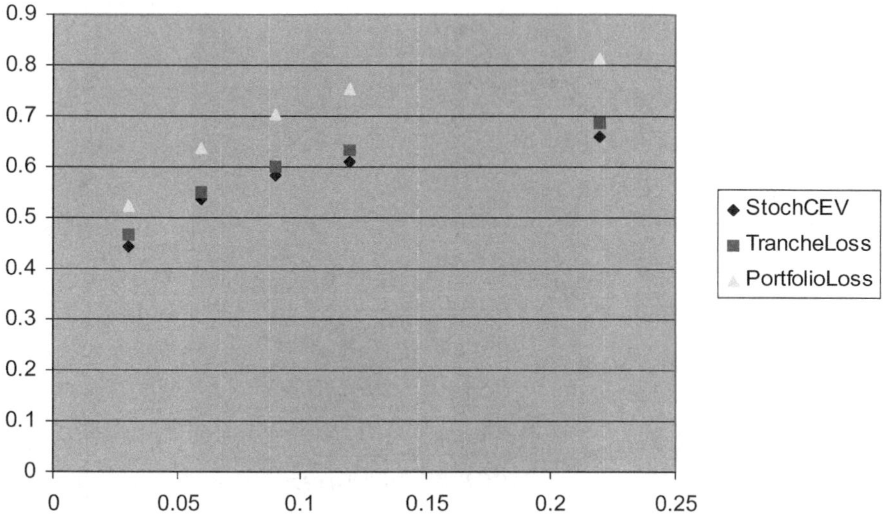

Fig. 6.5 Comparison of the rescaling properties of the StochCEV model with TrancheLoss and PortfolioLoss rescaling for a spread-tightening scenario. All curves are multiplied by a factor of 0.5

Again, the Stochastic CEV model behaves similarly to Tranche Loss rescaling. In this case, we do not have the difference due to the smile effect since the ATM rate becomes 0.9959% (which is lower than the 3% tranche strike). In summary, as long as the ATM is lower than the smallest tranche strike of 3%, the two methods would give comparable results. They would differ, however, for strikes that are lower than the ATM rate.

6.6 Risk Management

The Black-Scholes analogy offers also a very intuitive framework for managing the various risks of a correlation book. By using the Black volatility surface for parameterizing the loss distribution, we can express all our Greeks in the more familiar option pricing language. The typical risk measures would include: Delta, Gamma, Vega, Vanna and Volga,

$$\begin{aligned}
\text{Delta} &= \frac{\partial BS}{\partial S_{0,T}};\\
\text{Gamma} &= \frac{\partial^2 BS}{\left(\partial S_{0,T}\right)^2};\\
\text{Vega} &= \frac{\partial BS}{\partial \sigma_{BS}};\\
\text{Vanna} &= \frac{\partial \text{Vega}}{\partial S_{0,T}} = \frac{\partial^2 BS}{\partial \sigma_{BS} \partial S_{0,T}};\\
\text{Volga} &= \frac{\partial \text{Vega}}{\partial \sigma_{BS}} = \frac{\partial^2 BS}{(\partial \sigma_{BS})^2}.
\end{aligned}$$

Hedging the credit spread risk would then consist in flattening the Delta position; hedging the correlation risk would translate into flattening the Vega position. Similarly, to hedge the convexity in the book, we would need to flatten the higher order derivatives, such as Gamma, Volga (Vol Gamma) and Vanna (cross-derivatives).

In the Stochastic CEV model, the Delta and Gamma measures would be the same, but would also need to include the skew dynamics built-in the model; in other words, the model delta with be equal to the BS delta plus a correlation skew correction:

$$\text{ModelDelta} = \frac{\partial V}{\partial S_{0,T}} = \frac{\partial BS}{\partial S_{0,T}} + \frac{\partial BS}{\partial \sigma_{BS}} \cdot \frac{\partial \sigma_{BS}}{\partial S_{0,T}}.$$

And finally, the correlation skew risk could be represented by the following model specific risk measures:

$$\begin{aligned}
\text{ModelVega} &= \frac{\partial V}{\partial \alpha} = \frac{\partial V}{\partial \sigma_{BS}} \cdot \frac{\partial \sigma_{BS}}{\partial \alpha};\\
\text{ModelVanna} &= \frac{\partial V}{\partial \rho} = \frac{\partial V}{\partial \sigma_{BS}} \cdot \frac{\partial \sigma_{BS}}{\partial \rho};\\
\text{ModelVolga} &= \frac{\partial V}{\partial \nu} = \frac{\partial V}{\partial \sigma_{BS}} \cdot \frac{\partial \sigma_{BS}}{\partial \nu}.
\end{aligned}$$

The ModelVega is computed with respect to the explanatory vol $\alpha_0 = \alpha$. The ModelVanna, which captures the dependence between Vol and Spot, is computed with respect to the correlation parameter ρ. And the ModelVolga, which captures the vol gamma, is computed with respect to the Vol of Vol parameter ν.

References

P. Hagan, D. Woodward, Equivalent black volatilities. Appl. Math. Financ. **6**, 147–157 (1999)

P. Hagan, D. Kumar, A. Lesniewski, D. Woodward, Managing smile risk. Wilmott Magazine, September, 2002, 84–108

P. Hagan, D. Kumar, A. Lesniewski, D. Woodward, Arbitrage free SABR. Wilmott Magazine, January, 2014, 60–75

7

An Introduction to the Marshall-Olkin Copula

In this chapter, we present the Marshall-Olkin copula model where the correlation profile is constructed via a set of common shocks, which can trigger joint defaults in the basket. We calibrate the model through a systematic formulation that captures the salient features of the observable measures of correlation. The analysis of the natural consequences of spread dynamics leads to a neat expression of the market factor loadings. The actual levels of the factors are then calibrated on benchmark basket instruments traded at mid-market levels. We study the phenomenon of "correlation regimes" in this universe and we give a quantitative comparison between the Marshall-Olkin copula and the well-known Gaussian copula. We also show that the MO model can be used to reproduce the observed correlation skew in the CDO market. More recently, there has been renewed interest in the MO copula in the context of model risk management (see Morini 2011) and systemic risk modelling (see Gatarek and Jablecki 2016).

7.1 Introduction

As usual, we work on a filtered probability space $(\Omega, \mathcal{G}, \mathbb{G}, P)$, on which is given a set of n non-negative random variables $(\tau_1, \ldots, \tau_n)$ representing the default times of a basket of obligors. We have seen in Chap. 2 a general description of the Marshall-Olkin copula where multiple joint defaults are triggered by $2^n - 1$ independent Poisson processes. This is referred to as the fatal shock representation.

Here we give a parsimonious construction of the MO copula, which is based on a set of m common Poisson shocks. The default of each single-name can then be triggered conditional on one of the common Poisson shocks. This model

Y. Elouerkhaoui, *Credit Correlation*, Applied Quantitative Finance,
DOI 10.1007/978-3-319-60973-7_7

specification is more useful for practical applications than the fatal shock one, which has an exponentially explosive number of parameters to specify.

7.2 Genesis of the Marshall-Olkin Model

As we have shown in Chap. 1, assuming instantaneous default independence and correlated intensities is not the right answer for constructing correlated defaults. This introduces a convexity correction in the expectations, which is not directly relevant for the default correlation problem and does not produce proper default events correlation. In this section, we will build the default dependence by relaxing the first assumption and restricting the second one. In other words, we will assume that joint defaults are possible and that they can be triggered by a set of common Poisson drivers; and we will assume that these factors are independent. The copula function implied by this model is known as the Marshall-Olkin copula.

7.2.1 Construction of Correlation

In a nutshell, the Marshall-Olkin model decomposes the default process of each obligor on a basis of independent Poisson processes. These independent processes can either be common market factors or issuer specific idiosyncratic events. Each market factor event can trigger one or multiple issuer defaults. The idiosyncratic events, on the other hand, will only trigger issuer specific defaults. The default probability of an issuer conditional on the occurrence of the common factor event is referred to as the loading on the market factor. It is a measure of the leverage of a particular credit with this market factor. A market factor for investment grade names, for example, can be an industry-specific event that affects all companies in the sector; for emerging market credits, a market factor could correspond to a region specific or a sovereign event.

The default time of each obligor is viewed as the first jump time of a (doubly-stochastic) Poisson process $\tau_i = \inf\left\{t : N_t^i > 0\right\}$. We denote the associated default process by $D_t^i = \mathbf{1}_{\{\tau_i \leq t\}}$. The conditional survival probability is given by

$$\mathbb{P}\left(\tau_i > T \,|\mathcal{G}_t\right) = \mathbb{P}\left(N_T^i = 0\,|\mathcal{G}_t\right) = \mathbf{1}_{\{N_t^i=0\}}\mathbb{E}\left[\exp\left(-\int_t^T \lambda_s^i ds\right)|\mathcal{G}_t\right].$$

We denote by $\left(\lambda_t^{C_j}, N_t^{C_j}\right)_{1\leq j\leq m}$ the intensities and Poisson counting processes for the common market factors; and we denote by $\left(\lambda_t^{0,i}, N_t^{0,i}\right)_{1\leq i\leq n}$

the intensities and Poisson counting processes for the idiosyncratic events. We also denote their cumulative hazard rates by $\Lambda_T^{C_j} = \int_0^T \lambda_t^{C_j} dt$ and $\Lambda_T^{0,i} = \int_0^T \lambda_t^{0,i} dt$ respectively. Each Poisson counting process $N_t^{C_j}$ is equivalently represented by its sequence of jump times $\left(\theta_r^{C_j}\right)$, for $r = 0, 1, 2, \ldots$. In total, we have $m + n$ Poisson processes: the first m factors are common market events and the last n factors are issuer specific idiosyncratic events.

The principle of the multivariate point process decomposition is as follows.

Each single-name (default) point process, $D_t^i = \mathbf{1}_{\{\tau_i \le t\}}$, is derived from a set of basic common market events and specific idiosyncratic events:

$$dD_t^i = \left(1 - D_{t-}^i\right)\left[\sum_{j=1}^{m} A_{i,j} dN_t^{C_j} + dN_t^{0,i}\right], \tag{7.1}$$

where $A_{i,j}$ are $\{0, 1\}$-valued random variables, which indicate, conditional on a common market event $\{j\}$, whether single-name $\{i\}$ has defaulted or not. The so-called market factor loadings in this representation are the conditional default probabilities $p_t^{i,j} = \mathbb{E}\left[A_{i,j} \,|\mathcal{G}_t\right]$.

An important implication of this representation is that the decomposition (7.1) in terms of independent Poisson events translates directly into a canonical decomposition of intensities.

Proposition 1 *The default intensity of each single-name obligor, λ_t^i, admits the following market factor decomposition*

$$\lambda_t^i = \sum_{j=1}^{m} p_t^{i,j} \lambda_t^{C_j} + \lambda_t^{0,i} \tag{7.2}$$

Similarly, the arrival rate of joint defaults can also be derived. The arrival intensity of the joint defaults (i, k) is given by

$$\lambda_t^{ik} = \sum_{j=1}^{m} p_t^{i,j} p_t^{k,j} \lambda_t^{C_j}. \tag{7.3}$$

More generally, for any subset $A \subset \{1, 2, \ldots, n\}$, the arrival intensity of the joint defaults of all references in A is

$$\lambda_t^A = \lim_{\epsilon \to 0} \mathbb{P}\left(\bigcap_{i \in A} \left\{ N_{t+dt}^i - N_t^i = 1 \right\} |\mathcal{G}_t \right),$$

which can be computed using the conditional independence property:

$$\lambda_t^A = \sum_{j=1}^{m} \left(\prod_{i \in A} p_t^{i,j} \right) \lambda_t^{C_j}. \tag{7.4}$$

Using Eq. (7.4), we can derive closed-form formulas to price basket credit derivative products.

7.2.2 Copula Function

To derive the copula function in this model, we use the k-variate process representation of $\left(N_t^1, \ldots, N_t^n\right)$, also known as the equivalent-fatal-shock model.

We define, for $\pi \in \Pi_n$, where Π_n is the set of non-empty subsets of $\{1, \ldots, n\}$, the point process N_t^π, which counts the number of simultaneous defaults in the subset π only:

$$dN_t^\pi = \sum_{j=1}^{m} \left(\prod_{i \in \pi} A_{i,j} \prod_{i \notin \pi} \left(1 - A_{i,j}\right) \right) dN_t^{C_j} + \sum_{i=1}^{n} \mathbf{1}_{\{\pi = \{i\}\}} dN_t^{0,i}.$$

The processes $\left\{N_t^\pi : t \geq 0\right\}_{\pi \in \Pi_n}$ are independent Poisson processes; and we can represent the individual single-name processes as

$$N_t^i = \sum_{\pi \in \Pi_n} \mathbf{1}_{\{i \in \pi\}} N_t^\pi.$$

Similarly, if we define the first arrival time of the instantaneous joint defaults in π, $\tau_\pi = \inf\left\{t : N_t^\pi > 0\right\}$, then the single-name default time τ_i is given by the first-to-default time of the basket $\{\tau_\pi : i \in \pi\} : \tau_i = \min\{\tau_\pi : i \in \pi\}$.

Using this representation with deterministic intensities, we can write the joint survival probability as

$$\mathbb{P}\left(\tau_1 > T_1, \ldots, \tau_n > T_n\right) = \prod_{\pi \in \Pi_n} \exp\left(- \int_0^{\max\{T_i : i \in \pi\}} \lambda_s^\pi ds \right).$$

This is the Multivariate Exponential Distribution (MVE) developed by Marshall and Olkin (1967).

More explicitly, if we have flat hazard rates $\lambda^{\{i\}}$, $\lambda^{\{ij\}}$, $\lambda^{\{ijk\}}$,...for the fatal shocks that trigger the defaults of $\{i\}$ only, $\{i, j\}$ only, $\{i, j, k\}$ only, ..., then the MVE survival probability[1] $\overline{F}(t_1, \ldots, t_n)$ is given by

$$\overline{F}(t_1, \ldots, t_n) = \exp\left(\begin{array}{c} -\sum_{i=1}^{n} \lambda^{\{i\}} t_i - \sum_{i<j} \lambda^{\{ij\}} \max(t_i, t_j) - \sum_{i<j<k} \lambda^{\{ijk\}} \max(t_i, t_j, t_k) \\ - \ldots - \lambda^{\{12 \ldots n\}} \max(t_1, \ldots, t_n) \end{array} \right).$$

For $n = 2$, if we set $u_1 = \mathbb{P}(\tau_1 > t_1)$, $u_2 = \mathbb{P}(\tau_2 > t_2)$ and $\alpha_1 = \frac{\lambda^{12}}{\lambda^1}$, $\alpha_2 = \frac{\lambda^{12}}{\lambda^2}$, we get the bivariate Marshall-Olkin survival copula[2]

$$\overline{\mathbb{C}}(u_1, u_2) = u_1 u_2 \min\left(u_1^{-\alpha_1}, u_2^{-\alpha_2}\right) = \min\left(u_1^{1-\alpha_1} u_2, u_1 u_2^{1-\alpha_2}\right). \quad (7.5)$$

It has an absolutely continuous part on the upper and lower triangles $\{u_1 < u_2\}$ and $\{u_2 < u_1\}$, and has a singular component on the diagonal $\{u_1 = u_2\}$.

The MVE distribution is also memoryless: i.e., for all $T_1 > t_1, \ldots, T_n > t_n$, we have

$$\mathbb{P}(\tau_1 > T_1, \ldots, \tau_n > T_n \,|\, \tau_1 > t_1, \ldots, \tau_n > t_n) = \mathbb{P}(\tau_1 > T_1 - t_1, \ldots, \tau_n > T_n - t_n).$$

This is the multidimensional version of the memoryless property for an exponential distribution.

7.2.3 Numerical Implementation

In this section, we discuss some useful numerical implementation approaches, which can be used for the Marshall-Olkin copula. This includes: the Monte-Carlo method, the Compound Poisson Process approximation, Panjer recursion and Duffie's approximation.

Monte-Carlo. An efficient simulation algorithm is based on the observation that the (events) process $N_t^e = \sum_{j=1}^{m} N_t^{C_j} + \sum_{i=1}^{n} N_t^{0,i}$ is a Poisson process with intensity $\lambda_t^e = \sum_{j=1}^{m} \lambda_t^{C_j} + \sum_{i=1}^{n} \lambda_t^{0,i}$.

Similarly, the probability of a market event of type $\{e_j\}_{1 \le j \le m+n}$, which can be either a common factor shock $N_t^{C_j}$ or an idiosyncratic shock $N_t^{0,i}$ is equal to

$$p_t^{e_j} = \frac{\lambda_t^{C_j}}{\lambda_t^e}, \text{ for } 1 \leq j \leq m,$$

$$p_t^{e_{m+i}} = \frac{\lambda_t^{0,i}}{\lambda_t^e}, \text{ for } 1 \leq i \leq n.$$

For the aggregate (events) process N_t^e, conditional on a factor jump at time θ_r^e, the identity of the market factor which triggered the jump follows a multinomial distribution with parameters $\left(\frac{\lambda_{\theta_r^e}^{e_1}}{\lambda_{\theta_r^e}^e}, \ldots, \frac{\lambda_{\theta_r^e}^{e_m}}{\lambda_{\theta_r^e}^e}, \ldots, \frac{\lambda_{\theta_r^e}^{e_{m+n}}}{\lambda_{\theta_r^e}^e}\right)$.

Thus, the Monte-Carlo algorithm proceeds as follows.

1. Simulate the jump times $\{\theta_r^e\}$ for any type of market factor event in the interval $[0, T]$.
2. For each jump time θ_r^e, simulate the identity of the market factor which triggered the jump J_r^e.
3. For each pair (θ_r^e, J_r^e), simulate the conditional single-name Bernoulli variables $\left(A_{\theta_r^e}^{1,J_r^e}, \ldots, A_{\theta_r^e}^{n,J_r^e}\right)$.
4. Set the single-name default times:

$$\tau_i = \min\left\{A_{\theta_r^e}^{i,J_r^e}\theta_r^e + \left(1 - A_{\theta_r^e}^{i,J_r^e}\right)T : \theta_r^e \leq T\right\}.$$

Compound Poisson Process. In order to study the aggregate default distribution $X_t = \sum_{i=1}^n D_t^i$, we can use the following approximation

$$X_t = Z_t \triangleq \sum_{i=1}^n N_t^i,$$

where the single-name default indicators D_t^i have been replaced with their associated Poisson processes N_t^i. By doing so, we have, implicitly, neglected the probability of multiple jumps in the Poisson process.

The process Z_t is a compound Poisson process, which is obtained as the sum of $m + n$ independent compound Poisson processes

$$Z_t = \sum_{j=1}^m Z_t^{C_j} + \sum_{i=1}^n N_t^{0,i},$$

where the process $Z_t^{C_j}$ is defined as

$$Z_t^{C_j} = \sum_{r=1}^{N_t^{C_j}} X_r^{C_j} = \sum_{r=1}^{N_t^{C_j}} \left[\sum_{i=1}^{n} A_{\theta_r^{C_j}}^{i,j} \right].$$

The distribution of Z_t is not available in closed form, but we can compute its moment generating function $\mathcal{L}_{Z_t}(\alpha) = \mathbb{E}\left[e^{-\alpha Z_t}\right]$:

$$\begin{aligned}
\mathcal{L}_{Z_t}(\alpha) &= \prod_{j=1}^{m} \mathcal{L}_{Z_t^{C_j}}(\alpha) \prod_{i=1}^{n} \mathcal{L}_{N_t^{0,i}}(\alpha) \\
&= \prod_{j=1}^{m} \exp\left(-t \sum_{k=0}^{n} \left(1 - e^{-\alpha k}\right) \lambda^{C_j} \mathbb{P}\left(X^{C_j} = k\right) \right) \\
&\quad \times \prod_{i=1}^{n} \exp\left(-t \left(1 - e^{-\alpha}\right) \lambda^{0,i} \right).
\end{aligned}$$

For each market factor, the distribution of $X^{C_j} = \sum_{i=1}^{n} A^{i,j}$ is generated by inverting its Fourier transform

$$\mathcal{F}_{X^{C_j}}(\alpha) = \mathbb{E}\left[e^{-i\alpha X^{C_j}}\right] = \prod_{i=1}^{n} \left(p^{i,j} e^{-i\alpha} + \left(1 - p^{i,j}\right) \right).$$

Panjer Recursion. Using the following representation (in Lindskog and McNeil (2003)): $Z_t = \sum_{r=1}^{\widetilde{N}_t} \widetilde{X}_r$, where $\widetilde{N}_t$ is a Poisson process with intensity $\widetilde{\lambda}$, and $\widetilde{X}_r$, $r = 0, 1, 2, \ldots$, are i.i.d. random variables whose distribution is given by:

$$\begin{aligned}
&\mathbb{P}\left(\widetilde{X} = 0\right) = 0, \\
&\mathbb{P}\left(\widetilde{X} = 1\right) = \frac{1}{\widetilde{\lambda}} \left[\sum_{j=1}^{m} \lambda^{C_j} \mathbb{P}\left(X^{C_j} = 1\right) + \sum_{i=1}^{n} \lambda^{0,i} \right], \\
&\mathbb{P}\left(\widetilde{X} = k\right) = \frac{1}{\widetilde{\lambda}} \left[\sum_{j=1}^{m} \lambda^{C_j} \mathbb{P}\left(X^{C_j} = k\right) \right], \text{ for } k \geq 2,
\end{aligned}$$

we can compute Z_t with Panjer's algorithm

$$\mathbb{P}(Z_t = 0) = \exp\left(-\widetilde{\lambda} t\right),$$
$$\mathbb{P}(Z_t = l) = \frac{\widetilde{\lambda} t}{l} \sum_{k=1}^{l} k \mathbb{P}\left(\widetilde{X} = k\right) \mathbb{P}(Z_t = l - k), \text{ for } l > 0.$$

Duffie's Approximation. In Duffie and Pan (2001), the Poisson shock model SDE is integrated as

$$D_t^i = \sum_{j=1}^{m} A_{i,j} D_t^{C_j} + D_t^{0,i},$$

where $D_t^{C_j}$ and $D_t^{0,i}$ are the default indicators for the Poisson processes $N_t^{C_j}$ and $N_t^{0,i}$, i.e., $D_t^{C_j} = \mathbf{1}_{\left\{N_t^{C_j} > 0\right\}}$, and $D_t^{0,i} = \mathbf{1}_{\left\{N_t^{0,i} > 0\right\}}$. With this representation, we are neglecting the probability of multiple jumps of the common market factors. The Fourier transform of the default distribution X_T can then be computed directly:

$$\mathcal{F}_{X_T}(\alpha) = \varphi^0(\alpha) \prod_{j=1}^{m} \left[\exp\left(-\Lambda_T^{C_j}\right) + \left(1 - \exp\left(-\Lambda_T^{C_j}\right)\right) \varphi^{C_j}(\alpha)\right],$$

where $\varphi^0(\alpha)$ and $\varphi^{C_j}(\alpha)$ are defined as

$$\varphi^0(\alpha) = \prod_{i=1}^{n} \left[\exp\left(-\Lambda_T^{0,i}\right) + \left(1 - \exp\left(-\Lambda_T^{0,i}\right)\right) e^{-i\alpha}\right],$$
$$\varphi^{C_j}(\alpha) = \prod_{i=1}^{n} \left[\left(1 - p^{i,j}\right) + p^{i,j} e^{-i\alpha}\right].$$

7.3 Calibration

In this section, we present a methodology to calibrate this rich correlation structure to "observable" measures of interdependence.

The aim of the calibration process is threefold:

1. define the common market factor events that constitute the backbone of the correlation structure;
2. calibrate the market factors loadings—the p_{ij}'s;
3. calibrate the intensity levels of the specified market factors in the model.

We shall discuss each item in turn.

7.3.1 Background Radiation

The choice of market factors, which drive the likelihood of joint defaults, is the first step in our construction. Clearly, this choice is going to be market specific. For instance, the set of economic conditions, which explain the default correlation "sentiment" for investment grade names, would be different from the ones that affect emerging markets or high-yield credits. Investment grade markets are mostly sector driven, whereas emerging market interdependencies can be explained by region and country factors.

In this calibration procedure, we will focus mainly on investment grade names.

To exhibit the existence of our market factors, we will zoom in on three segments of the market.

- **Intra-sector segment**. The credit spreads for obligors in the same industry have a tendency to move together. This would imply the existence of a sector factor, which is fairly stable and jumps occasionally. The sector factor shocks can then be observed through the joint co-movements of the credit spreads in that sector. Obviously, the sector factor itself cannot be observed but we can observe its effect. If this causality principle holds then the existence of a sector factor would be justified.
- **Inter-sector segment**. If we look at a basket of names in different industries, we can observe historically that they also have a tendency to move together to some extent. The interdependence between names in different sectors is much lower than the one observed intra-sector, nevertheless it cannot be neglected. We will refer to the inter-sector driver as the "Beta" Factor—to use an obvious analogy with equity markets.
- **Super Senior tranches**. If we believe in the Gaussian copula model, then the value of a super senior tranche would be equal to zero. The market, however, has a different opinion. Super AAA tranches are trading at a few basis points; this suggests that the market is pricing the "highly unlikely"

global Armageddon risk, i.e., a scenario where everybody defaults at the same time. Unlike the Gaussian copula pricer, the MO model is capable of capturing this effect. This is what we call the background radiation effect. There exists a global "World" driver trading at a few basis points, and everybody in the (credit) universe has a loading of 1 on that factor. In other words, the probability of the Armageddon event is very low, but if the event actually happens then everybody will default without a doubt. The world driver sits in the background silently and will probably never be "active", but its effect is factored into the market prices. Including it in our construction ensures that the price of super senior tranches will always be floored at the world driver level.

Summary 2 *In summary, for investment grade credits, we will have typically a decomposition of the form:*

$$\lambda_t^i = \lambda_t^{0,i} + \left[p_t^{i,B} \lambda_t^B + \sum_{k=1}^{m_S} p_t^{i,S_k} \lambda_t^{S_k} + \lambda_t^W \right], \tag{7.6}$$

where λ_t^W *is the intensity of the "World" driver;*
λ_t^B *is the intensity of the "Beta" driver, and* $p^{i,B}$ *is the loading on that driver;*
$\lambda_t^{S_k}$ *is the intensity of the "Sector" driver* S_k, *and* p^{i,S_k} *is the loading on that sector—if* $i \in S_k$ *then* $p^{i,S_k} > 0$ *otherwise* $p^{i,S_k} = 0$;
$\lambda_t^{0,i}$ *is the intensity of the idiosyncratic events.*

7.3.2 The Expanding Universe

The next step is to formulate a systematic definition of the loadings on the Beta and Sector drivers. To do so, we want to capture two basic features imposed by the dynamics of our credit universe: a) idiosyncratic moves, b) market-wide moves. If we look at a particular sector and observe the day-to-day spread changes in that sector, then two scenarios could happen.

- **Scenario 1**. Figure 7.1 depicts the situation where one credit spread goes up or down and all the other spreads in the sector are unchanged. In this configuration, we can actually state that this is not a market or sector move. And therefore, the intensity of the market factor will remain constant and the spread change of the outlier is explained either by a change in the loading of that name or a change in the idiosyncratic term or both. In fact, all three explanations are possible. In the extreme case where we keep increasing the

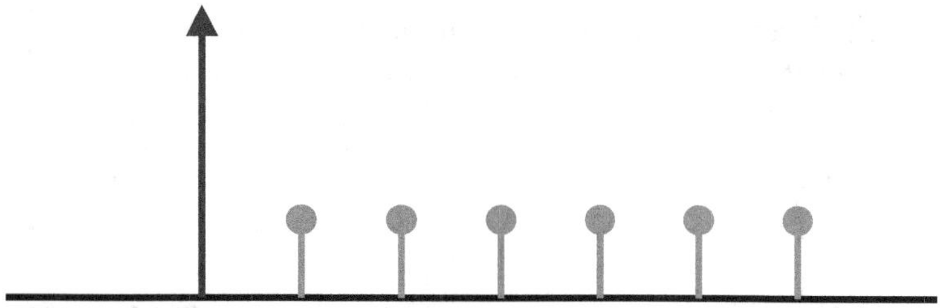

Fig. 7.1 Scenario 1—Idiosyncratic move

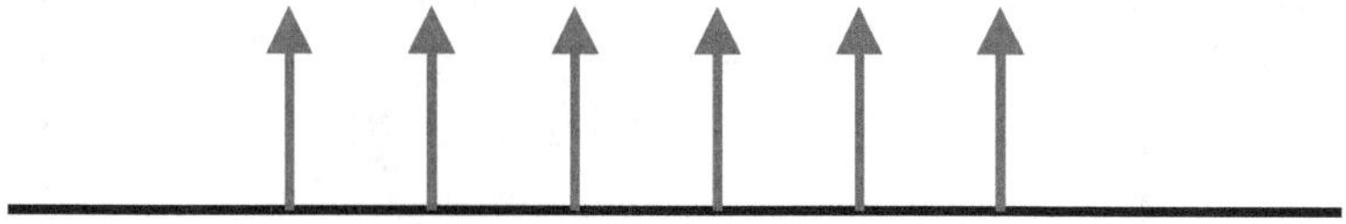

Fig. 7.2 Scenrio 2—Market-wide move

spread of the outlier, we can see that at some point the loading will reach its maximum value of 1, and every incremental increase thereafter will be an idiosyncratic shock. In the mirror situation where we keep decreasing the spread, the idiosyncratic term will reach its minimal value of 0, and all subsequent negative shocks are due to a decrease in the factor loading. For all situations in between, the spread change will be explained by a combination of the changes in the factor loading and the idiosyncratic term.

- **Scenario 2**. Figure 7.2 shows a situation where all the single-name credits move in tandem. If all the obligors move together proportionally, then we can state that this, in fact, is a market factor move. This implies, in turn, that the loadings and the idiosyncratic terms are unchanged.

These two pictures are extreme configurations that may not be realistic, but they do exhibit clearly the qualitative behaviour of our expanding (credit) universe.

The statistical measure, which quantifies the effect of the credit spreads joint co-movements, is: spread correlation. This is going to be our starting point. We will ask the question: what should the factor loadings (the p_{ij}'s) be such that we can reproduce a given spread correlation matrix?

In all the rest, we assume that the spread correlation matrix is specified by two numbers: the correlation intra-sector ρ_{intra} and the correlation inter-sector ρ_{inter}.

To derive the formulas of the loading on the Beta driver $p_{i,B}$, we will consider the spread correlation between two names in different sectors:

$$Cov\left(d\lambda^{i}, d\lambda^{j}\right) = Cov\left(d\lambda^{W}, d\lambda^{W}\right) + p^{i,B} p^{j,B} Cov\left(d\lambda^{B}, d\lambda^{B}\right) + \sum_{k=1}^{m_S} p^{i,S_k} p^{j,S_k} Cov\left(d\lambda^{S_k}, d\lambda^{S_k}\right).$$

We assume that the world driver has no volatility and is a fixed constant at all times: $Cov\left(d\lambda^{W}, d\lambda^{W}\right) = 0$.

Since the chosen single-names are in different sectors, we have, by construction,

$$p^{i,S_k} p^{j,S_k} = 0, \text{ for } 1 \leq k \leq m_S.$$

Hence, the covariance simplifies to

$$Cov\left(d\lambda^{i}, d\lambda^{j}\right) = p^{i,B} p^{j,B} \left(\sigma^{B}\right)^{2} dt.$$

On the other hand, using the definition of the inter-sector spread correlation, we have

$$Cov\left(d\lambda^{i}, d\lambda^{j}\right) = \rho_{inter} \sigma^{i} \sigma^{j} dt.$$

Thus, for all pairs (i, j) in different sectors, the product of the factor loadings is given by the inter-sector correlation and the ratios of the single-name and market factor volatilities

$$p^{i,B} p^{j,B} = \rho_{inter} \left(\frac{\sigma^{i}}{\sigma^{B}}\right) \left(\frac{\sigma^{j}}{\sigma^{B}}\right), \text{ for all } 1 \leq i, j \leq n.$$

This set of equations has a unique solution given by:

$$p^{i,B} = \sqrt{\rho_{inter}} \left(\frac{\sigma^{i}}{\sigma^{B}}\right). \tag{7.7}$$

We can do the same analysis for two names in the same sector S_k. In that case, we obtain

$$Cov\left(d\lambda^{i}, d\lambda^{j}\right) = p^{i,B} p^{j,B} Cov\left(d\lambda^{B}, d\lambda^{B}\right) + p^{i,S_k} p^{j,S_k} Cov\left(d\lambda^{S_k}, d\lambda^{S_k}\right),$$

which implies

$$p^{i,S_k} = \sqrt{\rho_{intra} - \rho_{inter}} \left(\frac{\sigma^{i}}{\sigma^{S_k}}\right). \tag{7.8}$$

We will assume that the credit spread (normal) volatilities are proportional to the spread levels:

$$\begin{aligned}\sigma^{i} &= \alpha\lambda^{i},\\ \sigma^{B} &= \alpha_{B}\lambda^{B},\\ \sigma^{S_k} &= \alpha_{S}\lambda^{S_k},\end{aligned}$$

which would express the final factor loading formulas as a function of the single-name and market factor intensities.

The problem with these $\left[p^{ij}\right]$-formulas is that the market factor loadings are not bounded between 0 and 1. Moreover, as $\lambda^{i} \longrightarrow \infty$, $p^{ij} \longrightarrow \infty$ as well. A simple solution is to make the $\left[p^{ij}\right]$ capped at 1.

Proposition 3 *"Capped" formulation of the market factor loadings* $\left[p_{ij}\right]$*:*

$$p^{i,B} = \min\left(1, \sqrt{\rho_{inter}}\frac{\alpha}{\alpha_{B}}\left(\frac{\lambda^{i}}{\lambda^{B}}\right)\right), \tag{7.9}$$

$$p^{i,S_k} = \min\left(1, \sqrt{\rho_{intra}-\rho_{inter}}\frac{\alpha}{\alpha_{S}}\left(\frac{\lambda^{i}}{\lambda^{S_k}}\right)\right).$$

This simple adjustment, however, introduces discontinuities in the first derivative at $\lambda^{i} \in \left\{\lambda^{B}\frac{\alpha_{B}}{\alpha}\frac{1}{\sqrt{\rho_{inter}}}, \lambda^{S_k}\frac{\alpha_{S}}{\alpha}\frac{1}{\sqrt{\rho_{intra}-\rho_{inter}}}\right\}$. This might lead to jumps in the hedge parameters when we cross this barrier.

One approach to address this issue is to make the market factor loadings $\left[p_{ij}\right]$ well-behaved decaying functions that converge to 1 at infinity:

$$p^{i,B} = 1-\exp\left(-\sqrt{\rho_{inter}}\left(\frac{\sigma^{i}}{\sigma^{B}}\right)\right),$$

$$p^{i,S_k} = 1-\exp\left(-\sqrt{\rho_{intra}-\rho_{inter}}\left(\frac{\sigma^{i}}{\sigma^{S_k}}\right)\right).$$

Replacing the volatility terms by their expressions gives a very neat formulation of the factor loadings $\left[p_{ij}\right]$.

Proposition 4 *"Exponential-Decay" formulation of the market factor loadings* $[p_{ij}]$*:*

$$p^{i,B} = 1 - \exp\left(-\sqrt{\rho_{inter}}\frac{\alpha}{\alpha_B}\left(\frac{\lambda^i}{\lambda^B}\right)\right), \quad (7.10)$$

$$p^{i,S_k} = 1 - \exp\left(-\sqrt{\rho_{intra} - \rho_{inter}}\frac{\alpha}{\alpha_S}\left(\frac{\lambda^i}{\lambda^{S_k}}\right)\right).$$

There are a few observations to make at this stage.

- ρ_{inter} and ρ_{intra} in the previous formulas correspond to the correlation numbers in the "standard" correlation regime. In particular, we have the following asymptotic behaviour:
  - if $\frac{\lambda^i}{\lambda^B} \ll 1$ (respectively $\frac{\lambda^i}{\lambda^{S_k}} \ll 1$), we can expand the exponential and get $p^{i,B} \simeq \sqrt{\rho_{inter}}\frac{\alpha}{\alpha_B}\left(\frac{\lambda^i}{\lambda^B}\right)$ (respectively $p^{i,S_k} \simeq \sqrt{\rho_{intra} - \rho_{inter}}$ $\frac{\alpha}{\alpha_S}\left(\frac{\lambda^i}{\lambda^{S_k}}\right)$);
  - if $\frac{\lambda^i}{\lambda^B} \gg 1$ (respectively $\frac{\lambda^i}{\lambda^{S_k}} \gg 1$), we converge asymptotically to 1, $p^{i,B} \simeq 1$ (respectively $p^{i,S_k} \simeq 1$).
- By construction, the idiosyncratic volatility is positive: $\sigma_{i,0}^2 = \sigma_i^2 - \left(\left(p^{i,B}\right)^2\sigma_B^2 + \left(p^{i,S_k}\right)^2\sigma_{S_k}^2\right)$. Indeed, we have the following two extremes:
  - when $\frac{\sigma_i}{\sigma_B} \gg 1$ (and $\frac{\sigma_i}{\sigma_{S_k}} \gg 1$), we have

$$\sigma_{i,0}^2 = \sigma_i^2 - \left(\sigma_B^2 + \sigma_{S_k}^2\right) \gg 0;$$

  - in the other case where $\frac{\sigma_i}{\sigma_B} \ll 1$ (and $\frac{\sigma_i}{\sigma_{S_k}} \ll 1$), we have

$$\begin{aligned}\sigma_{i,0}^2 &\simeq \sigma_i^2 - \left(\left[\sqrt{\rho_{inter}}\left(\frac{\sigma_i}{\sigma_B}\right)\right]^2\sigma_B^2 + \left[\sqrt{\rho_{intra} - \rho_{inter}}\left(\frac{\sigma_i}{\sigma_{S_k}}\right)\right]^2\sigma_{S_k}^2\right)\\ &\simeq \sigma_i^2 - \left(\rho_{intra}\sigma_i^2\right) > 0.\end{aligned}$$

- There is no guarantee that the idiosyncratic spread is positive: $\lambda^{i,0} = \lambda^i - \lambda^W - \left(p^{i,B}\lambda^B + p^{i,S_k}\lambda_{S_k}\right)$. In particular, in the tight spreads case, where

$\frac{\sigma_i}{\sigma_B} \ll 1$ (and $\frac{\sigma_i}{\sigma_{S_k}} \ll 1$), we have

$$\lambda^{i,0} \simeq \lambda^i - \lambda^W - \left(\left[\sqrt{\rho_{inter}}\frac{\alpha}{\alpha_B}\left(\frac{\lambda^i}{\lambda^B}\right)\lambda^B\right] + \left[\sqrt{\rho_{intra} - \rho_{inter}}\frac{\alpha}{\alpha_S}\left(\frac{\lambda^i}{\lambda^{S_k}}\right)\lambda^{S_k}\right]\right);$$

thus, to ensure that the positivity constraint is satisfied, we must have

$$1 - \left(\frac{\lambda^W}{\lambda^i}\right) \geq \sqrt{\rho_{inter}}\frac{\alpha}{\alpha_B} + \sqrt{\rho_{intra} - \rho_{inter}}\frac{\alpha}{\alpha_S}.$$

- The relative change in the factor loadings $\left[p^{ij}\right]$ is proportional to the difference between the credit spread and market factor relative changes

$$\frac{dp^{i,B}}{p^{i,B}} \propto \left(\frac{d\lambda^i}{\lambda^i} - \frac{d\lambda^B}{\lambda^B}\right),$$
$$\frac{dp^{i,S_k}}{p^{i,S_k}} \propto \left(\frac{d\lambda^i}{\lambda^i} - \frac{d\lambda^{S_k}}{\lambda^{S_k}}\right).$$

This formulation of the factor loadings gives the right qualitative behaviour that we expect. However, because of the highly decaying nature of the exponential we could get some undesirable features in the price and risk asymptotics. The speed of decay also implies that the actual spread correlation calculated by the model can be substantially lower than the two parameters that we have used as input in the $\left[p_{ij}\right]$ formula.

To resolve these limitations of the decaying formulation, we can use a smoothing method known as the "Corner-Layer expansion".[3]

Consider the following function of two variables $u_0(x,t)$:

$$u_0(x,t) = \begin{cases} -1, & \text{if } x < -t, \\ \frac{x}{t}, & \text{if } -t < x < t, \\ 1, & \text{if } x > t. \end{cases} \tag{7.11}$$

The discontinuities at t and $-t$ can be smoothed out via the corner-layer expansion as follows:

$$u(x,t;\epsilon) = 1 - 2\sqrt{\frac{\epsilon}{\pi t}}\frac{\exp\left(-\frac{x_c^2}{4t}\right)}{1 - \mathrm{erf}\left(-\frac{x_c}{2\sqrt{t}}\right)}, \tag{7.12}$$

where $x_c = \frac{x-t}{\sqrt{\epsilon}}$, and erf(.) is the error function

$$\operatorname{erf}(z) = \frac{2}{\sqrt{\pi}} \int_0^z e^{-s^2} ds.$$

ϵ is the expansion factor. The smoothing is done in the neighborhood of the discontinuity on an interval of the order of $\sqrt{\epsilon}$.

Proposition 5 *"Corner-Layer" formulation of the market factor loadings* $[p_{ij}]$*:*

$$p^{i,B} = u\left(\sqrt{\rho_{inter}}\frac{\alpha}{\alpha_B}\lambda^i, \lambda^B; \epsilon\right), \tag{7.13}$$

$$p^{i,S_k} = u\left(\sqrt{\rho_{intra} - \rho_{inter}}\frac{\alpha}{\alpha_S}\lambda^i, \lambda^{S_k}; \epsilon\right).$$

Note that the capped $[p_{ij}]$-formulation is similar to the parameterisation proposed in Lindskog and McNeil (2003), where the Beta component and the Sector component are assumed to be fixed percentages of the spread. To see this, it suffices to decompose the issuer spread into its Beta, Sector and idiosyncratic parts:

$$\lambda^i = \lambda^W + \lambda^{i,B} + \lambda^{i,S_k} + \lambda^{0,i}, \tag{7.14}$$

and to replace $p^{i,B}$ and the p^{i,S_k} by their expressions

$$\lambda^{i,B} = p^{i,B}\lambda^B = \sqrt{\rho_{inter}}\frac{\alpha}{\alpha_B}\left(\frac{\lambda^i}{\lambda^B}\right)\lambda^B = \left(\sqrt{\rho_{inter}}\frac{\alpha}{\alpha_B}\right)\lambda^i,$$

$$\lambda^{i,S_k} = p^{i,S_k}\lambda^{S_k} = \left(\sqrt{\rho_{intra} - \rho_{inter}}\frac{\alpha}{\alpha_S}\right)\lambda^i.$$

7.3.3 The Big Bang State

The restrictions imposed by spread correlations imply a natural formulation of the market factor loadings, which is consistent with the credit spread dynamics. However, this piece of information does not provide any insights into the "Big-Bang" state i.e. the actual levels of the market factors that we start at and subsequently keep drifting from. The market factor levels can be recovered from a second "observable" measure of interdependence such as benchmark first-to-default prices.

In order to calibrate the market factor levels, we set the intensities of the Beta driver and the Sector drivers to match "mid-market" prices of benchmark first-to-default instruments. This will be conducted in two steps.

Step 1: calibrate the Beta driver on a "diversified" benchmark basket FTD.
Step 2: calibrate the Sector drivers on "undiversified" sector specific benchmark basket FTDs.

A "diversified" basket in this context means a basket of multiple obligors in different sectors. An "undiversified" basket refers to a basket of single-name credits in the same sector.

This two-step procedure is made possible by the choice of $\left[p^{ij}\right]$ parametrisation explained in the previous section. Recall that the intensity of a first-to-default for two names is given by

$$\begin{aligned}\lambda^{ftd} &= \lambda^1 + \lambda^2 - \lambda^{12} \\ &= \lambda^1 + \lambda^2 - \left[\sum_{j=1}^{m} \left(p^{1,j} p^{2,j}\right) \lambda^{C_j}\right].\end{aligned}$$

For a diversified two-name-FTD, it becomes

$$\lambda^{div} = \lambda^1 + \lambda^2 - \left[\lambda^W + \left(p^{1,B} p^{2,B}\right) \lambda^B\right].$$

And since the factor loading $p_{i,B}$ is a function of the issuer intensity λ^i and the market factor λ^B, the value of the diversified basket FTD is then given as a function of the Beta intensity (and the single-name intensities)

$$s^{div} \simeq (1 - recovery)\, \lambda^{div} = f\left(\lambda^1, \lambda^2, \lambda^B\right);$$

this relationship can be inverted directly to obtain the Beta intensity which matches the diversified FTD price.

Similarly, for a basket of two obligors in the same sector S_k, we have

$$\lambda^{undiv} = \lambda^1 + \lambda^2 - \left[\lambda^W + \left(p^{1,B} p^{2,B}\right) \lambda^B + \left(p^{1,S_k} p^{2,S_k}\right) \lambda^{S_k}\right].$$

The sector loading p_{i,S_k} is a function of the issuer intensity and the sector intensity. Hence, the price of an FTD on a single-sector basket will depend on the Beta intensity and the Sector intensity

$$s^{undiv} \simeq (1 - recovery)\, \lambda^{undiv} = g\left(\lambda^1, \lambda^2, \lambda^B, \lambda^{S_k}\right).$$

Since, we have already determined the value of the Beta driver in the first step, the intensity of the sector is the only unknown left to solve for. And we repeat this operation for each sector separately.

Note that the calibration method presented in this section is similar, in spirit, to the Heath-Jarrow-Morton model calibration for yield curve dynamics

$$df(t,T) = \mu_i(t,T)\, dt + \sum_{i=1}^{n} \Sigma_i(t,T)\, dW_t^i,$$

which is also done in three steps:

1. define the number of drivers, which explain the dynamics of the yield curve, e.g., $n = 3$;
2. specify a parametrisation of the volatility curve for each driver $F_i(T - t)$, which is consistent with the results of a historical Principal Component Analysis (PCA):

$$\Sigma_i(t,T) = \sigma_i(t)\, F_i(T - t);$$

3. calibrate the instantaneous volatility levels $\sigma_i(t)$ on a set of benchmark fixed income instruments such as caps or swaptions.

Another possibility for calibrating the market factor drivers is to use empirical default correlations. For example, we can use the average inter-sector default correlation to fit the Beta driver, then for each sector driver use the average intra-sector default correlations. The default correlation in the Marshall-Olkin model is given by

$$\rho_{ij}^{I} = \frac{\exp\left(-\int_0^T \left(\lambda_t^i + \lambda_t^j - \lambda_t^{ij}\right) dt\right) - Q_{0,T}^i Q_{0,T}^j}{\sqrt{Q_{0,T}^i\left(1 - Q_{0,T}^i\right) Q_{0,T}^j\left(1 - Q_{0,T}^j\right)}},$$

where $Q^i_{0,T} = \exp\left(-\int_0^T \lambda^i_t dt\right)$ is the single-name survival probability for time horizon T.

For two names in different sectors, the pairwise default correlation will only depend on the Beta driver; and for names in the same sector, it will depend on the Beta and the Sector drivers. Thus, we can fit the drivers in two steps as explained earlier.

7.3.4 Correlation Regimes

Correlation regimes refer to the intuitive idea that correlations are not constant over time. Correlations are, generally, fairly stable but they do depend on the particular market configuration that we are in at that point in time. The market configuration defines the actual "regime" of correlations that we are exposed to. As before, this can be explained with a few examples.

In a standard correlation regime, typical names in the same sector would be trading at comparable levels and then moving together in roughly the same proportion, thereby exhibiting highly correlated dynamics. In this case, as the Beta or Sector drivers increase, the single-name intensities change in tandem and the idiosyncratic component remains unchanged.

In the opposite situation where all the spreads in a sector are stable and one of the names keeps drifting away from the rest, we can state that this name is becoming more and more "de-correlated" from the rest and is behaving in a very idiosyncratic manner. In this case, the market factor levels are unchanged but this single-name's factor loadings and idiosyncratic spreads change. It is another way of saying that this specific obligor's behaviour does not reflect the general dynamics of that sector anymore.

Our market factor loading formulation describes this phenomenon very naturally. Indeed, for a fixed value of the market factor level, we have the following asymptotics of the credit spread dynamics: as the single-name intensity goes to infinity, $\lambda^i \to \infty$, the market factor loadings start increasing until they get maxed at 1, $p^{i,B} \to 1$, $p^{i,S_k} \to 1$, and the idiosyncratic intensity goes to infinity, $\lambda^{0,i} \to \infty$; this, in turn, implies that the pairwise default correlation goes to zero, $\rho^I_{ij} \to 0$.

7.4 Marshall-Olkin Vs Gaussian Copula

In this section, we compare the Marshall-Olkin copula with the well-known elliptical copulas: the Gaussian and the t-copula.

The standard Gaussian copula function is the most commonly used copula in the credit literature. With Gaussian copula dependence, the multivariate distribution of default times takes the following form

$$F(t_1, \ldots, t_n) = \Phi_n\left(\Phi^{-1}(F_1(t_1)), \ldots, \Phi^{-1}(F_n(t_n))\right),$$

where Φ_n denotes the n-dimensional normal cumulative distribution function, Φ^{-1} is the inverse normal distribution and F_i is the marginal exponential distribution of the default time τ_i. The Gaussian copula is the copula function of a set of correlated Gaussian variables transposed back into "uniform" space with the inverse normal.

This specification of the correlation happens to be the same as the one implied in a (structural) Firm Value model. And as such, the correlation matrix Σ used in the n-dimensional normal distribution has the interpretation of an asset correlation matrix.

Similarly, the Student-t copula is an extension of the Gaussian copula where the normal distributions are replaced with t-distributions. The multivariate default times' distribution, in this case, is given by

$$F(t_1, \ldots, t_n) = t_\nu^n\left(t_\nu^{-1}(F_1(t_1)), \ldots, t_\nu^{-1}(F_n(t_n))\right),$$

where t_ν^n is the n-dimensional t-distribution with parameter ν, and t_ν^{-1} is the inverse of the univariate t-distribution. The t-copula is the copula function of a set of correlated t-variables transposed back into uniform space with the inverse student function.

Some of the main differences between MO and the Gaussian and Student-t copulas can be summarised as follows.

- Elliptical copulas do not allow for multiple instantaneous defaults, i.e., $\mathbb{P}(\tau_i = \tau_j) = 0$, for $i \neq j$. The probability of instantaneous joint defaults can be non-zero for MO.
- Elliptical copulas are absolutely continuous; MO has a continuous part and a singular part (on the diagonal).
- MO can be implemented in closed form. The Gaussian and t-copula, in general, need a Monte-Carlo simulation except in the one-factor version where semi-analytic results are available.

- Tail dependence: the upper tail dependence[4] is equal to zero for the Gaussian copula; whereas MO and the t-copula have a non-zero tail dependence:

$$\lambda_U^{Gaussian} = 0, \lambda_U^{T-copula} = 2\left(1 - t_{\nu+1}\left(\sqrt{\nu+1}\frac{\sqrt{1-\rho}}{\sqrt{1+\rho}}\right)\right),$$

$$\lambda_U^{M.O.} = \min\left(\frac{\lambda^{12}}{\lambda^1}, \frac{\lambda^{12}}{\lambda^2}\right) \geq \lambda^W \min\left(\frac{1}{\lambda^1}, \frac{1}{\lambda^2}\right).$$

7.4.1 Multimodal Default Distribution

To study the shape of the loss distribution in the MO model, we construct a portfolio of 100 single-names, split across 10 sectors, with a 10% concentration in each sector. All the single-name intensities are set to $\lambda^i = 100$ bps. The intensities of the World driver, Beta driver and Sector drivers are set to $\lambda^W = 5$ bps, $\lambda^B = 500$ bps, $\lambda^{S_k} = 250$ bps respectively.

We use a Marshall-Olkin decomposition where 60% of the spread is due to Beta, 20% is from the sector contribution and the remaining 20% corresponds to the pure idiosyncratic component. This implies the following market factor loadings $p^{i,B} = 0.24$ and $p^{i,S_k} = 0.16$. The implied 5-year default correlation in this model is 19.25% intra-sector and 16.16% inter-sector. The numerical values in this example are chosen to exhibit the shape of the distributions and compare the various copulas; for empirical studies of default correlation, we refer the reader to the papers by Nagpal and Bahar (2001) and Servigny and Renault (2002).

To start with, we look at the default distribution at the 5-year time horizon for this correlated basket (with $p^{i,B} = 0.24$ and $p^{i,S_k} = 0.16$) and we compare it with a similar portfolio where we have the same marginals but with no Beta or Sector contributions: $p^{i,B} = 0$ and $p^{i,S_k} = 0$.

Figure 7.3 depicts typical shapes of the multi-modal Marshall-Olkin default distribution. In the high correlation case, the default distribution has four modes mirroring the MO decomposition on the Beta, Sector, Idiosyncratic and World drivers. The first big hump corresponds to the idiosyncratic term, the second hump corresponds to the Beta contribution, the third hump is from the sector contribution and the last spike at the very end is due to the world driver. In the low correlation case where the Beta and Sector contributions have been turned off, we only have two modes: the idiosyncratic mode and the World driver mode at the end. The idiosyncratic mode is further away to the right compared to the high correlation distribution since the idiosyncratic default probabilities are higher in this case.

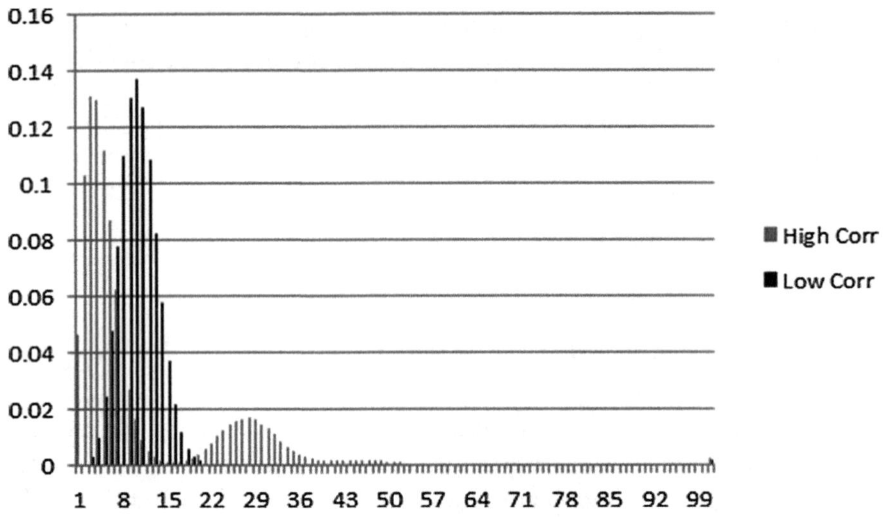

Fig. 7.3 Marshall-Olkin default distributions with high and low correlations

To compare MO with the elliptical copulas, we use the 5-year default correlation matrix implied by the Marshall-Olkin model as input for calibrating the Gaussian copula and the t-copula. Calibrating the asset correlation for the Gaussian copula, we get $\rho_{asset} = 41.68\%$ intra-sector and $\rho_{asset} = 36.39\%$ inter-sector. Doing the same calibration for a t-copula with $\nu = 9$, we get $\rho_{asset} = 35.92\%$ intra-sector and $\rho_{asset} = 30.12\%$ inter-sector.

Figure 7.4 shows the default distributions for the three calibrated copulas.

The main take-away point is that the M.O. default distribution admits multiple humps, whereas the Gaussian copula and the t-copula produce unimodal default distributions. To explain this fundamental difference, we analyze the following simplified example:

1. we consider a homogeneous portfolio;
2. we assume that we have one common market factor in the M.O. copula

$$dD_t^i = \left(1 - D_{t-}^i\right) \left[A^{i,B} dN_t^B + dN_t^{0,i}\right];$$

3. we assume that the asset correlation matrix in the Gaussian copula is built from one common factor

$$Z_i = \beta Y + \sqrt{1 - \beta^2} \epsilon_i, \, \rho_{ij}^{asset} = Cov\left(Z_i, Z_j\right),$$

where Y and $\epsilon_1, \ldots, \epsilon_n$ are independent standard Gaussian variables.

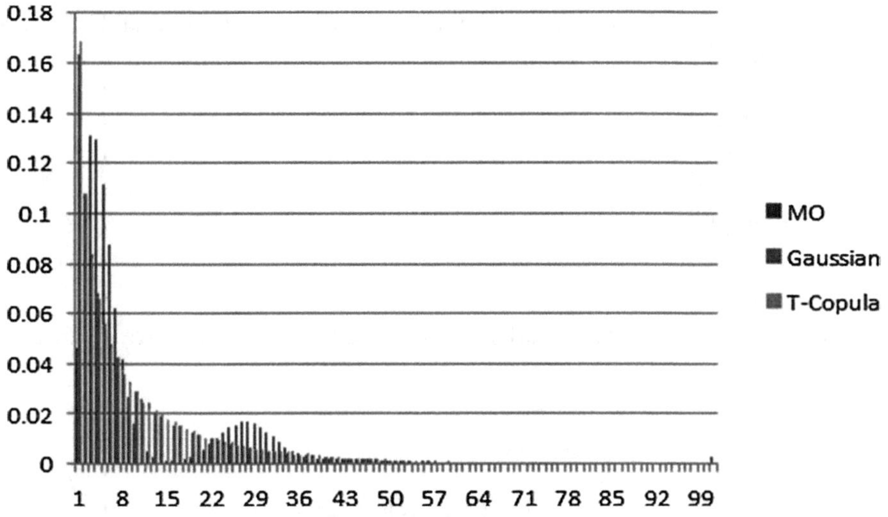

Fig. 7.4 Comparison of the default distributions for the MO, Gaussian and T-Copula

MO Loss Distribution. To derive the M.O. default distribution in closed form, we borrow the following approximation from Duffie and Pan (2001)

$$D_t^i \approx A^{i,B} D_t^B + D_t^{0,i},$$

where the SDE has been linearalised and the double-counting of common factor defaults and idiosyncratic defaults ignored.

To ensure that we do not have any double counting, we will use an improved version of the approximation: in order to remove the non-linearity, we replace the term $\left(1 - D_t^i\right)$ by $\left(1 - D_t^{0,i}\right)$ in the original SDE, then we integrate

$$D_t^i \approx A^{i,B} \left(1 - D_t^{0,i}\right) D_t^B + D_t^{0,i}. \tag{7.15}$$

The default probabilities of the common factor and the idiosyncratic events, at time horizon T, are given by

$$p_T^{0,i} = 1 - \mathbb{E}\left[\exp\left(-\int_0^T \lambda_t^{0,i} dt\right)\right],$$
$$p_T^B = 1 - \mathbb{E}\left[\exp\left(-\int_0^T \lambda_t^B dt\right)\right].$$

Thus, the Fourier transform of the defaults-counting random variable X_T can be approximated as

$$\begin{aligned}\psi^{X_T}(u) &= \mathbb{E}\left[\exp\left(-iuX_T\right)\right] \\ &\approx \mathbb{E}\left[\exp\left(-iuX_T\right)\middle|D_T^B=1\right]p_T^B+\mathbb{E}\left[\exp\left(-iuX_T\right)\middle|D_T^B=0\right]\left(1-p_T^B\right) \\ &= w_{S_1}\left[\left(1-p_{S_1}+p_{S_1}\exp\left(-iu\right)\right)^n\right] \\ &\quad+w_{S_2}\left[\left(1-p_{S_2}+p_{S_2}\exp\left(-iu\right)\right)^n\right], \end{aligned} \tag{7.16}$$

where the weights, w_{S_1}, w_{S_2}, and state probabilities, p_{S_1}, p_{S_2}, are given by

$$\begin{aligned} w_{S_1} &= p_T^B,\ w_{S_2}=1-p_T^B, \\ p_{S_1} &= p_T^{0,i}, \\ p_{S_2} &= p^{i,B}\left(1-p_T^{0,i}\right)+p_T^{0,i}. \end{aligned}$$

Equation (7.16) is the Fourier transform of a Bernoulli mixture. Thus, the distribution of defaults, at a time horizon T, is a weighted average of Bernoulli distributions

$$\mathbb{P}\left(X_T\le k\right)=w_{S_1}B\left(k,n,p_{s_1}\right)+w_{S_2}B\left(k,n,p_{s_2}\right), \tag{7.17}$$

where $B\left(k,n,p\right)=\sum_{l=0}^{k}\binom{n}{l}p^l\left(1-p\right)^{n-l}$.

The first mode of the distribution corresponds to the pure idiosyncratic events. The second mode is due to the common market factor. More generally, the number of different modes in the loss distribution corresponds to the number of Poisson market factor shocks in the MO decomposition.

Gaussian Loss Distribution. On the other hand, the distribution of defaults for the Gaussian copula is obtained by conditioning on the possible outcomes of the common factor Y

$$\mathbb{P}\left(X_T\le k\right)=\int_{-\infty}^{\infty}\mathbb{P}\left(X\le k\,|Y=y\right)\phi(y)dy.$$

Conditional on a realisation of the common factor, all the random variables D_T^i are independent. Hence

$$\mathbb{P}\left(X_T\le k\,|Y=y\right)=B\left(k,n,p_T(y)\right),$$

where the conditional single-name default probability $p_T(y)$ is defined as

$$p_T(y) = \mathbb{P}\left(D_T^i = 1 \,|Y = y\right) = \mathbb{P}\left(Z_i \le \Phi^{-1}\left(p_T^i\right) |Y = y\right)$$
$$= \mathbb{P}\left(\epsilon_i \le \frac{\Phi^{-1}\left(p_T^i\right) - \beta Y}{\sqrt{1-\beta^2}} \,|Y = y\right) = \Phi\left(\frac{\Phi^{-1}\left(p_T^i\right) - \beta y}{\sqrt{1-\beta^2}}\right).$$

Thus, the default distribution is an integral of Bernoulli distributions.

$$\mathbb{P}(X_T \le k) = \int_{-\infty}^{\infty} B(k, n, p_T(y))\,\phi(y)\,dy. \tag{7.18}$$

This continuum of Bernoulli mixtures degenerates to a single mode distribution. This is also the case for most absolutely continuous copula functions. MO is one of the very few copulas with a multi-modal distribution.

7.4.2 Correlation Term Structure

Another major difference between the M.O. and the Gaussian copulas is their "time-dependence" behaviour. We consider an example with one Beta market factor

$$\lambda^i = p^{i,B}\lambda^B + \lambda^{0,i}.$$

We set the single-name and the market factor intensities to $\lambda^i = 100$ bps, $\lambda^B = 100$ bps. And we set the factor loadings to $p^{i,B} = 0.3915$ so that the pairwise default correlation, at the 5-year time horizon, is equal to 15%. Similarly, we set the Gaussian asset correlation to $\rho^{asset} = 41.04\%$ in order to match the 15% default correlation. Now, having calibrated both copulas to the same 5-year default correlation, we compare the term structure of default correlations implied by the two models.

As expected, Fig. 7.5 shows that, for the Marshall-Olkin model, the implied default correlation is very stable, at around 15%, as we roll forward on the curve. The Gaussian copula, on the other hand, is highly time-dependent: its implied default correlation term structure decays rapidly as the duration of the trade shortens. At time $t = 0$, a 5-year first-to-default basket would be priced at 15% default correlation. In one year's time, the FTD becomes a 4-year trade and gets marked at 13.8% default correlation. And at time $t = 4$ year, with one year left to maturity, we will be marking the same basket at 8% default correlation. Since the implied default correlation in the Gaussian copula decays

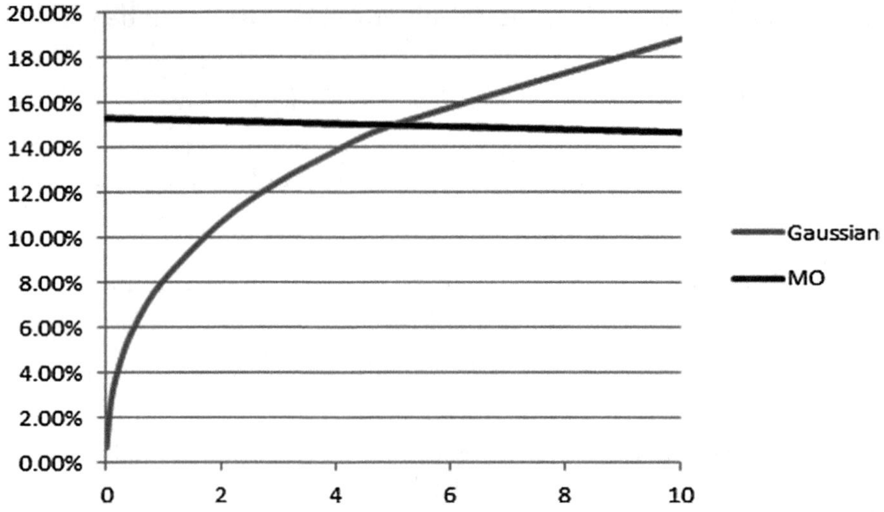

Fig. 7.5 Default correlation term structure for the Gaussian and MO Copulas

rapidly with time, the price of an FTD basket will become cheaper as time goes by even if the underlying credit spreads are unchanged.

7.4.3 Correlation Skew

In this sub-section, we illustrate schematically how Marshall-Olkin can be used to reproduce the observed market correlation skew. Using the multi-modality offered by the MO loss distribution, it is possible to use each mode to match various parts of the capital structure. Typically, the World driver is used to calibrate the super senior tranche, the idiosyncratic terms are used to calibrate the equity piece, and the Beta driver is used to match the Mezzanine tranches. Additional refinement of the calibration can be done with the sector drivers. Figures 7.6 and 7.7 show typical calibration results, for the two examples below, where the calibrated MO skew is compared with the quoted market bid/offer correlation skews.

Index level 31		
0–3%	16.8	18.0
3–6%	86	92
6–9%	30.5	36.5
9–12%	22.5	27.5
12–22%	11	12.3

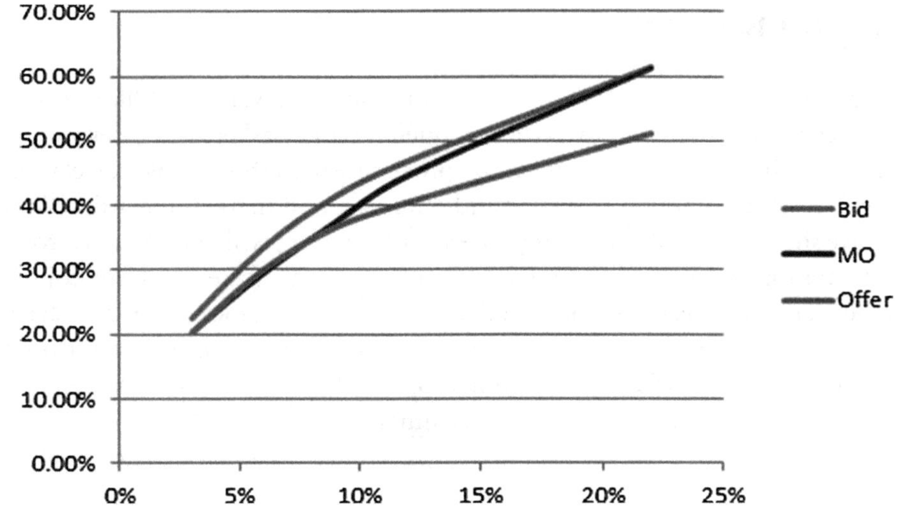

Fig. 7.6 Calibrated MO base correlation skew—index level = 31 bps

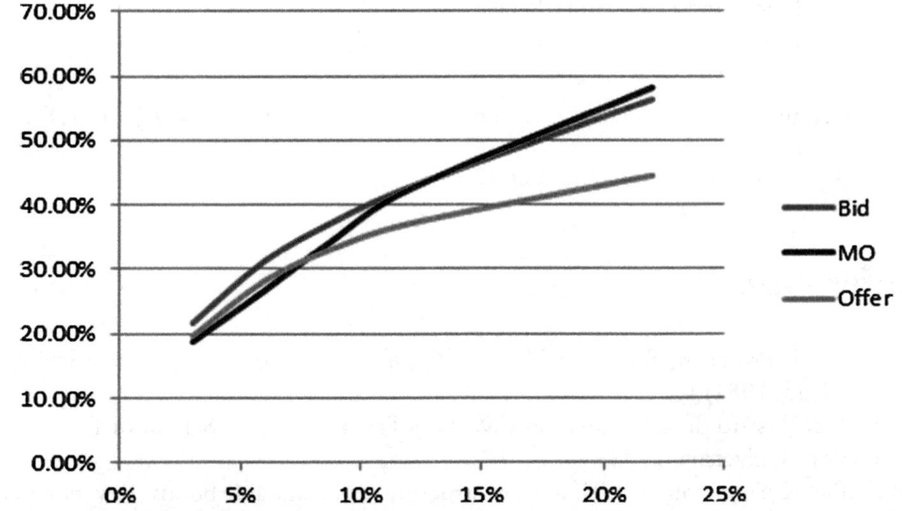

Fig. 7.7 Calibrated MO base correlation skew—index level = 44 bps

Index level 44		
0–3%	28.5	30.0
3–6%	176	186
6–9%	74	80
9–12%	43	50
12–22%	19	24

7.5 Conclusion

We have presented in this chapter an introduction to the Marshall-Olkin copula in the context of default correlation modelling and basket credit derivatives pricing. We have discussed a calibration procedure to fit this flexible correlation structure, which is both pragmatic and reflects sound market dynamics. And we have shown that MO has many interesting features, which make it a strong candidate for replacing the Gaussian copula. Indeed, if the Gaussian copula can be seen as the Black-Scholes model for default correlation, then Marshall-Olkin would be akin to a Heath-Jarrow-Morton framework: it is built on a pre-defined set of market factors, which once calibrated can be used to price consistently all correlation sensitive instruments.

Notes

1. See Barlow and Proschan (1981).
2. We refer the reader to Nelsen (1999) for details.
3. See Kevorkian and Cole (1996) for example.
4. The upper tail dependence is defined as $\lambda_U = \lim_{u \nearrow 1} \mathbb{P}\left(X > F_X^{-1}(u) \,|\, Y > F_Y^{-1}(u)\right)$. See Embrechts et al. (2003).

References

R. Barlow, F. Proschan, *Statistical Theory of Reliability and Life Testing* (Silver Spring, Maryland, 1981)

D. Duffie, First-to-default valuation (Working Paper, Graduate School of Business, Stanford University, 1998)

D. Duffie, K. Singleton, Modeling term structures of defaultable bonds. Rev. Financ. Stud. **12**(4), 687–720 (1999a)

D. Duffie, K. Singleton, Simulating correlated defaults (Working Paper, Graduate School of Business, Stanford University, 1999b)

D. Duffie, J. Pan, Analytical Value-At-Risk with jumps and credit risk. Financ. Stoch. **5**, 155–180 (2001)

P. Embrechts, F. Lindskog, A. McNeil, Modelling dependence with copulas and applications to risk management, in *Handbook of Heavy Tailed Distributions in Finance*, ed. by S.T. Rachev (Elsevier, Amsterdam, North-Holland, 2003)

R. Frey, A. McNeil, Dependent defaults in models of portfolio credit risk. J. Risk **6**(1), 59–92 (2003)

D. Gatarek, J. Jablecki, Modeling joint defaults in correlation-sensitive instruments. J. Credit Risk **12**(3), 2016 (2016)

H. Joe, *Multivariate Models and Dependence Concepts* (Chapman & Hall, London, 1997)

J. Kevorkian, J.D. Cole, *Multiple Scale and Singular Perturbation Methods* (Springer, New York, 1996)

F. Lindskog, A. McNeil, Common Poisson Shock models: applications to insurance and credit risk modelling. ASTIN Bullet. **33**(2), 209–238 (2003)

A.W. Marshall, I. Olkin, A multivariate exponential distribution. J. Am. Stat. Assoc. (1967)

M. Morini, *Understanding and Managing Model Risk: A Practical Guide for Quants, Traders and Validators* (Wiley, Chischester, 2011)

K. Nagpal, R. Bahar, measuring default correlation. Risk (March 2001), 129–132

R. Nelsen, *An Introduction to Copulas* (Springer, New York, 1999)

H. Panjer, Recursive evaluation of a family of compound distributions. ASTIN Bullet. **12**, 22–26 (1981)

P.J. Schönbucher, Factor models for portfolio credit risk. J. Risk Financ. **3**(1), 45–56 (2001)

A. Servigny, O. Renault, Default correlation: empirical evidence (Working Paper, Standard and Poors, 2002)

8

Numerical Tools: Basket Expansions

In the next few chapters, we provide some efficient numerical methods for the valuation of large basket credit derivatives. While the approaches are presented in the Marshall-Olkin copula model, most of the numerical techniques are generic and could be used with other copulas as well. The methods presented span a large spectrum of applied mathematics: Fourier transforms, changes of probability measure, numerical stable schemes, high-dimensional Sobol integration, recursive convolution algorithms.

8.1 Introduction

The valuation of large basket products was usually done with Monte-Carlo simulations. The advantage of Monte-Carlo is its simplicity and generality. Its main drawback, however, is the quality of the convergence, especially when one computes sensitivities, such as deltas and gammas. A good convergence is particularly hard to achieve for credit products since default events are usually rare, and probabilities in the tail of the distribution are difficult to estimate. On the other hand, the direct implementation in closed form is very accurate but is less trivial to implement; it is also very expensive computationally. Indeed, this method is based on enumerating the 2^n default configurations of the basket and computing the probability of each configuration. This algorithm explodes exponentially as the number of credits increases. To address this problem, we use a collection of techniques from numerical analysis and actuarial mathematics, and we develop a suite of semi-analytic numerical methods based on the asymptotic behaviour of the portfolio. Although the implementation is done in

Y. Elouerkhaoui, *Credit Correlation*, Applied Quantitative Finance,
DOI 10.1007/978-3-319-60973-7_8

the Marshall-Olkin framework, most of the numerical techniques are generic and could be used with other copula models. Other analytical approximations of the Marshall-Olkin model have been developed by Duffie and Pan (2001), and by Lindskog and McNeil (2003).

8.2 Set-Up and Notations

The Marshall-Olkin copula is based on the Multivariate Poisson Process construction. We give here a brief overview of the model (see a detailed description in Chap. 7).

We consider a set of non-negative random variables $(\tau_1, \ldots, \tau_n)$, defined on a probability space $(\Omega, \mathcal{G}, P)$, representing the default times of n obligors. For each firm i, we denote by $D_t^i \triangleq \mathbf{1}_{\{\tau_i \leq t\}}$ the default indicator process.

We assume that there exists a set of $(m+n)$ independent Poisson processes $\left(N_t^{c_j}\right)_{t\geq 0}$ with intensities $\lambda^{c_j}(t)$, which can trigger joint defaults.

For every event type c_j, and for all $t \geq 0$, we define a set of independent Bernoulli variables $\left(A_t^{1,j}, \ldots, A_t^{n,j}\right)$ with probabilities $\left(p_{1,j}, \ldots, p_{n,j}\right)$, $p_{i,j} \in [0, 1]$.

We assume that for $j \neq k$, the vectors $\mathbf{A}_t^j = \left(A_t^{1,j}, \ldots, A_t^{n,j}\right)$ and $\mathbf{A}_t^k = \left(A_t^{1,k}, \ldots, A_t^{n,k}\right)$ are independent.

We assume that for $t \neq s$, the vectors $\mathbf{A}_t^j = \left(A_t^{1,j}, \ldots, A_t^{n,j}\right)$ and $\mathbf{A}_s^j = \left(A_s^{1,j}, \ldots, A_s^{n,j}\right)$ are independent.

At the rth occurrence of the common market event of type j, at time $\theta_r^{c_j}$, we draw the set of n independent Bernoulli variables $\left(A_{\theta_r^{c_j}}^{1,j}, \ldots, A_{\theta_r^{c_j}}^{n,j}\right)$. The variable $A_{\theta_r^{c_j}}^{i,j}$ indicates if obligor i has defaulted or not.

The process $\left(N_t^i\right)_{t\geq 0}$, defined as

$$N_t^i \triangleq \sum_{j=1}^{m+n} \sum_{\theta_r^{c_j} \leq t} A_{\theta_r^{c_j}}^{i,j}, \tag{8.1}$$

is also a Poisson process obtained by superpositionning a set of independent (thinned) Poisson processes. Its intensity is given by

$$\lambda_i(t) = \sum_{j=1}^{m+n} p_{i,j} \lambda^{c_j}(t). \tag{8.2}$$

We define the default time τ_i as the first jump time of the Poisson process N_t^i:

$$\tau_i \triangleq \min \left\{ t : N_t^i > 0 \right\}. \tag{8.3}$$

We denote by $Q_i(T)$ the survival probability of τ_i :

$$Q_i(T) \triangleq \mathbb{P}(\tau_i > T) = \mathbb{P}\left(N_T^i = 0\right) = \exp\left(-\int_0^T \lambda_i(s)\, ds\right), \tag{8.4}$$

which will be referred to as the Q-factor associated with τ_i.

To specify the model further, we assume that we have two types of market factors: (1) the first m drivers are common market factors which affect more than one obligor; they can represent global economic factors, industry sector factors, or regional and country factors; (2) the n remaining ones are idiosyncratic issuer-specific shocks; they will be denoted as: $N_t^{0,i} \triangleq N_t^{c_{m+i}}$ (and $\lambda^{0,i}(t) \triangleq \lambda^{c_{m+i}}(t)$), for $1 \leq i \leq n$. Their corresponding factor loadings are: $p_{i,m+i} = 1$ and $p_{i,m+k} = 0$, for $1 \leq k \neq i \leq n$.

Equation (8.2) then becomes

$$\lambda_i(t) = \sum_{j=1}^{m} p_{i,j} \lambda^{c_j}(t) + \lambda^{0,i}(t). \tag{8.5}$$

The copula function of the default times $(\tau_1, ..., \tau_n)$, implied by this multivariate Poisson process, is known as the Marshall-Olkin copula (see Proposition 6 in Lindskog and McNeil (2003)).

Furthermore, we associate with the set $(\tau_1, ..., \tau_n)$ the ordered sequence of random times $\tau^{[1]} \leq \tau^{[2]} \leq ... \leq \tau^{[n]}$, defined as $\tau^{[1]} = \min(\tau_1, ..., \tau_n)$, and for $k = 2, ..., n$,

$$\tau^{[k]} = \min\left(\tau_i : i = 1, ..., n,\ \tau_i > \tau^{[k-1]}\right). \tag{8.6}$$

Now, we consider a kth-to-default swap (also called a k out of n default swap in Laurent and Gregory (2005)), which matures at time T, and is specified by the following contractual obligations:

- if k default events occur before maturity, $\tau^{[k]} \leq T$, and credit i is the one that has last defaulted, $\tau^{[k]} = \tau_i$, then the protection seller makes a payment equal to $(1 - \delta_i)$, where δ_i is the recovery value of issuer i;
- in return, the buyer makes a series of periodic premium payments $(C_1, ..., C_N)$,on the cash-flow dates $(T_1, ..., T_N)$, as long as the total number of defaults is less than k. Each cash flow C_i is equal to the product of the premium and the day-count fraction $(T_i - T_{i-1})$.

The value of the premium leg is given by the expectation of the discounted payoff

$$\text{premium\_leg} = \sum_{i=1}^{N} p_{0,T_i} C_i \mathbb{P}\left(\tau^{[k]} > T_i\right),$$

where $p_{0,T}$ is the risk-free zero-coupon bond maturing at time T. For simplicity, interest rates and credit processes are assumed to be independent.

The value of the protection leg is given by the following recovery integral:

$$\text{protection\_leg} = \sum_{i=1}^{n} (1 - \delta_i) \left[\int_0^T p_{0,t} \mathbb{P}\left(\tau^{[k]} = \tau_i, \tau^{[k]} \in dt\right) dt\right]. \tag{8.7}$$

$Q^{[k]}(T)$, the kth-to-default Q-factor, is the survival probability of the default time $\tau^{[k]}$; it can be written as:

$$Q^{[k]}(T) \triangleq \mathbb{P}\left(\tau^{[k]} > T\right) = \mathbb{P}(X_T < k),$$

where X_T counts the portfolio aggregate defaults at time T

$$X_T \triangleq \sum_{i=1}^{n} D_T^i. \tag{8.8}$$

If we have a basket with homogeneous recoveries, the value of the protection leg simplifies to

$$\text{protection\_leg} = -(1 - \delta) \int_0^T p_{0,t} dQ^{[k]}(t);$$

otherwise, one needs to compute the density:

$$\mathbb{P}\left(\tau^{[k]} = \tau_i, \tau^{[k]} \in dt\right).$$

We refer to Laurent and Gregory (2005) or Bielecki and Rutkowski (2002b) for more details.

It should now be clear that to evaluate basket default swaps, we need to generate the survival probability curve $Q^{[k]}(t)$, for $0 \leq t \leq T$. This latter is discretized on a fine mesh and used to estimate the recovery integral of the protection leg.

We show, in this chapter, how to compute this curve with a direct method.

Then, we shall study the properties of the aggregate default distribution X_T by using its probability generating function

$$\varphi(x) = \sum_{k=0}^{n} \mathbb{P}(X_T = k)\ x^k;$$

the "Homogeneous Transformation", the "Asymptotic Homogeneous Expansion", and the "Asymptotic Expansion" methods, presented in the next chapters, are all based on the fundamental "probability generating function representation" of Theorem 85 (in Chap. 10).

8.3 Direct Approach

To motivate the numerical methods developed in the next chapters, we show, in this section, how to compute the key component $Q^{[k]}(T)$ with the direct approach, which is based on enumerating all default combinations in the basket, and mixing their probabilities by using some simple algebraic rules. This combinatorial recipe is only needed for higher-order baskets, where $k \geq 2$. For first-to-default swaps, the algebra is very simple, and closed form solutions are readily available. We begin with the simple FTD case; then we build up the algorithm for the more burdensome cases. We shall use the equivalent fatal shock representation of Lindskog and McNeil

8.3.1 Equivalent Fatal Shock Representation

Let $\Pi_\mathbf{n}$ be the set of all subsets of $\{1, ..., n\}$. For each $\pi \in \Pi_\mathbf{n}$, we introduce the point process N_t^π, which counts the number of shocks in $(0, t]$ resulting in joint defaults of the obligors in π only:

$$N_t^{\pi} \triangleq \sum_{j=1}^{m+n} \sum_{r=1}^{N_t^{c_j}} A_{\theta_r^{c_j}}^{\pi,j}, \tag{8.9}$$

where, for each trigger time $\theta_r^{c_j}$, $A_{\theta_r^{c_j}}^{\pi,j}$ is a Bernoulli variable, which is equal to 1 if all obligors $i \in \pi$ default and all the others, $i \notin \pi$, survive:

$$A_t^{\pi,j} \triangleq \prod_{i \in \pi} A_t^{i,j} \prod_{i \notin \pi} \left(1 - A_t^{i,j}\right). \tag{8.10}$$

For example, if $\pi = \{1, 2\}$, then the process $N_t^{\{1,2\}}$ counts the shocks, which trigger simultaneous defaults of obligors 1 and 2 but not the other obligors 3 to n.

Further, let $\widetilde{N}_t$ count all shocks which results in any kind of loss, i.e.,

$$\widetilde{N}_t \triangleq \sum_{\substack{\pi \in \Pi_{\mathbf{n}} \\ \pi \neq \emptyset}} N_t^{\pi}. \tag{8.11}$$

We have the following fatal shock representation key result. We refer to Lindskog and McNeil (2003) for details (see Proposition 4).

Proposition 77 (Fatal shock representation)

1. *The processes* $(N^{\pi})_{\pi \in \Pi_{\mathbf{n}}}$ *are independent Poisson processes with intensities*

$$\lambda^{\pi}(t) = \sum_{j=1}^{m+n} p_{\pi,j} \lambda^{c_j}(t),$$

where

$$p_{\pi,j} = \prod_{i \in \pi} p_{i,j} \prod_{i \notin \pi} \left(1 - p_{i,j}\right).$$

2. $\widetilde{N}$ *is a Poisson process with intensity*

$$\widetilde{\lambda}(t) = \sum_{j=1}^{m+n} \left[1 - \prod_{i=1}^{n} \left(1 - p_{i,j}\right)\right] \lambda^{c_j}(t).$$

This provides a fatal shock representation of the original not-necessarily-fatal shock set-up. Each obligor i can be represented as

$$N_t^i = \sum_{\pi \in \Pi_{\mathbf{n}}} \mathbf{1}_{\{i \in \pi\}} N_t^{\pi}.$$

Each default configuration, represented by the subset π, can be alternatively be defined with a n-dimensional vector $\widetilde{x_s} = \left(x_{s,1}, ..., x_{s,2}\right)$ of zeros and ones such as

$$i \in \pi \Longleftrightarrow x_{s,i} = 1.$$

If $n = 3$, for instance, we have the following mappings

$$\begin{aligned} \{1, 2\} &\Longleftrightarrow (1, 1, 0) \\ \{1, 3\} &\Longleftrightarrow (1, 0, 1) \\ \{2\} &\Longleftrightarrow (0, 1, 0) \end{aligned}$$

Denoting the fatal shocks N_t^{π} (and its intensity $\lambda^{\pi}(t)$) by

$$N_t^{[\widetilde{x_s}]} \triangleq N_t^{\pi} \text{ and } \lambda^{[\widetilde{x_s}]}(t) \triangleq \lambda^{\pi}(t),$$

we can express each obligor i as

$$N_t^i = \sum_{s=1}^{2^n} x_{s,i} N_t^{[\widetilde{x_s}]}. \tag{8.12}$$

We shall see that the notations of (8.12) will be useful in the derivation of pricing formulas for basket default swaps.

8.3.2 First-to-Default Swap: $k = 1$

To derive the formula of the FTD Q-factor $Q^{[1]}(T)$, it suffices to observe

$$\begin{aligned} Q^{[1]}(T) &\triangleq \mathbb{P}\left(\tau^{[1]} > T\right) = \mathbb{P}\left(\tau_1 > T, ..., \tau_n > T\right) \\ &= \mathbb{P}\left(\widetilde{N}_T = 0\right) = \exp\left(-\int_0^T \widetilde{\lambda}(t)\, dt\right). \end{aligned}$$

Proposition 78 *The survival probability of the first-to-default time is given by*

$$\mathbb{P}\left(\tau^{[1]} > T\right) = \exp\left(-\int_0^T \lambda^{[1]}(t)\, dt\right),$$

where

$$\lambda^{[1]}(t) = \sum_{j=1}^{m}\left[1 - \prod_{i=1}^{n}\left(1 - p_{i,j}\right)\right]\left[\lambda^{c_j}(t)\right] + \left[\sum_{i=1}^{n}\lambda^{0,i}(t)\right]. \quad (8.13)$$

Now, we turn to evaluating the density $\mathbb{P}\left(\tau^{[1]} = \tau_i, \tau^{[1]} \in dt\right)$, which is needed for the protection leg when the recovery rates are non-homogeneous. The formulas in Laurent and Gregory (2005), are based on the assumption that the probability of simultaneous defaults is equal to zero; in the Marshall-Olkin model, this is not the case; we need to examine the instantaneous joint defaults case a bit closer. Indeed, if joint defaults occur, we can choose which reference obligation to deliver. One market convention is to deliver the bond with the lowest recovery value. We assume, without loss of generality, that the underlying references in the basket are ordered such that $\delta_1 \leq \delta_2 \leq ... \leq \delta_n$. This means that the ith reference is delivered if $\tau_i \in (t - \epsilon, t]$ and all the other references with lower recoveries $(\tau_1, \tau_2, ..., \tau_{i-1})$ are alive. In other words, the density is given by

$$\begin{aligned}
\mathbb{P}\left(\tau_i = \tau^{[1]}, \tau^{[1]} \in dt\right) &= \lim_{\epsilon \to 0^+}\frac{1}{\epsilon}\mathbb{P}\left(\tau^{[1]} > t - \epsilon, \tau_i \in (t - \epsilon, t], \bigcap_{l=1}^{i-1}\tau_l \notin (t - \epsilon, t]\right) \\
&= \lim_{\epsilon \to 0^+}\frac{1}{\epsilon}\mathbb{P}\left(\widetilde{N}_{t-\epsilon} = 0, N_t^i - N_{t-\epsilon}^i = 1, \bigcap_{l=1}^{i-1} N_t^l - N_{t-\epsilon}^l = 0\right) \\
&= \lim_{\epsilon \to 0^+}\frac{1}{\epsilon}\mathbb{P}\left(\widetilde{N}_{t-\epsilon} = 0\right)\mathbb{P}\left(N_t^i - N_{t-\epsilon}^i = 1, \bigcap_{l=1}^{i-1} N_t^l - N_{t-\epsilon}^l = 0\right).
\end{aligned}$$

To show that

$$\begin{aligned}
&\lim_{\epsilon \to 0^+}\frac{1}{\epsilon}\mathbb{P}\left(N_t^i - N_{t-\epsilon}^i = 1, \bigcap_{l=1}^{i-1} N_t^l - N_{t-\epsilon}^l = 0\right) \\
&= \sum_{j=1}^{m}\left[p_{i,j}\prod_{l=1}^{i-1}\left(1 - p_{l,j}\right)\right]\left[\lambda^{c_j}(t)\right] + \left[\lambda^{0,i}(t)\right],
\end{aligned}$$

we argue as follows. The probability to have more than one common market event in $(t-\epsilon, t]$ is of order $o(\epsilon)$. Hence

$$\mathbb{P}\left(N_t^i - N_{t-\epsilon}^i = 1, \bigcap_{l=1}^{i-1} N_t^l - N_{t-\epsilon}^l = 0\right)$$

$$= \sum_{j=1}^{m+n} \left[\begin{array}{c} \mathbb{P}\left(N_t^i - N_{t-\epsilon}^i = 1, \bigcap_{l=1}^{i-1} N_t^l - N_{t-\epsilon}^l = 0 \,\middle|\, N_t^{c_j} - N_{t-\epsilon}^{c_j} = 1, \right. \\ \left. \bigcap_{k\neq j} N_t^{c_k} - N_{t-\epsilon}^{c_k} = 0\right) \\ \times \mathbb{P}\left(N_t^{c_j} - N_{t-\epsilon}^{c_j} = 1, \bigcap_{k\neq j} N_t^{c_k} - N_{t-\epsilon}^{c_k} = 0\right) \end{array}\right] + o(\epsilon).$$

But, since we have

$$N_t^i - N_{t-\epsilon}^i = \sum_{j=1}^{m+n} \sum_{\theta_r^{c_j} \in (t-\epsilon, t]} A_{\theta_r^{c_j}}^{i,j},$$

Then,

$$\mathbb{P}\left(N_t^i - N_{t-\epsilon}^i = 1, \bigcap_{l=1}^{i-1} N_t^l - N_{t-\epsilon}^l = 0\right)$$

$$= \sum_{j=1}^{m+n} \mathbb{P}\left(A_{t-\epsilon}^{i,j} = 1, \bigcap_{l=1}^{i-1} A_{t-\epsilon}^{l,j} = 0\right) \mathbb{P}\left(N_t^{c_j} - N_{t-\epsilon}^{c_j} = 1\right) + o(\epsilon)$$

$$= \sum_{j=1}^{m+n} \left[p^{i,j} \prod_{l=1}^{i-1} \left(1 - p^{l,j}\right)\right] \left[\epsilon \lambda^{c_j}(t)\right] + o(\epsilon).$$

Taking the limit, we arrive at the following expression:

$$\frac{\mathbb{P}\left(\tau_i = \tau^{[1]}, \tau^{[1]} \in dt\right)}{Q^{[1]}(t)} = \sum_{j=1}^{m} \left[p_{i,j} \prod_{l=1}^{i-1} \left(1 - p_{l,j}\right)\right] \left[\lambda^{c_j}(t)\right] + \left[\lambda^{0,i}(t)\right]. \tag{8.14}$$

If we had a different convention for the bonds to be delivered in the case of joint defaults, we would need to re-order the default times according to the recovery rate delivery rule, and formula (8.14) would hold for the re-ordered basket.

8.3.3 kth-to-Default Swap: $k > 1$

In order to evaluate the Q-factor of the random time $\tau^{[k]}$, for $k > 1$, we have to enumerate all the possible basket default configurations, and to compute their corresponding probabilities. We shall represent each default configuration with a n-dimensional vector of zeros and ones, $\widetilde{x_s} = (x_{s,1}, ..., x_{s,2})$ for $s = 1, ..., 2^n$: if $x_{s,i} = 1$, then the ith reference has defaulted. By $d(\widetilde{x_s}) = \sum_{i=1}^{n} x_{s,i}$, we denote the number of defaults in this configuration. For example, if we have $n = 3$ underlying credits, the basket default configurations are: $\widetilde{x_1} = (0, 0, 0)$, $\widetilde{x_2} = (1, 0, 0)$, $\widetilde{x_3} = (0, 1, 0)$, $\widetilde{x_4} = (1, 1, 0)$, $\widetilde{x_5} = (0, 0, 1)$, $\widetilde{x_6} = (1, 0, 1)$, $\widetilde{x_7} = (0, 1, 1)$, $\widetilde{x_8} = (1, 1, 1)$.

Let $Q^{[\widetilde{x_s}]}(T)$ denote the probability of the default configuration $\widetilde{x_s}$ on the interval $(0, T]$:

$$Q^{[\widetilde{x_s}]}(T) \triangleq \mathbb{E}\left[\prod_{i=1}^{n} \left(D_T^i\right)^{x_{s,i}} \left(1 - D_T^i\right)^{1-x_{s,i}}\right], \tag{8.15}$$

with the convention $0^0 = 1$. By definition, summing up the probabilities of all configurations, such that $d(\widetilde{x_s}) < k$, gives the value of the Q-factor $Q^{[k]}(T)$:

$$Q^{[k]}(T) = \sum_{\{s:d(\widetilde{x_s})<k\}} Q^{[\widetilde{x_s}]}(T).$$

It is easy to see, from Eq. (8.15), that we can write $Q^{[\widetilde{x_s}]}(T)$ as a linear combination of FTD Q-factors on subsets of $\{\tau_1, ..., \tau_n\}$. We introduce the notation $\Pi_{\mathbf{n}}$ to represent the set of all subsets of $\{1, ..., n\}$. For each subset $\pi \in \Pi_{\mathbf{n}}$, we define the first-to-default random time $\tau_\pi^{[1]} = \min\{\tau_i : i \in \pi\}$, and its Q-factor $Q_\pi^{[1]}(T)$:

$$Q_\pi^{[1]}(T) = \mathbb{P}\left(\tau_\pi^{[1]} > T\right) = \mathbb{E}\left[\prod_{i\in\pi} \left(1 - D_T^i\right)\right].$$

In the example with $n = 3$, we have the following *basis* of first-to-default Q-factors: $\left\{1, Q_{\{1\}}^{[1]}(T), Q_{\{2\}}^{[1]}(T), Q_{\{3\}}^{[1]}(T), Q_{\{1,2\}}^{[1]}(T), Q_{\{1,3\}}^{[1]}(T), Q_{\{2,3\}}^{[1]}(T), Q_{\{1,2,3\}}^{[1]}(T)\right\}$. Note that $Q_\pi^{[1]}(T)$ for the empty set $\pi = \emptyset$ is equal to

1. For the first configuration $\widetilde{x_1} = (0, 0, 0)$, the Q-factor $Q^{[\widetilde{x_1}]}(T)$ is given by

$$\mathbb{P}((0,0,0)) = \mathbb{E}\left[\left(1 - D_T^1\right)\left(1 - D_T^2\right)\left(1 - D_T^3\right)\right] = Q^{[1]}_{\{1,2,3\}}(T).$$

For the second configuration $\widetilde{x_2} = (1, 0, 0)$, we have

$$\mathbb{P}((0,0,0)) = \mathbb{E}\left[D_T^1\left(1 - D_T^2\right)\left(1 - D_T^3\right)\right] = Q^{[1]}_{\{2,3\}}(T) - Q^{[1]}_{\{1,2,3\}}(T).$$

In general, the default configuration Q-factors can be written as

$$Q^{[\widetilde{x_s}]}(T) = \sum_{\pi \in \Pi_\mathbf{n}} \alpha_\pi^{\widetilde{x_s}} Q^{[1]}_\pi(T), \tag{8.16}$$

where the coefficients $\alpha_\pi^{\widetilde{x_s}}$ take values in $\{-1, 0, 1\}$. For $n = 3$, the α_π-coefficients for each configuration are:

	1	$Q^{[1]}_{\{1\}}$	$Q^{[1]}_{\{2\}}$	$Q^{[1]}_{\{3\}}$	$Q^{[1]}_{\{1,2\}}$	$Q^{[1]}_{\{1,3\}}$	$Q^{[1]}_{\{1,3\}}$	$Q^{[1]}_{\{1,2,3\}}$
0 0 0	0	0	0	0	0	0	0	1
1 0 0	0	0	0	0	0	0	1	−1
0 1 0	0	0	0	0	0	1	0	−1
1 1 0 $\rightarrow$	0	0	0	0	1	−1	−1	1
0 0 1	0	0	0	0	1	0	0	−1
1 0 1	0	0	1	0	−1	0	−1	1
0 1 1	0	1	0	0	−1	−1	0	1
1 1 1	1	−1	−1	−1	1	1	1	−1

In the general case, this table can be constructed recursively as follows.

We define a partition of $\Pi_\mathbf{n+1}$: $\Pi_\mathbf{n+1} = \Pi^+_\mathbf{n+1} \cup \Pi^-_\mathbf{n+1}$, and $\Pi^+_\mathbf{n+1} \cap \Pi^-_\mathbf{n+1} = \emptyset$. By $\Pi^+_\mathbf{n+1}$, we denote the set of all subsets that contain $(n+1)$, and by $\Pi^-_\mathbf{n+1}$, the set of all subsets that do not contain $(n+1)$:

$$\Pi^+_\mathbf{n+1} = \{\pi_{n+1} : \pi_{n+1} \in \Pi_\mathbf{n+1},\ (n+1) \in \pi_{n+1}\},$$
$$\Pi^-_\mathbf{n+1} = \{\pi_{n+1} : \pi_{n+1} \in \Pi_\mathbf{n+1},\ (n+1) \notin \pi_{n+1}\}.$$

For a $(n+1)$-dimensional basket, the default configurations are either $\left(\widetilde{x_s^n}, 1\right)$ or $\left(\widetilde{x_s^n}, 0\right)$, where $\widetilde{x_s^n}$ is the default configuration for a n-dimensional basket:

$$
\begin{aligned}
Q^{[(\widetilde{x_s^n},0)]}(T) &= \mathbb{E}\left[\left(1 - D_T^{n+1}\right)\prod_{i=1}^{n}\left(D_T^i\right)^{x_{s,i}^n}\left(1 - D_T^i\right)^{1-x_{s,i}^n}\right] \\
&= \mathbb{E}\left[\left(1 - D_T^{n+1}\right)\sum_{\pi_n\in\Pi_{\mathbf{n}}}\alpha_{\pi_n}\prod_{i\in\pi_n}\left(1 - D_T^i\right)\right] \\
&= \mathbb{E}\left[\sum_{\pi_{n+1}\in\Pi_{\mathbf{n+1}}^{+}}\alpha_{\pi_n}\prod_{i\in\pi_{n+1}}\left(1 - D_T^i\right)\right],
\end{aligned}
$$

$$
\begin{aligned}
Q^{[(\widetilde{x_s^n},1)]}(T) &= \mathbb{E}\left[\left(D_T^{n+1}\right)\prod_{i=1}^{n}\left(D_T^i\right)^{x_{s,i}^n}\left(1 - D_T^i\right)^{1-x_{s,i}^n}\right] \\
&= \mathbb{E}\left[\left(D_T^{n+1}\right)\sum_{\pi_n\in\Pi_{\mathbf{n}}}\alpha_{\pi_n}\prod_{i\in\pi_n}\left(1 - D_T^i\right)\right] \\
&= \mathbb{E}\left[\sum_{\pi_n\in\Pi_{\mathbf{n}}}\alpha_{\pi_n}\prod_{i\in\pi_n}\left(1 - D_T^i\right)\right] - \mathbb{E}\left[\left(1 - D_T^{n+1}\right)\sum_{\pi_n\in\Pi_{\mathbf{n}}}\alpha_{\pi_n}\prod_{i\in\pi_n}\left(1 - D_T^i\right)\right] \\
&= \mathbb{E}\left[\sum_{\pi_{n+1}\in\Pi_{\mathbf{n+1}}^{-}}\alpha_{\pi_n}\prod_{i\in\pi_{n+1}}\left(1 - D_T^i\right)\right] - \mathbb{E}\left[\sum_{\pi_{n+1}\in\Pi_{\mathbf{n+1}}^{+}}\alpha_{\pi_n}\prod_{i\in\pi_{n+1}}\left(1 - D_T^i\right)\right].
\end{aligned}
$$

In summary, we have the following recursion:

$$
\begin{aligned}
\alpha_{\pi_{n+1}}^{(\widetilde{x_s^n},0)} &= 0, \text{ for } \pi_{n+1} \in \Pi_{\mathbf{n+1}}^{-}, \\
\alpha_{\pi_{n+1}}^{(\widetilde{x_s^n},0)} &= \alpha_{\pi_n}^{\widetilde{x_s^n}}, \text{ for } \pi_{n+1} \in \Pi_{\mathbf{n+1}}^{+} \text{ and } \pi_{n+1} = \pi_n \cup \{n+1\},
\end{aligned}
$$

and

$$
\begin{aligned}
\alpha_{\pi_{n+1}}^{(\widetilde{x_s^n},1)} &= \alpha_{\pi_n}^{\widetilde{x_s^n}}, \text{ for } \pi_{n+1} \in \Pi_{\mathbf{n+1}}^{-} \text{ and } \pi_{n+1} = \pi_n \cup \emptyset, \\
\alpha_{\pi_{n+1}}^{(\widetilde{x_s^n},1)} &= -\alpha_{\pi_n}^{\widetilde{x_s^n}}, \text{ for } \pi_{n+1} \in \Pi_{\mathbf{n+1}}^{+} \text{ and } \pi_{n+1} = \pi_n \cup \{n+1\}.
\end{aligned}
$$

Once we have generated the α_π-representation (Eq. 8.16) for each default configuration, and we have computed the subset FTD Q-factors, we are in a position to evaluate the kth -to-default Q-factor $Q^{[k]}(T)$.

Proposition 79 *Using the notations in this section, the survival probability of the kth-to-default random time is given by*

$$\mathbb{P}\left(\tau^{[k]} > T\right) = \sum_{\{s:d(\widetilde{x_s})<k\}} \sum_{\pi\in\Pi_\mathbf{n}} \alpha_\pi^{\widetilde{x_s}} Q_\pi^{[1]}(T),$$

where the $\{-1, 0, 1\}$-valued coefficients $\alpha_\pi^{\widetilde{x_s}}$ are computed recursively.

As mentioned in Sect. 8.2, for non-homogeneous recovery rates, we need to evaluate the density $\mathbb{P}\left(\tau_i = \tau^{[k]}, \tau^{[k]} \in dt\right)$. First, let us remark that credit i would be the kth-to-default obligor in an infinitesimal interval $(t-\epsilon, t]$ iff:

$$\left\{\tau_i = \tau^{[k]}, \tau^{[k]} \in (t-\epsilon, t]\right\} \Leftrightarrow \left\{D_t^i - D_{t-\epsilon}^i = 1, \sum_{l=1}^{n} D_{t-\epsilon}^l + \sum_{l=1}^{i} D_t^l - D_{t-\epsilon}^l = k\right\}.$$

This corresponds to the default of credit i being in the interval $(t-\epsilon, t]$, and the total number of defaults is equal to k. If there are instantaneous joint defaults, only the credits in the set $\{1, ..., i\}$ define the default-rank for credit i (the references are assumed to be ordered according to the pre-specified recovery-delivery rule). In other words, a credit l would be considered to be in default before credit i if $\tau_l < \tau_i$, or $\tau_l = \tau_i$ and $l < i$. Thus, we need to compute

$$\begin{aligned}\mathbb{P}\left(\tau_i = \tau^{[k]}, \tau^{[k]} \in dt\right) &= \lim_{\epsilon\to 0^+} \frac{1}{\epsilon}\mathbb{P}\left(D_t^i - D_{t-\epsilon}^i = 1, \sum_{l=1}^{n} D_{t-\epsilon}^l + \sum_{l=1}^{i} D_t^l - D_{t-\epsilon}^l = k\right)\\ &= \lim_{\epsilon\to 0^+} \frac{1}{\epsilon}\mathbb{E}\left[\left(D_t^i - D_{t-\epsilon}^i\right)\mathbf{1}_{\left\{\sum_{l=1}^{n} D_{t-\epsilon}^l + \sum_{l=1}^{i} D_t^l - D_{t-\epsilon}^l = k\right\}}\right]\end{aligned}$$

But we have

$$\mathbb{P}\left(D_t^i - D_{t-\epsilon}^i = 1\right) = \left(1 - D_{t-\epsilon}^i\right)\mathbb{P}\left(N_t^i - N_{t-\epsilon}^i > 0\right).$$

Hence

$$\begin{aligned}&\mathbb{E}\left[\left(D_t^i - D_{t-\epsilon}^i\right)\mathbf{1}_{\left\{\sum_{l=1}^{n} D_{t-\epsilon}^l + \sum_{l=1}^{i} D_t^l - D_{t-\epsilon}^l = k\right\}}\right]\\ &= \mathbb{E}\left[\left(1 - D_{t-\epsilon}^i\right)\left(N_t^i - N_{t-\epsilon}^i\right)\mathbf{1}_{\left\{\sum_{l=1}^{n} D_{t-\epsilon}^l + \sum_{l=1}^{i}\left(1 - D_{t-\epsilon}^l\right)\left(N_t^l - N_{t-\epsilon}^l\right) = k\right\}}\right] + o(\epsilon).\end{aligned}$$

Using the equivalent fatal shock representation, we know that

$$N_t^i - N_{t-\epsilon}^i = \sum_{\pi\in\Pi_\mathbf{n}} x_{s,i}\left(N_t^{[\widetilde{x_s}]} - N_{t-\epsilon}^{[\widetilde{x_s}]}\right),$$

and

$$D_t^i = \sum_{s=1}^{2^n} x_{s,i} D_t^{[\widetilde{x_s}]}.$$

Combining the probabilities of the default configurations $\widetilde{x_{s_1}}$, in the interval $(0, t - \epsilon]$, and the probabilities of the configurations $\widetilde{x_{s_2}}$, in the instantaneous interval $(t - \epsilon, t]$, we arrive at

$$\mathbb{E}\left[\left(1 - D_{t-\epsilon}^i\right)\left(N_t^i - N_{t-\epsilon}^i\right)\mathbf{1}_{\left\{\sum_{l=1}^n D_{t-\epsilon}^l + \sum_{l=1}^i (1-D_{t-\epsilon}^l)(N_t^l - N_{t-\epsilon}^l) = k\right\}}\right]$$
$$= \sum_{s_1=1}^{2^n}\sum_{s_2=1}^{2^n}\left[\left(1 - x_{s_1,i}\right)x_{s_2,i}\mathbf{1}_{\left\{d(\widetilde{x_{s_1}}) + \sum_{l=1}^i (1 - x_{s_1,l})x_{s_2,l} = k\right\}}\right] Q^{[\widetilde{x_{s_1}}]}(t)\left[\epsilon\lambda^{[\widetilde{x_{s_2}}]}(t)\right] + o(\epsilon),$$

which is a weighted average over all possible scenarios. This is the convolution between the basket default distributions on the intervals $(0, t - \epsilon]$ and $(t - \epsilon, t]$, weighted by the indicator function of the default scenarios that contribute to the recovery payoff $(1 - \delta_i)$.

The density follows by taking the limit

$$\mathbb{P}\left(\tau_i = \tau^{[k]}, \tau^{[k]} \in dt\right) = \sum_{s_1=1}^{2^n}\sum_{s_2=1}^{2^n} Q^{[\widetilde{x_{s_1}}]}(t)\,\lambda^{[\widetilde{x_{s_2}}]}(t)$$
$$\left[\left(1 - x_{s_1,i}\right)x_{s_2,i}\mathbf{1}_{\left\{d(\widetilde{x_{s_1}}) + \sum_{l=1}^i (1 - x_{s_1,l})x_{s_2,l} = k\right\}}\right]. \tag{8.17}$$

References

T. Bielecki, M. Rutkowski, *Credit Risk: Modeling, Valuation and Hedging* (Springer, New York, 2002a)

T. Bielecki, M. Rutkowski, Intensity-based valuation of basket credit derivatives, in *Mathematical Finance*, ed. by J. Yong (World Scientific, Singapore, 2002b), pp. 12–27

D. Duffie, J. Pan, Analytical value-At-Risk with jumps and credit risk. Financ. Stochast. **5**, 155–180 (2001)

J.P. Laurent, J. Gregory, Basket default swaps, CDOs and factor copulas. J. Risk **7**(4), 103–122 (2005)

F. Lindskog, A. McNeil, Common Poisson Shock models: applications to insurance and credit risk modelling. ASTIN Bullet. **33**(2), 209–238 (2003)

9

Static Replication

In principle, the direct method requires, for each time step, 2^{n+1} values, corresponding to the possible combinations of s_1 and s_2 in Eq. (8.17); as the size of the underlying basket increases, the number of default configurations explodes exponentially. This significant limitation restricts the applicability of the method to baskets under 10 or 11 credits. As an alternative, we describe here a different approach, which is based on a static replication idea. We show that each kth-to-default swap can be synthetically generated with a portfolio of first-to-default swaps referenced to subsets of the original basket. This result is not very surprising, since sub-FTDs may be viewed as a type of Arrow-Debreu prices of the multivariate default space. Given the set of all sub-FTD prices, one can compute the value of any higher-order basket default swap. The key building block is the FTD evaluator; a fast implementation of FTD prices ensures, in turn, a reasonable run time for kth -to-default swaps.

In this chapter, we describe how this static FTD replication is done: first, we show the relationship between kth-to-default and $(k-1)$th-to-default swaps; then, we apply this recursion step-by-step until we arrive at the complete FTD expansion.

9.1 The Recursive Formula

To illustrate the recursive relationship, we start with a simple example, where we have a basket of $n = 3$ credits: A, B and C. We observe that the $\{A, B, C\}$-second-to-default swap can be replicated with the first-to-default swaps
referenced to the sub-baskets $\{A, B\}$, $\{A, C\}$, $\{B, C\}$ and $\{A, B, C\}$ as:

Y. Elouerkhaoui, *Credit Correlation*, Applied Quantitative Finance,
DOI 10.1007/978-3-319-60973-7_9

$$STD(A,B,C) = FTD(A,B) + FTD(A,C) + FTD(B,C) - 2FTD(A,B,C). \quad (9.1)$$

To see this, it suffices to verify that the payoffs of the two sides of the equation match in all default scenarios. Since Eq. (9.1) is symmetric in A, B and C, we suppose, for instance, that credit A defaults, and we denote its recovery rate by δ_A. On the left-hand side, $STD(A,B,C)$ becomes $FTD(B,C)$, and makes no cash payments; on the right-hand side, $FTD(A,B)$ pays $(1-\delta_A)$ and terminates, $FTD(A,C)$ pays $(1-\delta_A)$ and terminates, $FTD(B,C)$ is unchanged and does not make any cash payments, and finally $FTD(A,B,C)$ pays $(1-\delta_A)$ and terminates. In summary, we have

PV		PV	Cash balance
$FTD(A,B)$	$\rightarrow$	0	$(1-\delta_A)$
$FTD(A,C)$	$\rightarrow$	0	$(1-\delta_A)$
$FTD(B,C)$	$\rightarrow$	$FTD(B,C)$	0
$-2FTD(A,B,C)$	$\rightarrow$	0	$-2(1-\delta_A)$
Total	$\rightarrow$	$FTD(B,C)$	0
PV		PV	cash balance
$STD(A,B,C)$	$\rightarrow$	$FTD(B,C)$	0

For the premium leg, all the trades pay one unit of premium before default. If A defaults, $STD(A,B,C)$ and $FTD(B,C)$ keep paying one unit of premium each, and the premium payments stop for all the other FTDs which contain the reference A: $FTD(A,B)$, $FTD(A,C)$, $FTD(A,B,C)$,

	Before default	After A defaults
$FTD(A,B)$	1	0
$FTD(A,C)$	1	0
$FTD(B,C)$	1	1
$-2FTD(A,B,C)$	-2	0
Total	1	1
	Before default	After A defaults
$STD(A,B,C)$	1	1

A similar analysis shows that we can replicate the $\{A,B,C\}$-third-to-default by second-to-defaults on the sub-baskets $\{A,B\}$, $\{A,C\}$, $\{B,C\}$ and $\{A,B,C\}$ as follows:

$$TTD(A,B,C) = \frac{1}{2}\left[STD(A,B) + STD(A,C) + STD(B,C) - STD(A,B,C)\right].$$

Now, we turn to the general case, and we establish the result for a basket of n credit. We denote by $V_n^{[k]}(A_1, \ldots, A_n)$ the value of a kth-to-default swap referenced to the n-dimensional basket $\{A_1, \ldots, A_n\}$. We also define the l-subsets of $\{1, 2, \ldots, n\}$ (i.e. subsets containing exactly l elements) by the mappings $\pi_s^l(.)$, for $1 \leq s \leq \binom{n}{l}$:

$$\pi_s^l(.) : \{1, 2, \ldots, k\} \to \{1, 2, \ldots, n\}.$$

Furthermore, it is notationaly convenient to assume that each l-subset π_s^l is an ordered sequence of indices $\pi_s^l(1) \leq \pi_s^l(2) \leq \cdots \leq \pi_s^l(k)$.

Example 80 We consider $n = 3$ credits; for subsets of $l = 2$ elements, we have:

$$\left\{\pi_1^2 : \pi_1^2(1) = 1; \pi_1^2(2) = 2\right\}, \left\{\pi_2^2 : \pi_2^2(1) = 1; \pi_2^2(2) = 3\right\}, \left\{\pi_3^2 : \pi_1^2(1) = 2; \pi_3^2(2) = 3\right\}.$$

For a second-to-default $V_n^{[2]}$, we can easily check that the replicating portfolio of FTD swaps contains the $\{A_1, \ldots, A_n\}$-FTD and all the $(n-1)$ -element-sub-FTDs,

$$V_n^{[2]}(A_1, \ldots, A_n) = \sum_{s=1}^{n} V_{n-1}^{[1]}\left(A_{\pi_s^{n-1}(1)}, \ldots, A_{\pi_s^{n-1}(n-1)}\right) - (n-1)\, V_n^{[1]}(A_1, \ldots, A_n).$$

For higher-orders $k \geq 2$, we have a recursion of the form

$$a_n^{[k]} V_n^{[k]}(A_1, \ldots, A_n) = \underbrace{\sum_{s=1}^{n} V_{n-1}^{[k-1]}\left(A_{\pi_s^{n-1}(1)}, \ldots, A_{\pi_s^{n-1}(n-1)}\right)}_{S} - b_n^{[k]} V_n^{[k-1]}(A_1, \ldots, A_n). \tag{9.2}$$

Equation (9.2) is symmetric in $A_1, \ldots, A_n$. Suppose, for example, that A_1 goes into default. The subsets $\left(\pi_s^{n-1}(.)\right)_{1 \leq s \leq n}$ are ordered such that the last set is $\pi_n^{n-1} = \{A_2, \ldots, A_n\}$ and all the other sets contain credit A_1. In the sum S, the first $(n-1)$ subsets contain the defaulted reference A_1, hence each sub-basket $V_{n-1}^{[k-1]}\left(A_{\pi_s^{n-1}(1)}, \ldots, A_{\pi_s^{n-1}(n-1)}\right)$, for $1 \leq s \leq n-1$, becomes $V_{n-2}^{[k-2]}\left(A_{\pi_s^{n-2}(2)}, \ldots, A_{\pi_s^{n-2}(n-1)}\right)$. The sub-basket $V_{n-1}^{[k-1]}$ $(A_2, \ldots, A_n)$ remains unchanged and the last term $V_n^{[k-1]}(A_1, \ldots, A_n)$ becomes $V_{n-1}^{[k-2]}(A_2, \ldots, A_n)$. It follows that, after default, Eq. (9.2) becomes

$$a_n^{[k]} V_{n-1}^{[k-1]}(A_2, \ldots, A_n) = \left[\sum_{s=1}^{n-1} V_{n-2}^{[k-2]}\left(A_{\pi_s^{n-2}(2)}, \ldots, A_{\pi_s^{n-2}(n-1)}\right) + V_{n-1}^{[k-1]}(A_2, \ldots, A_n) \right]$$
$$-b_n^{[k]} V_{n-1}^{[k-2]}(A_2, \ldots, A_n);$$

re-arranging the terms, we obtain

$$\left(a_n^{[k]} - 1\right) V_{n-1}^{[k-1]}(A_2, \ldots, A_n) = \sum_{s=1}^{n-1} V_{n-2}^{[k-2]}\left(A_{\pi_s^{n-1}(2)}, \ldots, A_{\pi_s^{n-1}(n-1)}\right)$$
$$-b_n^{[k]} V_{n-1}^{[k-2]}(A_2, \ldots, A_n). \tag{9.3}$$

On the other hand, applying Eq. (9.2) to the $(n-1)$-element-basket $\{A_2, \ldots, A_n\}$ yields

$$a_{n-1}^{[k-1]} V_{n-1}^{[k-1]}(A_2, \ldots, A_n) = \sum_{s=1}^{n-1} V_{n-2}^{[k-2]}\left(A_{\pi_s^{n-1}(2)}, \ldots, A_{\pi_s^{n-1}(n-1)}\right)$$
$$-b_{n-1}^{[k-1]} V_{n-1}^{[k-2]}(A_2, \ldots, A_n), \tag{9.4}$$

which, compared with Eq. (9.3), gives

$$a_n^{[k]} - 1 = a_{n-1}^{[k-1]}, \text{ and } b_n^{[k]} = b_{n-1}^{[k-1]}. \tag{9.5}$$

Applying the double-recursion (9.5), we arrive at the values of $a_n^{[k]}$:

$$a_n^{[k]} = a_{n-1}^{[k-1]} + 1 = \cdots = a_{n-(k-2)}^{[2]} + (k-2) = 1 + (k-2) = k - 1,$$

and the values of $b_n^{[k]}$:

$$b_n^{[k]} = b_{n-1}^{[k-1]} = \cdots = b_{n-(k-2)}^{[2]} = n - (k-2) - 1 = n - k + 1.$$

Proposition 81 *The replication of a kth-to-default swap referencing a basket of n credits $\{A_1, \ldots, A_n\}$, is done with $(n+1)$ $(k-1)$ th-to-default swap instruments: the replicating swaps reference the original basket $\{A_1, \ldots, A_n\}$ and all its $(n-1)$-element-sub-baskets,*

$$V_n^{[k]}(A_1, \ldots, A_n) = \frac{1}{a_n^{[k]}} \left[\sum_{s=1}^{n} V_{n-1}^{[k-1]}\left(A_{\pi_s^{n-1}(1)}, \ldots, A_{\pi_s^{n-1}(n-1)}\right) - b_n^{[k]} V_n^{[k-1]}(A_1, \ldots, A_n) \right], \tag{9.6}$$

with $a_n^{[k]} = k - 1$ and $b_n^{[k]} = n - k + 1$.

9.2 The Complete Expansion

Using the relationship (9.6), it is easy to see that we can proceed recursively until we arrive at the complete expansion in terms of the basic FTD instruments, each FTD swap references a sub-basket of l elements, where l varies between $(n-k+1)$ and n:

$$V_n^{[k]}(A_1,\ldots,A_n)=\sum_{l=n-k+1}^{n}\sum_{s=1}^{\binom{n}{l}}\alpha_n^{[k]}(l)\,V_l^{[1]}\left(A_{\pi_s^l(1)},\ldots,A_{\pi_s^l(k)}\right). \tag{9.7}$$

Note that $\alpha_n^{[k]}(l)$, the instrument contribution to the replicating portfolio, depends solely on the size of the subset π_s^l. This is a consequence of the symmetry with respect to the references $A_1,\ldots,A_n$. In order to derive the expression of the coefficients $\alpha_n^{[k]}(l)$, we replace each term, on the right-hand side of the recursive relationship (9.6), by its FTD representation (9.7),

$$\begin{aligned}V_n^{[k]}(A_1,\ldots,A_n)&=\frac{1}{a_n^{[k]}}\left[\underbrace{\sum_{t=1}^{n}\sum_{l=n-k+1}^{n-1}\sum_{s=1}^{\binom{n-1}{l}}\alpha_{n-1}^{[k-1]}(l)\,V_l^{[1]}\left(A_{\pi_t^{n-1}(\pi_s^l(1))},\ldots,A_{\pi_t^{n-1}(\pi_s^l(l))}\right)}_{S_1}\right]\\&\quad-\frac{b_n^{[k]}}{a_n^{[k]}}\left[\underbrace{\sum_{l=n-k+2}^{n}\sum_{s=1}^{\binom{n}{l}}\alpha_n^{[k-1]}(l)\,V_l^{[1]}\left(A_{\pi_s^l(1)},\ldots,A_{\pi_s^l(l)}\right)}_{S_2}\right].\end{aligned} \tag{9.8}$$

We simplify the triple sum S_1 by interchanging the summation order and using the following lemma.

Lemma 82

$$\sum_{t=1}^{n}\sum_{s=1}^{\binom{n-1}{l}}V_l^{[1]}\left(A_{\pi_t^{n-1}(\pi_s^l(1))},\ldots,A_{\pi_t^{n-1}(\pi_s^l(l))}\right)=(n-l)\sum_{s=1}^{\binom{n}{l}}V_l^{[1]}\left(A_{\pi_s^l(1)},\ldots,A_{\pi_s^l(l)}\right).$$

Proof To generate all the subsets of the double mapping $\pi_t^{n-1}\left(\pi_s^l(.)\right)$, for $1\le t\le n$, $1\le s\le\binom{n-1}{l}$, first we enumerate all the $(n-1)$-subsets of $\{1,\ldots,n\}$, then for each $(n-1)$-subset, we enumerate all of its l-subsets; in total, we have generated $n\times\binom{n-1}{l}$ subsets of $\{1,\ldots,n\}$ containing exactly

l elements. This means that we have spanned the set of all l-subsets a number of times d, where d verifies $n \times \binom{n-1}{l} = d \times \binom{n}{l}$; hence $d = n - l$. □

The sum S_1, then, becomes

$$S_1 = \sum_{l=n-k+1}^{n-1} \alpha_{n-1}^{[k-1]}(l) \left[(n-l) \sum_{s=1}^{\binom{n}{l}} V_l^{[1]} \left(A_{\pi_s^l(1)}, \ldots, A_{\pi_s^l(l)} \right) \right].$$

On the other hand, we can re-write the double sum S_2 as

$$S_2 = \sum_{l=n-k+2}^{n} \alpha_n^{[k-1]}(l) \sum_{s=1}^{\binom{n}{l}} V_l^{[1]} \left(A_{\pi_s^l(1)}, \ldots, A_{\pi_s^l(l)} \right).$$

By splitting the k-sums in S_1 and S_2, Eq. (9.8) becomes

$$\begin{aligned} & V_n^{[k]}(A_1, \ldots, A_n) \\ &= \sum_{l=n-k+1}^{n-k+1} \left(\frac{n-l}{a_n^{[k]}} \right) \alpha_{n-1}^{[k-1]}(l) \sum_{s=1}^{\binom{n}{l}} V_l^{[1]} \left(A_{\pi_s^l(1)}, \ldots, A_{\pi_s^l(l)} \right) \\ &+ \sum_{l=n-k+2}^{n-1} \left(\left(\frac{n-l}{a_n^{[k]}} \right) \alpha_{n-1}^{[k-1]}(l) - \left(\frac{b_n^{[k]}}{a_n^{[k]}} \right) \alpha_n^{[k-1]}(l) \right) \sum_{s=1}^{\binom{n}{l}} V_l^{[1]} \left(A_{\pi_s^l(1)}, \ldots, A_{\pi_s^l(l)} \right) \\ &- \sum_{l=n}^{n} \left(\frac{b_n^{[k]}}{a_n^{[k]}} \right) \alpha_n^{[k-1]}(l) \sum_{s=1}^{\binom{n}{l}} V_l^{[1]} \left(A_{\pi_s^l(1)}, \ldots, A_{\pi_s^l(l)} \right). \end{aligned} \tag{9.9}$$

Comparing Eq. (9.9) with the original expansion (9.7), we obtain the following double-recursion:

$$\begin{cases} \alpha_n^{[k]}(l) = \left(-\frac{b_n^{[k]}}{a_n^{[k]}} \right) \alpha_n^{[k-1]}(l), \text{ for } l = n, \\ \alpha_n^{[k]}(l) = \left(\frac{n-l}{a_n^{[k]}} \right) \alpha_{n-1}^{[k-1]}(l) - \left(\frac{b_n^{[k]}}{a_n^{[k]}} \right) \alpha_n^{[k-1]}(l), \text{ for } n-k+1 < l < n, \\ \alpha_n^{[k]}(l) = \left(\frac{n-l}{a_n^{[m]}} \right) \alpha_{n-1}^{[k-1]}(l), \text{ for } l = n-k+1. \end{cases}$$

Let us solve this double-recursion step-by-step.

We start with the case $l = n$,

$$\alpha_n^{[k]}(n) = \left(-\frac{n-k+1}{k-1}\right)\left(-\frac{n-k}{k-2}\right)\cdots\left(-\frac{n-1}{1}\right)\underbrace{\alpha_n^{[1]}(n)}_{=1}$$
$$= (-1)^{k-1}\frac{(n-1)!}{(k-1)!\,(n-k)!} = (-1)^{k-1}\binom{n-k}{n-1}.$$

Then, for $l = n - k + 1$, we have

$$\alpha_n^{[k]}(n-k+1) = \frac{n-(n-k+1)}{k-1}\alpha_{n-1}^{[k-1]}(n-k+1)$$
$$= \alpha_{n-1}^{[k-1]}(n-k+1) = \cdots = \alpha_{n-k+1}^{[1]}(n-k+1) = 1.$$

And finally, for $n - k + 1 < l < n$,

$$\alpha_n^{[k]}(l) = \left(\frac{n-l}{k-1}\right)\alpha_{n-1}^{[k-1]}(l) - \left(\frac{n-k+1}{k-1}\right)\alpha_n^{[k-1]}(l). \tag{9.10}$$

From the shape of Eq. (9.10), we postulate that the solution is of the form

$$\alpha_n^{[k]}(l) = (-1)^{\beta}\binom{\gamma}{\delta},$$

and using some elementary algebra, we find that the solution is:

$$\alpha_n^{[k]}(l) = (-1)^{l-(n-k+1)}\binom{l-1}{n-k}.$$

To summarize, we have the following proposition.

Proposition 83 *The replication of a kth-to-default referencing a basket of n credits $\{A_1, \ldots, A_n\}$, is done with the first-to-default swaps that reference all the l-subsets of the original basket, where $n - k + 1 \leq l \leq n$:*

$$V_n^{[k]}(A_1, \ldots, A_n) = \sum_{l=n-k+1}^{n}\sum_{s=1}^{\binom{n}{l}}\alpha_n^{[k]}(l)\,V_l^{[1]}\left(A_{\pi_s^l(1)}, \ldots, A_{\pi_s^l(l)}\right),$$

where $\alpha_n^{[k]}(l) = (-1)^{l-(n-k+1)}\binom{l-1}{n-k}$, for $n - k + 1 \leq l \leq n$.

Remark 84 Note that the first-to-default expansion is generic and is applicable to all copula models. The replication itself is independent of the choice of the multivariate dependence.

In order to compare the FTD expansion with the direct approach, we apply Eq. (9.7) to an kth-to-default credit linked zero-coupon cash flow, and we establish the expression of the Q-factor associated with the random time $\tau^{[k]}$

$$\mathbb{P}\left(\tau^{[k]} > T\right) = \sum_{l=n-k+1}^{n} \sum_{s=1}^{\binom{n}{l}} \alpha_n^{[k]}(l)\, Q_{\pi_s^l}^{[1]}(T). \tag{9.11}$$

Recall the expression of $Q^{[k]}(T)$ with the direct method (Eq. 8.16)

$$\mathbb{P}\left(\tau^{[k]} > T\right) = \sum_{\{s:d(\widetilde{x_s})<k\}} \sum_{\pi \in \Pi_{\mathbf{n}}} \alpha_\pi^{\widetilde{x_s}}\, Q_\pi^{[1]}(T). \tag{9.12}$$

The key difference lies in the fact that instead of generating the entire set of subsets $\Pi_{\mathbf{n}}$, and add up many terms whose coefficients are $\alpha_\pi^{\widetilde{x_s}} = 0$, we restrict ourselves to those sub-FTDs $Q_{\pi_s^l}^{[1]}(T)$ that contribute to the value of the sum. Indeed, looking back at the 2^n-by-2^n matrix $\left[\alpha_\pi^{\widetilde{x_s}}\right]$, in the example of Chap. 8, one notices that it is actually very sparse. The only extreme case where all the sub-FTDs are required is for the last-to-default swap $k = n$.

10

The Homogeneous Transformation

In general, the number of sub-FTDs in the replication formula (9.7) is a function of n, the size of the basket $\{A_1, \ldots, A_n\}$, and k, the order of the basket default swap. For instance, a STD swap is replicated with $(1+n)$ sub-FTD instruments, a TTD swap is replicated with $\left(1+n+\frac{n(n+1)}{2}\right)$ sub-FTD instruments, and so on. A kth -to-default swap, in general, is replicated with all the FTD instruments referencing the l-sub-baskets, where l varies between $(n-m+1)$ and n; the size of this replicating portfolio is given by the truncated binomial expansion $N(k,n) = 1+\binom{n-1}{n}+\cdots+\binom{n-k+1}{n}$. Clearly, for small values of k and n, the total number of sub-FTDs is reasonable, but when one considers higher default orders, the combinatorial coefficients are larger, and may be impossible to handle numerically in extreme cases: a last-to-default swap, for example, requires a total of $N(n,n) = 2^n - 1$ sub-FTDs.

The most time-consuming step in the evaluation is the generation of the sub-FTDs $Q^{[1]}_{\pi^l_s}(T)$, for all combinations $\left\{\pi^l_s : \pi^l_s \in \Pi_{\mathbf{n}},\ \left|\pi^l_s\right| = l\right\}$. If we had a homogeneous basket, then, for a given subset size l, all the FTD instruments would have exactly the same value; and Eq. (9.11) would then collapse to

$$\mathbb{P}\left(\tau^{[k]} > T\right) = \sum_{l=n-k+1}^{n} \alpha^{[k]}_n(l)\binom{n}{l} Q^{[1]}_{\pi^l}(T),$$

where π^l is one l-subset of the homogeneous basket. The number of sub-FTDs to compute, would reduce to one evaluation per l-subset, hence a total of $N(k,n) = k$ FTD evaluations for the whole kth-to-default swap. So, the

Y. Elouerkhaoui, *Credit Correlation*, Applied Quantitative Finance,
DOI 10.1007/978-3-319-60973-7_10

first (natural) approximation that we consider is to transform the original non-homogeneous basket to a homogeneous one while preserving some properties of the aggregate default distribution. The homogeneous transformation idea is not new. It goes back to Moody's binomial expansion model and the diversity-score approach where they have used a transformation, which preserves the mean and the variance of the portfolio loss distribution. It is a well-known fact, however, that the first and second moments are not enough to represent heavy-tailed loss distributions. In the approach described here, for each default order, we consider the corresponding percentile of the aggregate default distribution, and we require that this quantity remains invariant with respect to the homogeneous approximation. We shall see that this transformation is exact for an FTD swap, and that, for higher-order defaults, the approximation gives very good results.

10.1 An Illustrative Example

We begin with a first-to-default swap, and we work out the homogeneous transformation that preserves its present value. We consider the non-homogeneous basket $\{A_1, \ldots, A_n\}$, which is represented with its MO decomposition: its market factor loadings matrix $\left[p_{i,j}\right]_{\substack{1\leq i\leq n\\ 1\leq j\leq m}}$, and its idiosyncratic intensities vector $\left(\lambda^{0,i}\right)_{1\leq i\leq n}$. Our objective is to find an equivalent portfolio $\left\{A_1^*, \ldots, A_n^*\right\}$, where all the underlying credits A_i^* have the same MO representation: $\left(\left(p_j^*\right)_{1\leq j\leq m}, \lambda^{0,*}\right)$, and the value of the FTD swap referencing this new portfolio is equal to the value of the original non-homogeneous FTD. As discussed previously, the FTD random time $\tau^{[1]}$, in a MO model, is exponentially distributed and its hazard rate $\lambda^{[1]}(t)$ is given by Eq. (8.14):

$$\lambda^{[1]}(\{A_1, \ldots, A_n\}) = \sum_{j=1}^{m}\left[1 - \prod_{i=1}^{n}\left(1 - p_{i,j}\right)\right]\left[\lambda^{c_j}(t)\right] + \left[\sum_{i=1}^{n}\lambda^{0,i}(t)\right].$$

To ensure that the two FTD swaps have the same value, we need to equate the hazard rates of the FTD times $\tau^{[1]}(\{A_1, \ldots, A_n\}) = \min(\tau_1, \ldots, \tau_n)$ and $\tau^{[1]}\left(\left\{A_1^*, \ldots, A_n^*\right\}\right) = \min\left(\tau_1^*, \ldots, \tau_n^*\right)$:

$$\lambda^{[1]}(\{A_1, \ldots, A_n\}) = \lambda^{[1]}\left(\left\{A_1^*, \ldots, A_n^*\right\}\right), \forall\left[\lambda^{c_j}(t)\right] \in \mathbb{R}^+, 1 \leq j \leq m. \tag{10.1}$$

Substituting the FTD hazard rates with their formulas, and because the equality (10.1) holds for all values of the factor intensities $\lambda_j^C(t)$, one gets

$$1-\prod_{i=1}^{n}\left(1-p_{i,j}\right)=1-\left(1-p_j^*\right)^n, \text{ for } 1\leq j\leq m; \text{ and } \sum_{i=1}^{n}\lambda^{0,i}(t)=n\lambda^{0,*}(t).$$

This completes the definition of the equivalent portfolio transformation. The FTD equivalent homogeneous portfolio is obtained by taking the arithmetic average of the idiosyncratic intensities, and the geometric average the market factor conditional survival probabilities $q_{i,j}=1-p_{i,j}$:

$$p_j^*=1-\exp\left(\frac{1}{n}\sum_{i=1}^{n}\log\left(1-p_{i,j}\right)\right),$$

$$\lambda^{0,*}(t)=\frac{1}{n}\sum_{i=1}^{n}\lambda^{0,i}(t).$$

10.2 The Homogeneous Transformation

Now, we turn to the general case of higher-order default swaps. We want to define the transformation

$$\left(\left[p_{i,j}\right]_{\substack{1\leq i\leq n\\1\leq j\leq m}},\left(\lambda^{0,i}\right)_{1\leq i\leq n}\right)\rightarrow\left(\left(p_j^*\right)_{1\leq j\leq m},\lambda^{0,*}\right),$$

such that the value of the kth-to-default is the same for the two portfolios $\{A_1,\ldots,A_n\}$ and $\left\{A_1^*,\ldots,A_n^*\right\}$. The key is to match the survival probabilities of the random default time $\tau^{[k]}$. Recall that the Q-factor $Q^{[k]}(T)$ and the portfolio aggregate defaults X_T are related as:

$$Q^{[k]}(T)=\mathbb{P}\left(\tau^{[k]}>T\right)=\mathbb{P}\left(X_T<k\right).$$

In order to study the properties of the random time $\tau^{[k]}$, we work with the aggregate default distribution X_T, and its probability generating function $\varphi(x)$. We show that $\varphi(x)$ can be conveniently expressed in terms of the basis FTD Q-factors $Q_\pi^{[1]}(T)$.

Theorem 85 *The probability generating function $\varphi(x)$ of the random variable X_T is given by:*

$$\varphi(x) = \sum_{\pi \in \Pi_\mathbf{n}} Q_\pi^{[1]}(T)\ x^{n-d(\pi)} (1-x)^{d(\pi)}. \tag{10.2}$$

Proof We proceed by induction: we assume that the property is verified for n, and we prove that it holds for $n+1$.

The probability generating function for a basket of $(n+1)$ underlying credits is defined by:

$$\varphi_{n+1}(x) = \sum_{k=0}^{n+1} \mathbb{P}\left(\sum_{i=1}^{n+1} D_T^i = k\right) x^k.$$

Conditioning on D_T^{n+1}, we can re-write the p.g.f. as:

$$\varphi_{n+1}(x) = \mathbb{E}\left[x^{D_T^1+\cdots+D_T^{n+1}}\right] = \mathbb{E}\left[\mathbb{E}\left[x^{D_T^1+\cdots+D_T^{n+1}} \middle| D_T^{n+1}\right]\right]. \tag{10.3}$$

D_T^{n+1} takes two values: 0 or 1, hence,

$$\begin{aligned}\varphi_{n+1}(x) = {} & \mathbb{P}\left(D_T^{n+1}=0\right)\mathbb{E}\left[x^{D_T^1+\cdots+D_T^n} \middle| D_T^{n+1}=0\right] \\ & + \mathbb{P}\left(D_T^{n+1}=1\right)\cdot x \cdot \mathbb{E}\left[x^{D_T^1+\cdots+D_T^n} \middle| D_T^{n+1}=1\right].\end{aligned} \tag{10.4}$$

Applying the induction relationship to the conditional p.g.f. of the n-sum $D_T^1 + \cdots + D_T^n$, we have

$$\mathbb{E}\left[x^{D_T^1+\cdots+D_T^n} \middle| D_T^{n+1}\right] = \sum_{\pi_n \in \Pi_\mathbf{n}} \mathbb{P}\left(\prod_{i\in\pi_n}\left(1-D_T^i\right) \middle| D_T^{n+1}\right) x^{n-d(\pi_n)} (1-x)^{d(\pi_n)}.$$

Replacing in (10.4), and observing that

$$\mathbb{P}\left(D_T^{n+1}=0\right)\mathbb{P}\left(\prod_{i\in\pi_n}\left(1-D_T^i\right) \middle| D_T^{n+1}=0\right) = \mathbb{P}\left(\left[\prod_{i\in\pi_n}\left(1-D_T^i\right)\right]\left(1-D_T^{n+1}\right)\right),$$

$$\mathbb{P}\left(D_T^{n+1}=1\right)\mathbb{P}\left(\prod_{i\in\pi_n}\left(1-D_T^i\right) \middle| D_T^{n+1}=1\right) = \mathbb{P}\left(\left[\prod_{i\in\pi_n}\left(1-D_T^i\right)\right] D_T^{n+1}\right),$$

we get

$$\varphi_{n+1}(x) = \sum_{\pi_n \in \mathbf{\Pi_n}} \mathbb{P}\left(\left[\prod_{i \in \pi_n}\left(1 - D_T^i\right)\right]\left(1 - D_T^{n+1}\right)\right) x^{n-d(\pi_n)}(1-x)^{d(\pi_n)}$$
$$+ \sum_{\pi_n \in \mathbf{\Pi_n}} \mathbb{P}\left(\left[\prod_{i \in \pi_n}\left(1 - D_T^i\right)\right] D_T^{n+1}\right) x^{1+n-d(\pi_n)}(1-x)^{d(\pi_n)}. \quad (10.5)$$

Observing that

$$\mathbb{P}\left(\left[\prod_{i \in \pi_n}\left(1 - D_T^i\right)\right] D_T^{n+1}\right) = \mathbb{P}\left(\prod_{i \in \pi_n}\left(1 - D_T^i\right)\right) - \mathbb{P}\left(\left[\prod_{i \in \pi_n}\left(1 - D_T^i\right)\right]\left(1 - D_T^{n+1}\right)\right), \quad (10.6)$$

Eq. (10.5) becomes

$$\varphi_{n+1}(x) = \sum_{\pi_n \in \mathbf{\Pi_n}} \mathbb{P}\left(\left[\prod_{i \in \pi_n}\left(1 - D_T^i\right)\right]\left(1 - D_T^{n+1}\right)\right) x^{n-d(\pi_n)}(1-x)^{1+d(\pi_n)}$$
$$+ \sum_{\pi_n \in \mathbf{\Pi_n}} \mathbb{P}\left(\prod_{i \in \pi_n}\left(1 - D_T^i\right)\right) x^{1+n-d(\pi_n)}(1-x)^{d(\pi_n)}. \quad (10.7)$$

Recall that $\mathbf{\Pi_{n+1}}$ is partitioned into two sets $\mathbf{\Pi^+_{n+1}}$ and $\mathbf{\Pi^-_{n+1}}$:

$$\mathbf{\Pi^+_{n+1}} = \left\{\pi_{n+1} : \pi_{n+1} \in \mathbf{\Pi_{n+1}},\ (n+1) \in \pi_{n+1}\right\},$$
$$\mathbf{\Pi^-_{n+1}} = \left\{\pi_{n+1} : \pi_{n+1} \in \mathbf{\Pi_{n+1}},\ (n+1) \notin \pi_{n+1}\right\}.$$

For $\pi_{n+1} \in \mathbf{\Pi^+_{n+1}}$: $\pi_{n+1} = \pi_n \cup \{n+1\}$, we have

$$Q^{[1]}_{\pi_{n+1}}(T) = \mathbb{E}\left(\prod_{i \in \pi_{n+1}}\left(1 - D_T^i\right)\right) = \mathbb{E}\left(\left[\prod_{i \in \pi_n}\left(1 - D_T^i\right)\right] \cdot \left(1 - D_T^{n+1}\right)\right),$$

and

$$d(\pi_{n+1}) = d(\pi_n) + 1.$$

For $\pi_{n+1} \in \mathbf{\Pi^-_{n+1}}$: $\pi_{n+1} = \pi_n \cup \emptyset$, we have

$$Q^{[1]}_{\pi_{n+1}}(T) = \mathbb{E}\left(\prod_{i \in \pi_{n+1}}\left(1 - D_T^i\right)\right) = \mathbb{E}\left(\prod_{i \in \pi_n}\left(1 - D_T^i\right)\right),$$

and

$$d(\pi_{n+1}) = d(\pi_{n+1}).$$

Substituting in Eq. (10.7) yields

$$\begin{aligned}\varphi_{n+1}(x) = &\sum_{\pi_{n+1}\in\Pi^{+}_{\mathbf{n+1}}} Q^{[1]}_{\pi_{n+1}}(T)\, x^{n+1-d(\pi_{n+1})}(1-x)^{d(\pi_{n+1})} \\ &+ \sum_{\pi_{n+1}\in\Pi^{-}_{\mathbf{n+1}}} Q^{[1]}_{\pi_{n+1}}(T)\, x^{n+1-d(\pi_{n+1})}(1-x)^{d(\pi_{n+1})},\end{aligned}$$

which ends the proof. □

Remark 86 Observing that the survival probability $Q^{[k]}(T)$ can be recovered from the probability generating function $\varphi(x)$, one gets

$$Q^{[k]}(T) = \sum_{l=0}^{k-1} \frac{\varphi^{(l)}(x)}{l!}.$$

Theorem 85 provides another proof of the FTD expansion formula (9.11).

Letting $\Lambda^{c_j}(T)$ and $\Lambda^{0,i}(T)$ denote the cumulative market factor and idiosyncratic intensities respectively, i.e., $\Lambda^{c_j}(T) = \int_0^T \lambda^{c_j}(t)\,dt$, for $j=1,\ldots,m$, and $\Lambda^{0,i}(T) = \int_0^T \lambda^{0,i}(t)\,dt$, for $i=1,\ldots,n$, the value of each sub-FTD Q-factor is given by Eq. (8.13),

$$Q^{[1]}_{\pi}(T) = \exp\left(-\left[\sum_{j=1}^{m}\left(1-\prod_{i\in\pi}(1-p_{i,j})\right)\Lambda^{c_j}(T)\right]-\left[\sum_{i\in\pi}\Lambda^{0,i}(T)\right]\right). \quad (10.8)$$

In order to remove the non-linear dependence on the market factor intensities, we expand the exponential function to first-order[1]: exp $\left(\sum_{j=1}^{m}\left(\prod_{i\in\pi}(1-p_{i,j})\right)\Lambda^{c_j}(T)\right) \simeq 1+\sum_{j=1}^{m}\left(\prod_{i\in\pi}(1-p_{i,j})\right)\Lambda^{c_j}(T)$, and we define a new set of adjusted market factor loadings $\widetilde{p_{i,j}} = 1-e^{-\Lambda^{0,i}(T)}(1-p_{i,j})$. Then, Eq. (10.8) becomes

$$Q^{[1]}_{\pi}(T) \simeq \exp\left(-\sum_{j=1}^{m}\Lambda^{c_j}(T)\right)\left[\exp\left(-\sum_{i\in\pi}\Lambda^{0,i}(T)\right)+\sum_{j=1}^{m}\left(\prod_{i\in\pi}(1-\widetilde{p_{i,j}})\right)\Lambda^{c_j}(T)\right]. \quad (10.9)$$

In the spirit of Eq. (8.14), used in the FTD case, this new formula expresses the survival probability as a linear combination of market factor intensities and a separate idiosyncratic contribution. Direct substitution of Eq. (10.9) in the expression of the probability generating function of Theorem 85 shows that we can write $\varphi(x)$ as a linear combination of functions $\varphi_j^c(x)$ and $\varphi_0(x)$, defined as:

$$\varphi_j^c(x) = \sum_{\pi \in \Pi_{\mathbf{n}}} \left(\prod_{i \in \pi} \left(1 - \widetilde{p_{i,j}}\right) \right) x^{n-d(\pi)} (1-x)^{d(\pi)},$$

$$\varphi_0(x) = \sum_{\pi \in \Pi_{\mathbf{n}}} \exp\left(-\sum_{i \in \pi} \Lambda^{0,i}(T) \right) x^{n-d(\pi)} (1-x)^{d(\pi)},$$

and which can be interpreted as the market factor conditional probability generating functions and the idiosyncratic probability generating function, respectively. Indeed, using Theorem 85, we can easily see that $\varphi_0(x)$ is the p.g.f. of a sum of independent Bernoulli variables X_i^0, where $\mathbb{P}\left(X_i^0 = 1\right) = 1 - e^{-\Lambda^{0,i}(T)}$, and $\varphi_j^c(x)$ is the p.g.f. of a sum of independent Bernoulli variables X_i^j, where $\mathbb{P}\left(X_i^j = 1\right) = \widetilde{p_{i,j}}$. Inverting the probability generating function $\varphi(x)$,

$$\varphi(x) \simeq \exp\left(-\sum_{j=1}^{m} \Lambda^{c_j}(T) \right) \left[\varphi_0(x) + \sum_{j=1}^{m} \Lambda^{c_j}(T) \left[\varphi_j^c(x) \right] \right],$$

gives the value of the kth-to-default Q-factor $\mathbb{P}(X_T < k)$ as a linear combination of market factor contributions and an idiosyncratic term, which allows a one-to-one mapping between the two portfolios,

$$\mathbb{P}(X_T < k) \simeq \exp\left(-\sum_{j=1}^{m} \Lambda^{c_j}(T) \right) \left[Q_0^{[k]}(T) + \sum_{j=1}^{m} Q_j^{[k]}(T) \Lambda^{c_j}(T) \right], \tag{10.10}$$

where

$$Q_0^{[k]}(T) = \mathbb{P}\left(\sum_{i=1}^{n} X_i^0 < k \right), \text{ and } Q_j^{[k]}(T) = \mathbb{P}\left(\sum_{i=1}^{n} X_i^j < k \right), \text{ for } j = 1, \ldots, m.$$

In light of Eq. (10.10), the homogeneous transformation for the kth-to-default payoff is defined as the one that keeps invariant the probability of the idiosyncratic mode $Q_0^{[k]}(T)$, and the probabilities of the market factor modes $Q_j^{[k]}(T)$. For the homogenous basket $\{A_1^*, \ldots, A_n^*\}$, the idiosyncratic and market factor probabilities $Q_0^{[k]}(T)$ and $Q_j^{[k]}(T)$ are computed directly as truncated sums of the binomial expansion: $Q_0^{[k]}(T) = B\left(k-1, n, 1-e^{-\Lambda^{0,i}(T)}\right)$, and $Q_j^{[k]}(T) = B\left(k-1, n, \widetilde{p_j^*}\right)$, where $B(k, n, p) = \sum_{i=0}^{k} \binom{n}{i} (1-p)^{n-i} (p)^i$ is the binomial distribution with parameter p. For the non-homogeneous basket $\{A_1, \ldots, A_n\}$, these probabilities can be computed by an inversion of the Fourier transform $\mathcal{F}_0(s) = \varphi_0(e^s)$ and $\mathcal{F}_j^c(s) = \varphi_j^c(e^s)$,

$$\mathbb{P}\left(\sum_{i=1}^{n} X_i^0 = k\right) = \mathcal{F}^{-1}\left(\mathcal{F}_0(s)\right)|_{s=k},$$

$$\mathbb{P}\left(\sum_{i=1}^{n} X_i^j = k\right) = \mathcal{F}^{-1}\left(\mathcal{F}_j^c(s)\right)|_{s=k}.$$

Since the variables $(X_i^0)_{1\le i\le n}$ and $\left(X_i^j\right)_{1\le i\le n}$ are independent, the p.g.f. $\varphi_0(x)$ and $\varphi_j^c(x)$ are given by the products $\varphi_0(x) = \prod_{i=1}^{n} \varphi_{X_i^0}(x)$ and $\varphi_j^c(x) = \prod_{i=1}^{n} \varphi_{X_i^j}(x)$, where $\varphi_X(x) = (px + (1-p))$ is the p.g.f. of a Bernoulli variable with parameter p. Other methods to generate the distribution of a sum of independent variables, such as the convolution recursion method, are available and are more efficient than the Fourier transform inversion.

In summary, the homogeneous transformation can be described algorithmically through the following steps:

1. Find $\Lambda^{0,*}(T)$:
   (a) generate the idiosyncratic default distribution $\left[\mathbb{P}\left(\sum_{i=1}^{n} X_i^0 = l\right)\right]_{0\le l\le n}$ by Fourier inversion;
   (b) use a Brent search to find $\Lambda^{0,*}(T)$ such that $B\left(1-e^{-\Lambda^{0,i}(T)}, k, n\right) = \mathbb{P}\left(\sum_{i=1}^{n} X_i^0 < k\right)$.
2. For each market factor, find p_j^*:
   (a) transform the loadings $p_{i,j}$ to the adjusted ones: $\widetilde{p_{i,j}} = 1 - e^{-\Lambda_i^0(T)}(1-p_{i,j})$;

(b) generate the conditional market factor default distribution $\left[\mathbb{P}\left(\sum_{i=1}^{n} X_i^j = l\right)\right]_{0\leq l\leq n}$ by Fourier inversion;

(c) use a Brent search to find $\widetilde{p_j^*}$ such that $B\left(\widetilde{p_j^*}, k, n\right) = P\left(\sum_{i=1}^{n} X_i^j < k\right)$;

(d) transform back $\widetilde{p_j^*}$ to p_j^*: $p_j^* = 1 - e^{\Lambda^{0,*}(T)}\left(1 - \widetilde{p_j^*}\right)$.

10.3 Fourier Inversion and Importance Sampling

One issue that arises in the Fourier inversion procedure is its failure to estimate accurately small probabilities in the tail of the distribution. Indeed, since underlying default probabilities are usually small, the portfolio aggregate default distribution is centred to the left, and the tail probabilities decay exponentially as we move further from the mean. When the tail probabilities are smaller that the machine double-precision $\sim 10^{-16}$, the numerical round-off errors dominate. This results in a well-documented oscillating behaviour, which is known in the numerical analysis literature as a round-off plateau (Fig. 10.1).[2]

To address this problem, we apply a technique inspired by the importance sampling method, which is often used in Monte Carlo simulations to improve the speed of the convergence. The idea is to consider the aggregate default distribution under a new probability measure. Provided that the change of measure is chosen appropriately, such that the new target distribution is centred in the tail, the percentile to be estimated will be significantly larger than the machine precision, and we are in a better position to estimate it accurately with the Fourier inversion algorithm. Once the percentile is evaluated, it can then be converted back to the original measure with the Radon-Nykodim derivative (which, in fact, is a fast-decaying-exponential).

To define the appropriate measure change, we use the concepts of tilted distributions, Esscher transforms and saddle points.[3]

For a given probability density function $f(x)$, we call tilted densities of f, the family of density functions $f_\theta(x)$ defined by

$$f_\theta(x) = f(x)\frac{e^{\theta x}}{M(\theta)} = f(x)\,e^{\theta x - K(\theta)}, \tag{10.11}$$

where $M(t) = \mathbb{E}\left[e^{tX}\right] = \int e^{tx} f(x)\,dx$ is its moment generating function, and $K(t) = \log(M(t))$ is its cumulant generating function. The mean and

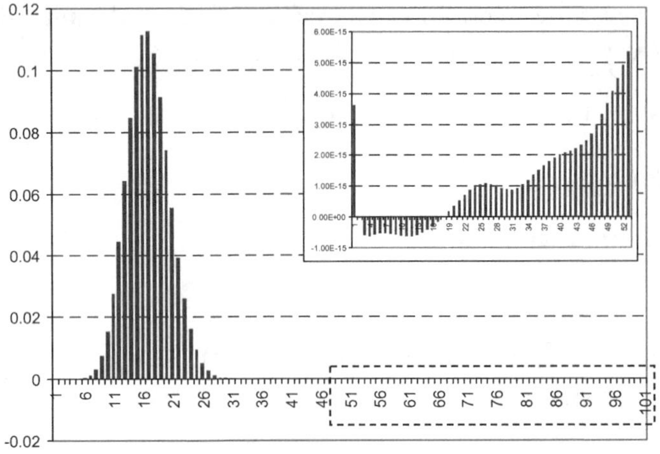

Fig. 10.1 Fourier inversion round-off plateau

the variance of the tilted distribution are given by: $\mu_\theta = K'_\theta(0) = K'(\theta)$, and $\sigma_\theta^2 = K''_\theta(0) = K''(\theta)$.

The change of probability measure corresponding to the unique solution of the equation

$$K'\left(\widehat{\theta}\right) = x \tag{10.12}$$

is called an Esscher transform. The parameter $\widehat{\theta}$ is known as the saddle-point. This, basically, defines a change of probability measure such that the new distribution is centred around x.

In our setting, we have a set of independent Bernoulli variables X_i with parameter p_i, and we need to estimate the percentile of the aggregate distribution $\mathbb{P}(X_1 + \cdots + X_n \geq k)$, for a given k.

The probability mass function of the Bernoulli variable X_i is given by:

$$f_i(x) = p_i^x (1 - p_i)^{1-x}, \; x = 0, 1,$$

its moment generating function is:

$$M_i(\theta) = p_i e^\theta + 1 - p_i,$$

and we can write its tilted probability mass as:

$$f_i^\theta(x) = \frac{e^{\theta x} p_i^x (1 - p_i)^{1-x}}{M_i(\theta)} = \widehat{p_i}^x \left(1 - \widehat{p_i}\right)^{1-x}, \text{ for } x = 0, 1,$$

with

$$\widehat{p_i} = \frac{p_i e^{\theta}}{p_i e^{\theta} + 1 - p_i}.$$

$f_i^{\theta}(x)$ corresponds to the probability mass function of a Bernoulli variable with the tilted probability $\widehat{p_i}$.

For the aggregate defaults r.v. $X_1 + \cdots + X_n$, the tilted probability mass is:

$$f^{\theta}(x) = \frac{e^{\theta x} f(x)}{M(\theta)} = \frac{e^{\theta x} f(x)}{\prod_{i=1}^{n} M_i(\theta)}, x = 0, 1, \ldots, n,$$

where, since the variables X_i are independent, the moment generating function $M(\theta)$ is given by the product of the individual m.g.f. $M_i(\theta)$. The Radon-Nykodim derivative used to transform $f^{\theta}(x)$ back to the original density $f(x)$ is:

$$f(x) = f^{\theta}(x) e^{K(\theta) - \theta x}.$$

Using the tilted probabilities $(\widehat{p_1}, \ldots, \widehat{p_n})$, we generate the distribution of the aggregate defaults under $\mathbb{P}^{\widehat{\theta}}$; then multiplying by the Radon-Nykodim derivative $\frac{d\mathbb{P}}{d\mathbb{P}^{\theta}} = e^{K(\theta)-\theta x}$, we arrive at the probability distribution of the random variable $X_1 + \cdots + X_n$ under the original probability measure $\mathbb{P}$.

To summarize, the algorithm to compute the tail probability $\mathbb{P}\left(\sum_{i=1}^{n} X_i < k\right)$ is as follows:

1. Step 1: use a Brent method to solve for the saddle point $\widehat{\theta}$, such that $K'\left(\widehat{\theta}\right) = k$. $K'(t)$ is given by

$$K'(t) = \sum_{i=1}^{n} \frac{p_i e^{t}}{p_i e^{t} + 1 - p_i}.$$

2. Step 2: compute the tilted probabilities $\widehat{p_i}$

$$\widehat{p_i} = \frac{p_i e^{\widehat{\theta}}}{p_i e^{\widehat{\theta}} + 1 - p_i}.$$

3. Step 3: generate the aggregate default distribution, in the new measure $\mathbb{P}^{\widehat{\theta}}$, $\left[\mathbb{P}^{\widehat{\theta}}\left(\sum_{i=1}^{n} X_i = j\right)\right]_{0 \le j \le n}$ by Fourier inversion.

4. Step 4: compute the default distribution in the original measure:

$$\mathbb{P}\left(\sum_{i=1}^{n} X_i = j\right) = \mathbb{P}^{\widehat{\theta}}\left(\sum_{i=1}^{n} X_i = j\right) \exp\left(K\left(\widehat{\theta}\right) - \widehat{\theta} j\right).$$

Notes

1. This expansion is a very good approximation since the values of the intensities $\left(\Lambda_j^c(T)\right)_{1 \leq j \leq m}$ are usually small.
2. See Boyd (2000) for a discussion on machine precision and round-off plateaus.
3. A good reference on exponential tilting can be found in Ross (1997).

References

J.P. Boyd, *Chebyshev and Fourier Spectral Methods* (Dover Publications, New York, 2000)

S.M. Ross, *Simulation* (Academic Press, Boston, 1997)

11

The Asymptotic Homogeneous Expansion

The transformation, described in the previous chapter, produces a homogeneous portfolio, which mimics some properties of the aggregate default distribution, and can be used pari-pasu for the purposes of basket default swap valuation. By using this homogeneous portfolio, the numerical burden that comes with the pricing of large baskets is eased, and the valuation algorithm is significantly speeded up. From Eq. (9.6), the kth-to-default survival probability $Q_n^{[k]} = \mathbb{P}\left(\tau^{[k]} > T\right)$ for the n -dimensional homogeneous portfolio $\left\{A_1^*, \ldots, A_n^*\right\}$ can be computed recursively as:

$$Q_n^{[k]} = \left(\frac{n}{k-1}\right)\left(Q_{n-1}^{[k-1]} - Q_n^{[k-1]}\right) + Q_n^{[k-1]}. \tag{11.1}$$

This simple-looking recursion hides a nasty numerical problem: it is numerically unstable. As one moves up the recursion tree, the numerical round-off errors (which originate mainly from the difference $\left(Q_{n-1}^{[k-1]} - Q_n^{[k-1]}\right)$) propagate rapidly, and the resulting prices are completely erroneous. In fact, a closer look at Eq. (11.1) shows that the latter resembles the discretized finite-difference scheme of the wave equation PDE, with time-dependent and space-dependent velocity:

$$\begin{aligned} Q_n^{[k]} - Q_n^{[k-1]} &= -\left(\frac{n}{k-1}\right)\left(Q_n^{[k-1]} - Q_{n-1}^{[k-1]}\right), \\ \frac{\partial Q_{x,t}}{\partial t} &= -v_{x,t}\frac{\partial Q_{x,t}}{\partial x}. \end{aligned} \tag{11.2}$$

Y. Elouerkhaoui, *Credit Correlation*, Applied Quantitative Finance,
DOI 10.1007/978-3-319-60973-7_11

Equation (11.2) is a well-known unstable difference scheme.[1]

To address this issue, we take a different route: rather than using the recursive approach, we study the asymptotic behaviour of the homogeneous portfolio. We show, in this chapter, that the solution $Q_n^{[k]}$ admits an asymptotic series expansion, and we explain how to compute each term in the expansion.

11.1 Asymptotic Series Expansion

We fix a time horizon T, and we consider the kth-to-default Q-factor: $Q^{[k]}(T) = \mathbb{P}\left(\tau^{[k]} > T\right)$.

Proposition 87 *Asymptotic Series Expansion*

The homogeneous portfolio kth-to-default Q-factor has the following series expansion:

$$Q^{[k]}(T) = e^{-\Lambda^c(T)} \left[\sum_{n_1=0}^{+\infty} \cdots \sum_{n_m=0}^{+\infty} \frac{\Lambda^{c_1}(T)^{n_1}}{n_1!} \cdots \frac{\Lambda^{c_m}(T)^{n_m}}{n_m!} B\left(k-1, n, p_{n_1,\ldots,n_m}\right) \right], \tag{11.3}$$

where

$$p_{n_1,\ldots,n_m} = 1 - e^{-\Lambda^{0,*}(T)} \left(1 - p_1^*\right)^{n_1} \cdots \left(1 - p_m^*\right)^{n_m},$$

and $B(k, n, p)$ is the cumulative Binomial probability with parameter p:

$$B(k, n, p) = \sum_{j=0}^{k} \binom{n}{j} p^j (1-p)^{n-j}.$$

Proof We consider the p.g.f. of the defaults counting variable X_T. By virtue of Theorem 85, $\varphi(x)$ is given by

$$\varphi(x) = \sum_{\pi \in \Pi_\mathbf{n}} Q_\pi^{[1]}(T)\ x^{n-d(\pi)} (1-x)^{d(\pi)},$$

where the sub-FTD Q-factors of the homogeneous basket are

$$Q_\pi^{[1]}(T) = \exp\left(- \left[\sum_{j=1}^{m} \left(1 - \left(1 - p_j^*\right)^{d(\pi)}\right) \Lambda^{c_j}(T) \right] - \left[d(\pi)\, \Lambda^{0,*}(T) \right] \right). \tag{11.4}$$

Replacing each exponential $\exp\left(\left(1-p_j^*\right)^{d(\pi)}\Lambda^{c_j}(T)\right)$ by its series expansion,

$$\exp\left(\left(1-p_j^*\right)^{d(\pi)}\Lambda^{c_j}(T)\right)=\sum_{n_j=0}^{+\infty}\frac{\left[\left(1-p_j^*\right)^{d(\pi)}\Lambda^{c_j}(T)\right]^{n_j}}{n_j!},$$

Eq. (11.4) becomes

$$\begin{aligned}
&Q\left(\widetilde{x_s}\right)\\
&=e^{-\Lambda^c(T)-\left[d(\pi)\Lambda^{0,*}(T)\right]}\sum_{n_1=0}^{+\infty}\cdots\sum_{n_m=0}^{+\infty}\frac{\left[\left(1-p_j^*\right)^{d(\pi)}\Lambda^{c_1}(T)\right]^{n_1}}{n_1!}\cdots\frac{\left[\left(1-p_j^*\right)^{d(\pi)}\Lambda^{c_m}(T)\right]^{n_m}}{n_m!}\\
&=e^{-\Lambda^c(T)}\sum_{n_1=0}^{+\infty}\cdots\sum_{n_m=0}^{+\infty}\frac{\Lambda^{c_1}(T)^{n_1}}{n_1!}\cdots\frac{\Lambda^{c_m}(T)^{n_m}}{n_m!}e^{-d(\pi)\Lambda^*_{n_1,\ldots,n_m}(T)},
\end{aligned}\tag{11.5}$$

where

$$\Lambda^*_{n_1,\ldots,n_m}(T)=\Lambda^{0,*}(T)+\log\left(\left(1-p_1^*\right)^{n_1}\right)+\cdots+\log\left(\left(1-p_m^*\right)^{n_m}\right).$$

Substituting (11.5) in the expression of the probability generating function gives

$$\varphi(x)=e^{-\Lambda^c(T)}\sum_{n_1=0}^{+\infty}\cdots\sum_{n_m=0}^{+\infty}\frac{\Lambda^{c_1}(T)^{n_1}}{n_1!}\cdots\frac{\Lambda^{c_m}(T)^{n_m}}{n_m!}\varphi_{n_1,\ldots,n_m}(x),$$

where

$$\begin{aligned}
\varphi_{n_1,\ldots,n_m}(x)&=\sum_{\pi\in\Pi_{\mathbf{n}}}Q_{n_1,\ldots,n_m}(\pi)\,x^{n-d(\pi)}(1-x)^{d(\pi)},\\
Q_{n_1,\ldots,n_m}(\pi)&=\exp\left(-d(\pi)\,\Lambda^*_{n_1,\ldots,n_m}(T)\right).
\end{aligned}$$

Using Theorem 85 again, it is easy to see that $\varphi_{n_1,\ldots,n_m}(x)$ is, in fact, the probability generating function of a set of n independent identically distributed variables $(Z_1,\ldots,Z_n)$, with an idiosyncratic intensity $\Lambda^*_{n_1,\ldots,n_m}(T)$, or a default probability $\mathbb{P}(Z_i=1)=1-e^{-\Lambda^*_{n_1,\ldots,n_m}(T)}$, for $i=1,\ldots,n$. Indeed, $Q_{n_1,\ldots,n_m}(\pi)$ is the sub-FTD Q-factor of the π-subset of $(Z_1,\ldots,Z_n)$,

$$Q_{n_1,\dots,n_m}(\pi) = \prod_{i\in\pi} \exp\left(-\Lambda^*_{n_1,\dots,n_m}(T)\right) = \prod_{i\in\pi} \mathbb{P}(Z_i = 0).$$

The p.g.f. $\varphi_{n_1,\dots,n_m}(x)$ is, then, given by the product of the p.g.f. of the individual independent variables Z_i:

$$\varphi_{n_1,\dots,n_m}(x) = \left(x + \left(e^{-\Lambda^0_*(T)}\left(1-p_1^*\right)^{n_1}\dots\left(1-p_m^*\right)^{n_m}\right)(1-x)\right)^n.$$

In other words, $\varphi_{n_1,\dots,n_m}(x)$ is the probability generating function of a Binomial distribution with parameter $p_{n_1,\dots,n_m} = 1 - e^{-\Lambda^{0,*}(T)}\left(1-p_1^*\right)^{n_1}\dots\left(1-p_m^*\right)^{n_m}$, and $\varphi(x)$ is a weighted average of the Binomial p.g.f. $\varphi_{n_1,\dots,n_m}(x)$:

$$\varphi(x) = \sum_{n_1=0}^{+\infty}\dots\sum_{n_m=0}^{+\infty} w_{n_1,\dots,n_m}\cdot\varphi_{n_1,\dots,n_m}(x),$$

$$w_{n_1,\dots,n_m} = e^{-\Lambda^c(T)}\frac{\Lambda^{c_1}(T)^{n_1}}{n_1!}\dots\frac{\Lambda^{c_m}(T)^{n_m}}{n_m!}.$$

A direct inversion of the probability generating function completes the proof. ■

Remark 88 Note that the weights $w_{n_1,\dots,n_m}$ in front of each Binomial function sum up to 1:

$$e^{-\Lambda^c(T)}\left[\sum_{n_1=0}^{+\infty}\sum_{n_2=0}^{+\infty}\dots\sum_{n_m=0}^{+\infty}\frac{\Lambda^{c_1}(T)^{n_1}}{n_1!}\frac{\Lambda^{c_2}(T)^{n_2}}{n_2!}\dots\frac{\Lambda^{c_m}(T)^{n_m}}{n_m!}\right] = 1.$$

Equation (11.3) can be seen as the formula of a "Binomial Mixture" i.e., a weighted average of Binomial distributions. The first mode $\left(w_{0,\dots,0}, p_{0,\dots,0}\right)$ is the "independent" mode, or the pure idiosyncratic mode; then the "correlation" modes are added on top. Each correlation mode corresponds to a market factor contribution, or a combination thereof.

As n goes to infinity, each Binomial mode converges to a Gaussian distribution, and Eq. (11.3) becomes the formula of a "Gaussian Mixture".

Figure 11.1 depicts an example of a typical default distribution in a M.O. model with three types of market factors: "World" driver, Beta driver and Sector driver. In Fig. 11.2, we give examples of the Binomial mixture modes with different values of the exponents (n_W, n_B, n_S).

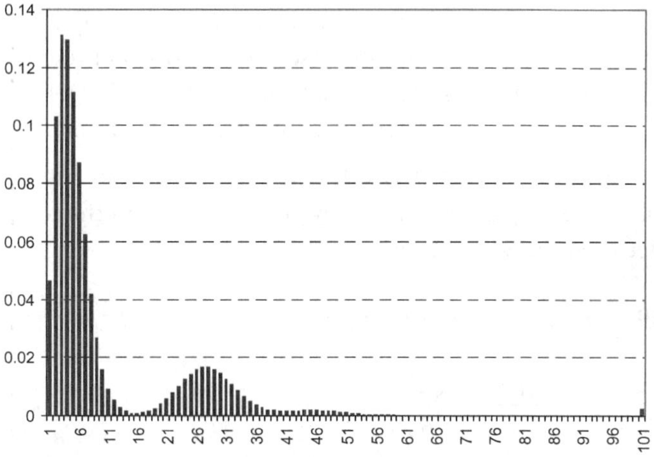

Fig. 11.1 Binomial mixture modes. There are four modes in this example: the first one corresponds to the idiosyncratic defaults; the second one is the Beta mode, representing joint defaults in different sectors; the third hump, which is less pronounced, is the sector mode; and finally, the last peak, at the end of the distribution, is a state of the world where all credits defaults

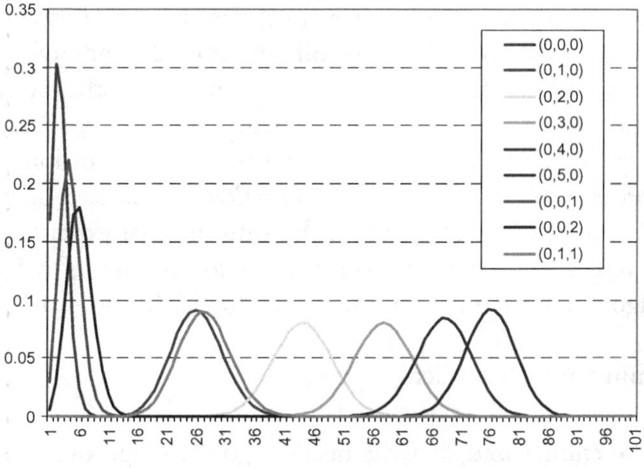

Fig. 11.2 Conditional binomial distributions in the asymptotic homogeneous expansion

In practice, rather than treating the market factor Poisson jumps separately, it is more efficient to consider the process $\{N^c(t), t \geq 0\}$:

$$N_t^c \triangleq \sum_{j=1}^{m} N_t^{c_j},$$

which counts the total number of market factor events of any type.[2] Since N_t^c is the sum of independent Poisson processes, it is a Poisson process as well, and its intensity is given by the sum of intensities: $\lambda^c(t) = \lambda^{c_1}(t) + \cdots + \lambda^{c_m}(t)$. For a fixed time horizon T, conditional on the total number of jumps, the probability of a jump of type j is given by $\pi_j = \frac{\Lambda^{c_j}(T)}{\Lambda^c(T)}$, for $1 \leq j \leq m$. Re-writing the series expansion (11.3) by conditioning on the total number of jumps, we get the following result.

Corollary 89 *The asymptotic series expansion (11.3) can be re-written as*

$$Q^{[k]}(T) = \sum_{n=0}^{+\infty} e^{-\Lambda^c(T)} \frac{\Lambda^c(T)^n}{n!} \left[\sum_{n_1+\cdots+n_m=n} \frac{n!}{n_1!\ldots n_m!} \pi_1^{n_1} \ldots \pi_m^{n_m} B\left(k-1, n, p_{n_1,\ldots,n_m}\right) \right]. \qquad (11.6)$$

The conditional terms $\frac{n!}{n_1!\ldots n_m!}\pi_1^{n_1} \ldots \pi_m^{n_m}$ are the probabilities of a multinomial distribution with parameters $(\pi_1, \ldots, \pi_m)$. For each expansion order n, the number of terms in the multinomial expansion (i.e., the number of sets $(n_1, \ldots, n_m)$ such that $n_1 + \cdots + n_m = n$) is equal to $\binom{n+m-1}{m-1}$. For high values of m and n, the number of terms in the multinomial expansion can be very large. For a typical diversified portfolio of 100–125 underlying credits, we find that the number of market factors, representing mostly various industry sectors, is of the order of $m \sim 25$. In order to reduce the truncation error in (11.6), this type of portfolios usually requires the series expansion cut off point to be of the order of $N \sim 15$. For $m = 25$ market factors, as soon as we exceed order $n = 6$, the number of terms in the multinomial expansions becomes significantly large, and very time-consuming to compute. To improve this part of the algorithm, we use a n-dimensional Quasi-Monte-Carlo integration technique.

A few comments are in order.

- It should be emphasized that the need to go into high orders is driven by the values of the cumulative intensity $\Lambda^c(T) = \Lambda^{c_1}(T) + \cdots + \Lambda^{c_m}(T)$; the truncation order is determined so that

$$\left| 1 - \sum_{n=0}^{N} e^{-\Lambda^c(T)} \frac{\Lambda^c(T)^n}{n!} \right| < \varepsilon_N,$$

where ε_N is a given accuracy level. Longer maturities, higher market factor intensities, and larger number of market factors require more orders in the expansion to achieve the same level of accuracy.

- The conditional jump type probabilities $(\pi_1, .., \pi_m)$ are, by construction, significantly higher than zero, therefore the convergence properties of the Quasi-Monte-Carlo sample are very good, which in turn implies that only a reduced number of paths is usually needed (typically, number of samples is $\sim$1000 paths).
- Estimating the probabilities of the multinomial distribution is a well-posed integration problem on the n-dimensional hypercube, which is particularly suited for Quasi-Monte-Carlo.

11.2 Quasi-Monte Carlo Integration of the Conditional Distribution

Re-writing Eq. (11.6) in terms of expectations, the kth-to-default Q-factor $Q^{[k]}(T)$ is:

$$Q^{[k]}(T) = \sum_{n=0}^{+\infty} e^{-\Lambda^c(T)} \frac{\Lambda^c(T)^n}{n!} \mathbb{P}\left(B\left(k-1, n, p_{N_T^{c1},\ldots,N_T^{cm}}\right) \middle| N_T^c = n\right).$$

The conditional expectation $\mathbb{P}\left(. \middle| N_T^c = n\right)$ is given analytically by summing over the weights of the multinomial distribution,

$$\begin{aligned}&\mathbb{P}\left(B\left(k-1, n, p_{N_T^{c1},\ldots,N_T^{cm}}\right) \middle| N_T^c = n\right)\\ &= \sum_{n_1+\cdots+n_m=n} \frac{n!}{n_1!\ldots n_m!} \pi_1^{n_1} \ldots \pi_m^{n_m} B\left(k-1, n, p_{n_1,\ldots,n_m}\right).\end{aligned}$$

Here, we estimate this expectation numerically using Quasi-Monte Carlo integration.[3]

We use an n-dimensional Sobol sequence to simulate a vector of n indices $(x_1, \ldots, x_n)$, where the x_i's are discrete random variables with the following distribution:

$$P(x_i = j) = \pi_j, \text{ for } 1 \leq j \leq m.$$

For each path, the parameter of the conditional Binomial distribution $p_{N_T^{c1},\ldots,N_T^{cm}}$ is given by:

$$p_{N_T^{c1},\ldots,N_T^{cm}} = 1 - e^{-\Lambda^{0,*}(T)} \left(1 - p_{x_1}^*\right) \ldots \left(1 - p_{x_n}^*\right) = p_{x_1,\ldots,x_n},$$

and the conditional expectation is evaluated as:

$$\begin{aligned}
&\mathbb{P}\left(B\left(k-1, n, p_{N_T^{c1},\ldots,N_T^{cm}}\right) \middle| N_T^c = n\right) \\
&= \mathbb{P}\left(B\left(k-1, n, p_{x_1,\ldots,x_n}\right)\right) \\
&\simeq \frac{1}{N}\sum_{path=1}^{N} B\left(k-1, n, p_{x_1^{path},\ldots,x_n^{path}}\right).
\end{aligned}$$

In practice, we compute the dominant terms of the series expansion in closed form, and we estimate the higher-order terms, needed for the convergence of the asymptotic series (i.e., to reduce the truncation error ε_n), by numerical integration. For a typical basket of 25 market factors, the orders 1 through 6 use the closed form formula; orders 7–15 use Quasi-Monte Carlo. Sobol integration is particularly appropriate for this type of high dimensional integration problems, it benefits from the space filling properties of this low-discrepancy sequence. We can also use other low-discrepancy sequences, such as Faure, Halton, Haselgrove.[4]

Notes

1. We refer the reader to Press et al. (1992).
2. See Duffie and Pan (2001).
3. A similar approach is used in Merino and Nyfeler (2002).
4. See Niederreiter (1992) for a discussion of low-discrepancy sequences and their application to high-dimensional integration.

References

D. Duffie, J. Pan, Analytical value-at-risk with jumps and credit risk. Financ. Stoch. **5**, 155–180 (2001)

S. Merino, M. Nyfeler, Calculating portfolio loss. Risk, 82–86 (2002)

H. Niederreiter, *Random Number Generation and Quasi-Monte Carlo Methods* (Society for Industrial and Applied Mathematics, Philadelphia, 1992)

W.H. Press, S.A. Teukolsky, W.T. Vetterling, B.P. Flannery, *Numerical Recipes in C* (Cambridge University Press, Cambridge, 1992)

12

The Asymptotic Expansion

In this chapter, we relax the homogeneous portfolio assumption, and we derive an asymptotic series expansion of the kth-to-default Q-factor in the non-homogeneous case. We also show how to compute the conditional aggregate default distributions that appear in the expansion using the convolution recursion algorithm. The latter and other recursive methods have been traditionally used in actuarial mathematics to evaluate ruin probabilities and insurance premia.

12.1 The Asymptotic Series Expansion

We consider the kth-to-default Q-factor, at a fixed time horizon T, for the non-homogeneous portfolio $\{A_1, ..., A_n\}$. By $B(k, \widetilde{p})$, we denote the cumulative probability of a random variable $Y_1 + ... + Y_n$,

$$B(k, \widetilde{p}) = \sum_{j=0}^{k} \mathbb{P}\left(\sum_{i=1}^{n} Y_i = j\right),$$

where $(Y_1, ..., Y_n)$ are n independent Bernoulli variables with parameters $\widetilde{p} = (p_1, ..., p_n)$.

Proposition 90 *Asymptotic Series Expansion*

Y. Elouerkhaoui, *Credit Correlation*, Applied Quantitative Finance,
DOI 10.1007/978-3-319-60973-7_12

The kth-to-default Q-factor $Q^{[k]}(T)$ for the non-homogeneous portfolio $\{A_1, ..., A_n\}$ has the following series expansion:

$$Q^{[k]}(T) = e^{-\Lambda^c(T)} \left[\sum_{n_1=0}^{+\infty} \cdots \sum_{n_m=0}^{+\infty} \frac{\Lambda^{c_1}(T)^{n_1}}{n_1!} \cdots \frac{\Lambda^{c_m}(T)^{n_m}}{n_m!} B\left(k-1, \widetilde{p_{n_1,...,n_m}}\right) \right], \tag{12.1}$$

where the probability vector , $\widetilde{p_{n_1,...,n_m}} = \left(p_{n_1,...,n_m}(1), ..., p_{n_1,...,n_m}(n)\right)$ *is given by:*

$$p_{n_1,...,n_m}(i) = 1 - e^{-\Lambda^{0,i}(T)} \left(1 - p_{i,1}\right)^{n_1} ... \left(1 - p_{i,m}\right)^{n_m}.$$

Proof As before, we proceed as follows:

1. we consider the probability generating function of the aggregate defaults counting variable $\varphi(x)$,
2. we expand the exponentials $\left[\exp\left(\prod_{i\in\pi}\left(1-p_{i,j}\right)\Lambda^{c_j}(T)\right)\right]$ that appear in the sub-FTD Q-factors $Q_\pi^{[1]}(T)$,
3. which is then substituted in the formula of the p.g.f. $\varphi(x)$,
4. and we arrive at the expression of $\varphi(x)$ as a weighted average of conditional independent probability generating functions $\varphi_{n_1,...,n_m}(x)$:

$$\varphi(x) = \sum_{n_1=0}^{+\infty} \cdots \sum_{n_m=0}^{+\infty} w_{n_1,...,n_m} \cdot \varphi_{n_1,...,n_m}(x),$$

$$w_{n_1,...,n_m} = e^{-\Lambda^c(T)} \frac{\Lambda^{c_1}(T)^{n_1}}{n_1!} \cdots \frac{\Lambda^{c_m}(T)^{n_m}}{n_m!},$$

where

$$\varphi_{n_1,...,n_m}(x) = \prod_{i=1}^{n} \left(x + \left(e^{-\Lambda^{0,i}(T)} \left(1 - p_{i,1}\right)^{n_1} ... \left(1 - p_{i,m}\right)^{n_m}\right)(1-x)\right),$$

is the probability generating function of the sum of n independent Bernoulli variables $(Y_1, ..., Y_n)$ with parameters $\widetilde{p_{n_1,...,n_m}} = \left(p_{n_1,...,n_m}(1), ..., p_{n_1,...,n_m}(n)\right)$:

$$p_{n_1,...,n_m}(i) = 1 - e^{-\Lambda^{0,i}(T)} \left(1 - p_{i,1}\right)^{n_1} ... \left(1 - p_{i,m}\right)^{n_m}.$$

Equation (12.1) follows from the inversion of the probability generating function $\varphi(x)$.

■

Conditioning on $N_t^c \triangleq N_t^{c1} + ... + N_t^{cm}$, the total number of market factor shocks in $(0, t]$, we obtain the compact version of Eq. (12.1).

Corollary 91 *The asymptotic series (12.1), for the non-homogeneous portfolio* $\{A_1, ..., A_n\}$, *can be written as the follows:*

$$Q^{[k]}(T) = \sum_{n=0}^{+\infty} e^{-\Lambda^c(T)} \frac{\Lambda^c(T)^n}{n!} \left[\sum_{n_1+...+n_m=n} \frac{n!}{n_1!...n_m!} \pi_1^{n_1} ... \pi_m^{n_m} B\left(k-1, \widetilde{p_{n_1,...,n_m}}\right) \right]. \tag{12.2}$$

As with the AHX method, the dominant terms in Eq. (12.2) are computed in closed form, and the higher-order terms are estimated with Sobol numerical integration.

12.2 Recursion Methods for the Computation of Aggregate Distributions

To compute the aggregate default distributions $B\left(k, \widetilde{p_{n_1,...,n_m}}\right)$ that appear in the series expansion (12.2), we can use the standard FFT algorithm; we can also use other recursive methods, which were studied extensively in the actuarial literature. Here, we present the convolution recursion and we refer the reader to the literature for other approximate recursions by Panjer (1981), Kornya (1983), Hipp (1986) and DePril (1986).

The general problem is the following: for a set of n independent discrete random variables $(Y_1, ..., Y_n)$, we want to derive the distribution of the sum

$$S_n = Y_1 + ... + Y_n. \tag{12.3}$$

The generating function of the random variable S_n,

$$\varphi_{S_n}(x) = \sum_{k=0}^{\infty} \mathbb{P}(S_n = k)\, x^k,$$

is given by the product of the generating functions of the variables Y_i (since they are all independent):

$$\varphi_S(x) = \prod_{i=1}^{n} \varphi_{Y_i}(x),$$

and its distribution $p_{S_n}(s) = \mathbb{P}(S_n = s)$ is given by the n-fold convolution of the distributions p_{Y_i}:

$$p_{S_n}(s) = \bigotimes_{i=1}^{n} p_{Y_i}(s) . \tag{12.4}$$

We compute this convolution product by applying the following recursion:

$$\begin{aligned} p_{S_{k+1}}(s) &= p_{S_k} \otimes p_{Y_{k+1}}(s) \\ &= \sum_{y=0}^{s} p_{Y_{k+1}}(y)\, p_{S_k}(s-y), \quad \text{for } 1 \le k \le n-1. \end{aligned} \tag{12.5}$$

In our set-up, the aggregate default distributions of Eq. (12.2) are sums of independent Bernoulli variables. The distribution density of the sum S_n, when $Y_1, \ldots, Y_n$ are Bernoulli variables with parameters $\widetilde{p} = (p_1, \ldots, p_n)$, is computed recursively as follows:

for $0 \le k \le n-1$,

$$p_{S_{k+1}}(s) = p_{k+1}\, p_{S_k}(s-1) + (1 - p_{k+1})\, p_{S_k}(s), \quad 0 \le s \le k+1,$$

with the convention

$$p_{S_k}(-1) = 0,$$

and the distribution of the empty sum S_0 being defined as:

$$\begin{cases} p_{S_0}(0) = 1 \\ p_{S_0}(s) = 0 \text{ for } s > 0 \end{cases}.$$

12.3 Numerical Comparisons

We consider a portfolio of $n = 100$ underlying credits, where we have 10 credits per sector, and all the individual intensities are equal to $\lambda_i = 200$

bps. We assume that the MO representation of each credit is given by:

$$\lambda_i = \lambda^W + p_{i,B}\lambda^B + p_{i,S_\cdot}\lambda^{S_\cdot} + \lambda^{0,i},$$

where the world driver is $\lambda^W = 5$ bps, the Beta driver is $\lambda^B = 500$ bps, and the sector drivers are $\lambda^{S_j} = 250$ bps. The loadings are fixed at $p_{i,B} = 0.24$ and $p_{i,S_j} = 0.16$, so that 60% of the intensity is due to the Beta component $\left[p_{i,B}\lambda^B\right]$ and 20% is due to the sector component $\left[p_{i,S_\cdot}\lambda^{S_\cdot}\right]$.

We compute the break-even spreads $\left(s^{[1]}, \ldots, s^{[n]}\right)$ of the kth-to-default swaps referencing this portfolio; and we compare the the accuracy of various approximations with the exact asymptotic solution (12.2). We estimate the accuracy of the asymptotic homogeneous expansion, as well as the approximation suggested by Duffie and Pan (2001), and the Panjer recursion based solution developed in Lindskog and McNeil (2003).

In Duffie and Pan (2001), the default indicator D_T^i is approximated by:

$$D_T^i \simeq \sum_{j=1}^{m} A_{i,j} D_T^{c_j} + D_T^{0,i}, \tag{12.6}$$

$$X_T = \sum_{i=1}^{n} D_T^i.$$

where $D_T^{c_j} \triangleq \mathbf{1}_{\left\{N_T^{c_j}>0\right\}}$ and $D_T^{0,i} \triangleq \mathbf{1}_{\left\{N_T^{0,i}>0\right\}}$.

In Lindskog and McNeil (2003), the aggregate defaults counter is approximated by the sum of Poisson variables, and then Panjer's recursion is used to derive its distribution:

$$N_T^i = \sum_{j=1}^{m} \sum_{\theta_r^{c_j} \le t} A_{\theta_r^{c_j}}^{i,j} + N_t^{0,i},$$

$$X_T \simeq \sum_{i=1}^{n} N_T^i. \tag{12.7}$$

By construction, approximation (12.6) underestimates the number of joint defaults since it considers only the first shock of the common Poisson processes $N_T^{c_j}$. On the other hand, approximation (12.7) overestimates the number of defaults since it accounts for all the Poisson events N_T^i.

Figure 12.1 shows the relative error of the three approximations.

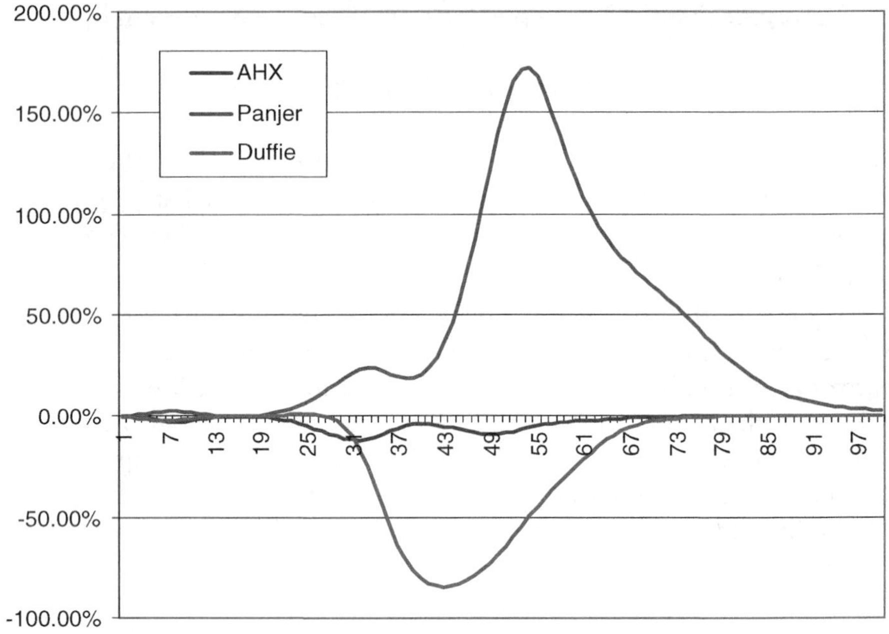

Fig. 12.1 Comparison between the accuracy of the AHX method, the Panjer approximation and Duffie's approximation

12.4 Concluding Remarks

We have considered the problem of pricing large basket credit derivatives in the Marshall-Okin framework with semi-analytic approaches. We have shown that the direct approach explodes exponentially as the basket size increases. As an alternative, we have explored a wide spectrum of methods that evolved from the replicating portfolio method to the asymptotic series expansion of the basket. we have presented how to compute each conditional distribution in the series expansion with a Fourier transform method, or a recursive convolution method. Recursion methods are very popular in insurance mathematics, and have been studied extensively in the actuarial literature. Combining the tools of stochastic calculus, numerical analysis and actuarial mathematics offers a very powerful platform for addressing the issues raised by large portfolio credit derivatives.

References

N. De Pril, On the exact computation of the aggregate claims distribution in the individual life model. ASTIN Bullet. **16**, 109–112 (1986)

D. Duffie, J. Pan, Analytical value-at-risk with jumps and credit risk. Financ. Stochast. **5**, 155–180 (2001)

C. Hipp, Improved approximations for the aggregate claims distribution in the individual model. ASTIN Bullet. **16**, 89–100 (1986)

P.S. Kornya, Distribution of the aggregate claims in the individual risk theory model. Trans. Soc. Actuar. **35**, 823–836. Discussion pp. 837–858

F. Lindskog, A. McNeil, Common Poisson shock models: applications to insurance and credit risk modelling. ASTIN Bullet. **33**(2), 209–238 (2003)

H. Panjer, Recursive evaluation of a family of compound distributions. ASTIN Bullet. **12**, 22–26 (1981)

13

CDO-Squared: Correlation of Correlation

In this chapter, we analyze the "correlation of correlation" risk in the Marshall-Olkin copula framework. The valuation of higher-order correlation products such as "CDOs of CDOs" (also known as "CDO-Squared") is mainly driven by correlation of correlation effects. First, We extend the first-to-default replication method to baskets of basket products. Then, we develop an intuitive methodology for analyzing this type of structures. The idea is to model each underlying basket security as a single name process, and to derive its equivalent intensity process and its decomposition on the MO common market factors. This, in turn, defines the multivariate dependence between the underlying basket securities in the portfolio.

13.1 Introduction

We study the default correlation risk in CDO-Squared structures, where the underlying portfolio is also a basket of CDO tranche securities. The payoff of the structure depends on the aggregate losses of the underlying tranches. For a universe of n credits, we can define p underlying portfolios with a matrix of notionals $\left[N_{ij}\right]_{\substack{1\leq j\leq p \\ 1\leq i\leq n}}$, if a credit i is not included in portfolio j, then $N_{ij} = 0$. For each portfolio, we define a CDO tranche with the lower and upper loss bounds $\left(\alpha_j\right)_{1\leq j\leq p}$ and $\left(\beta_j\right)_{1\leq j\leq p}$.

Y. Elouerkhaoui, *Credit Correlation*, Applied Quantitative Finance,
DOI 10.1007/978-3-319-60973-7_13

The loss variable L_t^j, for each portfolio $1 \leq j \leq m$, is

$$L_t^j \triangleq \sum_{i=1}^{n} N_{ij} (1 - \delta_i) \mathbf{1}_{\{\tau_i \leq t\}}.$$

Each CDO tranche $1 \leq j \leq p$, with attachment points $0 \leq K_1^j < K_2^j \leq 1$, covers the losses from $\alpha_j \triangleq K_1^j \sum_{i=1}^n N_{ij}$ to $\beta_j \triangleq K_2^j \sum_{i=1}^n N_{ij}$. The tranche loss variable is given by

$$M_t^{j,\alpha_j,\beta_j} \triangleq \min\left(\max\left(L_t^j - \alpha_j, 0\right), \beta_j - \alpha_j\right).$$

Now, we can define the aggregate loss on the portfolio of CDO tranches as the sum of the tranche losses

$$L_t \triangleq \sum_{j=1}^{p} M_t^{j,\alpha_j,\beta_j},$$

and the (α, β)-tranche of the CDO of CDOs is given by

$$M_t^{\alpha,\beta} \triangleq \min\left(\max\left(L_t - \alpha, 0\right), \beta - \alpha\right).$$

The valuation of CDOs of CDOs depends on a higher-order type of default correlation risk that we refer to as "correlation of correlation". In the same way that "options on options" are sensitive to a "volatility of volatility" parameter, CDO-Squareds are sensitive to correlation of correlation. The choice of the copula function used to model the default times' multivariate dependence and its calibration becomes particularly relevant to the pricing problem. While any mis-calibration of the copula model may have a first order effect on standard CDOs, this error can be much larger for CDO-Squareds. Similarly, we could also define baskets of baskets i.e., nth-to-default swaps referencing a basket of nth-to-default securities.

Typical CDO-Squareds use a master pool of 250–350 single-name credits. The names in the master pool are used to reference a number of CDO tranches (typically, 6–10 Mezzanine tranches). Then, these CDO tranches are used as the underlying portfolio for the CDO-Squared structure. Figure 13.1 illustrates the mechanics of the CDO-Squared.

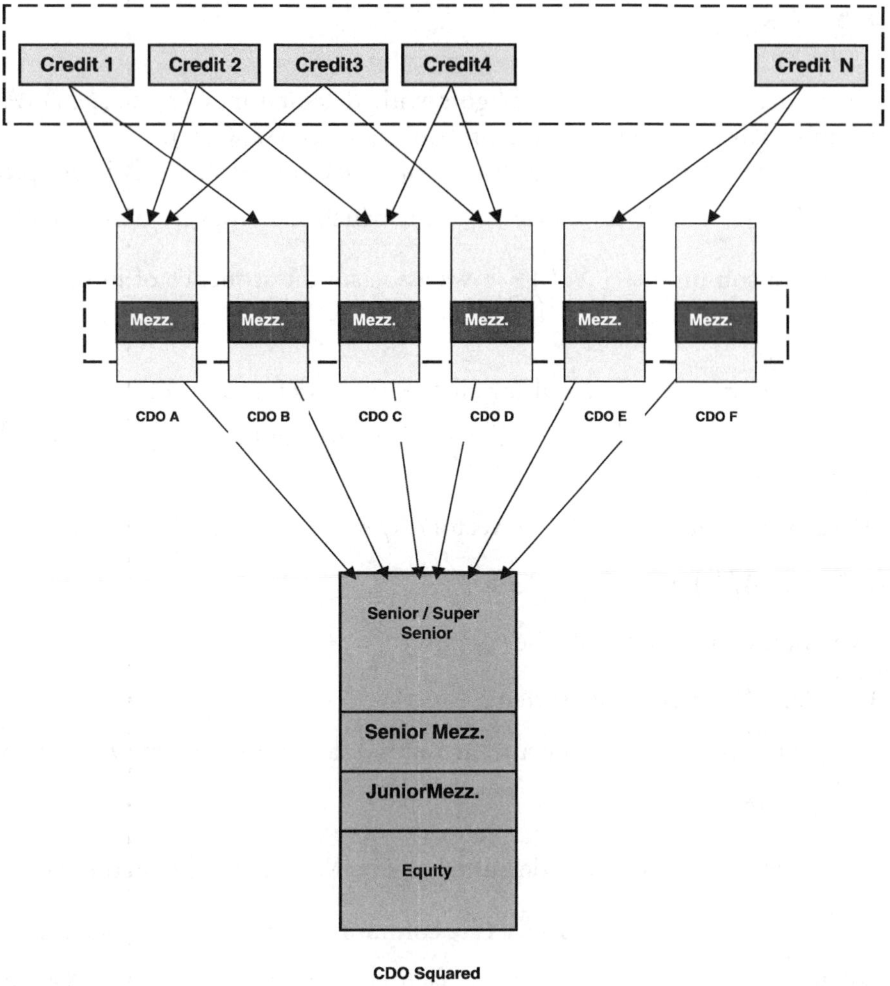

Fig. 13.1 Typical CDO squared structure

The aim of this chapter is to provide an intuitive method for analyzing the correlation of correlation risk in the Marshall-Olkin copula framework. In the first part, we extend the FTD replication method to basket of basket products. Then, in the second part, we develop a simple intuitive approach to modelling basket of basket structures: the idea is to look at each underlying basket as a single name security and to derive its intensity process and its multivariate representation.

13.2 Set-up

We consider a portfolio of n obligors with default times $(\tau_1, \ldots, \tau_n)$. All random variables are defined on a probability space $(\Omega, \mathcal{G}, \mathbb{P})$.

We assume that we have a set of $(m+n)$ independent Poisson processes $\left(N_t^{c_j}\right)_{t\geq 0}$ with intensities $\lambda^{c_j}(t)$, which can trigger joint defaults. To each Poisson process $\left(N_t^{c_j}\right)_{t\geq 0}$, we associate the sequence of jump times $\left\{\theta_r^{c_j}\right\}_{r\in\{1,2,\ldots\}}$.

For every event type c_j, and for all $t \geq 0$, we define a set of independent Bernoulli variables $\left(A_t^{1,j}, \ldots, A_t^{n,j}\right)$ with probabilities $\left(p_{1,j}, \ldots, p_{n,j}\right)$, $p_{i,j} \in [0, 1]$.

We assume that for $j \neq k$, the vectors $\mathbf{A}_t^j = \left(A_t^{1,j}, \ldots, A_t^{n,j}\right)$ and $\mathbf{A}_t^k = \left(A_t^{1,k}, \ldots, A_t^{n,k}\right)$ are independent.

We assume that for $t \neq s$, the vectors $\mathbf{A}_t^j = \left(A_t^{1,j}, \ldots, A_t^{n,j}\right)$ and $\mathbf{A}_s^j = \left(A_s^{1,j}, \ldots, A_s^{n,j}\right)$ are independent.

At time $\theta_r^{c_j}$ (i.e., at the rth occurrence of the market event of type j), we draw the set of n independent Bernoulli variables $\left(A_{\theta_r^{c_j}}^{1,j}, \ldots, A_{\theta_r^{c_j}}^{n,j}\right)$. The variable $A_{\theta_r^{c_j}}^{i,j}$ indicates if obligor i has defaulted or not. We classify the market factors $\left(N_t^{c_j}\right)_{t\geq 0}$ so that the first m factors are common market events, which impact more than two obligors, and the last n factors are idiosynratic events. We shall denote the latter by $N_t^{0,i} \triangleq N_t^{c_{m+i}}$ (and $\lambda^{0,i}(t) \triangleq \lambda^{c_{m+i}}(t)$), for $1 \leq i \leq n$. The corresponding factor loadings are: $p_{i,m+i} = 1$ and $p_{i,m+k} = 0$, for $1 \leq k \neq i \leq n$.

We define the process $\left(N_t^i\right)_{t\geq 0}$ as

$$N_t^i \triangleq \sum_{j=1}^{m+n} \sum_{r=1}^{N_t^{c_j}} A_{\theta_r^{c_j}}^{i,j}. \tag{13.1}$$

$\left(N_t^i\right)_{t\geq 0}$ is also a Poisson process since it is obtained as the superposition of independent (thinned) Poisson processes. Its intensity is given by

$$\lambda_i(t) = \sum_{j=1}^{m+n} p_{i,j}\lambda^{c_j}(t) = \sum_{j=1}^{m} p_{i,j}\lambda^{c_j}(t) + \lambda^{0,i}(t).$$

The single-name survival probability of obligor i is given by

$$Q_i(T) \triangleq \mathbb{P}(\tau_i > T) = \mathbb{P}\left(N_T^i = 0\right) = \exp\left(-\int_0^T \lambda_i(t)\, dt\right); \quad (13.2)$$

We denote by $D_t^i = \mathbf{1}_{\{\tau_i \leq t\}}$ the default indicator process of the ith credit. The multivariate dependence structure can be alternatively described by the following SDE

$$dD_t^i = \left(1 - D_{t-}^i\right)\sum_{j=1}^{m+n} A_t^{i,j}\, dN_t^{c_j}.$$

The copula function implied by this Poisson shock model is called a Marshall-Olkin copula.

13.3 Replication Method

In Chap. 9, we have shown that any nth-to-default payoff can be replicated with a set of first-to-defaults referencing sub-baskets of the original pool of credits. For example, the second-to-default on a three-name basket $\{A, B, C\}$ can be replicated by first-to-defaults on the sub-baskets $\{A, B\}, \{A, C\}, \{B, C\}$ and $\{A, B, C\}$ as:

$$STD(A, B, C) = FTD(A, B) + FTD(A, C) + FTD(B, C) - 2FTD(A, B, C).$$

The third-to-default on the basket $\{A, B, C\}$ can be replicated by second-to-defaults on the sub-baskets $\{A, B\}$, $\{A, C\}$, $\{B, C\}$ and $\{A, B, C\}$ as:

$$TTD(A, B, C) = \frac{1}{2}\left[STD(A, B) + STD(A, C) + STD(B, C) - STD(A, B, C)\right].$$

The STDs can then be decomposed into first-to-defaults on smaller sub-baskets

$$TTD(A,B,C)=\frac{1}{2}\left\{\begin{array}{c}[FTD(A)+FTD(B)-FTD(A,B)]\\+[FTD(A)+FTD(C)-FTD(A,C)]\\+[FTD(B)+FTD(C)-FTD(B,C)]\\-[FTD(A,B)+FTD(A,C)+FTD(B,C)-2FTD(A,B,C)]\end{array}\right\}.$$

In general, we can show that an nth-to-default payoff can be replicated by $(n-1)$-th-to-default payoffs, which in turn can be replicated by $(n-2)$-th-to-default payoffs, and so forth. Applying this replication recursively, we get the complete FTD expansion.

The generic FTD replication result is given in the following proposition.

Let $V_n^{[k]}(A_1,\ldots,A_n)$ denote the value of a kth-to-default on the n-name basket $\{A_1,\ldots,A_n\}$. And let us define the l-subsets of $\{1,2,\ldots,n\}$ (i.e. the subsets containing exactly l elements) by the mappings $\pi_s^l(.)$ for $1\leq s\leq\binom{n}{l}$:

$$\pi_s^l(.):\{1,2,\ldots,l\}\rightarrow\{1,2,\ldots,n\}.$$

We also assume, without loss of generality, that the elements of each l -subset are sorted in an increasing order

$$\pi_s^l(1)\leq\pi_s^l(2)\leq\cdots\leq\pi_s^l(l).$$

Proposition 92 (First-to-Default Expansion) *A kth-to-default on n names $V_n^{[k]}$ can be replicated with first-to-defaults on sub-baskets of l names $V_l^{[1]}$, for $n-k+1\leq l\leq n$, as follows*

$$V_n^{[k]}(A_1,\ldots,A_n)=\sum_{l=n-k+1}^{n}\sum_{s=1}^{\binom{n}{l}}\alpha_n^{[k]}(l)\,V_l^{[1]}\left(A_{\pi_s^l(1)},\ldots,A_{\pi_s^l(l)}\right),\tag{13.3}$$

where

$$\alpha_n^{[k]}(l)=(-1)^{l-(n-k+1)}\binom{l-1}{n-k},\ \textit{for}\ n-k+1\leq l\leq n.$$

To apply the FTD replication to baskets of baskets, we need to introduce some additional notations.

Suppose we have a master pool of credits $\{A_1,\ldots,A_N\}$ which contains all the names in the underlying basket securities. And we have n underlying basket securities $\{S_1,\ldots,S_n\}$ defined as

$$S_i=V_{n_i}^{k_i}\left(A_{\Sigma_i(1)},\ldots,A_{\Sigma_i(n_i)}\right),$$

where n_i specifies the number of credits in the underlying basket S_i, k_i specifies the order of the nth-to-default (e.g. FTD, STD, ...) and the mapping

$$\Sigma_i\left(.\right):\{1,2,\ldots,n_i\}\rightarrow\{1,2,\ldots,N\}$$

specifies the names included in the underlying basket S_i.

The objective is to find the FTD replication of the master basket of baskets $V_n^{[k]}\left(S_1,\ldots,S_n\right)$.

To start with, we apply the expansion (13.3) formally to the basket $V_n^{[k]}$

$$V_n^{[k]}\left(S_1,\ldots,S_n\right)=\sum_{l=n-k+1}^{n}\sum_{s=1}^{\binom{n}{l}}\alpha_n^{[k]}\left(l\right)V_l^{[1]}\left(S_{\pi_s^l(1)},\ldots,S_{\pi_s^l(l)}\right);$$

then, we compute each first-to-default term

$$V_l^{[1]}\left(S_{i_i},\ldots,S_{i_l}\right).$$

To this end, we expand the securities S_i into their FTD representation and we apply the following algebraic rules to the FTD operator $V^{[1]}=V_l^{[1]}(\ldots)$

$$V^{[1]}\left(S_1,S_2,\ldots,S_k\right)=V^{[1]}\left(S_1,V^{[1]}\left(S_2,\ldots,S_k\right)\right),\qquad(13.4)$$

$$V^{[1]}\left(\alpha S_1+\beta S_2,S_3\right)=\alpha V^{[1]}\left(S_1,S_3\right)+\beta V^{[1]}\left(S_2,S_3\right),\qquad(13.5)$$

$$V^{[1]}\left(S_1,S_2\right)=V^{[1]}\left(S_2,S_1\right).\qquad(13.6)$$

For example, consider two baskets $\{A,B,C\}$ and $\{A,D,E\}$. And let us generate the FTD expansions of the following baskets of baskets

1. $V^{[1]}\left(V^{[1]}\left(A,B,C\right),V^{[1]}\left(A,D,E\right)\right)$
2. $V^{[2]}\left(V^{[1]}\left(A,B,C\right),V^{[1]}\left(A,D,E\right)\right)$
3. $V^{[1]}\left(V^{[2]}\left(A,B,C\right),V^{[2]}\left(A,D,E\right)\right)$

The first example is simple. It suffices to observe that a first-to-default of first-to-defaults is a first-to-default on the union of the two baskets

$$V^{[1]}\left(V^{[1]}\left(A,B,C\right),V^{[1]}\left(A,D,E\right)\right)=V^{[1]}\left(A,B,C,D,E\right).$$

To do the second example, we expand the master STD, then we apply our algebra

$$\begin{aligned}
& V^{[2]}\left(V^{[1]}(A, B, C), V^{[1]}(A, D, E)\right) \\
&= V^{[1]}\left(V^{[1]}(A, B, C)\right) + V^{[1]}\left(V^{[1]}(A, D, E)\right) \\
&\quad -V^{[1]}\left(V^{[1]}(A, B, C), V^{[1]}(A, D, E)\right) \\
&= V^{[1]}(A, B, C) + V^{[1]}(A, D, E) - V^{[1]}(A, B, C, D, E).
\end{aligned}$$

To expand the third example, we apply the FTD expansion to the STD sub-securities, then do the algebra

$$\begin{aligned}
& V^{[1]}\left(V^{[2]}(A, B, C), V^{[2]}(A, D, E)\right) \\
&= V^{[1]}\begin{pmatrix} V^{[1]}(A, B) + V^{[1]}(B, C) + V^{[1]}(A, C) - V^{[1]}(A, B, C), \\ V^{[1]}(A, D) + V^{[1]}(D, E) + V^{[1]}(A, E) - V^{[1]}(A, D, E) \end{pmatrix} \\
&= V^{[1]}(A, B, D) + V^{[1]}(A, B, D, E) + V^{[1]}(A, B, E) - V^{[1]}(A, B, D, E) \\
&\quad +V^{[1]}(A, B, C, D) + V^{[1]}(B, C, D, E) + V^{[1]}(A, B, C, E) - V^{[1]}(A, B, C, D, E) \\
&\quad +V^{[1]}(A, C, D) + V^{[1]}(A, C, D, E) + V^{[1]}(A, C, E) - V^{[1]}(A, C, D, E) \\
&\quad -V^{[1]}(A, B, C, D) - V^{[1]}(A, B, C, D, E) - V^{[1]}(A, B, C, E) + V^{[1]}(A, B, C, D, E).
\end{aligned}$$

For CDO-Squared structures, where we have the same notional n and same recovery δ for all the names in the master pool, we can easily convert each CDO payoff to a portfolio of NTD baskets.

The portfolio loss is given by

$$L_t \triangleq \sum_{i=1}^{n} n(1-\delta)\,\mathbf{1}_{\{\tau_i \leq t\}},$$

and the portfolio aggregated defaults counter is

$$X_t \triangleq \sum_{i=1}^{n} \mathbf{1}_{\{\tau_i \leq t\}}.$$

A CDO tranche that covers the losses between a lower bound α and an upper bound β, is equivalent to the default slices between

$$\frac{\alpha}{n(1-\delta)} \leq X_t \leq \frac{\beta}{n(1-\delta)}.$$

If we define $m_\alpha = \left[\frac{\alpha}{n(1-\delta)}\right]$ and $m_\beta = \left[\frac{\beta}{n(1-\delta)}\right]$, where $[a]$ denotes the integer part of a, then the value of the CDO tranche $V^{(\alpha,\beta)}$ will be expressed in terms of the default slices $V^{[k]}$ as follows

$$V^{(\alpha,\beta)} = \left(1 - \frac{\alpha}{n\,(1-\delta)} + m_\alpha\right) V^{[m_\alpha]} + \sum_{k=m_\alpha+1}^{m_\beta} V^{[k]} + \left(\frac{\beta}{n\,(1-\delta)} - m_\beta\right) V^{[m_\beta]}. \tag{13.7}$$

We can then apply the replication for baskets of baskets as explained before.

13.4 Equivalent Single Name Process

The idea of the "Equivalent Single Name Process" method is to consider each basket security in the underlying portfolio as a single name security whose default time is driven by a Poisson process, and to derive the probabilistic properties of this process. We shall use a similar technique to our "Homogeneous Portfolio Approach" (see Chap. 10).

We start with a first-to-default payoff to illustrate the method, then we tackle the general kth-to-default case, where $k > 1$.

13.4.1 First-to-Default Case

We consider a portfolio of n names $\{A_1, \ldots, A_n\}$ with intensities $(\lambda_i\,(t))_{1\le i\le n}$.

The decomposition of each credit $\{A_i\}$ in the Marshall-Olkin model is given by a vector of loadings $\left[p_{i,j}\right]_{1\le j\le m}$ and an idiosyncratic intensity $\left(\lambda^{0,i}\,(t)\right)$. We need to find an equivalent single name process (τ^{eq}), which has the same intensity and the same MO multivariate representation as the FTD basket.

In the first-to-default case, the default time $\left(\tau^{ftd} \triangleq \min\,(\tau_1, \ldots, \tau_n)\right)$ is a Poisson process and its intensity is given by

$$\lambda^{ftd}\,(\{\tau_1, \ldots, \tau_n\}) = \sum_{j=1}^{m}\left[1 - \prod_{i=1}^{n}\left(1 - p_{i,j}\right)\right]\left[\lambda^{c_j}\,(t)\right] + \left[\sum_{i=1}^{n}\lambda^{0,i}\,(t)\right].$$

On the other hand, the intensity of the equivalent process is

$$\lambda^{eq}(t) = \sum_{j=1}^{m} p_j^{eq}\left[\lambda^{c_j}(t)\right] + \left[\lambda^{0,eq}(t)\right].$$

To ensure that the values of the FTD basket and the single name security are the same, we need to have

$$\lambda^{ftd}(t)(\{\tau_1, \ldots, \tau_n\}) = \lambda^{eq}(t), \forall \left[\lambda^{c_j}(t)\right] \in \mathbb{R}^+, 1 \leq j \leq m,$$

which implies for each market factor, $1 \leq j \leq m$

$$p_j^{eq} = 1 - \prod_{i=1}^{n}\left(1 - p_{i,j}\right), \tag{13.8}$$

and for the idiosyncratic term

$$\lambda^{0,eq}(t) = \sum_{i=1}^{n} \lambda^{0,i}(t). \tag{13.9}$$

This defines the equivalence transformation for first-to-default payoffs.

13.4.2 The Equivalence Transformation

In the general case, we want to define the transformation

$$\left(\left[p_{i,j}\right]_{\substack{1\leq i\leq n\\1\leq j\leq m}}, \left(\lambda^{0,i}\right)_{1\leq i\leq n}\right) \to \left(\left(p_j^{eq}\right)_{1\leq j\leq m}, \lambda^{0,eq}\right),$$

such that the intensity of a kth-to-default time $\tau^{[k]}$ of the portfolio $(\tau_1, \ldots, \tau_n)$ and its multivariate properties are the same as the equivalent single name default time τ^{eq}.

As usual, we fix a time horizon T and we consider the distribution of the random variable

$$X_T \triangleq \sum_{i=1}^{n} D_T^i.$$

D_T^i is the default indicator for issuer i, and X_T counts the number of defaults in the basket that occurred before time T. The Q-factor associated with the

kth-to-default time $\tau^{[k]}$ is given by

$$Q^{[k]}(T) \triangleq \mathbb{P}\left(\tau^{[k]} > T\right) = \mathbb{P}(X_T < k). \tag{13.10}$$

For fixed T, we consider the probability generating function of the r.v. X_T

$$\varphi(x) \triangleq \sum_{k=0}^{n} \mathbb{P}(X_T = k)\, x^k. \tag{13.11}$$

We denote by $\Pi_{\mathbf{n}}$ the set of all subsets of $\{1, \ldots, n\}$. For each subset $\pi \in \Pi_{\mathbf{n}}$, we define the first-to-default random time $\tau_\pi^{[1]} = \min\{\tau_i : i \in \pi\}$, and its associated Q-factor $Q_\pi^{[1]}(T)$:

$$Q_\pi^{[1]}(T) \triangleq \mathbb{P}\left(\tau_\pi^{[1]} > T\right) = \mathbb{E}\left[\prod_{i\in\pi}\left(1 - D_T^i\right)\right]. \tag{13.12}$$

We know that the probability generating function $\varphi(x)$ can be expressed in terms of the first-to-default Q-factors $\left[Q_\pi^{[1]}(T)\right]_{\pi\in\Pi_{\mathbf{n}}}$ as follows (see Chap. 10).

Proposition 93 *The probability generating function $\varphi(x)$ of the random variable X_T is given by*

$$\varphi(x) = \sum_{\pi\in\Pi_{\mathbf{n}}} Q_\pi^{[1]}(T)\, x^{n-d(\pi)}(1-x)^{d(\pi)}. \tag{13.13}$$

We also know that the FTD Q-factor $Q_\pi^{[1]}(T)$ is given by

$$Q_\pi^{[1]}(T) = \exp\left(-\left[\sum_{j=1}^{m}\left(1 - \prod_{i\in\pi}\left(1 - p_{i,j}\right)\right)\Lambda_j^c(T)\right] - \left[\sum_{i\in\pi}\Lambda^{0,i}(T)\right]\right), \tag{13.14}$$

where $\Lambda^{c_j}(T)$ and $\Lambda^{0,i}(T)$ are the cumulative intensities

$$\Lambda^{c_j}(T) \triangleq \int_0^T \lambda^{c_j}(t)\,dt,$$

$$\Lambda^{0,i}(T) \triangleq \int_0^T \lambda_i^0(t)\,dt.$$

We expand the exponential to first order and we define a new set of loadings $\widetilde{p_{i,j}} = 1 - e^{-\Lambda^{0,i}(T)} \left(1 - p_{i,j}\right)$

$$Q_{\pi}^{[1]}(T) \simeq \exp\left(-\sum_{j=1}^{m} \Lambda^{c_j}(T)\right) \left[\exp\left(-\sum_{i\in\pi} \Lambda^{0,i}(T)\right) + \sum_{j=1}^{m} \left(\prod_{i\in\pi} \left(1 - \widetilde{p_{i,j}}\right)\right) \Lambda^{c_j}(T)\right]. \tag{13.15}$$

Substituting Eq. (13.15) in the expression of the p.g.f. yields

$$\exp\left(\sum_{j=1}^{m} \Lambda^{c_j}(T)\right) \varphi(x) = \left[\sum_{s=1}^{2^n} \exp\left(-\sum_{i\in\pi} \Lambda^{i,0}(T)\right) x^{n-d(\pi)} (1-x)^{d(\pi)}\right] + \sum_{j=1}^{m} \Lambda^{c_j}(T) \left[\sum_{s=1}^{2^n} \left(\prod_{i\in\pi} \left(1 - \widetilde{p_{i,j}}\right)\right) x^{n-d(\pi)} (1-x)^{d(\pi)}\right].$$

We define the idiosyncratic p.g.f. $\varphi_0(x)$ and the market factor conditional p.g.f. $\varphi_j^c(x)$ as

$$\varphi_0(x) \triangleq \sum_{s=1}^{2^n} \exp\left(-\sum_{i\in\pi} \Lambda^{i,0}(T)\right) x^{n-d(\pi)} (1-x)^{d(\pi)}, \tag{13.16}$$

$$\varphi_j^c(x) \triangleq \sum_{s=1}^{2^n} \left(\prod_{i\in\pi} \left(1 - \widetilde{p_{i,j}}\right)\right) x^{n-d(\pi)} (1-x)^{d(\pi)}. \tag{13.17}$$

Using the result in Proposition 93, we know that $\varphi_0(x)$ is the characteristic function of a sum of independent Bernoulli variables X_i^0, where $\mathbb{P}\left(X_i^0 = 1\right) = 1 - \exp\left(-\Lambda^{i,0}(T)\right)$ and $\mathbb{P}\left(X_i^0 = 0\right) = \exp\left(-\Lambda^{i,0}(T)\right)$. Similarly, $\varphi_j^c(x)$ is the characteristic function of a sum of independent Bernoulli variables X_i^j, where $\mathbb{P}\left(X_i^j = 1\right) = \widetilde{p_{i,j}}$ and $\mathbb{P}\left(X_i^j = 0\right) = 1 - \widetilde{p_{i,j}}$. Thus

$$\varphi_0(x) = \prod_{i=1}^{n} \left(\left(1 - \exp\left(-\Lambda^{i,0}(T)\right)\right) x + \exp\left(-\Lambda^{i,0}(T)\right)\right), \tag{13.18}$$

$$\varphi_j^c(x) = \prod_{i=1}^{n} \left(\widetilde{p_{i,j}} x + \left(1 - \widetilde{p_{i,j}}\right)\right). \tag{13.19}$$

Using the fact that

$$\mathbb{P}(X = l) = \frac{\varphi^{(l)}(x)}{l!},$$

the kth-to-default Q-factor $\mathbb{P}(X_T < k)$ can be expressed as a linear combination of market factor contributions and the idiosyncratic term

$$\mathbb{P}(X_T < k) \simeq \exp\left(-\sum_{j=1}^{m} \Lambda^{c_j}(T)\right)\left[\mathbb{P}\left(\sum_{i=1}^{n} X_i^0 < k\right) + \sum_{j=1}^{m} \mathbb{P}\left(\sum_{i=1}^{n} X_i^j < k\right)\Lambda^{c_j}(T)\right]. \tag{13.20}$$

On the other hand, the Q-factor of the equivalent single name default time τ^{eq} is given by

$$Q^{eq}(T) \triangleq \mathbb{P}\left(\tau^{eq} > T\right) = \exp\left(-\left[\sum_{j=1}^{m} p_j^{eq}\Lambda^{c_j}(T)\right] - \left[\Lambda^{0,eq}(T)\right]\right), \tag{13.21}$$

where $\Lambda^{0,eq}(T)$ is the cumulative idiosyncratic intensity

$$\Lambda^{0,eq}(T) \triangleq \int_0^T \lambda_{eq}^0(t)\,dt.$$

We expand the exponential and we replace with the adjusted set of loadings $\widetilde{p_j^{eq}} = 1 - e^{-\Lambda_{eq}^0(T)}\left(1 - p_j^{eq}\right)$

$$Q^{eq}(T) \simeq \exp\left(-\sum_{j=1}^{m} \Lambda^{c_j}(T)\right)\left[\exp\left(-\Lambda^{0,eq}(T)\right) + \sum_{j=1}^{m}\left(1 - \widetilde{p_j^{eq}}\right)\Lambda^{c_j}(T)\right]. \tag{13.22}$$

The equivalence transformation will be defined as the one that keeps the Q-factors invariant

$$Q^{[k]}(T)(\{\tau_1,\ldots,\tau_n\}) = Q^{eq}(T), \forall \left[\Lambda^{c_j}(T)\right] \in \mathbb{R}^+, 1 \le j \le m. \tag{13.23}$$

We substitute (13.20) and (13.22) in (13.23). This implies that the market factor components and idiosyncratic terms are equal

$$\mathbb{P}\left(\sum_{i=1}^{n} X_i^0 < k\right) = \exp\left(-\Lambda^{0,eq}(T)\right),$$

$$\mathbb{P}\left(\sum_{i=1}^{n} X_i^j < k\right) = 1 - \widetilde{p_j^{eq}}.$$

The probabilities $\mathbb{P}\left(\sum_{i=1}^{n} X_i^0 < k\right)$ and $\mathbb{P}\left(\sum_{i=1}^{n} X_i^j < k\right)$ can be computed as usual with the standard methods: Fourier Inversion or Convolution Recursion.

The equivalent single name process is therefore completely defined by its idiosyncratic intensity and its market factor loadings as

$$\Lambda^{0,eq}(T) = -\log\left[\mathbb{P}\left(\sum_{i=1}^{n} X_i^0 < k\right)\right]. \tag{13.24}$$

$$p_j^{eq} = 1 - \left[\exp\left(\Lambda^{0,eq}(T)\right)\mathbb{P}\left(\sum_{i=1}^{n} X_i^j < k\right)\right]. \tag{13.25}$$

In summary, the equivalence transformation can be described algorithmically by the following steps:

1. Find $\lambda^{0,eq}$
    (a) Generate the idiosyncratic default distribution $\left[\mathbb{P}\left(\sum_{i=1}^{n} X_i^0 = j\right)\right]_{0\le j\le n}$ by Fourier inversion or Convolution recursion
    (b) Solve for $\lambda^{0,eq}$ using Eq. (13.24)

2. For each market factor, find p_j^{eq}

   (a) Transform the $p_{i,j}$'s to the $\widetilde{p_{i,j}}$'s: $\widetilde{p_{i,j}} = 1 - e^{-\Lambda^{0,i}(T)}\left(1 - p_{i,j}\right)$
   (b) Generate the conditional market factor default distribution $\left[\mathbb{P}\left(\sum_{i=1}^{n} X_i^j = j\right)\right]_{0 \leq j \leq n}$ by Fourier inversion or Convolution recursion
   (c) Solve for p_j^{eq} using Eq. (13.25)

13.4.3 Equivalence Transformation for CDOs

In this section, we construct the "Equivalence Transformation" for CDO tranches.

We fix a time horizon T and we consider the aggregate portfolio loss variable L_T

$$L_T \triangleq \sum_{i=1}^{n} L_i D_T^i.$$

D_T^i and L_i are the default indicator and the loss variable for issuer i.

The Q-factor associated with the (α, β)-tranche is given by

$$Q^{\alpha,\beta}(T) \triangleq \mathbb{E}\left[1 - \frac{M_T^{\alpha,\beta}}{\beta - \alpha}\right], \tag{13.26}$$

where $M_T^{\alpha,\beta}$ is the tranche loss

$$M_T^{\alpha,\beta} \triangleq \min\left(\max\left(L_T - \alpha, 0\right), \beta - \alpha\right).$$

We consider the Fourier transform $\phi : \mathbb{R} \to \mathbb{C}$ of the r.v. L_T, defined by

$$\phi(u) = \mathbb{E}\left[\exp\left(iuL_T\right)\right]. \tag{13.27}$$

First, we derive a similar result to Proposition 93. We express $\phi(u)$ in terms of the first-to-default Q-factors $\left[Q_\pi^{[1]}(T)\right]_{\pi \in \Pi_\mathbf{n}}$.

Proposition 94 *The Fourier transform $\phi(u)$ of the random variable L_T is given by*

$$\phi(u) = \sum_{\pi \in \Pi_{\mathbf{n}}} Q_{\pi}^{[1]}(T) \left[\prod_{i \notin \pi} \psi_i \prod_{i \in \pi} (1 - \psi_i) \right], \tag{13.28}$$

where $\psi_i = \exp(iuL_i)$.

Proof We proceed by induction. We assume the property is verified for n, and we prove that it holds for $n + 1$.

For $n + 1$ names, the Fourier transform of the loss variable

$$L_T^{n+1} = \sum_{i=1}^{n+1} L_i D_T^i, \tag{13.29}$$

is given by

$$\phi^{n+1}(u) = \mathbb{E}\left[\exp\left(-iuL_T^{n+1}\right)\right].$$

Conditioning on D_T^{n+1}, we have

$$\begin{aligned} \phi^{n+1}(u) &= \mathbb{E}\left[\mathbb{E}\left[\exp\left(-iuL_T^{n+1}\right) \middle| D_T^{n+1}\right]\right] \\ &= \mathbb{E}\left[\exp\left(-iuL^{n+1}D_T^{n+1}\right) \mathbb{E}\left[\exp\left(-iuL_T^{n}\right) \middle| D_T^{n+1}\right]\right]. \end{aligned} \tag{13.30}$$

Using the induction relationship, we can write the conditional characteristic function as

$$\begin{aligned} \mathbb{E}\left[\exp\left(-iuL_T^{n}\right) \middle| D_T^{n+1}\right] = \sum_{\pi_n \in \Pi_{\mathbf{n}}} \mathbb{E}\left(\prod_{i \in \pi_n} \left(1 - D_T^i\right) \middle| D_T^{n+1}\right) \\ \left[\prod_{i \notin \pi_n} \psi_i \prod_{i \in \pi_n} (1 - \psi_i)\right], \end{aligned}$$

and Eq. (13.30) becomes

$$\phi^{n+1}(u) = \mathbb{E}\left[\psi_{n+1}^{D_T^{n+1}}\left[\sum_{\pi_n\in\Pi_{\mathbf{n}}}\mathbb{E}\left(\prod_{i\in\pi_n}\left(1-D_T^i\right)\middle|D_T^{n+1}\right)\right.\right.$$
$$\left.\left.\left[\prod_{i\notin\pi_n}\psi_i\prod_{i\in\pi_n}(1-\psi_i)\right]\right]\right].$$

Writing the expectation explicitly, we get

$$\begin{aligned}
&\phi^{n+1}(u)\\
&=\mathbb{P}\left(D_T^{n+1}=0\right)\sum_{\pi_n\in\Pi_{\mathbf{n}}}\mathbb{E}\left(\prod_{i\in\pi_n}\left(1-D_T^i\right)\middle|\left\{D_T^{n+1}=0\right\}\right)\\
&\quad\left[\prod_{i\notin\pi_n}\psi_i\prod_{i\in\pi_n}(1-\psi_i)\right] \qquad (13.31)\\
&+\mathbb{P}\left(D_T^{n+1}=1\right)\psi_{n+1}\sum_{\pi_n\in\Pi_{\mathbf{n}}}\mathbb{E}\left(\prod_{i\in\pi_n}\left(1-D_T^i\right)\middle|\left\{D_T^{n+1}=1\right\}\right)\\
&\quad\left[\prod_{i\notin\pi_n}\psi_i\prod_{i\in\pi_n}(1-\psi_i)\right].
\end{aligned}$$

Observing that

$$\begin{aligned}
&\mathbb{P}\left(D_T^{n+1}=0\right)\mathbb{E}\left(\prod_{i\in\pi_n}\left(1-D_T^i\right)\middle|\left\{D_T^{n+1}=0\right\}\right)\\
&=\mathbb{E}\left(\left[\prod_{i\in\pi_n}\left(1-D_T^i\right)\right]\left(1-D_T^{n+1}\right)\right),
\end{aligned}$$

and

$$\begin{aligned}
&\mathbb{P}\left(D_T^{n+1}=1\right)\mathbb{E}\left(\prod_{i\in\pi_n}\left(1-D_T^i\right)\Big|\left\{D_T^{n+1}=1\right\}\right)\\
&=\mathbb{E}\left(\left[\prod_{i\in\pi_n}\left(1-D_T^i\right)\right]D_T^{n+1}\right)\\
&=\mathbb{E}\left(\left[\prod_{i\in\pi_n}\left(1-D_T^i\right)\right]\right)-\mathbb{E}\left(\left[\prod_{i\in\pi_n}\left(1-D_T^i\right)\right]\left(1-D_T^{n+1}\right)\right),
\end{aligned}$$

Eq. (13.31) becomes

$$\begin{aligned}
\phi^{n+1}(u)=&\sum_{\pi_n\in\boldsymbol{\Pi}_\mathbf{n}}\mathbb{E}\left(\left[\prod_{i\in\pi_n}\left(1-D_T^i\right)\right]\left(1-D_T^{n+1}\right)\right)(1-\psi_{n+1})\\
&\left[\prod_{i\notin\pi_n}\psi_i\prod_{i\in\pi_n}(1-\psi_i)\right]\\
&+\sum_{\pi_n\in\boldsymbol{\Pi}_\mathbf{n}}\mathbb{E}\left(\left[\prod_{i\in\pi_n}\left(1-D_T^i\right)\right]\right)\psi_{n+1}\\
&\left[\prod_{i\notin\pi_n}\psi_i\prod_{i\in\pi_n}(1-\psi_i)\right].
\end{aligned}\qquad(13.32)$$

The set of subsets $\boldsymbol{\Pi}_{\mathbf{n+1}}$ is partitioned into $\boldsymbol{\Pi}_{\mathbf{n+1}}^+$ and $\boldsymbol{\Pi}_{\mathbf{n+1}}^-$:

$$\begin{aligned}
\boldsymbol{\Pi}_{\mathbf{n+1}}^+&=\left\{\pi_{n+1}:\pi_{n+1}\in\boldsymbol{\Pi}_{\mathbf{n+1}},\ (n+1)\in\pi_{n+1}\right\},\\
\boldsymbol{\Pi}_{\mathbf{n+1}}^-&=\left\{\pi_{n+1}:\pi_{n+1}\in\boldsymbol{\Pi}_{\mathbf{n+1}},\ (n+1)\notin\pi_{n+1}\right\}.
\end{aligned}$$

For $\pi_{n+1}\in\boldsymbol{\Pi}_{\mathbf{n+1}}^+$: $\pi_{n+1}=\pi_n\cup\{n+1\}$, we have

$$Q_{\pi_{n+1}}^{[1]}(T)=\mathbb{E}\left(\prod_{i\in\pi_{n+1}}\left(1-D_T^i\right)\right)=\mathbb{E}\left(\left[\prod_{i\in\pi_n}\left(1-D_T^i\right)\right]\cdot\left(1-D_T^{n+1}\right)\right),$$

and

$$n + 1 \in \pi_{n+1}.$$

For $\pi_{n+1} \in \Pi^-_{\mathbf{n+1}}$: $\pi_{n+1} = \pi_n \cup \emptyset$, we have

$$Q^{[1]}_{\pi_{n+1}}(T) = \mathbb{E}\left(\prod_{i \in \pi_{n+1}} \left(1 - D^i_T\right)\right) = \mathbb{E}\left(\prod_{i \in \pi_n} \left(1 - D^i_T\right)\right),$$

and

$$n + 1 \notin \pi_{n+1}.$$

Substituting in Eq. (13.32) yields

$$\begin{aligned}\phi^{n+1}(u) = &\sum_{\pi_{n+1} \in \Pi^+_{\mathbf{n+1}}} Q^{[1]}_{\pi_{n+1}}(T) \left[\prod_{i \notin \pi_{n+1}} \psi_i \prod_{i \in \pi_{n+1}} (1 - \psi_i)\right] \\ &+ \sum_{\pi_{n+1} \in \Pi^-_{\mathbf{n+1}}} Q^{[1]}_{\pi_{n+1}}(T) \left[\prod_{i \notin \pi_{n+1}} \psi_i \prod_{i \in \pi_{n+1}} (1 - \psi_i)\right],\end{aligned}$$

which ends the proof. ■

If we denote by $\phi\left(u, \widetilde{p_{n_1,\ldots,n_m}}\right)$ the Fourier transform of the random variable $\sum_{i=1}^n L_i Y_i$, where $(Y_1, \ldots, Y_n)$ are n independent Bernoulli variables with parameters $\widetilde{p} = (p_1, \ldots, p_n)$

$$\phi\left(u, \widetilde{p_{n_1,\ldots,n_m}}\right) \triangleq \prod_{i=1}^{n} \left((1 - p_i) + p_i \exp\left(iuL_i\right)\right), \tag{13.33}$$

we can show that $\phi(u)$ admits the following asymptotic series expansion.

Proposition 95 (Asymptotic Series Expansion) *The loss Fourier transform function $\phi(u)$ has the following series expansion*

$$\phi(u) = e^{-\Lambda^c(T)} \left[\sum_{n_1=0}^{+\infty} \cdots \sum_{n_m=0}^{+\infty} \frac{\Lambda^{c_1}(T)^{n_1}}{n_1!} \cdots \frac{\Lambda^{c_m}(T)^{n_m}}{n_m!} \phi\left(u, \widetilde{p_{n_1,\ldots,n_m}}\right)\right],$$

where $\Lambda^c(T) \triangleq \sum_{i=1}^{n} \Lambda^{c_j}(T)$, *and the probability vector* $\widetilde{p_{n_1,\ldots,n_m}} = \left(p_{n_1,\ldots,n_m}(1), \ldots, p_{n_1,\ldots,n_m}(n)\right)$ *is given by*

$$p_{n_1,\ldots,n_m}(i) = 1 - e^{-\Lambda_i^0(T)}\left(1 - p_{i,1}\right)^{n_1} \ldots \left(1 - p_{i,m}\right)^{n_m}.$$

Proof Using Proposition 94, we express the probability generating function of the r.v. L_T in terms of the FTD Q-factors $Q_\pi^{[1]}(T)$

$$\phi(u) = \sum_{\pi \in \Pi_{\mathbf{n}}} Q_\pi^{[1]}(T)\left[\prod_{i \notin \pi} \psi_i \prod_{i \in \pi}(1 - \psi_i)\right]. \tag{13.34}$$

Each FTD Q-factor $Q_\pi^{[1]}(T)$ is given by

$$Q_\pi^{[1]}(T) = \exp\left(-\left[\sum_{j=1}^{m}\left(1 - \prod_{i \in \pi}\left(1 - p_{i,j}\right)\right)\Lambda^{c_j}(T)\right] - \left[\sum_{i \in \pi} \Lambda^{0,i}(T)\right]\right). \tag{13.35}$$

We expand the exponentials

$$\exp\left(\prod_{i \in \pi}\left(1 - p_{i,j}\right)\Lambda^{c_j}(T)\right) = \sum_{n_j=0}^{+\infty} \frac{\left[\prod_{i \in \pi}\left(1 - p_{i,j}\right)\Lambda^{c_j}(T)\right]^{n_j}}{n_j!}.$$

We substitute in Eq. (13.35)

$$\begin{aligned}
&Q_\pi^{[1]}(T) \\
&= e^{-\Lambda^c(T) - \left[\sum_{i \in \pi} \Lambda^{0,i}(T)\right]} \sum_{n_1=0}^{+\infty} \ldots \\
&\quad \sum_{n_m=0}^{+\infty} \frac{\left[\prod_{i \in \pi}\left(1 - p_{i,1}\right)\Lambda^{c_1}(T)\right]^{n_1}}{n_1!} \frac{\left[\prod_{i \in \pi}\left(1 - p_{i,m}\right)\Lambda^{c_m}(T)\right]^{n_m}}{n_m!} \\
&= e^{-\Lambda^c(T)} \sum_{n_1=0}^{+\infty} \ldots \sum_{n_m=0}^{+\infty} \frac{\Lambda^{c_1}(T)^{n_1}}{n_1!} \ldots \\
&\quad \frac{\Lambda^{c_m}(T)^{n_m}}{n_m!} e^{-\sum_{i \in \pi}\left(\Lambda^{0,i}(T) + \log\left(\left(1 - p_{i,1}\right)^{n_1}\right) + \cdots + \log\left(\left(1 - p_{i,m}\right)^{n_m}\right)\right)}.
\end{aligned}$$

Substituting in the expression of the characteristic function yields

$$\phi(u) = e^{-\Lambda^{c}(T)} \sum_{n_1=0}^{+\infty} \cdots \sum_{n_m=0}^{+\infty} \frac{\Lambda^{c_1}(T)^{n_1}}{n_1!} \cdots \frac{\Lambda^{c_m}(T)^{n_m}}{n_m!} \sum_{\pi \in \Pi_{\mathbf{n}}} Q_{n_1,\ldots,n_m}(\pi) \left[\prod_{i \notin \pi} \psi_i \prod_{i \in \pi} (1-\psi_i) \right],$$

where

$$Q_{n_1,\ldots,n_m}\left(\pi\right) = \exp\left(-\sum_{i \in \pi} \left(\Lambda^{0,i}\left(T\right) + \log\left(\left(1 - p_{i,1}\right)^{n_1} \right) + \cdots + \log\left(\left(1 - p_{i,m}\right)^{n_m} \right) \right) \right).$$

We define the function $\phi_{n_1,\ldots,n_m}(u)$

$$\phi_{n_1,\ldots,n_m}(u) = \sum_{\pi \in \Pi_{\mathbf{n}}} Q_{n_1,\ldots,n_m}(\pi) \left[\prod_{i \notin \pi} \psi_i \prod_{i \in \pi} (1-\psi_i) \right]. \tag{13.36}$$

$\phi_{n_1,\ldots,n_m}(u)$ is the Fourier transform of n independent idiosycratic terms with intensities $\left(\Lambda^{i}_{n_1,\ldots,n_m}(T)\right)_{1 \le i \le n}$

$$\Lambda^{i}_{n_1,\ldots,n_m}(T) = \Lambda^{0,i}(T) + \log\left((1-p_{i,1})^{n_1}\right) + \cdots + \log\left((1-p_{i,m})^{n_m}\right).$$

Hence

$$\phi_{n_1,\ldots,n_m}(u) = \prod_{i=1}^{n} \left(\psi_i + \left(e^{-\Lambda^{0,i}(T)} \left(1-p_{i,1}\right)^{n_1} \cdots \left(1-p_{i,m}\right)^{n_m} \right) (1-\psi_i) \right).$$

Thus, $\phi_{n_1,\ldots,n_m}(u)$ is the Fourier transform of the random variable $\sum_{i=1}^{n} Y_i$, where $(Y_1, \ldots, Y_n)$ are n independent Bernoulli variables with parameters $p_{n_1,\ldots,n_m} = \left(p_{n_1,\ldots,n_m}(1), \ldots, p_{n_1,\ldots,n_m}(n)\right)$

$$p_{n_1,\ldots,n_m}(i) = 1 - e^{-\Lambda^{0,i}(T)} \left(1-p_{i,1}\right)^{n_1} \cdots \left(1-p_{i,m}\right)^{n_m}.$$

$\phi(u)$ is a weighted average of the conditional independent characteristic functions $\varphi_{n_1,\dots,n_m}(x)$

$$\varphi(x) = \sum_{n_1=0}^{+\infty} \dots \sum_{n_m=0}^{+\infty} w_{n_1,\dots,n_m} \cdot \phi_{n_1,\dots,n_m}(u), \tag{13.37}$$
$$w_{n_1,\dots,n_m} = e^{-\Lambda^c(T)} \frac{\Lambda^{c_1}(T)^{n_1}}{n_1!} \dots \frac{\Lambda^{c_m}(T)^{n_m}}{n_m!},$$

which ends the proof. ■

We define the idiosyncratic loss variable $L_T^0 \triangleq \sum_{i=1}^n L_i X_i^0$, where $(X_i^0)_{1\le i\le n}$ is a vector of independent Bernoulli variables, given by $\mathbb{P}(X_i^0 = 1) = 1 - \exp(-\Lambda^{i,0}(T))$ and $\mathbb{P}(X_i^0 = 0) = \exp(-\Lambda^{i,0}(T))$. Similarly, for each market factor, we define the loss variable $L_T^{c_j} \triangleq \sum_{i=1}^n L_i X_i^j$, where $\left(X_i^j\right)_{1\le i\le n}$ is a vector of independent Bernoulli variables, given by $\mathbb{P}\left(X_i^j = 1\right) = \widetilde{p_{i,j}}$ and $\mathbb{P}\left(X_i^j = 0\right) = 1 - \widetilde{p_{i,j}}$. $\widetilde{p_{i,j}}$ is the duration adjusted loading $\widetilde{p_{i,j}} = 1 - e^{-\Lambda^{0,i}(T)}(1 - p_{i,j})$.

We also define the (α, β)-tranche loss variables M_T^0 and $M_T^{c_j}$

$$M_T^0 \triangleq \min\left(\max\left(L_T^0 - \alpha, 0\right), \beta - \alpha\right),$$
$$M_T^{c_j} \triangleq \min\left(\max\left(L_T^{c_j} - \alpha, 0\right), \beta - \alpha\right).$$

Using the order-one truncated expansion, we have

$$\mathbb{E}\left[M_T^{\alpha,\beta}\right] \simeq \exp\left(-\sum_{j=1}^m \Lambda^{c_j}(T)\right)\left[\mathbb{E}\left[M_T^0\right] + \sum_{j=1}^m \mathbb{E}\left[M_T^{c_j}\right]\Lambda^{c_j}(T)\right],$$

and the (α, β)-tranche Q-factor is given by

$$Q^{\alpha,\beta}(T) \simeq \exp\left(-\sum_{j=1}^m \Lambda^{c_j}(T)\right)\left[\mathbb{E}\left[1 - \frac{M_T^0}{\beta - \alpha}\right]\right.$$
$$\left. + \sum_{j=1}^m \mathbb{E}\left[1 - \frac{M_T^{c_j}}{\beta - \alpha}\right]\Lambda^{c_j}(T)\right]. \tag{13.38}$$

On the other hand, we have from Eq. (13.15) the expression of the Q-factor for the single name equivalent process

$$Q^{eq}(T) \simeq \exp\left(-\sum_{j=1}^{m} \Lambda^{c_j}(T)\right)\left[\exp\left(-\Lambda^{0,eq}(T)\right) + \sum_{j=1}^{m}\left(1-\widetilde{p_j^{eq}}\right)\Lambda^{c_j}(T)\right].$$

Hence, by matching the idiosyncratic term and the market factor components, we get

$$\exp\left(-\Lambda^{0,eq}(T)\right) = \mathbb{E}\left[1 - \frac{M_T^0}{\beta-\alpha}\right],$$

$$\left(1-\widetilde{p_j^{eq}}\right) = \mathbb{E}\left[1 - \frac{M_T^{c_j}}{\beta-\alpha}\right].$$

the (α, β)-tranche equivalence transformation is completely defined by

$$\Lambda^{0,eq}(T) = -\log\left[\mathbb{E}\left[1 - \frac{M_T^0}{\beta-\alpha}\right]\right]. \tag{13.39}$$

$$p_j^{eq} = 1 - \left[\exp\left(\Lambda^{0,eq}(T)\right)\mathbb{E}\left[1 - \frac{M_T^{c_j}}{\beta-\alpha}\right]\right]. \tag{13.40}$$

In summary, the (α, β)-tranche equivalence transformation can be described algorithmically as follows:

1. Find $\lambda^{0,eq}$
   (a) Generate the idiosyncratic loss distribution $\left[\mathbb{P}\left(L_T^0 = l\right)\right]$ by Fourier inversion or Convolution recursion
   (b) Compute the Expected value of the idiosyncratic (α, β)-tranche loss

$$\mathbb{E}\left[M_T^0\right] = \sum_{l=\alpha}^{\beta}(l-\alpha)\,\mathbb{P}\left(L_T^0 = l\right) + (\beta-\alpha)\,\mathbb{P}\left(L_T^0 > \beta\right).$$

   (c) Solve for $\lambda^{0,eq}$ using Eq. (13.39)

2. For each market factor, find p_j^{eq}

   (a) Transform the $p_{i,j}$'s to the $\widetilde{p_{i,j}}$'s: $\widetilde{p_{i,j}} = 1 - e^{-\Lambda^{0,i}(T)} \left(1 - p_{i,j}\right)$
   (b) Generate the conditional market factor default distribution $\left[\mathbb{P}\left(L_T^{c_j} = l\right)\right]$ by Fourier inversion or Convolution recursion
   (c) Compute the Expected value of the market factor (α, β)-tranche loss

$$\mathbb{E}\left[M_T^{c_j}\right] = \sum_{l=\alpha}^{\beta} (l - \alpha)\, \mathbb{P}\left(L_T^{c_j} = l\right) + (\beta - \alpha)\, \mathbb{P}\left(L_T^{c_j} > \beta\right).$$

   (d) Solve for p_j^{eq} using Eq. (13.40)

13.4.4 Numerical Examples

In this section, we use the results derived in the previous section to study the default correlation properties of basket securities.

We use a 100-name investment grade diversified portfolio. The average credit spread is 120 bps, the maximum spread is 500 bps and the minimum spread is 30 bps. The portfolio is diversified across 19 industry sectors, where the industry concentrations vary from 2% to 11%. The portfolio spread distribution is given in Fig. 13.2.

The portfolio industry concentrations is given in Fig. 13.3.

We assume that the intensity of each issuer is has the following Marshall-Olkin decomposition:

$$\lambda_i(t) = \left[\lambda^W(t)\right] + p_{i,B}\left[\lambda^B(t)\right] + \sum_{j=1}^{m} p_{i,S_j}\left[\lambda^{S_j}(t)\right] + \left[\lambda^{0,i}(t)\right],$$

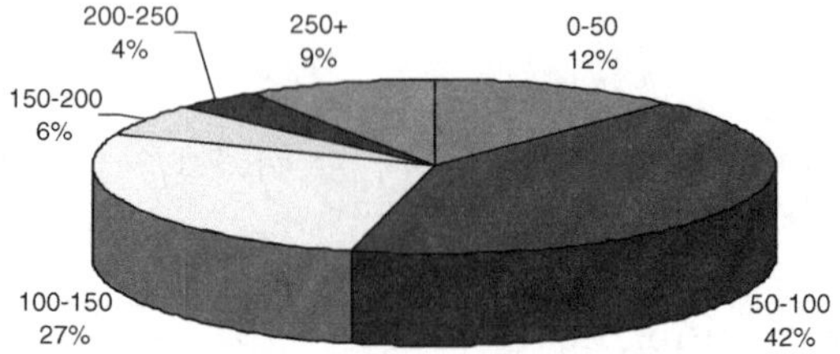

Fig. 13.2 Portfolio spread distribution

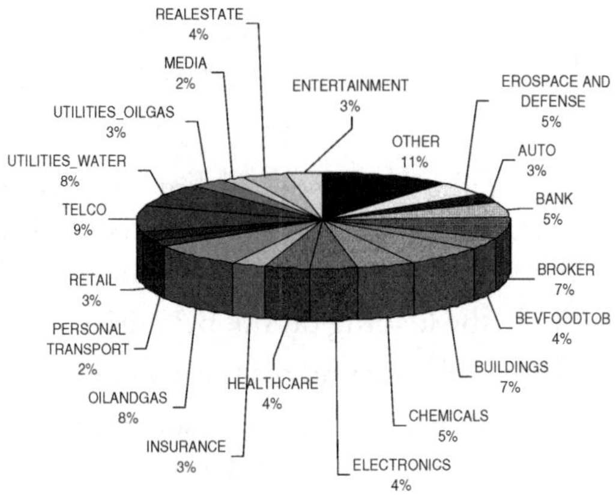

Fig. 13.3 Portfolio industry concentration

where

$\lambda_W(t)$ is the intensity of the "World" driver

$\lambda_B(t)$ is the intensity of the "Beta" driver, and $p_{i,B}$ is the loading on that driver

$\lambda_{S_j}(t)$ is the intensity of the "Sector" driver S_j, and p_{i,S_j} is the loading on that sector[1]

$\lambda_i^0(t)$ is the intensity of the idiosyncratic events

The "World" driver represents the global Armageddon risk, which triggers the joint defaults of all the credits in the universe. The world driver event is a very low-probability event. However, the loading of each credit on this driver is equal to 1. The world driver is used to calibrate to the super AAA risk in the CDO market. The "Beta" driver is responsible for the correlation between names in different sectors. And the "Sector" drivers make names in the same sector more correlated than the rest of the universe.

The intensity of the World driver is $\lambda^W = 2.5$ bps. The intensity of the Beta driver is $\lambda^B = 400$ bps. The Sector driver intensities vary from 100 bps to 300 bps. We assume that 50% of the spread is Beta, 25% is sector and 25% is idiosyncratic. On average, the implied 5 year default correlation in this model is 7% inter-sector and 14% intra-sector.

Using the equivalence transformation, we generate the equivalent single name process and its MO decomposition for all the default slices on this portfolio i.e. FTD, STD, ... The MO representation for the intensity of a kth-to-default time $\tau^{[k]}$ of the portfolio $(\tau_1, \ldots, \tau_n)$ is then given by

[1] If $i \in S_j$ then $p_{i,S_j} > 0$ otherwise $p_{i,S_j} = 0$.

$$\lambda^{[k]}(t) = \left[\lambda^{W}(t)\right] + p_{B}^{[k]}\left[\lambda^{B}(t)\right] + \sum_{j=1}^{m} p_{S_j}^{[k]}\left[\lambda^{S_j}(t)\right] + \left[\lambda^{[k],0}(t)\right]. \tag{13.41}$$

Obviously, the loading of the kth-to-default on the world driver will also be equal to 1. The world driver triggers the defaults of all the names in the portfolio, therefore $\tau^{[k]}$ will also trigger for all default slices $1 \leq k \leq n$.

Figure 13.4 shows how the loading on the Beta driver $p_{B}^{[k]}$ and the loading on one of the Sector drivers $p_{S_j}^{[k]}$ vary across the default slices $k = 1, 2, \ldots, n$.

Clearly, for the high-order tranches (i.e. super-senior risk) from $k = 30$ to $k = 100$, the default event $\tau^{[k]}$ becomes a pure World driver event. It is very unlikely that a super-senior tranche will be hit by defaults unless there is a global meltdown where everyone defaults. So, the intensity of $\tau^{[k]}$ reduces to $\lambda^{[k]}(t) = \lambda^{W}(t)$ and $p_{B}^{[k]} = 0$, $p_{S_j}^{[k]} = 0$.

On the other side of the default spectrum (low values of k), we know that for a FTD intensity the loadings are given by

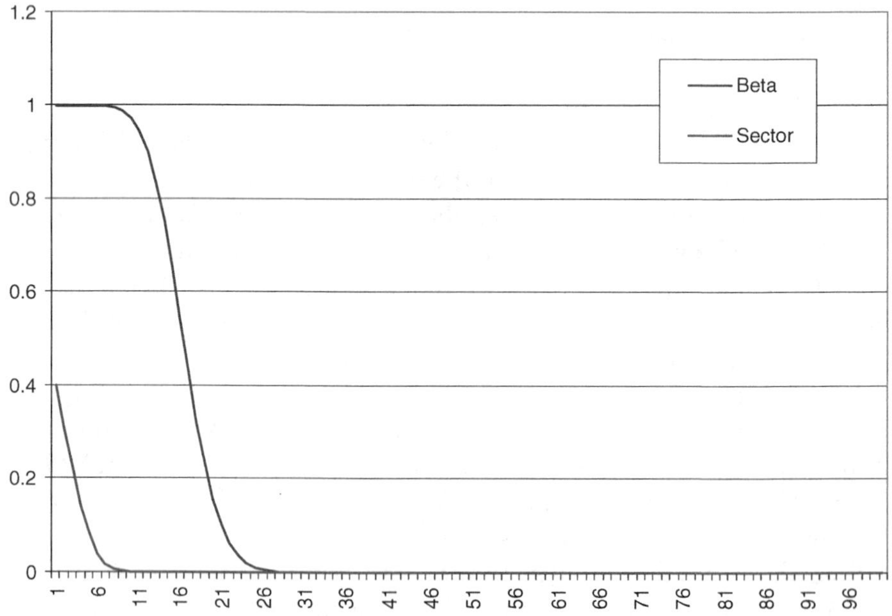

Fig. 13.4 Loading of the default time $\tau^{[k]}$ on the Beta driver and one of the sector drivers as a function of the slice index k

$$p_B^{[1]} = 1 - \prod_{i=1}^{n} \left(1 - p_{i,B}\right) \leq 1 - \left(1 - \max_{1 \leq i \leq n} \left(p_{i,B}\right)\right)^n .$$

For large portfolios $n \to \infty$,

$$\prod_{i=1}^{n} \left(1 - p_{i,B}\right) \to 0 \text{ as } n \to \infty,$$

and the order of magnitude of $p_B^{[1]}$ will be around 1

$$p_B^{[1]} \simeq 1.$$

For the loadings on the sector drivers, $p_{S_j}^{[1]}$ is equal to $1 - \prod_{i \in S_j} \left(1 - p_{i,S_j}\right)$. Since the portfolio is diversified, the loading $p_{S_j}^{[1]}$ will be of the order of

$$\begin{aligned} p_{S_j}^{[1]} &= 1 - \prod_{i \in S_j} \left(1 - p_{i,S_j}\right) \\ &\simeq 1 - \left(1 - \overline{p_{i,S_j}}\right)^{|S_j|} \\ &\simeq 1 - \left(1 - \left|S_j\right| * \overline{p_{i,S_j}}\right) \\ &\simeq \text{number of names in } S_j \ * \text{ average sector loading.} \end{aligned}$$

Now, if we look at the implied 5y default correlation of the default times $\left(\tau^{[1]}, \tau^{[2]}, \ldots, \tau^{[n]}\right)$, we get the surface depicted in Fig. 13.5.

The default correlation between the low-default slices is close to zero. The correlation between the super-senior slices is equal to 1. And we have a hump in the middle where the correlation increases to 0.95 and drops again to 0.15. This can be seen more easily on Fig. 13.6 where we plot the upper-diagonal (i.e. the pairs $\rho_{1,2}, \rho_{2,3}, \rho_{3,4}, \ldots, \rho_{n-1,n}$).

For the higher slices, the default event $\tau^{[k]}$ degenerates to a pure world driver event, therefore, by construction the default correlation between all senior slices will be a perfect 1. For the lowest slices, as one would expect we have exactly the opposite effect. Equity slices are mostly driven by idiosyncratic events, therefore the default correlation between these events is close to zero. The hump that we observe for the middle Mezzanine slices can be explained by the Beta driver. The slices $k = 8, 9, 10, 11, 12$ have a high probability of triggering almost simultaneously if a Beta event occurs. Therefore, their default correlation is exceptionally high. This effect can also be exhibited if we

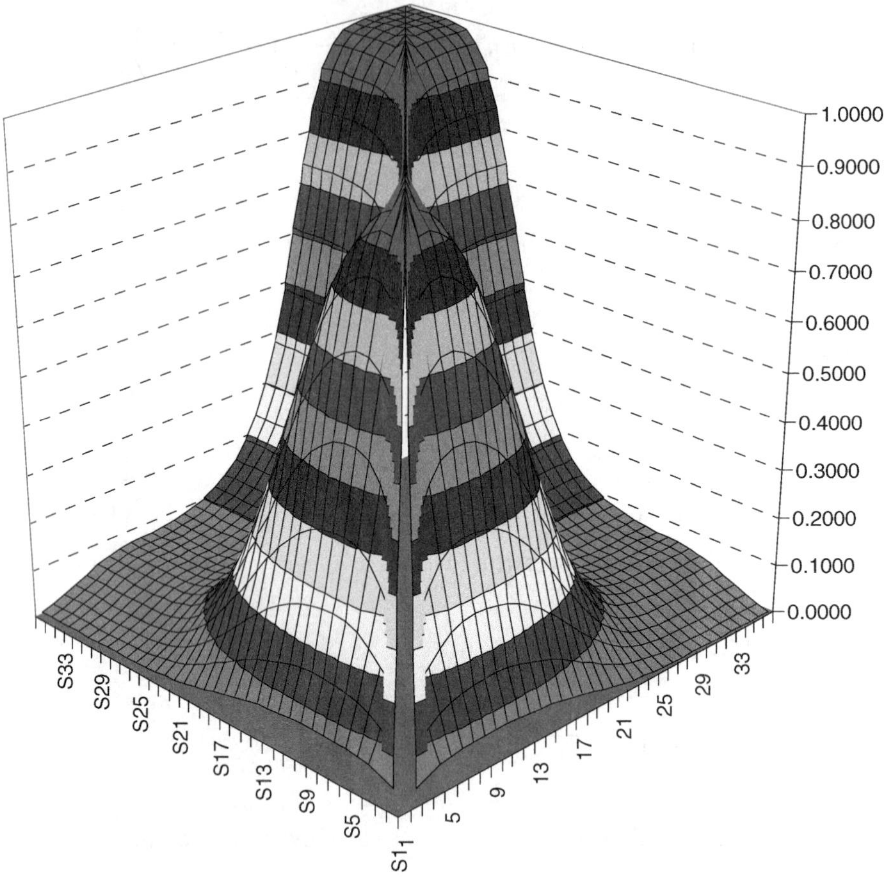

Fig. 13.5 5-year default correlation surface of the kth-to-default times $(\tau^{[k]})$

superpose on our correlation plot the ATM spreads of the corresponding slices (Fig. 13.7). We can immediately spot that there are two plateaus in the graph:

1. The super-senior plateau where all the slices $k \geq 30$ have an ATM spread of 2.5 bps i.e. the world driver spread.
2. The Beta plateau where the slices $k = 8, 9, 10, 11, 12$ have roughly the same spread $\simeq$ 400–450 bps.

If we remove the Beta driver and Sector driver dependencies, the Mezzanine hump will disappear and the default correlation plot will vary from 0 for low slices to 1 for high slices (Fig. 13.8). The speed of switching from the 0-correlation regime to the 1-correlation regime will depend on the level of the world driver.

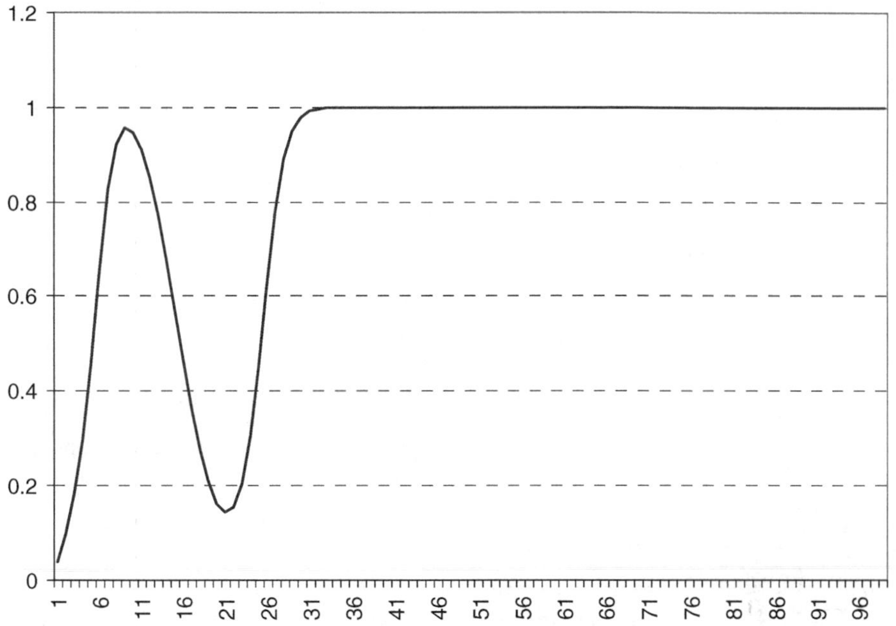

Fig. 13.6 Upper-diagonal correlation curve $\left[\rho_{1,2}, \rho_{2,3}, \rho_{3,4}, \ldots, \rho_{n-1,n}\right]$

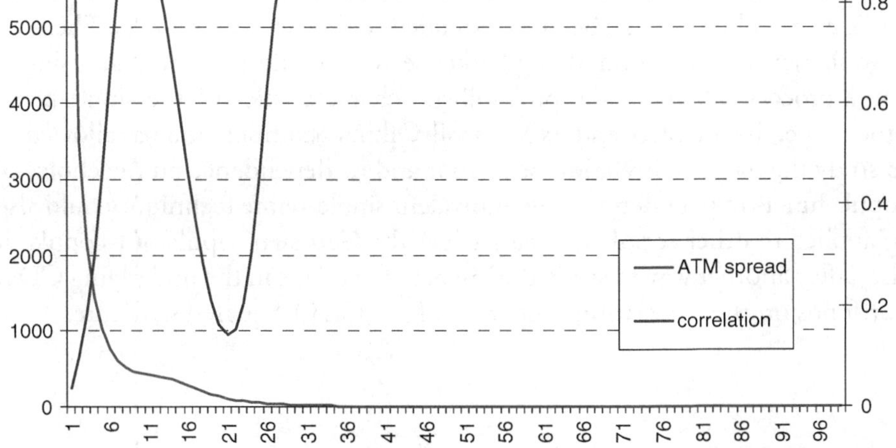

Fig. 13.7 Upper-diagonal correlation and ATM spreads for the corresponding tranches

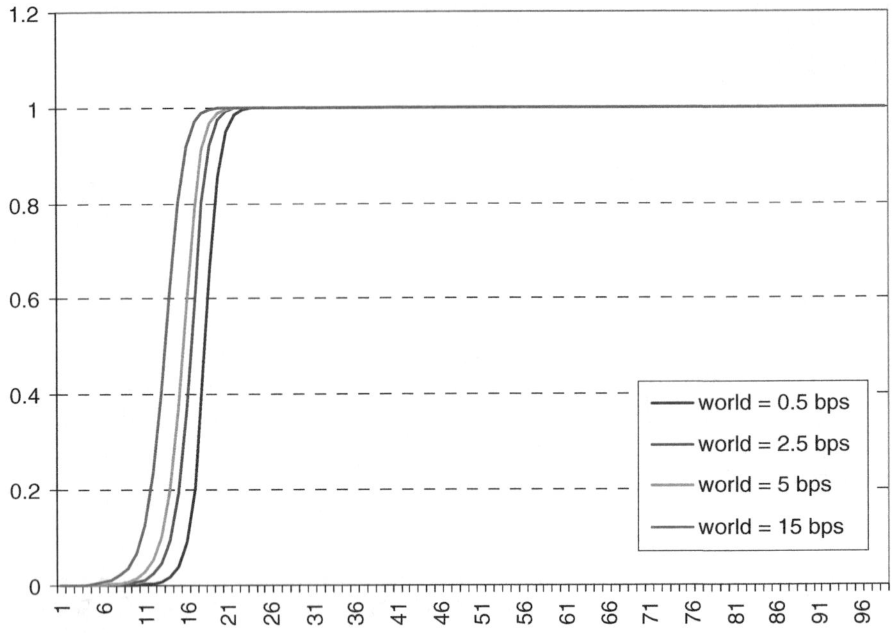

Fig. 13.8 Upper-diagonal correlation curves for different values of the world driver

13.5 Conclusion

In this chapter, we have presented a simple approach to modelling CDO-Squared structures, which provides a better intuitive understanding of the correlation of correlation effects. First, we have shown that the first-to-default replication techniques can be easily extended to this type of products. Then, we have shown that each underlying basket security could be viewed as a single-name process. Deriving the probabilistic characteristics of the single-name process, i.e., its intensity and its Marshall-Olkin decomposition has allowed us to study the basket correlation behaviour and its dependence on the choice of copula function parameters. This equivalent single-name technique could also be applied to other copula models such as the Gaussian copula or t-copula. It also offers another way to study the impact of overlaps in the underlying CDO portfolios on the correlation properties of the CDO-Squared structure.

14

Second Generation Models: From Flat to Correlation Skew

In this chapter, we review some popular correlation skew models. We give a brief description of each model and discuss the advantages and limitations of each modelling framework. This includes: the Stochastic Correlation model, Local Correlation (and Random Factor Loading), the Levy copula and the Implied (hazard rate) Copula.

The stochastic and local correlation models are so called "second generation" models, which extend the Gaussian copula in an attempt to model the tranche prices of the entire capital structure at a fixed time horizon. The Levy copula is also an extension of the Gaussian copula, which accounts for the correlation skew, but it also provides some dynamics for the expected loss process. And last but not least, the implied hazard rate copula is a non-parametric copula function, which constructs the implied distribution of the conditioning factor (and the associated conditional single-name probabilities) from the tranche prices directly.

To aid the description of the various correlation skew model extensions covered in this chapter—which are essentially very similar to the Gaussian copula approach—we briefly review the specification of the (basic) one-factor Gaussian copula model.

14.1 Gaussian Copula

In the standard Gaussian copula approach, we define the (default) latent variables X_i for each credit, with $i \in \{1, \ldots, n\}$, by

Y. Elouerkhaoui, *Credit Correlation*, Applied Quantitative Finance,
DOI 10.1007/978-3-319-60973-7_14

$$X_i = \rho Y + \sqrt{1-\rho^2}\varepsilon_i, \tag{14.1}$$

where Y and $\varepsilon_1, \ldots, \varepsilon_n$ are independent standard Gaussian random variables and ρ is some constant parameter in $[0, 1)$. If the default time for each credit i is given by τ_i, then we make the following connection between X_i and τ_i by choosing a constant default threshold u_i such that

$$p_T^i = \mathbb{P}(\tau_i \le T) = \mathbb{P}(X_i \le u_i)$$

holds. In this case, the threshold u_i is given by

$$u_i = \Phi^{-1}\left(p_T^i\right). \tag{14.2}$$

The correlation between the names is then specified by the parameter ρ , which determines each single-name's dependency on the central shock variable Y and its idiosyncratic factor ε_i. The level u_i is then selected in order to match the single name's default probabilities. The steps now to calculate the expected loss for a given tranche are as follows.

1. Evaluate the default thresholds u_i from the single name default credit curves, given by Eq. (14.2).
2. Obtain the conditional default probability given the central shock variable Y, which is in this case

$$p_T^i(y) = \mathbb{P}(\tau_i \le T \,|Y = y) = \Phi\left(\frac{\Phi^{-1}\left(p_T^i\right) - \rho y}{\sqrt{1-\rho^2}}\right). \tag{14.3}$$

3. We then compute the conditional loss distribution, either by using standard recursion techniques, or using one of the various approximations methods available, such as the normal proxy method. In the latter, based on the central limit theorem, we approximate the distribution of losses, conditional on the central shock Y, with a normal random variable, and the conditional probabilities given by (14.3) are used to calculate its mean and variance.
4. Finally, we integrate the payoff over the central shock variable. The density in this case is simply the standard Gaussian function

$$\varphi(y) = \frac{1}{\sqrt{2\pi}} \exp\left(-\frac{y^2}{2}\right).$$

The implementation of all the second generation models described in this chapter follows broadly the same steps with small changes required to compute the new single-name conditional probabilities and the (new) marginal distributions of the latent variables (if impacted).

14.2 Stochastic Correlation

In this section, we describe the "Stochastic Correlation" model, which basically, can be thought of as a Mixture of Gaussian copulas. This type of models is inspired by the stochastic volatility models, typically, used in equity or foreign exchange options pricing. One of the key features of the model is that it preserves the Gaussian nature of the underlying latent variables, which simplifies the numerical implementation substantially as it follows closely the standard implementation algorithm used for the Gaussian copula. We shall see later that this is not always the case for other correlation skew models; for example, in the local correlation model, the marginal distributions of the latent variables are not Gaussian anymore and we need to derive them explicitly before we can proceed to pricing. We refer, for example, to Burtschell et al. (2007, 2009) and Li and Liang (2005) for different specifications of stochastic correlation models. Here, we review two specifications: the Gaussian mixture and the stochastic de-correlated version.

14.2.1 Gaussian Mixture

In the Stochastic Correlation model we incorporate the correlation skew effect by making the Gaussian correlations themselves random variables.

More formally, we introduce a random (correlation) variable $\widetilde{\rho}\,(\omega)$, which takes some values $\{\alpha_1, \ldots, \alpha_m\}$ in the range $[0, 1)$, with probabilities p_j, $\mathbb{P}\left(\widetilde{\rho} = \alpha_j\right) = p_j$. We also assume that the correlation variable $\widetilde{\rho}\,(\omega)$ is independent from all the other underlying latent variables Y and $\varepsilon_1, \ldots, \varepsilon_n$. For each name i, we define the latent variable X_i as

$$X_i = \widetilde{\rho} Y + \sqrt{1 - \widetilde{\rho}^2} \varepsilon_i \ ;$$

thus, due to the independence of the random variables $\widetilde{\rho}$, Y and ε_i, the marginal distribution of the latent variable X_i still remains a standard Gaussian, and therefore the default threshold u_i can be calculated in the usual way with Eq. (14.2). The only difference from the Gaussian approach above, is the default probability conditioned on the central shock Y is different from (14.3).

In this case we need to condition on both the common factor Y and the random correlation $\widetilde{\rho}$; the single-name conditional probabilities for this stochastic correlation model are given by

$$p_T^i(y,\widetilde{\rho}) = \mathbb{P}(\tau_i \leq T \,|Y = y, \widetilde{\rho}) = \Phi\left(\frac{\Phi^{-1}(p_T^i) - \widetilde{\rho}y}{\sqrt{1-\widetilde{\rho}^2}}\right),$$

$$p_T^i(y) = \mathbb{E}\left[p_T^i(y,\widetilde{\rho})\right] = \sum_{j=1}^{m} p_j \Phi\left(\frac{\Phi^{-1}(p_T^i) - \alpha_j y}{\sqrt{1-\alpha_j^2}}\right).$$

This is closely related to the Gaussian Mixture model used, for example, in Li and Liang (2005). The copula function, in this case, is defined as a mixture of Gaussian copulas $\mathbb{C}(u_1, \ldots, u_n \,|\rho)$, where the mixing factor ρ has a distribution density function $v(\rho)$ of all possible correlation values

$$\mathbb{C}(u_1, \ldots, u_n) = \mathbb{E}\left[\mathbb{C}(u_1, \ldots, u_n \,|\rho)\right] = \int_\rho \mathbb{C}(u_1, \ldots, u_n \,|\rho)\, v(\rho) d\rho.$$

If we assume a discrete (correlations) distribution,

$$v(\rho) = \sum_{i=1}^{m} p_j \mathbf{1}_{\{\rho=\alpha_j\}},$$

we obtain the discrete Gaussian mixture copula function

$$\mathbb{C}(u_1, \ldots, u_n) = \sum_{i=1}^{m} p_j \mathbb{C}\left(u_1, \ldots, u_n \,|\rho = \alpha_j\right).$$

14.2.2 De-Correlated Stochastic Correlations

In Burtschell et al. (2007), a different specification of stochastic correlation is provided. Instead of assuming fully correlated stochastic correlations as in the Gaussian mixture model, they assume that the dynamics of the copula correlations are linked together with a two-factor model where we have a systemic component and an (independent) idiosyncratic component; in this set-up, the latent variables are defined as

$$X_i = \widetilde{\rho}_i Y + \sqrt{1 - \widetilde{\rho}_i^{\,2}} \varepsilon_i,$$
$$\widetilde{\rho}_i = (1 - B_s)(1 - B_i)\rho + B_s,$$

where the correlation state variables B_s, B_i are Bernoulli variables with probabilities q_s and q respectively, $\mathbb{P}(B_s = 1) = q_s$ and $\mathbb{P}(B_i = 1) = q$; we also assume independence between all the correlation state variables (B_s, B_i), on one hand, and the default state variables (Y, ε_i) on the other hand. The marginal distribution of stochastic correlation in this model is discrete and takes three values: 0 , ρ, 1, with probabilities $q(1 - q_s)$, $(1 - q)(1 - q_s)$, q_s.

The single-name conditional default probabilities are computed by conditioning on the random correlation system factor as well

$$\mathbb{P}(\tau_i \le T \,|Y = y, B_s) = (1 - B_s)\left[(1 - q)\,\Phi\left(\frac{\Phi^{-1}(p_T^i) - \rho y}{\sqrt{1 - \rho^2}}\right) + q p_T^i\right] + B_s \mathbf{1}_{\{Y \le \Phi^{-1}(p_T^i)\}}.$$

In practice, a 3-point Gaussian mixture tends to match the equity and senior tranches fairly well and struggles with the mezzanine tranches, or vice-versa. By adding the de-correlation between the random correlation factors, we introduce a hump in the distribution, which improves the fit across the capital structure.

14.2.3 LHP Approximation

As in the Gaussian copula case, to analyze the qualitative features of the model, we can derive useful Large Homogenous Portfolio approximations by taking the infinite portfolio size limit.

The (normalized) homogenous loss variable L_T is given by the portfolio aggregate default fraction $X_T \triangleq \frac{1}{n}\sum_{i=1}^{n} D_T^i$, rescaled with the loss-given-default (LGD)

$$L_T \triangleq \frac{(1 - R)}{n}\sum_{i=1}^{n} D_T^i = (1 - R) \cdot X_T.$$

We condition on the common factor Y and a given realization of the stochastic correlation $\widetilde{\rho}$, so that the single-name default indicators become independent; then, by virtue of the law of large numbers, we know that the fraction

$\frac{1}{n}\sum_{i=1}^{n}\mathbf{1}_{\{\tau_i\leq T\}}$ converges to its average $p_T(y,\widetilde{\rho})$,

$$X_{T|Y=y,\widetilde{\rho}}=\frac{1}{n}\sum_{i=1}^{n}\mathbf{1}_{\{\tau_i\leq T\}}\to\mathbb{E}\left[\mathbf{1}_{\{\tau_i\leq T\}}|Y=y,\widetilde{\rho}\right]=p_T(y,\widetilde{\rho}).$$

Thus, we can approximate the loss distribution L_T, for all $0\leq x\leq 1$, as

$$\begin{aligned}\mathbb{P}(L_T\leq x)&=\sum_{j=1}^{m}p_j\int_{-\infty}^{+\infty}\mathbb{P}\left(X_T\leq\frac{x}{1-R}|Y=y,\widetilde{\rho}=\alpha_j\right)\varphi(y)\,dy\\&\simeq\sum_{j=1}^{m}p_j\int_{-\infty}^{+\infty}\mathbb{P}\left(p_T(y,\alpha_j)\leq\frac{x}{1-R}|Y=y,\widetilde{\rho}=\alpha_j\right)\varphi(y)\,dy\\&=\sum_{j=1}^{m}p_j\int_{-\infty}^{+\infty}\mathbf{1}_{\left\{p_T(y,\alpha_j)\leq\frac{x}{1-R}\right\}}\varphi(y)\,dy=\sum_{j=1}^{m}p_j\Phi\left(-p_T^{-1}\left(\frac{x}{1-R},\alpha_j\right)\right).\end{aligned}$$

Replacing the inverse $p_T^{-1}(x,\alpha_j)$ with its expression yields the final result

$$\mathbb{P}(L_T\leq x)=\sum_{j=1}^{m}p_j\Phi\left(\frac{\sqrt{1-\alpha_j^2}\Phi^{-1}\left(\frac{x}{1-R}\right)-\Phi^{-1}(p_T)}{\alpha_j}\right).$$

Proposition 96 (Gaussian Mixture LHP Approximation) *In the Stochastic Correlation Gaussian Mixture model, the loss distribution can be computed with its Large Homogenous Portfolio (LHP) approximation as*

$$\mathbb{P}(L_T\leq x)=\sum_{j=1}^{m}p_j\Phi\left(\frac{\sqrt{1-\alpha_j^2}\Phi^{-1}\left(\frac{x}{1-R}\right)-\Phi^{-1}(p_T)}{\alpha_j}\right).$$

We follow the same steps and work out the LHP approximation for the De-correlated Stochastic Correlation model specification (of Burtschell et al. 2007) by conditioning on the common factor Y and the systemic correlation factor B_s. We have the following result.

Proposition 97 (De-correlated Stochastic Correlation LHP Approximation) *In the De-correlated Stochastic Correlation model, the loss distribution Large Homogenous Approximation (LHP) is given by*

$$\mathbb{P}\left(L_T \leq x\right) = \left(1 - q_s\right)\left[\mathbf{1}_{\{0<\widetilde{x}<1\}}\Phi\left(\frac{\sqrt{1-\rho^2}\Phi^{-1}\left(\widetilde{x}\right) - \Phi^{-1}\left(p_T\right)}{\rho}\right) + \mathbf{1}_{\{\widetilde{x}\geq 1\}}\right] + q_s\left(1 - p_T\right),$$

where $\widetilde{x}$ *is defined as* $\widetilde{x} = \frac{\frac{x}{1-R} - qp_T}{1-q}$.

So far, in the stochastic recovery models presented, we have assumed that the Gaussian copula correlations are stochastic: either 100% correlated (as in the Gaussian Mixture model) or with some amount of de-correlation between the single-name random correlations introduced via independent idiosyncratic and systemic terms. But throughout, we have always assumed independence between the random correlations and the common factor Y, which drives the correlation between defaults.

In the next section, we shall relax this hypothesis and assume instead that the Gaussian correlations are not only random, but the source of their randomness is also driven directly by the (defaults) common factor Y; in particular, we consider a specification of the random correlation as a functional form of Y: $\widetilde{\rho}_i\left(\omega\right) = \rho\left(Y\left(\omega\right)\right)$, for some real function $\rho\left(.\right) : \mathbb{R} \rightarrow [0, 1)$, thereby creating a link between defaults and correlations and vice-versa.

14.2.4 Calibration

The calibration of this type of models to market quotes is not always straightforward. For the Gaussian Mixture model, the best calibrations are usually obtained with $\widetilde{\rho}_i$ having a range of five values. In general, relatively close matches can be found for the equity and senior tranches but not for mezzanine levels. In particular, under normal market conditions, it is difficult to get enough weight in the 3–6% iTraxx and 3–7% CDX tranches, while maintaining the equity and senior tranche prices. This highlights the fact that the Mixture model, in its original form, lacks a suitable systemic shock, which would better control the shape of the distribution by allowing more weight to be shifted down to the middle part of the distribution. The improved stochastic recovery specification of Burtschell et al. (2007), which introduces a level of de-correlation between the random Gaussian correlations, provides a bit more flexibility by clustering the defaults in the equity and senior regions using the systemic correlation factor q_s, and controlling the support of the middle part with the idiosyncratic correlation probability q.

14.2.5 Model Comparison

Stochastic correlation models intend to provide a fit to the tranche prices of the whole capital structure, therefore providing a self-consistent, arbitrage-free interpolation for non-standard tranches. Their main advantage is the ease of implementation, with the only change being to the calculation of the conditional default probabilities. But the calibration remains particularly challenging with no real intuition for the choice of stochastic correlation parameters.

14.3 Local Correlation and RFL

In this section, we describe the class of "Local Correlation" models, originally introduced as the Random Factor Loading (RFL) model by Andersen and Sidenius (2005), then redefined and studied comprehensively by Turc et al. (2005). The local correlation terminology is borrowed from the "Local volatility" model of Dupire (1994), which was developed for equity options to fit the volatility smile surface. We use here the term Local Correlation loosely to refer to the general type of models where the Gaussian correlations or factor loadings depend on the common factor Y. However, the exact specification and implementation details differ significantly between the two model specifications described in Andersen and Sidenius (2005) and Turc et al. (2005).

14.3.1 Local Correlation

The Local Correlation model is another extension of the Gaussian copula, which attempts to fit the correlation skew curve observed in the market. In this model, the constant correlation parameter is replaced with a deterministic function $\rho(.)$ of the common factor Y,

$$\rho(y) : (-\infty, \infty) \to [0, 1);$$

the definition of the single-name latent variables X_i becomes

$$X_i = \rho(Y)\, Y + \sqrt{1 - \rho(Y)^2}\varepsilon_i,$$

where the random variables Y and ε_i's are again independent standard Gaussian variables.

However, unlike the stochastic correlation case, the latent variables X_i are no longer Gaussian. This means the formula (14.2) used to determine the default thresholds u_i is not valid anymore. The distribution of the single-name variable X_i is now given by

$$F_{X_i}(x) = \mathbb{P}(X_i \leq x) = \int_{-\infty}^{+\infty} \Phi\left(\frac{x - \rho(y)\, y}{\sqrt{1 - \rho(y)^2}}\right) \varphi(y)\, dy.$$

In the numerical implementation, we assume that $\rho(y)$ is a step function whose domain is defined by a finite partition over $\mathbb{R}$, i.e., $\rho(y)$ is of the form

$$\rho(y) = \sum_{j=1}^{m} c_j \mathbf{1}_{\{y \in [y_j, y_{j+1})\}}.$$

With this definition of the local correlation function $\rho(y)$, we obtain the following expression for the marginal probabilities $\mathbb{P}(X_i \leq x)$, which involves a set of bivariate Gaussian integrals,

$$F_{X_i}(x) = \mathbb{P}(X_i \leq x) = \sum_{j=1}^{m} \mathbb{P}\left(\{X_i \leq x\} \cap \{y_j \leq Y < y_{j+1}\}\right)$$

$$= \sum_{j=1}^{m} \left[\Phi_2\left(x, y_{j+1}; c_j\right) - \Phi_2\left(x, y_j; c_j\right)\right],$$

where we have used the fact that the bivariate distribution of the two variables Y and X_i, conditional on $Y \in [y_j, y_{j+1})$, is a standard bivariate normal with correlation parameter c_j.

The calculation for the conditional default distribution of τ_i, given Y, is simply

$$p_T^i(y) = \mathbb{P}(\tau_i \leq T \,|Y = y) = \Phi\left(\frac{F_{X_i}^{-1}\left(p_T^i\right) - \rho(y)\, y}{\sqrt{1 - \rho(y)^2}}\right).$$

14.3.2 Random Factor Loading

In the Random Factor Loading (RFL) specification, the set-up is slightly different; the factor loading on the common driver Y is a function of the

Y-factor, but the loading on the idiosyncratic term is not; and an additional constant drift is added to the equation to compensate for it:

$$X_i = a_i(Y) \cdot Y + \nu_i \varepsilon_i + m_i,$$

where the drift m_i and the volatility ν_i terms are chosen such that the latent variable X_i has zero mean and unit variance,

$$m_i = -\mathbb{E}[a_i(Y) \cdot Y] = -\int_{-\infty}^{+\infty} a_i(y) \cdot y\varphi(y)\, dy,$$

$$\nu_i = \sqrt{1 - Var[a_i(Y) \cdot Y]} = \sqrt{1 - \int_{-\infty}^{+\infty} (a_i(y) \cdot y)^2 \varphi(y)\, dy + m_i^2}.$$

With this specification, the conditional default probabilities are given by

$$p_T^i(y) = \mathbb{P}(\tau_i \leq T \,|Y = y) = \Phi\left(\frac{F_{X_i}^{-1}(p_T^i) - a_i(y)\, y - m_i}{\nu_i}\right).$$

Now, assuming a piece-wise constant function for the factor loadings

$$a_i(Y) = \sum_{j=1}^{m} \alpha_{i,j} \mathbf{1}_{\{y \in [y_j, y_{j+1})\}},$$

we can compute explicitly the drift and variance and work out the latent variables marginal distributions.

First, we start with the drift and the variance

$$m_i = -\sum_{j=1}^{m} \alpha_{i,j} \int_{-\infty}^{+\infty} \mathbf{1}_{\{y \in [y_j, y_{j+1})\}} \cdot y\varphi(y)\, dy = \sum_{j=1}^{m} \alpha_{i,j} (\varphi(y_{j+1}) - \varphi(y_j)),$$

$$\nu_i^2 - m_i^2 = 1 - \sum_{j=1}^{m} \alpha_{i,j}^2 \int_{-\infty}^{+\infty} \mathbf{1}_{\{y \in [y_j, y_{j+1})\}} y^2 \varphi(y)\, dy$$

$$= 1 - \sum_{j=1}^{m} \alpha_{i,j}^2 [(\Phi(y_{j+1}) - \Phi(y_j)) - (y_{j+1}\varphi(y_{j+1}) - y_j\varphi(y_j))];$$

then, we calculate the marginal distribution of the variable X_i,

$$
\begin{aligned}
F_{X_i}(x) &= \mathbb{P}(X_i \le x) = \mathbb{E}\left[\mathbb{P}\left(a_i(Y)\cdot Y + \nu_i\varepsilon_i + m_i \le x \,|Y\right)\right] \\
&= \sum_{j=1}^{m} \int_{-\infty}^{+\infty} \mathbf{1}_{\{y\in[y_j,y_{j+1})\}} \Phi\left(\frac{x - m_i - \alpha_{i,j}y}{\nu_i}\right) \varphi(y)\,dy \\
&= \sum_{j=1}^{m} \Phi_2\left(\frac{x-m_i}{\sqrt{\nu_i^2+\alpha_{i,j}^2}}, y_{j+1}; \frac{\alpha_{i,j}}{\sqrt{\nu_i^2+\alpha_{i,j}^2}}\right) \\
&\quad -\Phi_2\left(\frac{x-m_i}{\sqrt{\nu_i^2+\alpha_{i,j}^2}}, y_j; \frac{\alpha_{i,j}}{\sqrt{\nu_i^2+\alpha_{i,j}^2}}\right),
\end{aligned}
$$

where we have used some standard Gaussian integral results (see Appendix A in Andersen and Sidenius (2005)).

14.3.3 LHP Approximation

We proceed as before and generate the loss distribution in the large homogenous portfolio limit for the local correlation and RFL models.

Starting with the local correlation model, we have the following result.

Proposition 98 (Local Correlation LHP Approximation) *In the Local Correlation model, the Large Homogenous Portfolio (LHP) approximation of the loss distribution is given by*

$$
\begin{aligned}
\mathbb{P}(L_T \le x) = \sum_{j=1}^{m} \mathbf{1}_{\left\{y_{j+1} > \frac{z(x)\sqrt{1-c_j^2}}{c_j}\right\}} \\
\left[\Phi\left(y_{j+1}\right) - \Phi\left(\max\left(\frac{z(x)\sqrt{1-c_j^2}}{c_j}, y_j\right)\right)\right],
\end{aligned}
$$

where $z(x) = u - \Phi^{-1}\left(\frac{x}{1-R}\right)$.

Proof We substitute the conditional probability with its expression in the Local Correlation model

$$p_T(y) = \mathbb{P}(\tau_i \leq T \,|Y = y) = \Phi\left(\frac{u - \rho(y)\,y}{\sqrt{1 - \rho(y)^2}}\right);$$

and as usual, conditional on the common factor Y, we approximate the loss variable L_T with its mean $(1 - R)\,p_T(y)$,

$$\begin{aligned}
\mathbb{P}(L_T \leq x) &= \int_{-\infty}^{+\infty} \mathbf{1}_{\left\{\Phi\left(\frac{u-\rho(y)y}{\sqrt{1-\rho(y)^2}}\right)\leq \frac{x}{1-R}\right\}}\varphi(y)\,dy = \int_{-\infty}^{+\infty} \mathbf{1}_{\left\{u-\Phi^{-1}\left(\frac{x}{1-R}\right)\leq \frac{\rho(y)y}{\sqrt{1-\rho(y)^2}}\right\}}\varphi(y)\,dy \\
&= \sum_{j=1}^{m}\int_{-\infty}^{+\infty} \mathbf{1}_{\{y\in[y_j,y_{j+1})\}}\mathbf{1}_{\left\{u-\Phi^{-1}\left(\frac{x}{1-R}\right)\leq \frac{c_j}{\sqrt{1-c_j^2}}y\right\}}\varphi(y)\,dy \\
&= \sum_{j=1}^{m}\int_{-\infty}^{+\infty} \mathbf{1}_{\{y\in[y_j,y_{j+1})\}}\mathbf{1}_{\left\{\frac{\sqrt{1-c_j^2}}{c_j}\left(u-\Phi^{-1}\left(\frac{x}{1-R}\right)\right)\leq y\right\}}\varphi(y)\,dy \\
&= \sum_{j=1}^{m}\mathbf{1}_{\left\{y_{j+1}>\frac{z(x)\sqrt{1-c_j^2}}{c_j}\right\}}\left[\Phi(y_{j+1}) - \Phi\left(\max\left(\frac{z(x)\sqrt{1-c_j^2}}{c_j}, y_j\right)\right)\right].
\end{aligned}$$

■

Similarly, in the Random Factor Loading set-up, we have the following LHP approximation.

Proposition 99 (Random Factor Loading LHP Approximation) *In the Random Factor Loading model, the Large Homogenous Portfolio (LHP) approximation of the loss distribution is given by*

$$\mathbb{P}(L_T \leq x) = \sum_{j=1}^{m}\mathbf{1}_{\left\{y_{j+1}>\frac{z(x)}{\alpha_j}\right\}}\left[\Phi(y_{j+1}) - \Phi\left(\max\left(\frac{z(x)}{\alpha_j}, y_j\right)\right)\right],$$

where $z(x) = u - \nu\Phi^{-1}\left(\frac{x}{1-R}\right) - m.$

Proof Approximating the loss distribution L_T with its conditional mean $(1-R)\, p_T(y)$, we can write

$$\mathbb{P}(L_T \le x) \simeq \int_{-\infty}^{+\infty} \mathbb{P}\left(p_T(y) \le \frac{x}{1-R} |Y = y\right) \varphi(y)\, dy = \int_{-\infty}^{+\infty} \mathbf{1}_{\left\{p_T(y) \le \frac{x}{1-R}\right\}} \varphi(y)\, dy.$$

Replacing $p_T(y)$ with its formula

$$p_T(y) = \Phi\left(\frac{u - a(y)\, y - m}{\nu}\right),$$

and defining the quantity $z(x) = u - \nu\Phi^{-1}\left(\frac{x}{1-R}\right) - m$, we obtain

$$\begin{aligned}
\mathbb{P}(L_T \le x) &= \int_{-\infty}^{+\infty} \mathbf{1}_{\left\{\Phi\left(\frac{u-a(y)y-m}{\nu}\right) \le \frac{x}{1-R}\right\}} \varphi(y)\, dy = \int_{-\infty}^{+\infty} \mathbf{1}_{\{z(x) \le a(y) y\}} \varphi(y)\, dy \\
&= \sum_{j=1}^{m} \int_{-\infty}^{+\infty} \mathbf{1}_{\{y \in [y_j, y_{j+1})\}} \mathbf{1}_{\{z(x) \le \alpha_j y\}} \varphi(y)\, dy \\
&= \sum_{j=1}^{m} \int_{-\infty}^{+\infty} \mathbf{1}_{\{y \in [y_j, y_{j+1})\}} \mathbf{1}_{\left\{\frac{z(x)}{\alpha_j} \le y\right\}} \varphi(y)\, dy \\
&= \sum_{j=1}^{m} \mathbf{1}_{\left\{y_{j+1} > \frac{z(x)}{\alpha_j}\right\}} \left[\Phi\left(y_{j+1}\right) - \Phi\left(\max\left(\frac{z(x)}{\alpha_j}, y_j\right)\right)\right].
\end{aligned}$$

■

14.3.4 Calibration

As with the stochastic correlation model, the calibration is not easy; but compared with stochastic correlation both iTraxx and CDX 5 year levels can be calibrated to a much better degree of accuracy. Differences in the calibration at 5 years are typically below a root mean square error of a few basis points. However, the calibration tends to fail with similar issues to the stochastic correlation when pricing the mezzanine tranches for longer dated tenors (at 7 and 10 years). Getting these longer dated tranches to calibrate is quite challenging, both with and without term structure of the local correlation functions $\rho(y)$.

Typically, the best calibration results are obtained with the correlation function $\rho(y)$ having a 5 point partition size.

14.3.5 Model Comparison

The local correlation model attempts to provide a "proper" model for the expected losses over the whole capital structure at a set of discrete maturities. It does calibrate for reasonable levels at 5 years but later tenors tend to be problematic. The calibration is also quite unintuitive and not always very stable.

14.4 Levy Copula

In this section, we describe the "Levy Copula" model introduced by Baxter (2006). To create heavy-tailed loss distributions, the definition of latent variables in the model is based on the construction of a Multivariate Levy Process. With different choices of Levy processes, one can construct a large family of copula functions. The Gaussian copula is a particular case where the chosen Levy process is a standard Brownian motion.

14.4.1 Gamma Copula

This model is an extension of the Gaussian copula, but not only does it provide a model for expected losses across the capital structure, it also has some built-in spread dynamics.

The construction of the model is based on the following lemma.

Lemma 100 *If $Y_g(t)$ and $Y_i(t)$ are independent Levy processes, with the same marginal distributions, then the process $X_i(t)$ defined by*

$$X_i(t) = Y_g(\phi t) + Y_i((1-\phi)t),$$

has the same distribution as $Y_g(t)$ and $Y_i(t)$. If the second-order moments of $Y_g(t)$ and $Y_i(t)$ are finite, then ϕ represents the correlation between these two variables.

This is how we choose to define the latent variable X_i. The parameter ϕ is the proportion a single-name's entity movement due to the central (or global) shock factor; the proportion $(1-\phi)$ is due to the idiosyncratic factor.

In this framework, any Levy process can be used to specify a dynamic credit model; if we choose the Gamma process, we can write the single-name latent variables as

$$X_i(t) = -\Gamma_g(t; \phi\gamma, \lambda) - \Gamma_i(t; (1-\phi)\gamma, \lambda),$$

where $\Gamma(t; \gamma, \lambda)$ is a Gamma process, which has a Gamma density with parameters $k = \gamma t$ and $\theta = \frac{1}{\lambda}$; the Gamma density is given by

$$f(x; k, \theta) = x^{k-1} \frac{e^{-x/\theta}}{\theta^k \Gamma(k)}, \text{ for } x > 0,$$

with $\Gamma(x)$ being the standard Gamma function. We note that if $\Gamma_1(t; \gamma_1, \lambda)$ and $\Gamma_2(t; \gamma_2, \lambda)$ are two independent Gamma processes, then the sum is also a Gamma process:

$$\Gamma_1(t; \gamma_1, \lambda) + \Gamma_2(t; \gamma_2, \lambda) = \Gamma(t; \gamma_1 + \gamma_2, \lambda).$$

With this definition of the latent variables X_i, the defaults are caused by negative jumps from the Gamma process. The parameter λ is the inverse jump-size of these downward jumps; the γ-parameter is the jump intensity and ϕ is the correlation between the entity's idiosyncratic jumps and the global jumps.

Remark 101 Note that the Gaussian copula can be re-formulated as a Levy process copula model by choosing the Brownian motion as a special case

$$X_i(t) = W_g(\rho t) + W_i((1-\rho)t),$$

where are $W_g(t)$ and $W_i(t)$ independent Brownian motions representing the global and idiosyncratic factors; the factor ρ is the Gaussian copula correlation parameter.

14.4.2 Implementation

Although this is to some extent a dynamic credit model—not dissimilar, in spirit, from structural models– it can still be implemented using the same framework as the one-factor Gaussian copula. The main differences are:

- how we determine the single-name default thresholds u_i;
- the computation of the conditional default probabilities $p_T^i(y)$;
- and the density function of the central shock variable $Y_g(t)$.

We address each point in turn.
First, to find the thresholds u_i, we use the following distribution function of the latent variable $X_i(t)$

$$\mathbb{P}(X_i(t) \le x) = 1 - G\left(-x; \gamma t, \frac{1}{\lambda}\right),$$

where $G(-x; k, \theta)$ denotes the Gamma distribution function. We can then find the u_i with a simple bisection algorithm.

Next, to compute the conditional default probabilities in the Gamma case, we have a convenient analytical formula

$$p_T^i(y) = \mathbb{P}\left(X_i(t) \le u_i \left| Y_g(t) = y\right.\right) = 1 - G\left(-u_i + y; (1-\phi)\gamma t, \frac{1}{\lambda}\right).$$

And finally, the density of the central shock variable $f_{Y_g(t)}(x)$, is also known analytically

$$f_{Y_g(t)}(x) = (-x)^{\phi\gamma t - 1} \frac{e^{\lambda x}}{\left(\frac{1}{\lambda}\right)^{\phi\gamma t} \Gamma(\phi\gamma t)}.$$

Equipped with these three building blocks (the default thresholds u_i, the conditional default probabilities $p_T^i(y)$, and the density of the common factor $f_{Y_g(t)}(x)$), we can easily implement the model by, literally plugging those building blocks into the standard pricing algorithm used for the Gaussian copula. That is one of the most appealing features of this model; although seemingly complicated at first sight, it turns out that the implementation details do not require any substantial (incremental) efforts above and beyond the well-established Gaussian case.

14.4.3 LHP Approximation

With the Levy copula set-up, we can also derive a useful Large Homogenous Portfolio (LHP) Approximation similar to the Gaussian copula.

We have the following result.

Proposition 102 (Levy Copula LHP Approximation) *In the Levy Copula Gamma Model, the Large Homogenous Portfolio approximation of the loss distribution is given by*

$$\mathbb{P}\left(L_T \leq x\right) = 1 - G\left(u + G^{-1}\left(1 - \frac{x}{1-R}; (1-\phi)\,\gamma T, \frac{1}{\lambda}\right); \phi\gamma T, \frac{1}{\lambda}\right),$$

where $u = -G^{-1}\left(1 - p_T^i; \gamma T, \frac{1}{\lambda}\right)$ *is the single-name default threshold.*

Proof In the Levy copula (Gamma) model, the conditional default probabilities are given by the Gamma distribution as

$$p_T^i\left(y\right) = 1 - G\left(-u_i + y; (1-\phi)\,\gamma T, \frac{1}{\lambda}\right),$$

where the default threshold u_i is also obtained by inverting a Gamma distribution

$$u_i = -G^{-1}\left(1 - p_T^i; \gamma T, \frac{1}{\lambda}\right).$$

For a homogenous portfolios, conditional on the common factor $Y_g\left(t\right)$, the loss variable L_T converges to its mean $(1-R)\,p_T\left(y\right)$, and we can write

$$\begin{aligned}
\mathbb{P}\left(L_T \leq x\right) &= \int_{-\infty}^{+\infty} \mathbf{1}_{\left\{p_T(y) \leq \frac{x}{1-R}\right\}} f_{Y_g(T)}\left(y\right) dy \\
&= \int_{-\infty}^{+\infty} \mathbf{1}_{\left\{1-G\left(-u+y;(1-\phi)\gamma T,\frac{1}{\lambda}\right) \leq \frac{x}{1-R}\right\}} f_{Y_g(T)}\left(y\right) dy \\
&= \int_{-\infty}^{+\infty} \mathbf{1}_{\left\{G^{-1}\left(1-\frac{x}{1-R};(1-\phi)\gamma T,\frac{1}{\lambda}\right) \leq -u+y\right\}} f_{Y_g(T)}\left(y\right) dy \\
&= 1 - G\left(u + G^{-1}\left(1 - \frac{x}{1-R}; (1-\phi)\,\gamma T, \frac{1}{\lambda}\right); \phi\gamma T, \frac{1}{\lambda}\right).
\end{aligned}$$

■

14.4.4 Calibration

The calibration process for the gamma model is very fast and, generally, exhibits good parameter stability. While the model calibrates reasonably well across the junior part of the capital structure, it does, however, suffer from underpricing the senior tranches (15–30% CDX and 12–22% iTraxx) and overpricing the super-senior tranches (30–100% CDX and 22–100% iTraxx). In addition, as we move further into longer tenors, the quality of fit starts to deteriorate, especially for iTraxx 10 years.

Some of these limitations can be mitigated to some extent by adding a Poisson shock to the Gamma model, transforming it into a so-called Catastrophe Gamma model, which puts more weight in the tail of the distribution and makes the fit a bit more precise. But even with this additional Catastrophe Poisson factor, the calibration remains challenging.

14.4.5 Model Comparison

This model provides a dynamic approach for modelling the expected loss of the entire capital structure. It is suitable for pricing CDOs and forward starting CDOs consistently, while the individual single-name defaults are still being modelled. The calibration is fast and the three parameters of the Gamma process are, in general, relatively stable with few local minima on the surface. By adding, in a Monte-Carlo simulation, some dynamics for the spread process, we could also, in principle, be able to price portfolio products with optionality, such as options on tranches and leveraged super-senior tranches. There are however much better fully dynamic portfolio credit models, which we will study in detail in Chap. 15.

14.5 Implied Hazard Rate Copula

In this section, we describe the "Implied Hazard Rate Copula" introduced by Hull and White (2006). It is a non-parametric copula function built implicitly by matching the prices of CDO tranches. It was referred to in their original papers as the "Perfect Copula" or the "Implied Copula".

14.5.1 Model Description

Recall that the Gaussian copula (or any factor copula for that matter) can be specified by two main components:

- the distribution density of the common conditioning factor Y

$$f_Y(y) = \varphi(y) = \frac{1}{\sqrt{2\pi}} e^{-\frac{x^2}{2}};$$

- and the individual single-name conditional probabilities

$$p_T^i(y) = \mathbb{P}(\tau_i \leq T \mid Y = y) = \Phi\left(\frac{\Phi^{-1}(u_i) - \rho y}{\sqrt{1-\rho^2}}\right);$$

thus defining the whole multivariate (default times) copula function

$$\mathbb{C}(u_1, \ldots, u_n) = \int_{-\infty}^{+\infty} \left[\prod_{i=1}^{n} \Phi\left(\frac{\Phi^{-1}(u_i) - \rho y}{\sqrt{1-\rho^2}}\right)\right] \varphi(y)\, dy.$$

The idea of the implied hazard rate copula model is exactly that: instead of constructing the copula function from a set of well-defined random variables, representing the common and idiosyncratic factors, which are then combined to define the individual single-name latent variables, one could specify the copula function directly in terms of its conditional default probabilities $p_T^i(y)$, and the distribution of its systematic factor $f_Y(y)$. This would be the most general way of specifying a one-factor copula function; and vice-versa, all one-factor copula functions can be represented as such; so, we can always write the default times multivariate distribution as

$$\mathbb{P}(\tau_1 \leq T_1, \ldots, \tau_n \leq T_n) = \int_{-\infty}^{+\infty} \left[\prod_{i=1}^{n} p_T^i(y)\right] f_Y(y)\, dy.$$

In particular, we can choose a discrete set of conditional default probabilities $\left(p_T^i(y_j)\right)_{\substack{1\leq i\leq n \\ 1\leq j\leq m}}$, and assign a probability q_j to each state of the world represented by a given realization of the common factor $\{Y = y_j\}$: i.e., we want to express the density function of the common factor $f_Y(y)$, as

$$f_Y(y) = \sum_{j=1}^{m} q_j \delta(y - y_j);$$

in order to specify the conditional probabilities $p_T^i(y_j)$, in each state j, we postulate that they are directly related to the marginal default probabilities $p_T^i = \mathbb{P}(\tau_i \leq T)$, and can be written simply as rescaled default probabilities, which depend on the (economic) state that we are in

$$p_T^i(y_j) = w_{i,j} \cdot p_T^i.$$

In other words, the conditional default probability for name i, under scenario j, is equal to a given weight times the unconditional default probability p_T^i.

When the model is fully calibrated, we end up with a set of probabilities $(q_j)_{1\leq j\leq m}$, for each state of the world; and the corresponding weights matrix $[w_{i,j}]$, which defines the conditional default probabilities, for each name, in each state of the world. As in the factor copula model, conditional on each state of the world $\{Y = y_j\}$, the individual single-name defaults are independent of each other. That makes the calculation of conditional expectations of a loss variable payoff—and by extension unconditional expectations—completely trivial,

$$\mathbb{E}[H(L_T)] = \sum_{j=1}^{m} q_j \mathbb{E}[H(L_T) | Y = y_j];$$

this forms the basis for the valuation of any portfolio credit product.

In general, if we have a total of l tranches in the benchmark index that we need to re-price $(P_T^k)_{1\leq k\leq l}$, and we have more states than calibration instruments $m >> 1$, then we can solve the problem by minimizing the (target) calibration benchmark pricing errors

$$\min \sum_{k=1}^{l} \left\| \mathbb{E}\left[M_T^k\right] - P_T^k \right\|^2,$$

where the (model) expected tranche loss is

$$\mathbb{E}\left[M_T^k\right] = \sum_{j=1}^{m} q_j \mathbb{E}\left[M_T^k | Y = y_j\right], \text{ for } 1 \leq k \leq l,$$

subject to the following constraints

$$\sum_{j=1}^{m} q_j = 1, q_j \geq 0, \text{ for } 1 \leq j \leq m,$$

$$\sum_{j=1}^{m} q_j \mathbb{E}[L_T | Y = y_j] = \mathbb{E}[L_T].$$

14.5.2 Calibration

In order to avoid (potential) numerical instabilities in the calibration process, we can impose more structure on the optimization problem to constrain it further; for example, we can choose to parameterize the hazard rate copula such that we only have one unknown parameter for each calibration equation. We can then specify the conditional default probabilities and find the discrete probability distribution that matches all the calibration equations by a well-posed matrix inversion instead of a general optimization under (linear) constraints.

When calibrating the skew curve, we typically have the expected loss for 5 index tranches, the portfolio expected loss and the additional constraint that the state probabilities should sum up to 1—in total 7 equations. Now, we need to choose a suitable partition of the scenario space Y. A good choice that works well in practice is to specify 7 levels of conditional default probabilities: one at zero, one that results in a conditional portfolio expected loss in the middle of each tranche and one systemic-type scenario where the conditional default probability is equal to one. The idea behind this discretization is to allow for sufficient flexibility in the support of the distribution in order to re-price the individual index tranches and the expected portfolio loss.

For each one of these scenarios, $j = 1, \ldots, m$, we have a conditional portfolio expected loss, $\mathbb{E}[L_T|Y = y_j]$, and a set of conditional tranche expected losses, $\mathbb{E}[M_T^k|Y = y_j]$, for $k = 1, \ldots, m-2$ and $j = 1, \ldots, m$. We can now set up a linear system of equations in order to find the scenario probabilities $(q_j)_{1\le j\le m}$ that re-price the unconditional expected losses correctly:

$$\sum_{j=1}^{m} q_j \mathbb{E}[M_T^k|Y = y_j] = \mathbb{E}[M_T^k], \text{ for } k = 1, \ldots, m-2,$$

$$\sum_{j=1}^{m} q_j \mathbb{E}[L_T|Y = y_j] = \mathbb{E}[L_T],$$

$$\sum_{j=1}^{m} q_j = 1.$$

Since this linear system has full rank, we can find the probabilities q_j uniquely using standard matrix algebra. However, there is nothing constraining the probabilities q_j to be non-negative. Negative probabilities can arise, for instance, when the original loss distribution allows for arbitrage. In these circumstances we can constrain the offending probabilities to zero and solve the resulting linear system in a least square sense.

Alternatively, if one is interested in producing a "smooth" distribution of the conditioning factor, we could allow for a large number of scenarios $m >> 1$, and find the most unbiased distribution which satisfies the constraints by using standard Entropy Maximization techniques, or solving a Relative Entropy Minimization problem with respect to an a-priori distribution such as the one generated by the standard Gaussian copula.

14.5.3 Model Comparison

The first benefit of the implied hazard rate copula model is that it gives rise to arbitrage-free loss distributions by construction. This is contrary to the base correlation framework where we need to take special care when interpolating the correlation skew curve.

The hazard rate copula model also provides a consistent framework for pricing bespoke CDOs. The name-specific weights of the previous section are dependent on the credit spread of the name in a very intuitive way. We can therefore price bespoke portfolios by selecting name-specific weights and comparing the individual spread levels of the bespoke portfolio to spread levels of the index constituents. Thus, we could price bespoke portfolios, within the model, in a consistent manner, which would induce its own model-specific skew rescaling method.

A further benefit is the treatment of CDO-Squared type structures. Compared to the base correlation model, we no longer need to introduce an exogenous (correlation) parameter linking the marginal underlying CDO loss variables; it is all handled within the model coherently. There is also a large improvement in terms of computational time.

In benchmarking the model against base correlation, there are certain issues that can be highlighted. One clear difference when comparing the hazard rate copula to the standard base correlation framework is that the distribution of the central chock Y is continuous in the Gaussian copula but discrete in the hazard rate copula. During calibration, we have some freedom as to where we want to place the name-specific weights. As described above we have chosen the weights so that the conditional portfolio expected loss for each scenario falls in the middle of each tranche, but by choosing the weights differently we would calibrate different scenario probabilities q_j. This difference in calibration would not impact the pricing of the standard index tranches, since we match them by construction but would affect the prices of bespoke tranches.

This uncertainty in bespoke tranche prices is rather small under normal circumstances, but could be amplified for narrow bespoke tranches. There are several possibilities for tackling this issue. One suggestion would be to add more scenarios for the conditional default probabilities and in this way suppress the discrete nature of the model. The discrete points could also be chosen in a specific way to smoothen the loss distribution. As mentioned previously, Entropy Maximization methods offer one way of handling such ill-posed mathematical problems, with an underdetermined system of equations.

References

L. Andersen, J. Sidenius, Extensions to the Gaussian Copula: random recovery and random factor loadings. J. Credit Risk **1**, 29–70 (2005)

M. Baxter, Levy process dynamic modelling of single-name credits and CDO tranches (Working Paper, Nomura, 2006)

X. Burtschell, J. Gregory, J.P. Laurent, Beyond the Gaussian Copula: stochastic and local correlation. J. Credit Risk **3**(1), 31–62 (2007)

X. Burtschell, J. Gregory, J.P. Laurent, A comparative analysis of CDO pricing models. J. Deriv. **16**(4), 9–37 (2009)

B. Dupire, Pricing with a smile. Risk **7**, 18–20 (1994)

J. Hull, A. White, Valuing credit derivatives using an implied copula approach. J. Deriv. **14**(2), 8–28 (2006)

D.X. Li, M. Liang, CDO squared pricing using a gaussian mixture model with transformation of loss distribution (Working Paper, Barclays Capital, 2005)

J. Turc, P. Very, D. Benhamou, *Pricing CDOs with a smile* (Quantitative Strategy, Credit Research, Societe Generale, 2005)

15

Third Generation Models: From Static to Dynamic Models

In this chapter, we review some of the most important dynamic credit models in the literature. We give a brief description of each model and discuss the advantages and limitations of each modelling framework. We also comment on the usefulness of each model for a given family of correlation products. The models discussed include: the Top-Down model (of Giesecke and Goldberg 2005), the Dynamic Generalized Poisson Loss model (of Brigo et al. 2007), the N+ Model (of Longstaff and Rajan 2008), the Markov Chain Portfolio Loss model (of Schönbucher 2005), and the SPA model (of Sidenius et al. 2008).

15.1 Top-Down Approach

In the general Top-Down approach, introduced by Giesecke and Goldberg (2005), multi-name credit portfolios are modeled using a point process approach where the dynamics of the whole portfolio are considered instead of the individual single-name credits. The model specification is done through a choice of (portfolio) intensity process; then the single-name dynamics are inferred from it by projecting down the portfolio variable on the underlying single-names using a random thinning technique. This general description covers all possible choices of joint dependence specification, ranging from the classical doubly-stochastic models (typically, constructed through a bottom-up approach), to self-exciting models, where the portfolio intensity is impacted dynamically, by prior default events, hence creating a feedback loop between defaults and intensities. This is the so-called contagion effect observed in credit markets.

Y. Elouerkhaoui, *Credit Correlation*, Applied Quantitative Finance,
DOI 10.1007/978-3-319-60973-7_15

15.1.1 Model Set-Up

We work in a probability space $(\Omega, \mathcal{G}, \mathbb{P})$, where we have n default times $(\tau_i)_{1\leq i\leq n}$. $\left(T^{(i)}\right)_{1\leq i\leq n}$ is the sequence of ordered default times (i.e., the time of the first-to-default event, second-to-default, and so forth). $\mathbb{G} = (\mathcal{G}_t)_{t\geq 0}$ is the enlarged filtration containing the default events' filtrations and the economic state-variables background filtration $\mathbb{F}$.

The single-name default indicators are $D_t^i = \mathbf{1}_{\{\tau_i \leq t\}}$, and the kth-to-default indicators are denoted by $O_t^{(k)} = \mathbf{1}_{\{T^{(k)}\leq t\}}$. The portfolio default counting process is defined by the sum $D_t = \sum_{i=1}^n D_t^i = \sum_{i=1}^n O_t^{(i)}$. Under usual technical assumptions, the single-name default processes D^i and the portfolio default process D admit a Doob-Meyer decomposition into a non-decreasing predictable process A^i (and A respectively) and a local martingale M^i (and M respectively), $D_t^i = A_t^i + M_t^i$ (and $D_t = A_t + M_t$ respectively). A^i (and A) are the compensators of their respective counting processes with respect to the filtration $\mathbb{G}$, and we have, for all $\epsilon \geq 0$,

$$\mathbb{E}\left[D_{t+\epsilon} - D_t \left| \mathcal{G}_t \right.\right] = \mathbb{E}\left[A_{t+\epsilon} - A_t \left| \mathcal{G}_t \right.\right].$$

In the limit, we can write, almost surely,

$$A_t = \lim_{\epsilon\downarrow 0}\frac{1}{\epsilon}\int_0^t \mathbb{E}\left[D_{s+\epsilon} - D_s \left| \mathcal{G}_s \right.\right] ds.$$

We say that a point process D is intensity-based if its compensator can be represented as

$$A_t = \int_0^t \lambda_s ds,$$

where λ is a positive adapted process, equal to 0, almost surely on the set $\left\{t > T^{(n)}\right\}$. Thus, we have $\lambda_t = \lim_{\epsilon\downarrow 0}\frac{1}{\epsilon}\mathbb{E}\left[D_{s+\epsilon} - D_s \left| \mathcal{G}_s \right.\right] ds$, almost surely.

λ is called the $\mathbb{G}$-intensity of the point process D.

Summation. If the single-name compensators A^i are given, then the compensator of the portfolio counting process A can be constructed by adding up the individual compensators

$$A = \sum_{i=1}^n A^i.$$

This property, obviously, holds for independent defaults, but it is equally true when defaults are correlated as long as we are using the single-name compensators with respect to the (same) enlarged filtration $\mathbb{G}$ (and not the individual single-name filtrations $\mathbb{G}^i$).

If the single-name default processes are intensity-based, so that $A_t^i = \int_0^t \lambda_s^i ds$, then the intensity of the portfolio point process λ is also given by summation, $\lambda = \sum_{i=1}^n \lambda^i$.

15.1.2 Random Thinning

The opposite of summation is random thinning. If we are given a compensator A of the portfolio default counter D, then the single-name compensators A^i can be constructed by the random thinning technique.

Definition 103 (*Thinning Process*) A thinning process for the compensator A is a bounded, non-negative predictable process Z for which the Stieltjes integral $\int_0^t Z_s dD_s$ defines the indicator process of a totally inaccessible stopping time.

We say that Z^i replicates D^i, if Z^i is a thinning process, which satisfies $D_s^i = \int_0^t Z_s^i dD_s$, almost surely.

Now, the random thinning process can be described as follows.

Proposition 104 (Random Thinning) *Suppose the portfolio compensator A is a continuous process. If Z^i is a thinning process for A that replicates D^i, then the single-name compensator A^i is given by the Stieltjes integral*

$$A_t^i = \int_0^t Z_s^i dA_s.$$

Furthermore, if it is an intensity-based model, then Z^i thins the intensities as well, so that we have

$$\lambda^i = Z^i \lambda.$$

15.1.3 Doubly Stochastic Intensity Models

We have two definitions of a doubly stochastic intensity model. We can define a top-down version using the portfolio counting process, or a bottom-up version using the single-name default processes. Giesecke and Goldberg (2005) show that the two definitions are mutually exclusive

Definition 105 (*Doubly Stochastic Intensity Model Top-Down Specification*) We say that an intensity based model is doubly-stochastic with respect to a right continuous, complete sub-filtration $\mathbb{F} \subset \mathbb{G}$, if there exists a non-negative $\mathbb{F}$-predictable process h so that, for each $t \geq 0$, $u > 0$, we have

$$\mathbb{P}\left(D_{t+u} - D_t = k \,|\mathcal{G}_t \vee \mathcal{F}_{t+u}\right) = \frac{1}{k!}\left(\int_t^{t+u} h_s ds\right)^k \exp\left(-\int_t^{t+u} h_s ds\right),$$

for k ranging from 0 to $n(\omega, t)$, which is the number of names that are alive after time t, in state ω.

The bottom-up definition of doubly stochastic intensity model is given in terms of the single-name default indicators D^i. This is the framework used, for example, in Lando (1998).

Definition 106 (*Doubly Stochastic Intensity Model Bottom-Up Specification*) We say that an intensity based model is doubly-stochastic with respect to a right continuous, complete sub-filtration $\mathbb{F} \subset \mathbb{G}$, if there exists a non-negative $\mathbb{F}$-predictable process h^i so that, for each $t \geq 0$, $u > 0$, we have

$$\mathbb{P}\left(\tau_i > t + u \,|\mathcal{G}_t \vee \mathcal{F}_{t+u}\right) = \exp\left(-\int_t^{t+u} h_s^i ds\right),$$

on the set $\{\tau_i > t\}$.

15.1.4 Self-exciting Intensity Models

On the other side of the model spectrum, we have the so-called self-exciting processes: a rich class of models where the portfolio intensity reacts to realizations of the portfolio defaults, which, in turn, impact the next wave of future defaults, thus creating a feedback loop between defaults and intensities; this leads to natural contagion effects and generates some very interesting default clustering patterns. They are also called, sometimes, self-affecting or path-dependent point processes, since the intensity depends on the path of the underlying counting process.

Hawkes Process. A familiar example of a self-exciting process is the Hawkes (1971) process. Its $\mathbb{G}$-intensity is given by the following Stieltjes integral

$$\lambda_t = c + \int_0^t d\,(t-u)\, dD_u,$$

where c is a positive constant, and $d(t)$ is a non-negative function of time. In this model, the intensity of the portfolio counting process D is continuously updated as defaults occur along its path. Before the first default time $T^{(1)}$, the intensity is equal to the initial value c, then after $T^{(1)}$, it jumps to a new level $c + d\left(t - T^{(1)}\right)$; after k defaults, the intensity accumulates all the default-incurred jumps along the way: $c + \sum_{i=1}^{k} d\left(t - T^{(i)}\right)$. Hawkes and Oakes (1974) have shown that a Hawkes process can be represented as a Poisson cluster process. Each default event triggers a cluster of new defaults with intensity $d(u)$; the Hawkes process is then obtained as a superposition of the initial Poisson process, with intensity c, and the incremental default-triggered Poisson processes.

There are many possible choices for the intensity impact function $d(t)$; we give here two possible examples: Exponential Decay and Multiplicative Jumps.

Example 107 (Exponential Decay). It is given by the parametrization

$$d(u) = \sum_{j=1}^{k} \alpha_j e^{-\beta_j u},$$

where the sum $\sum_{j=1}^{k} \frac{\alpha_j}{\beta_j} < 1$ for some $k \geq 1$. The parameter α_j controls the size of the intensity jump, and β_j controls the impact decay over time.

Example 108 (Multiplicative Jumps). It is given by the parametrization

$$d(u) = g^i(u)\,\delta^i,$$

where $g^i(u)$ is a deterministic non negative function, and δ^i is an $\mathcal{F}_{T^{(i)}}$-measurable random variable. The parameter δ^i is the random jump at the time of the ith-default, and $g^i(u)$ controls the decay after default.

Next, we study some practical examples of Top-Down models that have been used successfully to calibrate to the index CDO tranche market and offer very realistic dynamics of the portfolio loss distribution over time.

15.2 Dynamic Generalized Poisson Loss Model

In this section, we review the Dynamic Generalized Poisson Loss (GPL) model of (Brigo et al. 2007). The main idea of this approach consists in representing the aggregate (portfolio) default distribution, at different time horizons, as a linear combination of independent Poisson processes. The jump of each (common) Poisson process represents the default of a sub-group of names in the portfolio. This is closely related to the Marshall-Olkin model studied, for example, in Lindskog and McNeil (2003) or Elouerkhaoui (2006). This is a top-down model in the sense that it does not explicitly model the single name defaults and focuses instead on modelling the aggregate portfolio distribution that reproduces the market quotes of (index) CDO tranches.

15.2.1 Basic GPL Model

In the basic GPL model, we have a set of independent Poisson processes $\left(N^1, ..., N^m\right)$, with time-dependent intensities $\left(\lambda^1, ..., \lambda^m\right)$, and we define the process

$$Z_t = \sum_{j=1}^{m} \alpha_j N_t^j,$$

where $(\alpha_1, ..., \alpha_m)$ is a set of integers, representing the number of defaulted names for each Poisson shock N^j. A possible choice is to set $\alpha_j = j$ so that we have

$$Z_t = N_t^1 + 2N_t^2 + 3N_t^3 .. + mN_t^m.$$

To ensure that the counting process Z_t is indeed a portfolio aggregate default process, we need to cap it at the total number of names in the portfolio. Thus, we model the portfolio default counter as

$$D_t = \min\left(Z_t, n\right) = Z_t \mathbf{1}_{\{Z_t < n\}} + n \mathbf{1}_{\{Z_t \geq n\}}.$$

Since we know explicitly the distribution of the auxiliary counting variable Z_t, we can derive easily the default distribution D_t,

$$\mathbb{P}\left(D_t \leq x\right) = \mathbb{P}\left(Z_t \leq x\right) \mathbf{1}_{\{x<n\}} + \mathbf{1}_{\{x \geq n\}},$$

and its associated (discrete) density function, $p_{D_t}(x) = \mathbb{P}\left(D_t = x\right)$,

$$p_{D_t}(x) = p_{Z_t}(x) \mathbf{1}_{\{x<n\}} + \mathbb{P}\left(Z_t \geq x\right) \mathbf{1}_{\{x \geq n\}}.$$

Similarly, we can compute its characteristic function through the (well-known) characteristic function of the variable Z_t

$$\varphi_{Z_t}(u) = \mathbb{E}\left[\exp\left(iuZ_t\right)\right] = \int_0^\infty \exp\left(iux\right) p_{Z_t}(x)\, dx.$$

Thanks to the independence of the Poisson processes N^j, the characteristic function $\varphi_{Z_t}(u)$ is simply given by the product of the individual characteristic functions of the Poisson processes $\varphi_{N_t^j}(u)$

$$\varphi_{Z_t}(u) = \prod_{j=1}^{m} \varphi_{N_t^j}(\alpha_j u) = \exp\left(\sum_{j=1}^{m} \Lambda_t^j \left(e^{i\alpha_j u} - 1 \right) \right),$$

where $\Lambda_t^j = \int_0^t \lambda_s^j ds$ is the cumulative intensity of the Poisson process N^j. The density $p_{Z_t}(x)$ can be obtained by standard inversion of the Fourier transform $\varphi_{Z_t}(u)$.

15.2.2 Portfolio Default Intensity

Consider A^D the compensator of the portfolio default counting process D, i.e., the non-decreasing predictable process satisfying the Doob-Meyer decomposition

$$\mathbb{E}\left[D_T - A_T^D \,|\mathcal{G}_t \right] = D_t - A_t^D.$$

Following Giesecke and Goldberg (2005), we can compute A^D as follows

$$\begin{aligned} A_T^D &= \lim_{\epsilon \downarrow 0} \frac{1}{\epsilon} \int_0^T \mathbb{E}\left[D_{t+\epsilon} - D_t \,|\mathcal{G}_t \right] dt \\ &= \lim_{\epsilon \downarrow 0} \frac{1}{\epsilon} \int_0^T \mathbb{E}\left[\min\left(Z_{t+\epsilon} - Z_t, n - Z_t \right) \mathbf{1}_{\{Z_t < n\}} \,|\mathcal{G}_t \right] dt \\ &= \lim_{\epsilon \downarrow 0} \frac{1}{\epsilon} \int_0^T \sum_{j=1}^{m} \min\left(\alpha_j, n - Z_t \right) \mathbf{1}_{\{Z_t < n\}} \mathbb{P}\left(Z_{t+\epsilon} - Z_t = \alpha_j \,|\mathcal{G}_t \right) dt, \end{aligned}$$

which gives the expression of the compensator and, therefore, the default intensity

$$A_T^D = \int_0^T \sum_{j=1}^{m} \min\left(\alpha_j, (n - Z_{t^-})^+ \right) \lambda_t^j dt,$$

$$\lambda_t^D = \sum_{j=1}^{m} \min\left(\alpha_j, (n - Z_{t^-})^+ \right) \lambda_t^j.$$

Thus the portfolio default intensity λ^D jumps in tandem with the default counting process D, and goes to zero when we have reached the total number of defaults n.

15.2.3 Calibration

During the calibration process, we need to find a parsimonious choice of both jump sizes α_j and Poisson intensities λ^j that match the index tranche quotes. The optimal values are found with the following incremental algorithm: for each maturity T_k,

1. start with the first Poisson process N^1, set $\alpha_1 = 1$ and calibrate $\Lambda^1_{T_k}$ by minimizing the market price errors' objective function;
2. add the second Poisson process N^2 and find the optimal value for α_2 in the range $[1, n]$ (typically, $n = 125$) by calibrating the intensities $\Lambda^1_{T_k}$ and $\Lambda^2_{T_k}$;
3. repeat the same process for α_j, for $j \geq 3$, by calibrating the intensities $\left(\Lambda^1_{T_k}, \Lambda^2_{T_k}, ..., \Lambda^j_{T_k}\right)$ until the calibration error is under its tolerance or the intensity $\Lambda^j_{T_k}$ is close to zero.

Typically, for index tranches, the best calibration results are obtained with a choice of 5 Poisson processes. And we get the typical multi-modal default distributions of the Marshall-Olkin model.

15.2.4 Dynamic Default Intensity

To make the model dynamic, the intensities are assumed to be stochastic (as opposed to time-dependent), and the Poisson processes N^j become independent Cox processes instead. In this case, the characteristic function $\varphi_{Z_t}(u)$ is given by the expectation

$$\varphi_{Z_t}(u) = \mathbb{E}\left[\exp\left(\sum_{j=1}^{m} \Lambda_t^j \left(e^{i\alpha_j u} - 1\right)\right)\right].$$

To compute it explicitly, we need to make a choice of intensity dynamics. We list here some possible examples.

Example 109 (Piecewise-Gamma Intensity). Assume the intensities' increments (on a fixed time grid T_l^Λ, for $l = 1, ..., g$) are independent and follow a Gamma distribution

$$\Lambda^j_{T_l^\Lambda} - \Lambda^j_{T_{l-1}^\Lambda} \backsim \Gamma\left(k_l^j, \theta_l^j\right),$$

with shape parameter $k_l^j > 0$, and scale parameter $\theta_l^j > 0$. The characteristic function is still given in closed form,

$$\varphi_{Z_t}^{GPW}(u) = \prod_{j=1}^{m}\prod_{l=1}^{g}\left(1 + \theta_l^j\left(1 - e^{i\alpha_j u}\right)\right)^{-k_l^j}.$$

Example 110 (Intensity Mixture). The intensities are specified on a set of different hypothetical "scenarios" where they take realized values $\Lambda_T^j = \Lambda^{j,k}(T)$ with probabilities q_k, for $k = 1, ..., h$; this leads naturally to a mixture model

$$\varphi_{Z_t}^{MIX}(u) = \sum_{k=1}^{h} q_k \exp\left(\sum_{j=1}^{m} \Lambda^{j,h}(t)\left(e^{i\alpha_j u} - 1\right)\right).$$

Example 111 (CIR Intensity). We can also assume that the Poisson intensities follow a standard CIR process

$$d\lambda_t^j = k_j\left(\theta_j - \lambda_t^j\right)dt + \sigma_j\sqrt{\lambda_t^j}\,dW_t^j,$$

which implies well-known closed form solutions similar to zero-coupon bond prices in the CIR interest rate model.

15.3 N+ Model

A very similar model, in spirit, to the GPL model is the common Poisson shock model of Longstaff and Rajan (2008). We will refer to it henceforth as the N+ model. One of its main advantages is that it works, directly, with the loss variable, as opposed to the default counting process in GPL, which then needs to be re-scaled back "appropriately" to represent a portfolio loss variable.

The N+ model could be viewed as a simplified version of the Marshall-Olkin model where joint defaults are triggered by a set of common Poisson shocks.

But instead of modelling the individual single-names, this approach focuses on the portfolio loss distribution and represents each mode of the distribution with a Cox process which triggers the joint default clusters. This simplifies the model dynamics and the calibration process.

15.3.1 Model Description

This approach models the loss variable L_t dynamically, in a similar fashion to the Marshall-Olkin copula, where the backbone of the correlation structure is built upon a set of independent (doubly-stochastic) Poisson processes representing possible market shock events.

The loss variable is specified as

$$L_t = 1 - \prod_{j=1}^{m} e^{-\gamma_j N_t^j},$$

where the factors $\left(N_t^j\right)_{1 \le j \le m}$ are m independent Cox processes, with (stochastic) intensities $\left(\lambda_t^j\right)_{1 \le j \le m}$. The terms γ_j represent the portfolio loss jump sizes occurring at each jump time of the process N_t^j. From this definition, it is clear that the loss function is monotonically decreasing and is consistent with a parametrization of the loss process. In particular, unlike the parametrization of the GPL model, this variable is bounded between 0 and 1 by construction.

It can seen as the solution of the equivalent simplified point process SDE of the Marshall-Olkin model transposed onto the loss variable

$$dL_t = (1 - L_{t^-}) \sum_{j=1}^{m} \overline{\gamma_j} dN_t^j, \text{ for } t \ge 0$$

$$L_t = 0,$$

where $\overline{\gamma_j} = 1 - e^{-\gamma_j}$. Thus it has all the natural properties of a portfolio loss variable.

We also assume that the intensities for the Cox processes follow some standard CIR dynamics

$$d\lambda_t^j = \left(\alpha_j - \beta_j \lambda_t^j\right) dt + \sigma_j \sqrt{\lambda_t^j} dW_t^j,$$

where $\left(W_t^j\right)_{1\le j\le m}$ are independent Brownian motions.

Given the choice of CIR dynamics for the (factor) intensity processes, a semi-analytic solution can be obtained for the expectation of any function H of the loss variable:

$$\mathbb{E}\left[H\left(L_T\right)\right] = \sum_{n_1=0}^{\infty}\sum_{n_2=0}^{\infty}\cdots\sum_{n_m=0}^{\infty}\left[\prod_{j=1}^{m}\frac{P_{n_j}\left(\lambda_0^j, T\right)}{n_j!}\right] H\left(1-\prod_{j=1}^{m}e^{-\gamma_j n_j}\right),$$

where $P_{n_j}\left(\lambda_0^j, T\right)$ is the probability that the Cox process is N_T^j is equal to n_j, conditional on today's intensity value λ_0^j, and rescaled by the factor $n_j!$,

$$P_{n_j}\left(\lambda_0^j, T\right) = \mathbb{P}\left(N_T^j = n_j\right)\cdot n_j! = \mathbb{E}\left[\exp\left(-\int_0^T \lambda_t^j dt\right)\left(-\int_0^T \lambda_t^j dt\right)^{n_j}\right].$$

The function $P_i\left(\lambda, T\right)$ satisfies the following recursive partial differential equation

$$\frac{\sigma^2\lambda}{2}\frac{\partial^2 P_i}{\partial\lambda^2} + \left(\alpha - \beta\lambda\right)\frac{\partial P_i}{\partial\lambda} - \lambda P_i + i\lambda P_{i-1} = \frac{\partial P_i}{\partial T},$$

whose solution is given by

$$P_i\left(\lambda, T\right) = A\left(T\right)e^{-B(T)\lambda}\sum_{j=0}^{i} C_{i,j}\left(T\right)\lambda^j,$$

where the functions $A\left(T\right)$ and $B\left(T\right)$ are defined as

$$A\left(T\right) = \exp\left(\frac{\alpha\left(\beta-\zeta\right)T}{\sigma^2}\right)\left(\frac{2\zeta}{\beta+\zeta-\left(\beta-\zeta\right)e^{-\zeta T}}\right)^{\frac{2\alpha}{\sigma^2}},$$

$$B\left(T\right) = \frac{2\zeta\left(\beta+\zeta\right)}{\sigma^2\left(\beta+\zeta-\left(\beta-\zeta\right)e^{-\zeta T}\right)} - \frac{\beta+\zeta}{\sigma^2},$$

and $\zeta = \sqrt{\beta^2 + 2\sigma^2}$. Finally, the functions $C_{i,j}(t)$ are solutions of a recursive system of first-order ordinary differential equations: for $1 \leq j \leq i - 1$,

$$\frac{dC_{i,i}(t)}{dt} = iC_{i-1,i-1}(t) - \left(\sigma^2 B(t) + \beta\right) iC_{i,i}(t),$$

$$\frac{dC_{i,j}(t)}{dt} = iC_{i-1,j-1}(t) - \left(\sigma^2 B(t) + \beta\right) jC_{i,j}(t) + (j+1)\left(\alpha + \frac{j\sigma^2}{2}\right) C_{i,j+1}(t),$$

$$\frac{dC_{i,0}(t)}{dt} = \alpha C_{i,1}(t),$$

with initial conditions $C_{0,0}(t) = 1$ and $C_{i,j}(0) = 0$, for all $i > 0$.

15.3.2 Calibration

Typically, the best (stable) calibration results are achieved with three factors $m = 3$. By choosing to include more factors in the model, $m > 3$, the calibration errors improve at the expense of a significant decrease in stability. Furthermore, the 3-factor model has a very natural market interpretation. When calibrated historically, one would find that the three jump sizes relate (approximately) to three types of market events: (a) an idiosyncratic event where one individual single name credit defaults, (b) a sector event where a few names from the same sector default and, (c) a major systemic shock event where approximately 70% of all firms (from different sectors) default. In the 3-factor model, the calibration is very accurate, especially for iTraxx and CDX 5 years. Typically, the root mean square error is below 1 bp for these tenors. It tends to struggle the most with senior tranches where the errors are often around 1–1.5 bp, but this is still very good and comparable to the market bid-offer spreads.

15.3.3 Model Comparison

Similar to the dynamic GPL model of Brigo et al. (2007), it offers another possible dynamic credit model specification, which gives very good calibration results for a single tenor (typically, the 5 year tenor); it does reasonably well with multiple tenor calibrations, but the quality of the fit deteriorates as we add more maturities to the calibration set.

It could be used, for example, as a consistent interpolation tool by calibrating its parameters to the base correlation skew and generating an arbitrage-free interpolation of the rest of the correlation surface. But since the calibration process does not match the skew perfectly for all tenors (and all parts of the capital structure), it could potentially produce some undesirable noise in both marks and risk if it is used on its own.

The model is more relevant for portfolio credit products with some path-dependency. It can be used to generate well-behaved forward skew dynamics that are needed to ensure consistent pricing and stable risk management of path-dependent products. We shall see later that we can also address some of the observed multi-tenor calibration errors by using the N+ model dynamics in conjunction with a Markov-Functional skew map overlay; by doing so, we ensure that the (skewed) marginal loss distributions, at each time horizon, are reproduced and that the implied time-dependence follows the multi-modal N+ skew dynamics.

15.4 Markov Chain Portfolio Loss

In this section, we review the Markov Chain Portfolio Loss model of Schönbucher (2005). In this framework, the dynamics of both the portfolio loss variable and the forward default probabilities are modelled concurrently. The portfolio loss distribution, for each time horizon, is represented with a time-inhomogeneous Markov chain and its evolution through time is described with the Markov chain forward transition rates. The model is calibrated, by construction, to the initial (spot) loss distribution across different strikes and maturities; then to evolve the loss distributions in time, the forward transition rates are assumed to follow some stochastic dynamics, which satisfy a drift condition similar to the Heath-Jarrow-Morton (HJM) drift condition for interest rates term structure models; this ensures that the martingale property is preserved and the (forward) implied model is arbitrage-free.

15.4.1 Loss Distribution

To simplify the presentation of the model, we assume that all credits in the portfolio have identical LGDs (loss-given-default) and they are all normalized to one, so that we can write the loss variable as the portfolio default counter D_t,

$$L_t = D_t = \sum_{i=1}^{n} D_t^i.$$

The (discrete) loss distribution, at time T, is represented by the probability vector $\mathbf{p}_{t,T} = \left(p_{t,T}^{(1)}, p_{t,T}^{(2)}, ..., p_{t,T}^{(n)}\right)$,

$$p_{t,T}^{(j)} \triangleq \mathbb{P}\left(L_T = j \,|\mathcal{G}_t\right).$$

At time $t = 0$, the (market) implied distribution is known and it is generated from a standard base correlation curve marked to observed index tranche prices: $p_{0,T}^{(j)} \triangleq \mathbb{P}\left(L_T = j\right)$.

Assumption (Regular Loss Distribution). The loss process L_t is a piece-wise constant, bounded non-decreasing process; we assume that standard regularity conditions are verified, namely

1. For all $j \leq n$, and $T \geq t$, $p_{t,T}^{(j)}$ is continuous in T;
2. $p_{t,T}^{(j)}$ is once-continuously differentiable in T, almost everywhere;
3. If $p_{t,T_0}^{(j)} > 0$ for some T_0, then $p_{t,T}^{(j)} > 0$ for all $T > T_0$.

It is also convenient to assume that the probability of simultaneous defaults is zero; in other words, the increments of the loss process are equal to one.

Furthermore, given the path structure of the loss process, the following properties are verified.

For all $0 \leq j \leq n$, and $t \leq T$, we have:

1. $p_{t,T}^{(j)} \geq 0$, and $\sum_{j=0}^{n} p_{t,T}^{(j)} = 1$;
2. $p_{t,t}^{(j)} = \mathbf{1}_{\{L_t = j\}}$, and $p_{t,T}^{(j)} = \mathbf{1}_{\{j \geq L_t\}}$;
3. The function $\sum_{k=0}^{j} p_{t,T}^{(k)} = \mathbb{P}\left(L_T \leq j \,|\mathcal{G}_t\right)$ is non-increasing in T.

Next, we represent the loss distribution with the transition rates of a Markov-chain process.

15.4.2 Time-Inhomogeneous Markov Chain

If the $A(t) = \left(a_{ij}(t)\right)_{0 \leq i, j \leq n}$ is the generator matrix of a time-inhomogeneous Markov chain $\widetilde{L}_t$, the transition rates $a_{ij}(t)$, from state $\widetilde{L}_t = i$ to $\widetilde{L}_t = j$, satisfy the following conditions

$$a_{ij}(t) \geq 0, \text{ for all } 0 \leq i, j \leq n,$$

$$\sum_{k=0,k\neq j}^{n} a_{jk}(t) = -a_{jj}(t) = a_j(t), \text{ for all } 0 \leq j \leq n.$$

And the transition probability matrix $P(t,T) = \left(P_{ij}(t,T)\right)_{0\leq i,j\leq n}$ (defined as $P_{ij}(t,T) \triangleq \mathbb{P}\left(\widetilde{L}_T = j \,\middle|\, \widetilde{L}_t = i\right)$) is the solution of a set of Kolmogorov equations

$$\frac{dP(t,T)}{dT} = P(t,T)\,A(t), \text{ with initial condition } P(t,t) = Id.$$

Since $\widetilde{L}_t$ is a non-decreasing process, the generator (and transition) matrix is upper-triangular with zeros in the last row (as $\widetilde{L}_t = n$ is an absorbing state). Integrating the Kolmogorov equations gives the transition probabilities as a function of the transition rates

$$P_{ij}(t,T) = 0, \text{ for } j < i,$$

$$P_{ij}(t,T) = \exp\left(-\int_t^T a_j(s)\,ds\right), \text{ for } i = j,$$

$$P_{ij}(t,T) = P_{jj}(t,T)\left[\sum_{k=i}^{j-1}\int_t^T \frac{P_{ik}(t,s)}{P_{jj}(t,s)} a_{kj}(s)\,ds\right], \text{ for } j > i.$$

The main result linking the loss process L_t to the Markov-chain probabilities is given in the following proposition.

Proposition 112 (Markov-Chain Representation of the Loss Distribution) *Let $t \geq 0$ and let the vector $\mathbf{p}_{t,T}$, for $T \geq t$, be a loss distribution satisfying the regularity properties above. Then, there exists a time-inhomogeneous Markov chain $\widetilde{L}_T$, for $T \geq t$, with generator matrix $A(t,T)$, such that its transition probability matrix $P(t,T)$ reproduces the loss distribution $\mathbf{p}_{t,T}$, i.e.,*

$$p_{t,T}^{(j)} = P_{\widetilde{L}_t,j}(t,T), \textit{ for } 0 \leq j \leq n, T \geq t.$$

In a nutshell, the proposition states that any arbitrage-free conditional loss distribution can be equivalently represented with the generator matrix of a time-inhomogeneous Markov chain. This is a very general representation, which does not restrict the set of possible loss distributions. We will refer to the elements of the generator matrix $A(t,T)$ as the *forward transition rates.*

So far, we have shown the existence of such a Markov chain representation. In the next result, we go one step further and show that, if we only consider one-step transition matrices, then the representation is unique.

Proposition 113 (One-Step Representation of the Loss Distribution) *Let $t \geq 0$ and let the vector $\mathbf{p}_{t,T}$, for $T \geq t$, be a loss distribution satisfying the regularity assumptions above. Define*

$$a_j(t,T) \triangleq -\frac{1}{p_{t,T}^{(j)}} \sum_{k=0}^{j} \frac{\partial}{\partial T} p_{t,T}^{(k)}, \textit{ for all } 0 \leq j \leq n,$$

and assume that $\lim_{T \to T} a_j(t,T) < \infty$. Then, there exists a unique one-step time-inhomogeneous Markov chain $\widetilde{L}_T$, for $T \geq t$, with generator matrix $A(t,T)$, and transition probability matrix $P(t,T)$, such that

$$p_{t,T}^{(j)} = P_{\widetilde{L}_t, j}(t,T), \textit{ for } 0 \leq j \leq n, T \geq t.$$

Now that we have represented uniquely the loss distributions with one-step Markov transition probabilities, the next step is to make them *stochastic* and assume some diffusion dynamics and work out the (HJM) drift conditions that would keep the model arbitrage-free.

15.4.3 Consistency

To start with, we make the following assumptions for the one-step Markov chain transition rates.

1. $A(t,T)$ is a bi-diagonal matrix, i.e., for all $0 \leq i, j < n$, we have $a_{ij}(t,T) = 0$ if $j \notin \{i, i+1\}$; $a_j(t,T) = -a_{jj}(t,T) = a_{j,j+1}(t,T) \geq 0$ and $a_n(t,T) = 0$.
2. The solution of the Markov chain Kolmogorov equations is well defined and given by: for all $0 \leq i, j \leq n$,

$$P_{ij}(t,T) = 0, \text{ for } j < i,$$

$$P_{ij}(t,T) = \exp\left(-\int_t^T a_j(t,s)\,ds\right), \text{ for } i = j,$$

$$P_{ij}(t,T) = \int_t^T P_{i,j-1}(t,s)\, a_{j-1}(t,s)\, e^{-\int_s^T a_j(t,u)du} ds, \text{ for } j > i.$$

3. There exists an integrable random variable $Y(\omega)$ such that $A(t, T; \omega) \leq Y(\omega)$, for all $t \leq T \leq T^*$ a.s.

Given a stochastic matrix $A(t, T)$, which satisfies the assumptions above, we need to check that it is consistent with the loss process L_t. The consistency between A and L is defined in the following sense.

Definition 114 (*Consistency*) Let $A(t, T)$ and $P(t, T)$ be two stochastic processes satisfying the conditions above. If L_t is a loss process with probability distributions $\mathbf{p}_{t,T}$, we say that A and L are consistent if for all $T \geq t$, and $0 \leq j \leq n$, we have

$$p_{t,T}^{(j)} = P(L_T = j \,|\mathcal{G}_t) = P_{L_t, j}(t, T).$$

This implies the following characterization of consistency where the predictable compensator (and intensity) of the loss (point) process L can be represented in terms of the forward transition rates $A(t, T)$.

Proposition 115 *Let $A(t, T)$ be consistent with L_t. Then, the point process L_t must have an arrival intensity $\lambda_L(t)$ given by*

$$\lambda_L(t) = a_{L_t}(t, t), \, a.e.$$

What this means in particular is that, in this framework, the default time is totally inaccessible and can be viewed as a generalized version of an intensity-based model. For structural Firm-Value models, it is still possible to represent (formally) the individual loss distributions at each time step t using forward transition rates, but they will tend to infinity as we get closer to the default time.

15.4.4 Specification of the Diffusion Dynamics

Now, we move on from the static version of the model and assume that the forward transition rates follow some diffusion dynamics:

$$da_n(t, T) = \mu_n(t, T)\, dt + \sigma_n(t, T)\, dW_t,$$

where W is a d-dimensional Brownian motion; the drift $\mu_n(t, T)$ and volatility $\sigma_n(t, T)$ are predictable stochastic processes in $\mathbb{R}$ and $\mathbb{R}^d$ respectively with sufficient regularity conditions.

The choice of diffusion dynamics here assumes implicitly that the forward rates intensities do not jump at default; they are dependent on the aggregate portfolio default state via the conditioning on the loss variable (since the loss intensity is given by $\lambda_L(t) = a_{L_t}(t,t)$) but they are not impacted by which name in the portfolio has defaulted. This is a standard assumption in top-down models in general where the conditioning on the default events in the bigger $\mathcal{G}_t$ filtration is replaced with conditioning on the loss variable $\sigma(L_t)$.

Similarly, we can write the dynamics of the transition probabilities $P_{ij}(t,T)$ as

$$dP_{ij}(t,T) = u_{ij}(t,T)\,dt + v_{ij}(t,T)\,dW,$$
$$dP_{L_t,j}(t,T) = u_j(t,T)\,dt + v_j(t,T)\,dW + \phi_j(t,T)\,dL_t,$$

where the dirft and volatility terms $u_j(t,T)$ and $v_j(t,T)$ are defined as $u_j(t,T) = u_{L_{t^-},j}(t,T)$ and $v_j(t,T) = v_{L_{t^-},j}(t,T)$.

Proposition 116 (Drift Condition) *The diffusion dynamics of the one-step transition rates $a_n(t,T)$ are consistent with the loss process L_t if and only if the following conditions are satisfied:*

- *the drift and volatility parameters are linked through the following relationship: for all $0 \le j \le n$, and $t \le T \le T^*$*

$$\mu_j(t,T) = -\sigma_j(t,T)\frac{v_{L_t,j}(t,T)}{P_{L_t,j}(t,T)};$$

- *the intensity of the loss process L_t is given by*

$$\lambda_L(t) = a_{L_t}(t,t).$$

The volatility term $v_{ij}(t,T)$ in the dynamics of the transition probabilities $P_{ij}(t,T)$ can be computed recursively as a weighted sum

$$v_{ij}(t,T) = \int_t^T w_{ij}(t,s)\left[\frac{v_{i,j-1}(t,T)}{P_{i,j-1}(t,T)} + \frac{\sigma_{j-1}(t,s)}{a_{j-1}(t,s)} - \int_s^T \sigma_i(t,u)\,du\right]ds, \text{ for } j > i,$$

where the weights $w_{ij}(t, s)$ are given by the ratios

$$w_{ij}(t, s) = \frac{P_{i,j-1}(t, s)\, a_{j-1}(t, s) \exp\left(-\int_s^T a_j(t, u)\, du\right)}{\int_t^T P_{i,j-1}(t, x)\, a_{j-1}(t, x) \exp\left(-\int_x^T a_j(t, u)\, du\right) dx},$$

and the initial conditions are $v_{ii}(t, T) = -\int_t^T \sigma_i(t, s)\, ds$, and $v_{ij}(t, T) = 0$ for $j < i$.

The drift condition in the previous proposition is the key result of the dynamic (portfolio) loss model in Schönbucher (2005). This is very similar to the drift condition of Heath-Jarrow-Morton used to model the dynamics of interest rate curves. Starting with a forward rates curve we ensure that we can price all the (spot) swap instruments, by construction, as of today (i.e., we match the initial rates term structure); then by imposing the HJM drift condition of the forward dynamics, we are free to choose any volatility process that we wish (in order to match options' market prices, for example) while ensuring that the forward interest rates curves generated by the model are arbitrage-free. The same principle applies here to forward loss distributions. At time 0, we start with a set of spot loss distributions built from market tranche prices at different time-horizons; we represent the loss distribution cube with a set of probability transition rates, which we then diffuse through time. The diffusion dynamics are specified via a choice of volatility process and the drift is determined from the HJM-type drift condition above. This ensures that we have arbitrage-free dynamics of the loss distribution while matching today's tranche market prices by construction.

As with HJM, the choice of the number of Brownian drivers and a particular volatility parametrization specifies completely the implied dynamics of the loss distribution.

A few examples are given below.

Example 117 Parallel Shift-Systemic Move: for a fixed positive constant a, we set

$$\sigma_i(t, s) = a > 0, \text{ for all } 0 \leq i < n.$$

A positive increment of the Brownian motion will increase all the individual transition rates in the same way, and a negative increment will similarly decrease all the transition rates in parallel. This parallel move of the individual transition probabilities will, in turn, impact the overall default risk in the portfolio and impact the portfolio index level.

Example 118 Correlation Tilt: for a given index $n^* < n$, and a set of positive and negative constants a_i that sum up to zero, $\sum_{i=0}^{n-1} a_i = 0$, we set

$$\sigma_i(t,s) = a_i > 0, \text{ for } i \leq n^*,$$
$$\sigma_i(t,s) = a_i < 0, \text{ for } i > n^*.$$

This would trigger a tilt in the loss distribution where probability mass is shifted to the right or to the left depending on the sign of the Brownian motion increment. By ensuring that the expected loss is unchanged, it would translate into a move of the base correlation curve, either a steepening or a flattening of base correlation leading to a different distribution of prices across the capital structure.

Example 119 Idiosyncratic Move: For a fixed (large) constant a, we set

$$\sigma_i(t,s) = a > 0, \text{ for } i = L_t,$$
$$\sigma_i(t,s) = 0, \text{ for } i > L_t.$$

This factor, by definition, impacts the transition rate of the next default only. This would represent an idiosyncratic event where one of the names in the portfolio blows up and dominates the default risk in the distribution.

There are other extensions and more general formulations of this model that could be used (see Schönbucher 2005). The main ones are: (a) allowing for jumps at default explicitly (which can be achieved by introducing a default event marker to reveal the identity of the defaulted name, and re-formulate the model within a marked point process framework) and, (b) using a multi-step Markov chain (to capture multiple simultaneous defaults and stochastic recovery assumptions).

15.5 SPA Model

In this section, we review the dynamic credit loss modelling framework of (Sidenius et al. 2008), referred to as the SPA model. In essence, it is very similar to the Markov Chain model of Schönbucher (2005), but the model construction is more "natural" as it operates on the loss distribution "discount factors" and gradually builds the no-arbitrage conditions and HJM-type dynamics on top. People familiar with the interest rates modelling mindset would probably find the SPA model construction more appealing although the results of both models are fairly identical.

15.5.1 Model Set-Up

In the SPA model, the main quantity of interest is the portfolio loss distribution; it is represented in terms of "Quantile Discount Factors" and then diffused conditional on realizations of the background filtration (which contains information on the economic state variables and excludes default events).

For a given portfolio with a set of notionals $(N_i)_{1\leq i\leq n}$, and recovery rates $(R_i)_{1\leq i\leq n}$, we define the (normalized) portfolio loss variable as

$$L_T \triangleq \frac{\sum_{i=1}^{n} N_i (1 - R_i) D_T^i}{\sum_{i=1}^{n} N_i}.$$

For all $x \in [0, 1]$, we define the stopping time τ_x as the first crossing time of the barrier x

$$\tau_x \triangleq \inf \{L_t > x, \text{ for } t \geq 0\},$$

and we denote its $\mathbb{F}$-intensity by λ_t^x. For each $x \in [0, 1]$, we can also define the τ_x-survival probability (or the loss-discount-factors) as

$$p_{t,T}^x \triangleq \mathbb{P}(L_T \leq x | \mathcal{F}_t) = \mathbb{P}(\tau_x > T | \mathcal{F}_t) = \mathbb{E}\left[\exp\left(-\int_0^T \lambda_s^x ds\right) | \mathcal{F}_t\right].$$

Each discount factor process $\left(p_{t,T}^x\right)_{t\geq 0}$ is an $\mathbb{F}$-martingale. The equivalent forward rates $f_{t,T}^x$ for the loss discount factors $p_{t,T}^x$ are defined by

$$f_{t,T}^x \triangleq -\frac{\frac{\partial}{\partial T} p_{t,T}^x}{p_{t,T}^x},$$

so that we can write

$$p_{t,T}^x = \exp\left(-\int_0^t f_{s,s}^x ds\right) \exp\left(-\int_t^T f_{t,s}^x ds\right), \text{ with } f_{s,s}^x = \lambda_s^x.$$

Thus, the loss distribution dynamics can be conveniently described either using the x-discount factors $p_{t,T}^x$ or the x-forward rates $f_{t,T}^x$. This is the equivalent of HJM term structure modelling in the portfolio loss (distribution) context.

To ensure that the dynamics of $p_{t,T}^x$ are feasible, we assume the following consistency conditions:

$$
\begin{aligned}
&p_{t,T}^1 = 1, \\
&p_{t,T_1}^x \geq p_{t,T_2}^x, \text{ for all } T_1 \leq T_2, \\
&p_{t,T}^x \leq p_{t,T}^y, \text{ for all } x \leq y.
\end{aligned}
$$

15.5.2 Loss Probability Diffusion

For each loss level x and maturity T, the process is $\left(p_{t,T}^x\right)_{t\geq 0}$ an $\mathbb{F}$-martingale. If the filtration $\mathbb{F}$ is generated from a set of independent Brownian motions W_t^α, $\alpha \geq 1$, then we can write the dynamics of $\left(p_{t,T}^x\right)_{t\geq 0}$ as

$$
\frac{dp_{t,T}^x}{dt} = \sum_{\alpha} \Sigma_x^\alpha (t, T) \cdot dW_t^\alpha,
$$

where $\Sigma_x^\alpha (t, T)$, $\alpha \geq 1$, is a set of (stochastic) volatility functions for each Brownian motion driver. This, in turn, yields the following dynamics for the x-forward rates $f_{t,T}^x$

$$
df_{t,T}^x = \sum_{\alpha} \sigma_x^\alpha (t, T) \, \Sigma_x^\alpha (t, T) \, dt + \sigma_x^\alpha (t, T) \cdot dW_t^\alpha,
$$

where

$$
\sigma_x^\alpha (t, T) = -\frac{\partial}{\partial T} \Sigma_x^\alpha (t, T) .
$$

Here we assume that the instantaneous volatility processes satisfy the necessary conditions so that the diffusion equations for $f_{t,T}^x$ and $p_{t,T}^x$ are well-defined (see, for example, Miltersen 1994). In addition, we need to ensure that the spatial consistency conditions are also satisfied, so that no-arbitrage properties in the forward loss distributions are still preserved during the diffusion, namely that $p_{t,T}^x \leq p_{t,T}^y$, for all $x \leq y$ (see Sidenius et al. 2008).

15.5.3 Loss Process as a One-Step Markov Chain

Given a set of (conditional) marginal loss distributions, $\mathbb{P} \left(L_T \leq x \,|\mathcal{F}_t\right)$, for all x, and $t < T$, (satisfying the usual consistency conditions) we can show that there exists a (conditional Markov) loss process, whose marginals match the input distributions. To prove the existence we construct such a process using a one-step Markov chain. In fact, in this particular case, the One-Step Markov chain solution is also unique.

We have the following result.

Proposition 120 *Assuming that the consistency conditions above are satisfied, then for any given discretization of losses* $x_0 = 0 < x_1 < \ldots < x_N = 1$, *for which* $p_{t,T}^{x_{i+1}} > p_{t,T}^{x_i}$, *there exists a non-decreasing, one-step loss process L on the grid* $\{x_i\}_{0 \le i \le N}$ *such that it is consistent with the loss distribution processes* $\left(p_{t,T}^x\right)_{t \ge 0}$ *for* $x \in \{x_i\}_{0 \le i \le N}$

$$\mathbb{P}\left(L_T \le x \,|\mathcal{F}_t\right) = p_{t,T}^x.$$

Proof Given the loss distributions $p_{t,T}^x$, at time t, for all losses x and maturities T, we can write the (conditional) density of the crossing time as

$$\mathbb{P}\left(\tau_x \in [T, T + dT)\,|\mathcal{F}_t\right) = -\frac{\partial}{\partial T} p_{t,T}^x dT.$$

Assume that we have a one-step loss process on the $\{x_i\}$ -grid (i.e., losses can only jump up from x_i to x_{i+1}), and let $\kappa_{t,T}^{x_i}$ be the intensity of a transition from x_i to x_{i+1}, at time horizon T, conditional on $\mathcal{F}_t$

$$\kappa_{t,T}^{x_i} dT = \mathbb{P}\left(L_{T+dT} = x_{i+1}\,|L_T = x_i, \mathcal{F}_t\right),$$

then we can write the τ_x-stopping time density in terms of the one-step intensities $\kappa_{t,T}^{x_i}$

$$\begin{aligned} \mathbb{P}\left(\tau_x \in [T, T + dT)\,|\mathcal{F}_t\right) &= \mathbb{P}\left(L_{T+dT} = x_{i+1}, L_T = x_i \,|\mathcal{F}_t\right) \\ &= \mathbb{P}\left(L_T = x_i \,|\mathcal{F}_t\right) \cdot \mathbb{P}\left(L_{T+dT} = x_{i+1}\,|L_T = x_i, \mathcal{F}_t\right) \\ &= \left(p_{t,T}^{x_{i+1}} - p_{t,T}^{x_i}\right) \cdot \kappa_{t,T}^{x_i} dT, \end{aligned}$$

which implies an explicit formula for the one-step transition rates

$$\kappa_{t,T}^{x_i} = \frac{-\frac{\partial}{\partial T} p_{t,T}^x dT}{p_{t,T}^{x_{i+1}} - p_{t,T}^{x_i}}.$$

From the loss distribution consistency conditions above, it follows that these rates are well-defined, non-negative one-step loss intensities; thus we have shown how to construct formally the associated one-step Markov chain. Furthermore, given the expression above, we have also shown that, when it exists, it is also unique. ■

15.5.4 Loss Process as a General Markov Chain

Now, we consider the general case and allow for a multi-step Markov chain. This is particularly relevant when we assume multiple simultaneous defaults and stochastic recovery rates.

First, we define the jump survival function on the continuous interval [0, 1].

Definition 121 The jump survival function $m_{t,T}^{z,x}$ is defined as

$$m_{t,T}^{z,x} dT = \mathbb{P}\left(L_{T+dT} > x \,|L_T = z, \mathcal{F}_t\right);$$

We have the following proposition.

Proposition 122 *If* L_t *is a non-decreasing pure-jump conditional Markov process on the state space* $[0, 1]$, *then we have*

$$m_{t,1}^{z,1} = 0, \textit{ for all } t, T, z,$$

and

$$\frac{\partial p_{t,T}^x}{\partial T} = -\int_0^x \left(m_{t,T}^{z,x} \cdot \frac{\partial p_{t,T}^z}{\partial z} \right) dz.$$

Proof L_t is restricted to the interval [0, 1] and we have

$$\begin{aligned}
\frac{\partial p_{t,T}^x}{\partial T} dT &= -\mathbb{P}\left(L_T \leq x, L_{T+dT} > x \,|\mathcal{F}_t\right) \\
&= -\int_0^z \mathbb{P}\left(L_T \in dz, L_{T+dT} > x \,|\mathcal{F}_t\right) \\
&= -\int_0^z \mathbb{P}\left(L_{T+dT} > x \,|L_T = z, \mathcal{F}_t\right) \mathbb{P}\left(L_T \in dz \,|\mathcal{F}_t\right) \\
&= -\int_0^x \left(m_{t,T}^{z,x} \cdot \frac{\partial p_{t,T}^z}{\partial z} \right) dz.
\end{aligned}$$

■

For a given set of "consistent" loss probabilities $p^x_{t,T}$ (i.e., satisfying the consistency conditions),to find a loss process whose implied dynamics (represented by the conditional survival function $m^{z,x}_{t,T}$) match the marginals $p^x_{t,T}$, we need to solve the forward Kolmogorov equation above. This is clearly an ill-posed problem, which could have multiple solutions (unlike the one-step Markov chain case where the solution is unique). In general, this type of underdetermined problems could be solved with a Relative Entropy Minimization approach. Typically, we would start with an initial well-behaved loss process $\tilde{L}_t$, then we find the "smallest" deformation of that process that satisfies the Kolmogorov equation above. The smallest deformation is given by minimizing the relative entropy function subject to the Kolmogorov equation constraint.

Sidenius et al. (2008) came up with a simpler approach and solved this problem by choosing a suitable "separable" parametrization of the survival function $m^{z,x}_{t,T}$, thus limiting the set of possible solutions, and deriving a well-posed analytical solution of the Kolmogorov equation (similar, in essence, to the one-step case). Observing that for a (spatially)-homogeneous Markov chain, the survival function $m^{z,x}_{t,T}$ depends only on the difference $(x - z)$, they postulate a solution of the form

$$m^{z,x}_{t,T} = \theta(T, x - z)\, v^x_{t,T},$$

where $\theta(T, y)$ is a non-negative time and space function given exogenously as an input; $v^x_{t,T}$ is a (rescaled) conditional survival process, which solves the Kolmogorov equation. The choice of the normalizing shape-function $\theta(T, y)$ is driven by the (stochastic) recovery distributions of the underlying portfolio, and would be centered around the average single-name recovery rate.

The solution with this chosen parametrization is unique in this case, and is given by the following proposition.

Proposition 123 *Assuming that the consistency conditions are satisfied, and that we have a normalizing shape function $\theta(T, y) \geq 0$ such that*

$$\int_0^x \theta(T, x - z)\frac{\partial p^z_{t,T}}{\partial z}dz > 0, \textit{ for all } t, T, x.$$

Then, there exists a non-decreasing loss process L, on the state space $[0, 1]$, $L_0 = 0$, such that it is consistent with the loss distributions $p^x_{t,T}$,

$$\mathbb{P}(L_T \leq x \,|\mathcal{F}_t) = p^x_{t,T}, \textit{ for } x \in [0, 1], 0 \leq t \leq T,$$

and the associated conditional survival function $m_{t,T}^{z,x}$ is given by

$$m_{t,T}^{z,x} = \theta(T, x - z)\, \nu_{t,T}^{x},$$

$$\nu_{t,T}^{x} = \frac{-\frac{\partial p_{t,T}^{x}}{\partial T}}{\int_0^x \theta(T, x - z) \frac{\partial p_{t,T}^{z}}{\partial z} dz} \geq 0.$$

Proof It suffices to substitute the parametric form in the Kolmogorov equation

$$\frac{\partial p_{t,T}^{x}}{\partial T} = -\nu_{t,T}^{x} \int_0^x \left(\theta(T, x - z) \cdot \frac{\partial p_{t,T}^{z}}{\partial z} \right) dz,$$

and to solve for $\nu_{t,T}^{x}$. ■

Remark 124 The one-step Markov chain construction, on the grid $\left\{\frac{i}{n}\right\}_{0 \leq i \leq n}$, can be recovered by setting

$$\theta(T, y) = \mathbf{1}_{\left\{y \in \left[0, \frac{1}{n}\right]\right\}}.$$

15.5.5 Implementation

Suppose we want to price a portfolio (loss) credit derivative instrument, whose payoff depends only on realizations of the loss variable and not the defaulted single names; then, we need to compute conditional expectations of functionals of the loss variable at a given time horizon:

$$\mathbb{E}\left[f(L_T) \,|\mathcal{G}_t\right] \simeq \mathbb{E}\left[f(L_T) \,|\mathcal{F}_t \vee \sigma(L_t)\right];$$

here we have assumed that our default filtration is generated by the aggregate loss variable and does not depend on which specific single names have defaulted. The expectation can be written as

$$\mathbb{E}\left[f(L_T) \,|\mathcal{F}_t, L_t = y\right] = \int \mathbb{E}\left[f(x) \,|\mathcal{F}_t, L_t = y\right] \frac{\partial p_{t,T}^{y,x}}{\partial x} dx,$$

where $p_{t,T}^{y,x}$ is the L_T-probability loss distribution conditional on a given realization of the loss L_t at time t, $\{L_t = y\}$,

$$p_{t,T}^{y,x} \triangleq \mathbb{P}(L_T \leq x \,|\mathcal{F}_t, L_t = y).$$

The main quantities needed for valuation are these probability distributions conditional jointly on the Brownian motion (spread) diffusions and the loss variable L_t. These can also be computed solving the same forward Kolmogorov equations as the model primitives $p^x_{t,T}$.

Proposition 125 *Fix $t \geq 0$ and $y \in [0, 1]$, and let the jump survival function $m^{x,z}_{t,T}$ be consistent with $p^x_{t,T}$, for all $x, z \in [0, 1]$, $T \geq t$, then all the conditional probabilities $p^{y,x}_{t,T}$ can be obtained by solving the system of forward Kolmogorov equations in T and x,*

$$\frac{\partial p^{y,x}_{t,T}}{\partial T} = -\int_0^x \left(m^{z,x}_{t,T} \cdot \frac{\partial p^{y,z}_{t,T}}{\partial z} \right) dz \ , T \geq t, \, x \in [0, 1],$$

with the initial condition at time $T = t$

$$p^{y,x}_{t,t} = \mathbf{1}_{\{x \geq y\}}.$$

These are the same Kolmogorov equations satisfied by the model primitives $p^x_{t,T}$, but with a different initial condition. As always, when discretized on the interval [0, 1], they become a set of ODEs easy to solve numerically. This can be done in one forward sweep for all x concurrently, so that only one Kolmogorov equation needs to be solved each time.

15.5.6 Dynamics

As with interest rate HJM models, we can specify the model dynamics via a suitable choice of volatility function.

Gaussian SPA Model. The simplest one-factor Markovian HJM model is the Gaussian model. As with interest rate models, it is simple to implement and has tractable closed form solutions but the forward rates can be negative. In our portfolio loss modelling context, negative forward rates mean that the loss process is not always non-decreasing in time.

We define the Gaussian one-factor SPA model with the following volatility structure

$$\begin{aligned} df^x_{t,T} &= \sigma_x(t, T)\, \Sigma_x(t, T)\, dt + \sigma_x(t, T) \cdot dW_t, \\ \sigma_x(t, T) &= \sigma e^{-a(T-t)}, \\ \Sigma_x(t, T) &= -\sigma \frac{1 - e^{-a(T-t)}}{a}. \end{aligned}$$

This can be conveniently re-written in terms of the dynamics of the (short rate) intensity λ_t^x as

$$f_{t,T}^x = f_{0,T}^x + e^{-a(T-t)} \left(\lambda_t^x - f_{0,t}^x + b(T-t)\,\theta_x(t)\right),$$
$$b(\tau) = \frac{1-e^{-a\tau}}{a},$$

where the dynamics of the intensity λ_t^x are given by

$$d\lambda_t^x = \left(\theta_t^x - a\lambda_t^x\right) dt + \sigma dW_t, \lambda_0^x = 0,$$

the drift $\theta_x(t)$ is a deterministic time-dependent function chosen to match the initial term structure of loss probabilities $p_{0,T}^x$.

Quasi-Gaussian SPA Model. Again, as in interest rates, to address some of their inherent limitations, Markov Gaussian models have been extended to include a local volatility function. Those are referred to as Quasi-Gaussian models (see Jamshidian 1991; Babbs 1990).

The dynamics of the (short rate) intensity are altered to use a local volatility function $\sigma\left(t, \lambda_t^x\right)$,

$$d\lambda_t^x = \left(\theta_x(t) - a\lambda_t^x\right) dt + \sigma\left(t, \lambda_t^x\right) dW_t,$$
$$\theta_t^x = \frac{\partial f_{0,t}^x}{\partial t} + af_{0,t}^x + \int_0^t e^{-2a(t-s)}\sigma^2\left(s, \lambda_s^x\right) ds;$$

the drift term θ_t^x is now a stochastic process through its dependence on the realized volatility $\sigma\left(s, \lambda_s^x\right)$, which, in turn, depends on the realized path of the intensity process λ_s^x.

Thus, the loss probabilities $p_{t,T}^x$ can be represented with the two state variables λ_t^x and θ_t^x,

$$p_{t,T}^x = \frac{p_{0,T}^x}{p_{0,t}^x} \exp\left(-b(T-t)\left[\lambda_t^x - f_{0,t}^x + \frac{1}{2}b(T-t)\,\theta_t^x\right]\right).$$

We can show that this model specification satisfies the required consistency conditions with very mild assumptions on the local volatility function.

Proposition 126 *If the local volatility function satisfies*

$$\sigma(t, 0) = 0,$$
$$\lambda^1 \geq \lambda^2 \Longrightarrow \sigma\left(t, \lambda^1\right) \geq \sigma\left(t, \lambda^2\right),$$

and the mean-reversion parameter is greater than the minimal value $a_{\min}$

$$a \geq a_{\min} = \max\left\{-\frac{\partial}{\partial t}\log\left(f_{0,t}^{y} - f_{0,t}^{x}\right), \textit{for all } t \in [0, T_{\max}], 0 \leq x \leq y \leq 1\right\},$$

then the loss distributions (no-arbitrage) consistency conditions are satisfied.

15.6 Concluding Remarks

The big advantage of the so-called third generation models presented in this chapter is their dynamic nature. Since they are directly linked to the portfolio loss dynamics, they can explicitly handle payoffs dependent on both portfolio losses and credit spreads. This enables consistent pricing of structured credit products with forward starting features and optionality triggered by tranche prices. Simplified valuation models are also possible, but only a fully dynamic model will give accurate and consistent pricing results.

Although dynamic credit modelling approaches are vital for structured credit products, they do not, however, answer all the outstanding questions with the current set of (vanilla) products. One of their main drawbacks is the lack of dispersion; since the dispersion in the underlying portfolio is not captured explicitly, they would not be suitable as a (correlation) skew rescaling tool. Rather, a rescaled skew needs to be derived from traditional methods and fed as an input to the dynamic loss model. Further, the calibration to standard index tranches could be challenging depending on the choice of loss parametrization; ensuring that no-arbitrage conditions are preserved at all times is not always easy with richer model dynamics. This underlines the fact that third generation dynamic models are best suited as a complement to the more traditional static models (used by the industry) and should only be used in situations where the dynamic nature of the product is the dominant factor in pricing.

In summary, while third generation dynamic models are needed for (sophisticated) structured credit products with forward-starting features and optionality, static models will not become obsolete nonetheless since 3G models are not naturally suitable for simple index tranches, bespoke CDOs or CDO-squared type structures.

References

S. Babbs, The term structure of interest rates: stochastic processes and contingent claims (Ph.D. Thesis, University of London, 1990)

D. Brigo, A. Pallavicini, R. Torresetti, Calibration of CDO tranches with the dynamical generalized-Poisson loss model. Risk Magazine, May (2007)

Y. Elouerkhaoui, Etude des problèmes de corrélation and d'incomplétude dans les marchés de crédit (Ph.D. Thesis, Université Paris-Dauphine, 2006)

K. Giesecke, L. Goldberg, A top down approach to multi-name credit (Working Paper, 2005)

A. Hawkes, Spectra of some self-exciting and mutually exciting point processes. Biometrika **58**(1), 83–90 (1971)

A. Hawkes, D. Oakes, A cluster process representation of a self-exciting process. J. Appl. Probab. **11**, 493–503 (1974)

F. Jamshidian, Bond and option evaluation in the Gaussian interest rate model. Res. Financ. **9**, 131–710 (1991)

D. Lando, On Cox processes and credit risky securities. Rev. Deriv. Res. **2**(2/3), 99–120 (1998)

F. Lindskog, A. McNeil, Common Poisson Shock models: applications to insurance and credit risk modelling. ASTIN Bullet. **33**(2), 209–238 (2003)

F. Longstaff, A. Rajan, An empirical analysis of the pricing of collateralized debt obligations. J. Financ. **63**(2), 529–563 (2008)

K.R. Miltersen, An arbitrage theory of the term structure of interest rates. Ann. Appl. Probab. **4**(4), 953–967 (1994)

P.J. Schönbucher, Portfolio losses and the term structure of loss transition rates: a new methodology for the pricing of portfolio credit derivatives (Working Paper, ETH Zurich, 2005)

J. Sidenius, V. Piterbarg, L. Andersen, A new framework for dynamic credit portfolio loss modeling. Int. J. Theor. Appl. Financ. **11**(2), 163–197 (2008)

Part III

Advanced Topics in Pricing and Risk Management

16

Pricing Path-Dependent Credit Products

This chapter addresses the problem of pricing (soft) path-dependent portfolio credit derivatives whose payoff depends on the loss variable at different time horizons. We review the general theory of copulas and Markov processes, and we establish the link between the copula approach and the Markov-Functional paradigm used in interest rates modelling. Equipped with these theoretical foundations, we show how one can construct a dynamic credit model, which matches the correlation skew at each tenor, by construction, and follows an exogenously specified choice of dynamics. Finally, we discuss the details of the numerical implementation and we give some pricing examples in this framework.

16.1 Introduction

Before the credit crisis, we used to price a number of exotic path-dependent portfolio credit derivatives such as Forward Starting CDOs, Step-up subordination CDOs, TARNs, Multi-maturity CDO-Squared. What these structures have in common is that their payoff can be written as

$$f\left(L_{T_1},\ldots,L_{T_n}\right);$$

this is a so-called "soft path-dependency" since the pricing depends only on the dynamics of the loss variables as opposed to the dynamics of each single name in the portfolio—this type of (strong) path-dependency is, typically, found in Cash CDO structures with waterfalls, OC/IC triggers, re-investments and so on. These structures can be priced in any of the standard second generation

Y. Elouerkhaoui, *Credit Correlation*, Applied Quantitative Finance,
DOI 10.1007/978-3-319-60973-7_16

models that allow for a sensible calibration of the market correlation skew, such as Random Factor Loading (RFL), Stochastic Correlation or the Levy Copula. They do, however, suffer from a number of limitations:

- the skew calibration is not perfect;
- the joint dependence of the loss variables at different time horizons L_{T_i} is built in model and may not be always intuitive;
- we do not have enough flexibility to separate the market correlation skew risk and the model risk (i.e., the choice of transition probabilities).

The Pricing Problem. The most general method to price these structures is to: a) use the marginal (loss) distributions as input for the (market) correlation skew surface, b) specify a copula function that links the marginals at each time horizon. With the copula approach, the risk can be split into two parts: market risk (i.e., individual risks of vanillas that we observe and calibrate to), model risk (i.e., the dependence structure between the vanilla (univariate) distributions, which drives the value of structured products).

The pricing problem then becomes

$$\begin{aligned}
&\mathbb{E}\left[f\left(L_{T_1},\ldots,L_{T_n}\right)\right] \\
&= \int\cdots\int f\left(x_1,\ldots,x_n\right)\mathbb{P}\left(L_{T_1}\in dx_1,\ldots,L_{T_n}\in dx_n\right) \\
&= \int\cdots\int f\left(x_1,\ldots,x_n\right)\frac{\partial^n}{\partial x_1\ldots\partial x_n}\mathbb{P}\left(L_{T_1}\leq x_1,\ldots,L_{T_n}\leq x_n\right)dx_1\ldots dx_n \\
&= \int\cdots\int f\left(x_1,\ldots,x_n\right)\frac{\partial^n}{\partial x_1\ldots\partial x_n}C\left(\phi_{L_{T_1}}\left(x_1\right),\ldots,\phi_{L_{Tn}}\left(x_n\right)\right)dx_1\ldots dx_n \\
&= \int\cdots\int f\left(\phi_{L_{T_1}}^{-1}\left(u_1\right),\ldots,\phi_{L_{Tn}}^{-1}\left(u_n\right)\right)\frac{\partial^n}{\partial u_1\ldots\partial u_n}C\left(u_1,\ldots,u_n\right)du_1\ldots du_n.
\end{aligned}$$

Now, the question is how do we choose this copula function?

The objective of this chapter is to address this question both theoretically and in practice.

The structure of the chapter is as follows.

In Sect. 16.2, we describe some of the most popular path-dependent portfolio credit products and their economic rationale.

In Sect. 16.3, we review the theory of copula and Markov processes. Then, in Sect. 16.4, we establish the link with the Markov-Functional approach used in interest rate modelling.

Using those principles, we build, in Sect. 16.5, a dynamic credit model, with suitable time-dependence, which fits the term structure of correlation skew by construction. And finally, in Sect. 16.6, we price sample path-dependent credit exotics in this framework and analyze their behaviour.

16.2 Product Description

In this section, we give a brief overview of some of most popular path-dependent portfolio credit derivatives and their economic rationale.

16.2.1 Forward Tranches

Forward Tranches are forward-starting trades in which the credit protection takes effect only after a given time. The investor is immune to all defaults before the trade's forward start date. All the names that have defaulted before the start date are removed from the portfolio but the Dollar subordination amount remains constant, which results in an increased subordination in percentage terms.

If we denote by S the forward start date of the trade, we define the forward portfolio loss variable, for $T \geq S$, as

$$L_{S,T} \triangleq L_T - L_S.$$

The payoff of forward tranche, with attachment and detachment points (K_1, K_2), at each coupon date $T_i > S$, is

$$\widehat{L}_{S,T_i} \triangleq \min\left(\max\left(L_{S,T_i} - K_1, 0\right), K_2 - K_1\right).$$

Rationale. In a market environment with steep credit curves, they offer a good spread pick-up compared to standard tranches.

16.2.2 Step-Up Subordination

A Step-up Subordination CDO is a CDO structure where the attachment and detachment points increase automatically according to a pre-defined schedule. In general, the attachment points are chosen to achieve a given rating for each time horizon. Since the credit enhancement increases over time, it provides a build-up of subordination, which protects the note holders in the later stages of the trade.

Rationale. In a market with steep credit curves, the implied default risk is mostly in the final years of the trade; that is the reason why it is important to maintain a high level of subordination in the back-end of the trade. This, in turn, offers a better yield compared to similarly rated standard tranches.

The payoff of the trade can be represented with a series of forward-starting tranches with different subordinations according to the step-up schedule.

16.2.3 Multi-Maturity CDO-Squared

This is a generic CDO-Squared structure where the underlying CDOs can have different maturities and can be both long and short.

Rationale. It is a more general structure that can offer the managers the flexibility to trade in and out of credits at different tenors and take advantage of curve trades.

If we denote by $\left(\widehat{L}_T^k\right)_{1 \le k \le m}$ the tranche losses, at time T, for each underlying CDO, then the CDO-Squared loss variable is defined as

$$L_T \triangleq \sum_{k=1}^{m} a_k \widehat{L}^k_{(T \wedge U_k)},$$

where $(U_k)_{1 \le k \le m}$ are the final maturities of the underlying CDOs, and $a_k \in \{-1, 1\}$ indicates if we are long or short the underlying CDO.

Forward tranches and Step-up subordination CDOs can be viewed as a special case of this more generic structure.

16.2.4 TARN

A TARN (Target Return Note or Target Redemption Note) is a Structured Note that pays a structured coupon (e.g., the coupon of an Equity tranche) and terminates when a pre-specified target return is reached. The knock-out time of the TARN is defined as the first coupon date when the return of the note is greater than the target. The principal is paid back at the knock-out time or the final maturity of the note whichever is first. A TARN can also be structured as a swap, in which case both the premium leg and protection leg terminate at the knock-out time. The Target Return feature ensures automatic redemption of the notes when the target return is reached, and the notes are principal protected.

Rationale. Typically for a 7-year TARN, the target return is chosen so that the note redeems in 5 years in the zero-defaults case. For investors with a view on back-loaded defaults, the TARN offers an attractive fixed coupon relative to both 5y and 7y bullet maturity principal protected notes.

Let $T_1, \ldots, T_n$ be a set of coupon dates. The coupon payments of a standard CDO tranche are given by

$$C_{T_i} \triangleq \left(N - \widehat{L}_{T_i}\right) \delta_i S = \widehat{N}_{T_i} \delta_i S,$$

where S is a fixed spread, δ_i is the daycount fraction, N is the notional of the trade, $\widehat{L}_{T_i}$ is the tranche loss at time T_i, and $\widehat{N}_{T_i}$ is the remaining tranche notional at time T_i.

The cumulative return of the note at time T_i is defined as

$$R_{T_i} \triangleq \frac{1}{N} \sum_{j=1}^{i} C_{T_j}.$$

The total remaining coupon after time T_i is given by

$$\Lambda_{T_i} \triangleq \left(R - R_{T_i}\right) N.$$

The coupon payment of the TARN is defined as

$$\widehat{C}_{T_i} \triangleq \min\left(C_{T_i}, \Lambda_{T_{i-1}}\right) \mathbf{1}_{\left\{R_{T_{i-1}} < R\right\}}.$$

The knock-out time of the note is

$$\tau^{ko} \triangleq \begin{cases} \min\left\{T_i : R_{T_i} \geq R\right\}, \text{if } R_{T_n} \geq R \\ T_n, \text{otherwise.} \end{cases}$$

The value of the TARN note, at time 0, is then given by

$$\begin{aligned} V &= \mathbb{E}\left[\sum_{i=1}^{n} p_{0,T_i} \widehat{C}_{T_i} + p_{0,\tau^{ko}} N\right] \\ &= \sum_{i=1}^{n} p_{0,T_i} \mathbb{E}\left[\widehat{C}_{T_i}\right] + N \sum_{i=1}^{n} p_{0,T_i} \mathbb{P}\left(\tau^{ko} = T_i\right), \end{aligned}$$

where $p_{0,T}$ denotes the value at time 0 of the zero-coupon discount factor maturing at time T.

There are many other variations of the TARN structure that can be offered to suit investors preferences. It can be structured as note or a swap, and the coupon payments and returns can be fixed or floating and all possible combinations thereof.

- **Coupon and Loss**. The definition of the return includes the tranche losses incurred to date

$$R_{T_i} \triangleq \frac{1}{N}\left[\sum_{j=1}^{i} C_{T_j} - \widehat{L}_{T_i}\right].$$

- **Coupon balance paid**. If the target return is not achieved, then the difference is paid out in the last coupon

$$V = \mathbb{E}\left[\sum_{i=1}^{n} D_{0,T_i}\widehat{C}_{T_i} + D_{0,\tau^{ko}}N + D_{0,T_n}\max\left(\Lambda_{T_n}, 0\right)\right].$$

- **Floating coupon**. The coupon rate is Libor plus spread

$$C_{T_i} \triangleq \widehat{N}_{T_i}\delta_i\left(l_{T_i} + S\right).$$

- **Floating payment**. The return is based on a fixed spread, but the coupon payment is floating

$$R_{T_i} \triangleq \frac{1}{N}\sum_{j=1}^{i} C_{T_i} = \frac{1}{N}\sum_{j=1}^{i} \widehat{N}_{T_i}\alpha_i S,$$

$$\widehat{C}_{T_i} \triangleq \left(\frac{l_{T_i} + S}{S}\right)\min\left(C_{T_i}, \Lambda_{T_{i-1}}\right)\mathbf{1}_{\left\{R_{T_{i-1}}<R\right\}}.$$

- **Floating return**. The return is based on a floating rate, but the coupon payment is fixed

$$R_{T_i} \triangleq \frac{1}{N}\sum_{j=1}^{i} C_{T_i} = \frac{1}{N}\sum_{j=1}^{i} \widehat{N}_{T_i}\alpha_i\left(l_{T_i} + S\right),$$

$$\widehat{C}_{T_i} \triangleq \left(\frac{S}{l_{T_i} + S}\right)\min\left(C_{T_i}, \Lambda_{T_{i-1}}\right)\mathbf{1}_{\left\{R_{T_{i-1}}<R\right\}}.$$

The TARN can also be structured as a swap, where the TARN coupons are paid out and the protection leg terminates at the knock-out time of the TARN

$$V = \mathbb{E}\left[\sum_{i=1}^{n} D_{0,T_i}\widehat{C}_{T_i} - \int_0^{\tau^{ko}} D_{0,t}d\widehat{L}_t\right].$$

16.3 Theory of Copulas and Markov Processes

The dependence of random variables can be classified into spatial and temporal dependence. Spatial dependence refers to the dependence between random variables at the same time, and temporal dependence refers to the dependence structure of a process. Most of the current research in copulas is done for the modelling of spatial dependence. However, the analysis of temporal dependence via copulas is also possible. The first paper in this field is Darsow et al. (1992) where they established the connection between copulas and Markov processes.

The basic idea is that many properties of stochastic processes can be characterized by their finite-dimensional distributions, and therefore copulas can be used for this analysis. Although many concepts in the theory of stochastic processes (such as martingales, SDEs, etc.) are stronger than the finite-dimensional distribution approach, it is possible and useful, in some special cases, to use the copula structure when constructing stochastic processes.

Howard Sherwood summarizes this view in the AMS-IMS-SIAM Conference of 1993:

"*The subject matter of these conference proceedings comes in many guises. Some view it as the study of probability distributions with fixed marginals; those coming to the subject from probabilistic geometry see it as the study of copulas; experts in real analysis think of it as the study of doubly stochastic measures; functional analysts think of it as the study of Markov operators; and statisticians say it is the study of possible dependence relations between pairs of random variables. All are right since all these topics are isomorphic.*"

16.3.1 Stochastic Processes and Copulas

The link between stochastic processes and copulas is provided by Kolmogorov's theorem (see Shiryaev 1996, Theorem II.9.1, pp. 246).

Theorem 127 (Kolmogorov) *Let* $\left\{F_{t_1,\ldots,t_n} : t_i \in T,\ t_1 < t_2 < \cdots < t_n\right\}$ *be a given family of finite-dimensional distribution functions, satisfying the following consistency condition*

$$\lim_{x_k \nearrow \infty} F_{t_1,\ldots,t_n}(x_1,\ldots,x_n) = F_{t_1,\ldots,t_{k-1},t_k,\ldots,t_n}(x_1,\ldots,x_{k-1},x_k,\ldots,x_n),$$

for all $1 \le k \le n$, $(x_1,\ldots,x_n) \in \mathbb{R}^n$, $(t_1,\ldots,t_n) \in T^n$. *Then, there exists a filtered probability space* $(\Omega,\mathcal{F},\mathbb{F},\mathbb{P})$ *and a stochastic process* $X = (X_t)_{t\in T}$ *on* $(\Omega,\mathcal{F},\mathbb{F},P)$ *such that*

$$\mathbb{P}\left(X_{t_1} \le x_1,\ldots,X_{t_n} \le x_n\right) = F_{t_1,\ldots,t_n}(x_1,\ldots,x_n),$$

for all $x_i \in \mathbb{R}$, $t_i \in T$, $1 \le i \le n$, $n \in \mathbb{N}$; *and* X_t *is* $\mathcal{F}_t$*-measurable for all* $t \in T$.

As a corollary, we can also define a stochastic process via a collection of copulas and marginal distributions.

Corollary 128 *Let* $\{C_{t_1,\ldots,t_n} : t_i \in T,\ n \in \mathbb{N},\ 1 \le i \le n,\ t_1 < t_2 < \cdots < t_n\}$ *be a collection of copulas satisfying the following consistency condition*

$$\lim_{u_k \nearrow 1} C_{t_1,\ldots,t_n}(u_1,\ldots,u_n) = C_{t_1,\ldots,t_{k-1},t_k,\ldots,t_n}(u_1,\ldots,u_{k-1},u_k,\ldots,u_n),$$

for all $1 \le k \le n$, $u_i \in [0,1]$; *and let* $\{F_t : t \in T\}$ *be a collection of one-dimensional distribution functions. Then, there exists a filtered probability space* $(\Omega,\mathcal{F},\mathbb{F},\mathbb{P})$ *and a stochastic process* $X = (X_t)_{t\in T}$ *on* $(\Omega,\mathcal{F},\mathbb{F},P)$ *such that*

$$\mathbb{P}\left(X_{t_1} \le x_1,\ldots,X_{t_n} \le x_n\right) = C_{t_1,\ldots,t_n}\left(F_{t_1}(x_1),\ldots,F_{t_n}(x_n)\right),$$

for all $x_i \in \mathbb{R}$, $t_i \in T$, $1 \le i \le n$, $n \in \mathbb{N}$; *and* X_t *is* $\mathcal{F}_t$*-measurable for all* $t \in T$.

In the copula approach, a stochastic process can be constructed with a process of uniform variables transformed by some given marginals. In other words, we have a process $(U_t)_{t\in T}$ where for each time t, the variable U_t is uniformly distributed on $[0,1]$, with a temporal dependence structure defined by a family of copulas. This process is referred to as the "uniform process". The original process $(X_t)_{t\in T}$ is obtained via the quantile transformation $X_t = F_t^{-1}(U_t)$, for all $t \in T$.

Darsow et al. (1992) studied the relationship between copulas and Markov processes, and derived necessary and sufficient conditions for a process to be Markovian based on its copula family.

16.3.2 Brownian Copula

As an example, we start with the copula function of a Brownian motion process.

The transition probability for a standard Brownian motion is given by

$$P(x, s; y, t) \triangleq \mathbb{P}(B_t \leq y \,|\, B_s = x) = \Phi\left(\frac{y - x}{\sqrt{t - s}}\right), \text{ for } t > s, x, y \in \mathbb{R}.$$

This process is Markovian, therefore we only need to extract its bivariate copula family $C_{t,s}$.

Lemma 129 *Let X and Y be two real-valued random variables on the same probability space $(\Omega, \mathcal{A}, P)$ with corresponding copula C and continuous marginals F_X and F_Y. Define the conditional probabilities*

$$\mathbb{P}(X \leq x \,|Y) \triangleq \mathbb{E}\left[\mathbf{1}_{\{X \leq x\}} \,|Y\right] \text{ and } \mathbb{P}(Y \leq y \,|X) \triangleq \mathbb{E}\left[\mathbf{1}_{\{Y \leq y\}} \,|X\right], \text{ for } x, y \in \mathbb{R}.$$

Then,

$$\frac{\partial}{\partial u_2} C(F_X(x), F_Y(Y)) \text{ is a version of } \mathbb{P}(X \leq x \,|Y), \text{ and}$$
$$\frac{\partial}{\partial u_1} C(F_X(X), F_Y(y)) \text{ is a version of } \mathbb{P}(Y \leq y \,|X).$$

From the previous lemma, we have a relationship between the copula function and the transition probability:

$$\mathbb{P}(B_t \leq y \,|B_s = x) = \frac{\partial}{\partial u_1} C^B_{s,t}(F_s(x), F_t(y)),$$

which yields

$$C^B_{s,t}(F_s(x), F_t(y)) = \int_{-\infty}^{x} \frac{\partial}{\partial u_1} C^B_{s,t}(F_s(z), F_t(y))\, dF_s(z)$$
$$= \int_{-\infty}^{x} \Phi\left(\frac{y - z}{\sqrt{t - s}}\right) dF_s(z).$$

Substituting the marginals of the Brownian motion, $F_t(x) = \Phi\left(\frac{x}{\sqrt{t}}\right)$, we obtain

$$C^B_{s,t}(u, v) = \int_{-\infty}^{u} \Phi\left(\frac{\sqrt{t}\Phi^{-1}(v) - \sqrt{s}\Phi^{-1}(w)}{\sqrt{t - s}}\right) dw.$$

The pairwise correlation implied by the Brownian copula (Fig.16.1) is given by:

$$\rho^B\left(L_{T_i}, L_{T_j}\right) = \sqrt{\frac{T_i}{T_j}}, \text{ for } T_i \le T_j.$$

16.3.3 Copulas and Markov Processes

We should note, at this junction, that the copula function of any strictly monotone increasing transformation of a Brownian motion is also a Brownian copula. For example, a geometric Brownian motion $X_t = X_0 + \mu \int_0^t X_s ds + \sigma \int_0^t X_s dB_s$ will have the same copula function, but different marginals. In fact, this is a special case of a more general question: how can we derive the family of copulas for general transformations of diffusions or local semi-martingales?

The answer to this question is given by the seminal work of Darsow et al. (1992). We start with some definitions and notations.

Definition 130 (*Markov process*) A real-valued process $(X_t)_{t \in T}$ is called a Markov process if for any $t_1 < \cdots < t_n \in T$, and $B \in \mathcal{B}$, we have

$$\mathbb{P}\left(X_{t_n} \in B \left| X_{t_1}, \ldots, X_{t_{n-1}}\right.\right) = \mathbb{P}\left(X_{t_n} \in B \left| X_{t_{n-1}}\right.\right).$$

As a consequence of the Markovian property, the conditional probabilities

$$P(x, s; y, t) \triangleq \mathbb{P}(X_t \le y \,|X_s = x),$$

satisfy the Chapman-Kolmogorov equation

$$P(x, s; y, t) = \int_{-\infty}^{\infty} P(x, s; dz, \theta)\, P(z, \theta; y, t).$$

If the conditional densities, $p(x, s; y, t) \triangleq \mathbb{P}(X_t = y \,|X_s = x)$, exist, this can also be written as

$$p(x, s; y, t) = \int_{-\infty}^{\infty} p(x, s; z, \theta)\, p(z, \theta; y, t)\, dz.$$

Darsow et al. (1992) were the first to relate Copulas to Markov processes. To do so, they defined a multiplication product for 2-copula functions, which

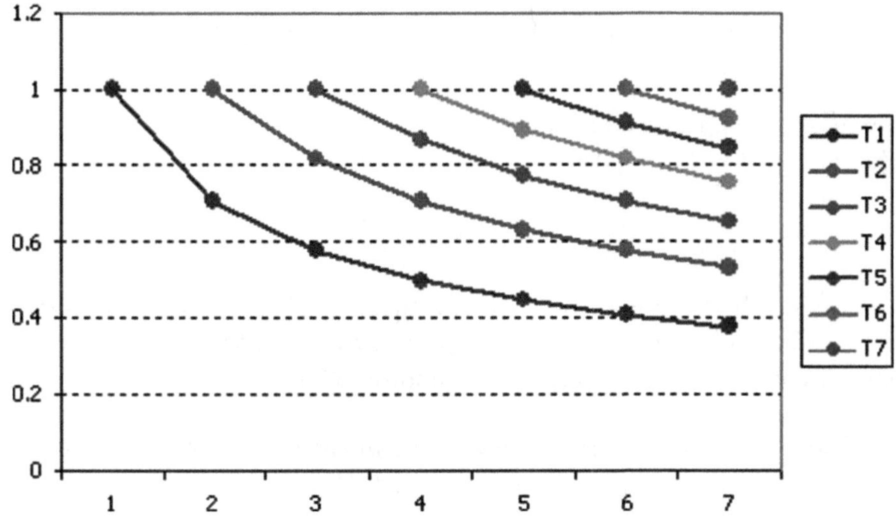

Fig. 16.1 Correlation matrix for the vector $(L_{T_1}, \ldots, L_{T_n})$ for a Brownian copula

essentially captures the dependence structure of the Chapman-Kolmogorov equation.

Definition 131 (*2-copula functions multiplication*) Let C_1 and C_2 be two bivariate copula functions, then we define the product $C_1 * C_2$ as

$$C_1 * C_2 : \begin{cases} [0,1]^2 \to [0,1] \\ (x, y) \to (C_1 * C_2)(x, y) = \int_0^1 \partial_2 C_1(x, s)\, \partial_1 C_2(s, y)\, ds \end{cases}$$

The relationship between Markov processes and the copula multiplication product is given in the following theorem (see Darsow et al. 1992, Theorem 3.2).

Theorem 132 *Let* $X = (X_t)_{t \in T}$ *be a stochastic process and let* $C_{s,t}$ *denote the copula function of the random variables* X_s *and* X_t. *Then, the following statements are equivalent:*

- *The transition probabilities* $P(x, s; y, t)$ *satisfy the Chapman-Kolmogorov equations*

$$P(x, s; y, t) = \int_{-\infty}^{\infty} P(x, s; dz, \theta) P(z, \theta; y, t),$$

for all $s < \theta < t$ *and almost all* $x \in \mathbb{R}$.

- *For all* $s < \theta < t$, *we have*

$$C_{s,t} = C_{s,\theta} * C_{\theta,t}.$$

Satisfying the Chapman-Kolmogorov equations is a necessary condition, but it is not a sufficient condition for a process to be Markovian. To derive the converse condition, they introduced an additional operator.

Definition 133 (*n-copula functions multiplication*) Let $C_1 \in \mathcal{C}_m$, $C_2 \in \mathcal{C}_n$, $m, n \geq 2$. Define the copula $C_1 \star C_2 : [0, 1]^{m+n-1} \to [0, 1]$ as

$$C_1 \star C_2 (u_1, \ldots, u_{m+n-1}) \triangleq \int_0^{u_m} \partial_m C_1 (u_1, \ldots, u_{m-1}, z) \, \partial_1 C_2 (z, u_{m+1}, \ldots, u_{m+n-1}) \, dz.$$

Remark 134 For $m = n = 2$, the relationship between the product $*$ and $\star$ is given by

$$C_1 * C_2 (u, v) = C_1 \star C_2 (u, 1, v) \text{ , for } u, v \in [0, 1].$$

Now, the characterization of a Markov process can be done fully using the $\star$-product (see Darsow et al. 1992, Theorem 3.3).

Theorem 135 *A real-valued process* $(X_t)_{t \in T}$ *is a Markov process if and only if, for all* $n \geq 3$, $t_1 < \cdots < t_n \in T$,

$$C_{t_1,\ldots,t_n} = C_{t_1,t_2} \star C_{t_2,t_3} \star \ldots \star C_{t_{n-1},t_n},$$

where $C_{t_1,\ldots,t_n} = C \langle X_{t_1}, \ldots, X_{t_n} \rangle$ *and* $C_{t_k,t_{k+1}} = C \langle X_{t_k}, X_{t_{k+1}} \rangle$.

This is a new approach to the theory of Markov processes and provides a new technique for constructing such processes. We quote here Darsow et al. (1992) who summarized the benefits of this powerful new approach as:

"*In the conventional approach, one specifies a Markov process by giving the initial distribution* μ *and a family of transition probabilities* $P_{s,t}(x, \mathcal{A})$ *satisfying the Chapman-Kolmogorov equations. In our approach, one specifies a Markov process by giving all of the marginal distributions and a family of 2-copulas... Ours is accordingly an alternative approach to the study of Markov processes which is different in principle from the conventional one. Holding the transition probabilities of a Markov process fixed and varying the initial distribution necessarily varies all of the marginal distributions, but holding the copulas of the process fixed and varying the initial distribution does not affect any other marginal distribution."*

16.3.4 Copulas of Local Martingales

In general, using the time transformation technique, one can derive the copula function for all continuous local martingales (see Schmitz 2003).

Theorem 136 (Dambis 1965; Dubins and Schwarz 1965) *If M is a continuous $(\mathcal{F}_t, P)$-local martingale vanishing at 0, such that $\langle M\rangle_t \overset{t\to\infty}{\to} \infty$ a.s., and if we set*

$$T_t \triangleq \inf\left\{s : \langle M\rangle_s > t\right\},$$

then the process $W_t \triangleq M_{T_t}$ is a $\left(\mathcal{F}_{T_t}\right)$-Brownian motion and $M_t = W_{\langle M\rangle_t}$, for all $t \in T$.

Since any continuous local martingale can be written as a time-changed Brownian motion, to derive its copula function it suffices to determine its quadratic variation and to use the copula function of a Brownian motion.

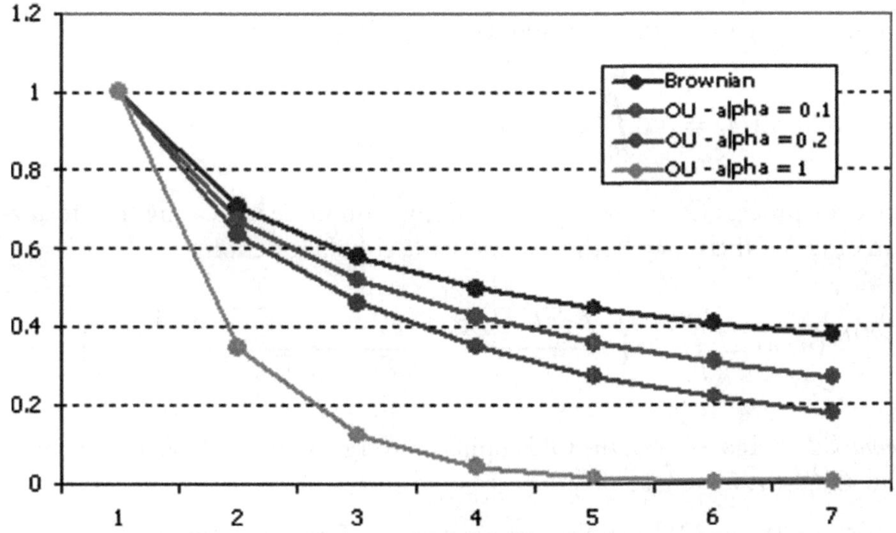

Fig. 16.2 Impact of the mean-reversion parameter on the covariance structure

Corollary 137 *If M is a continuous $(\mathcal{F}_t, P)$-local martingale vanishing at 0, such that $h(t) \triangleq \langle M \rangle_t \overset{t\to\infty}{\rightarrow} \infty$ a.s., then the bivariate copulas of the process are given by*

$$C_{s,t}(u, v) = \int_0^u \Phi\left(\frac{\sqrt{h(t)}\Phi^{-1}(v) - \sqrt{h(s)}\Phi^{-1}(w)}{\sqrt{h(t) - h(s)}}\right) dw.$$

16.3.5 Ornstein-Uhlenbeck Copula

As an example, we consider the Ornstein-Uhlenbeck process described by the following SDE

$$dX_t = -\alpha X_t dt + \sigma dB_t, t > 0.$$

The solution is given by

$$X_t = X_0 \exp(-\alpha t) + \sigma \int_0^t \exp(-\alpha(t - s)) dB_s, \text{ for } t \geq 0.$$

The process defined as

$$M_t \triangleq \exp(\alpha t) X_t - X_0 = \sigma \int_0^t \exp(\alpha s) dB_s,$$

is a martingale vanishing at 0, and its quadratic variation is

$$\langle M \rangle_t = \sigma^2 \int_0^t \exp(2\alpha s) ds = \frac{\sigma^2}{2\alpha}(\exp(2\alpha t) - 1).$$

Since the process (X_t) is a displaced strictly monotone increasing transformation of (M_t), then they both have the same copula function

$$C_{s,t}^{OU}(u, v) = \int_0^u \Phi\left(\frac{\sqrt{e^{2\alpha t} - 1}\Phi^{-1}(v) - \sqrt{e^{2\alpha s} - 1}\Phi^{-1}(w)}{\sqrt{e^{2\alpha t} - e^{2\alpha s}}}\right) dw.$$

Remark 138 If $\alpha \to \infty$, the OU copula converges to the independent copula, $\lim_{\alpha\to\infty} C_{s,t}^{OU}(u, v) = uv$.

If $\alpha \to 0$, the OU copula converges to the Brownian copula, $\lim_{\alpha\to\infty} C_{s,t}^{OU}(u, v) = C_{s,t}^{B}(u, v)$

The correlations for the Ornstein-Uhlenbeck copula (Fig. 16.2) are given by:

$$\rho^{OU}\left(L_{T_i}, L_{T_j}\right) = \sqrt{\frac{e^{2\alpha T_i} - 1}{e^{2\alpha T_j} - 1}}, \text{ for } T_i \leq T_j.$$

16.4 Link with Markov-Functional Models

By choosing a temporal copula function, which satisfies the Markovian characterization of Theorem 135, we have effectively specified a uniform process $(U_T)_{T\geq 0}$, which is Markovian and consistent with the time-dependence specified by the copula function; and our original process $(L_T)_{T\geq 0}$ is given by the inverse transformation

$$L_T = \phi_{L_T}^{-1}(U_T)$$

This is directly related to the Markov-Functional approach used in interest rate modelling (see Hunt et al. 2000), where one chooses a low-dimensional Markovian process $(x_T)_{T\geq 0}$ (under some martingale measure), and all the financial quantities required are represented as a functional form of this underlying process

$$L_T = f_T(x_T).$$

Indeed, it suffices to observe that

$$U_T = \phi_{x_T}(x_T),$$
$$f_T = \phi_{L_T}^{-1} \circ \phi_{x_T}.$$

This offers a recipe for building new dynamic credit models, which are:

- calibrated, by construction, on the term structure of correlation skew;
- flexible enough to allow for a sensible and intuitive time-dependence;
- numerically tractable thanks to the Markovian nature of the time-copula.

16.4.1 The A-Priori Model: Marshall-Olkin

To construct the low-dimensional Markov process $(x_T)_{T\geq 0}$, we shall use, as an a-priori model, a top-down version of the Marshall-Olkin model, which we refer to as the N+ model.

The main reasons for this choice are:

- the bottom-up construction of the model has an natural and intuitive interpretation;
- the MO copula can be described in SDE format, which, in turn, is very useful for deriving the dynamics of the loss variable;
- the multi-modal nature of its loss distribution captures, fundamentally, the three types of risk in the correlation market: Equity, Mezzanine, Senior;
- a (self-consistent) top-down version of the model can be derived, which captures its correlation dynamics and is numerically tractable.

We start with the MO single-name default dynamics:

$$dD_t^i = \left(1 - D_{t^-}^i\right) \sum_{j=1}^{m_C+n} A_t^{i,j} dN_t^{C_j},$$

where $N_t^{C_j}$ are independent Poisson processes, and $A_t^{i,j}$ are independent Bernoulli variables with probabilities $p^{i,j} \in [0, 1]$. In general, we use the convention that the first m_C factors are common and the remaining n factors are idiosyncratic. To derive the dynamics of the loss variable $L_t = \frac{1}{n}\sum_{i=1}^{n} D_t^i$, we can combine the individual single-name SDEs, and then approximate the (stochastic) weights that we obtain on each market factor with their deterministic projections (see Elouerkhaoui (2007) for details).

Thus, we end up with a top-down version of the model described by the following dynamics of the loss variable

$$dL_t = (1 - L_{t^-}) \left[\sum_{j=1}^{m} \gamma^j dN_t^j\right], \; L_0 = 0.$$

This can be easily calibrated and preserves the intuition of the original MO framework. This is the SDE studied empirically in Longstaff and Rajan (2008). In their paper, they have investigated how this description of the loss variable performs against realized historical data.

Some of their findings include:

- the dynamics of the correlation market are best explained with a 3-factor model, which essentially captures firm-specific, industry and economy-wide default events (Fig. 16.3);

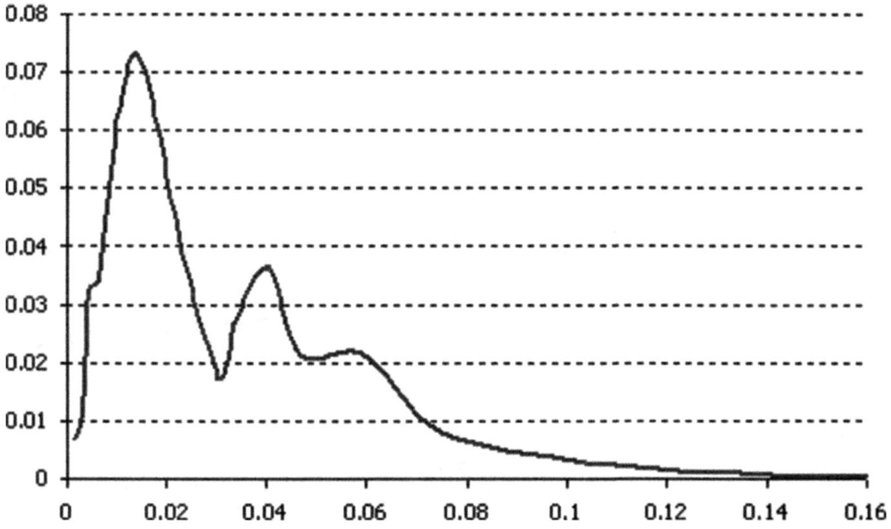

Fig. 16.3 Loss density function of the N+ model

- the estimated jump sizes for each factor are: 0.4%, 6%, 35%, which correspond to 1 default, 15 defaults and 87.5 defaults if the recovery rate is 50%;
- the expected default frequencies are: 1.2 years for the firm-specific events, 41.5 years for the industry factor and 763 years for the systemic risk, which is equivalent to default intensities of 0.8333, 0.0241 and 0.0013;
- on average, 65% of the spread is explained by idiosyncratic risk, 27% is due to industry or sector factor, and 8% represents systemic default risk.

16.4.2 Skewed N+ Model

Our dynamic MF credit model is constructed by the Markovian process $(x_t)_{t\geq 0}$ and the family of skewed marginal distributions $\left(\phi_{L_t}\right)_{t\geq 0}$ as follows

$$dx_t = (1 - x_t)\left[\sum_{j=1}^{m}\left(e^{\gamma^j} - 1\right) dN_t^j\right], \; x_0 = 0,$$

$$L_t = \phi_{L_t}^{-1}\left(\phi_{x_t}(x_t)\right), \text{ for all } t \geq 0.$$

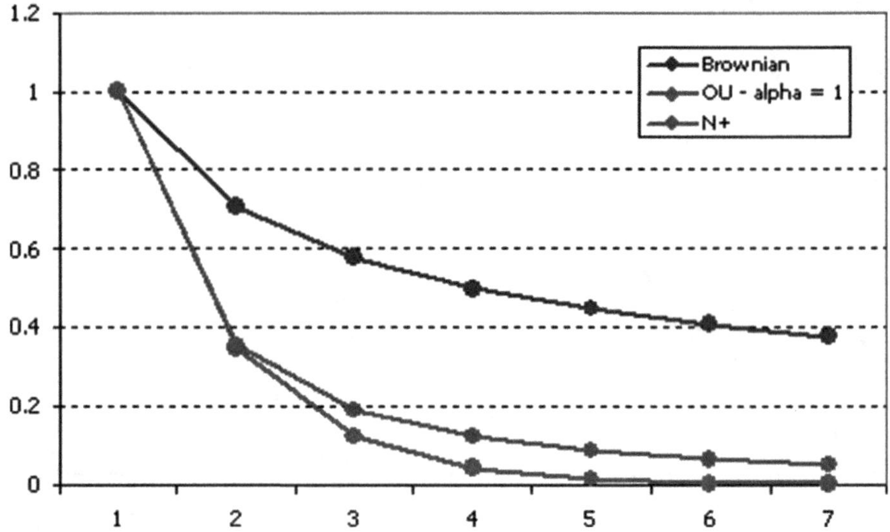

Fig. 16.4 Correlation structure for the N+ copula

The pairwise correlation for the N+ copula (Fig. 16.4) is given by:

$$\rho^{N+}\left(L_{T_i}, L_{T_j}\right) = \prod_{k=1}^{m} \sqrt{\frac{e^{\lambda^k T_i \left(e^{\gamma^k}-1\right)^2} - 1}{e^{\lambda^k T_j \left(e^{\gamma^k}-1\right)^2} - 1}}, \text{ for } T_i \leq T_j.$$

16.5 Numerical Implementation

In this section, we look at the numerical implementation details for two example payoffs: Forward CDOs and TARNs.

16.5.1 Forward CDO

To price a forward tranche, we need to evaluate the following expectation

$$\mathbb{E}\left[M^{(K_1,K_2)}\left(L_{T_2} - L_{T_1}\right)\right],$$

where $M^{(K_1,K_2)}$ is the payoff of a CDO tranche with attachment and detachment points (K_1, K_2),

$$M^{(K_1,K_2)}(x) = \min\left(\max\left(x - K_1, 0\right), K_2 - K_1\right).$$

The forward (non-skewed) loss variable is given by

$$x_{T_2} - x_{T_1} = \left(1 - x_{T_1}\right)\left[1 - \exp\left(-\sum_{j=1}^{m}\gamma^j\left(N_{T_2}^j - N_{T_1}^j\right)\right)\right].$$

To compute the value of the forward tranche, we need to map the skewed and unskewed loss variables

$$L_T = f_T\left(x_T\right) = \phi_{L_T}^{-1}\left(\phi_{x_T}\left(x_T\right)\right),$$

and to calculate two expectations: first the conditional expectation on x_{T_1}

$$\begin{aligned}\mathbb{E}\left[H\left(L_{T_2} - L_{T_1}\right)|x_{T_1}\right] &= \mathbb{E}\left[H\left(f_{T_2}\left(x_{T_2}\right) - f_{T_1}\left(x_{T_1}\right)\right)|x_{T_1}\right]\\ &= \sum_{n_1=0}^{+\infty}\cdots\sum_{n_m=0}^{+\infty}\left[\prod_{j=1}^{m} e^{-\left(\Lambda_{T_2}^{c_j}-\Lambda_{T_1}^{c_j}\right)}\frac{\left(\Lambda_{T_2}^{c_j} - \Lambda_{T_1}^{c_j}\right)^{n_j}}{n_j!}\right]\\ &\quad\times H\left(f_{T_2}\left(\left(1 - x_{T_1}\right)\left[1 - e^{-\sum_{j=1}^{m}\gamma^j n_j}\right]\right) - f_{T_1}\left(x_{T_1}\right)\right);\end{aligned}$$

then, we integrate with respect to the x_{T_1}-density

$$\begin{aligned}\mathbb{E}\left[H\left(L_{T_2} - L_{T_1}\right)\right] &= \sum_{n_1=0}^{+\infty}\cdots\sum_{n_m=0}^{+\infty}\left[\prod_{j=1}^{m} e^{-\Lambda_{T_1}^{c_j}}\frac{\left(\Lambda_{T_1}^{c_j}\right)^{n_j}}{n_j!}\right]\\ &\quad\times\mathbb{E}\left[H\left(L_{T_2} - L_{T_1}\right)\Big|x_{T_1} = 1 - e^{-\sum_{j=1}^{m}\gamma^j n_j}\right].\end{aligned}$$

16.5.2 TARN

For TARNs, in principle, for each coupon at time T_n, we need to compute an n-dimensional integral

$$\mathbb{E}\left[C^{TARN}\left(L_{T_1}, \ldots, L_{T_n}\right)\right] = \mathbb{E}\left[C^{TARN}\left(f_{T_1}\left(x_{T_1}\right), \ldots, f_{T_n}\left(x_{T_n}\right)\right)\right].$$

This can be implemented analytically by using a set of successive conditionings on $\left(x_{T_1}, \ldots, x_{T_n}\right)$ in a forward-induction algorithm, but it can quickly become very time-consuming for long dated structures; it can also be slow when the number of driving Poisson processes is large. For $m = 1$ or $m = 2$, it can be implemented on a 1D or a 2D recombining tree.

A faster implementation can be done using Monte-Carlo integration.

The algorithm is as follows.

1. For each coupon date $(T_1, \ldots, T_n)$, pre-compute the skewed and unskewed distributions $\left(\phi_{L_{T_i}}, \phi_{x_{T_i}}\right)$ and store the Markov-functional maps $f_{T_i} = \phi_{L_{T_i}}^{-1} \circ \phi_{x_{T_i}}$.
2. Simulate the driving Poisson processes $\left(N_{T_1}^j(\omega), \ldots, N_{T_n}^j(\omega)\right)_{1 \leq j \leq m}$, at the coupon dates $(T_1, \ldots, T_n)$.
3. For each path, compute the value of the unskewed losses $\left(x_{T_1}(\omega), \ldots, x_{T_n}(\omega)\right)$
$$x_{T_i}(\omega) = 1 - e^{-\sum_{j=1}^m \gamma^j N_{T_i}^j(\omega)}.$$
4. Use the Markov-functional maps to convert to the skewed losses $\left(L_{T_1}(\omega), \ldots, L_{T_n}(\omega)\right)$
$$L_{T_i}(\omega) = f_{T_i}\left(x_{T_i}(\omega)\right) = \phi_{L_{T_i}}^{-1}\left(\phi_{x_{T_i}}\left(x_{T_i}(\omega)\right)\right).$$
5. Compute the value of the TARN payoff: $C^{TARN}\left(L_{T_1}(\omega), \ldots, L_{T_n}(\omega)\right)$.

16.5.3 The Mapping Function

The two main ingredients of the model are: the mapping function and the copula process. In Fig. 16.5, we look at the skewed and unskewed distributions, and combine the two to get the implied mapping function depicted in Fig. 16.6.

16.6 Numerical Examples

In this section, we give some numerical pricing examples for a 7-year TARN structure, both in the N+ model and the Skewed N+ model.

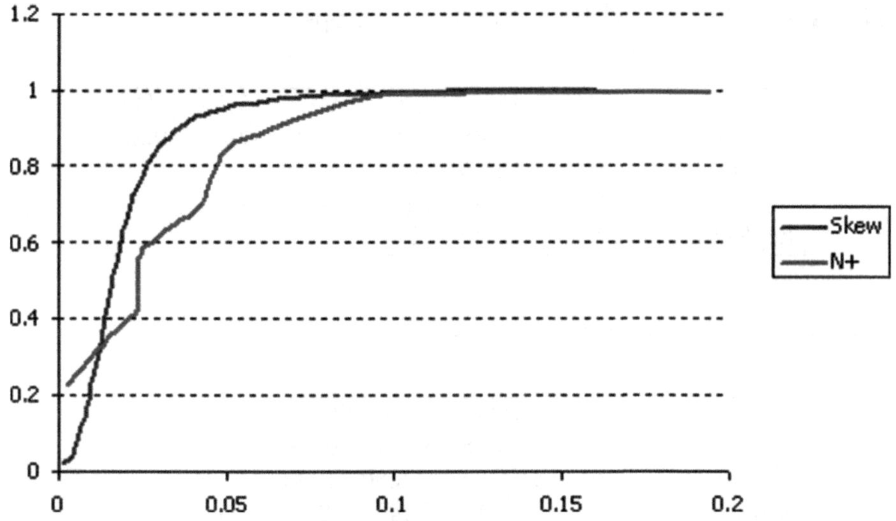

Fig. 16.5 The skewed and unskewed marginal distributions

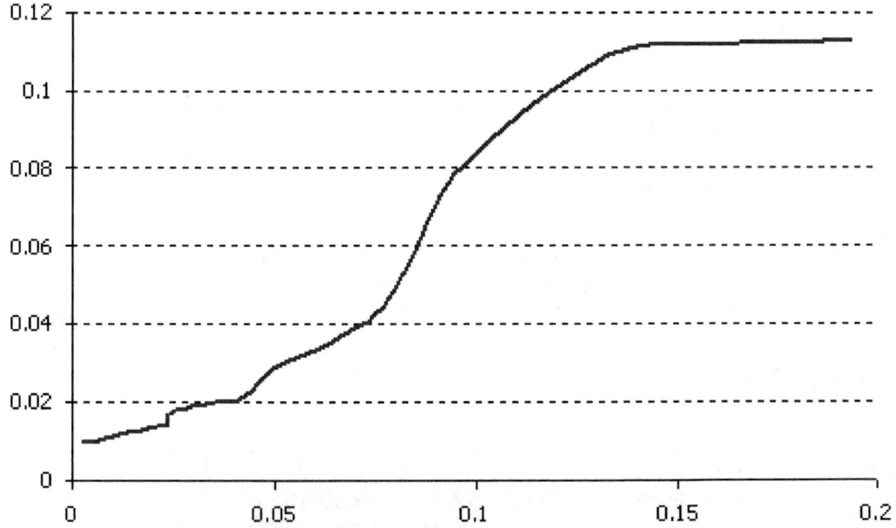

Fig. 16.6 The mapping function

N+ Model. Base Case—Index Level = 52.34 bps.

Target	5%	15%	35%	50%	75%	99%
PremiumPV	4.72	13.88	28.96	35.85	39.21	39.28
NotePV	100.91	105.66	111.11	113.08	114.89	114.97
ExpKOTime	1.14	2.38	5.11	6.54	7.00	7.00
ExpKOTimeDF	96.19	91.79	82.15	77.24	75.69	75.69

Skewed N+ Model. Base Case—Index Level = 52.34 bps.

Target	5%	15%	35%	50%	75%	99%
PremiumPV	4.83	14.38	31.71	40.75	45.66	45.66
NotePV	101.53	107.45	116.50	119.00	121.37	121.55
ExpKOTime	1.00	2.01	4.34	6.24	7.00	7.00
ExpKOTimeDF	96.69	93.07	84.79	78.25	75.71	75.69

N+ Model. Base Case—Target = 35%.

Index multiplier	0.5	1	3	5	10
Index level	35.67	52.34	119.00	185.67	352.33
PremiumPV	30.65	28.96	13.28	6.52	1.25
NotePV	116.05	111.11	89.14	82.21	76.94
ExpKOTime	4.17	5.11	6.95	7.00	7.00
ExpKOTimeDF	85.40	82.15	75.87	75.69	75.69

Skewed N+ Model. Base Case—Target = 35%.

Index multiplier	0.5	1	3	5	10
Index level	35.67	52.34	119.00	185.67	352.33
PremiumPV	31.58	31.71	31.70	31.73	31.79
NotePV	115.65	116.50	116.71	116.97	117.08
ExpKOTime	4.55	4.34	4.28	4.22	4.20
ExpKOTimeDF	84.07	84.79	85.01	85.23	85.29

16.7 Summary and Conclusion

In this chapter, we have addressed the problem of pricing path-dependent portfolio credit derivatives. To do so, we started with presenting the general theory of copulas and Markov processes, and in particular the necessary and sufficient conditions for a copula to be Markovian. Then, we have established the link between the copula approach and the Markov-Functional paradigm used in interest rates modelling. Using these theoretical foundations, we have built a dynamic credit model, which matches the correlation skew at each tenor, by construction, and follows the multi-modal Marshall-Olkin dynamics. And finally, we have reviewed the numerical implementation of the model and analyzed the pricing of TARN structures in this framework.

References

K. Dambis, On the decomposition of continuous sub-martingales. Theor. Probab. Appl. **10**, 401–410 (1965)

W. Darsow, B. Nguyen, E. Olsen, Copulas and Markov processes. Ill. J. Math. **36**(4), 600–642 (1992)

L. Dubins, G. Schwarz, On continuous Martingales. Proc. Nat. Acad. Sci. USA **53**, 913–916 (1965)

Y. Elouerkhaoui, Pricing and Hedging in a Dynamic Credit Model. International Journal of Theoretical and Applied Finance. **10**(4), 703–731 (2007)

P. Hunt, J. Kennedy, A. Pelsser, Markov-functional interest rate models. Financ. Stochast. **4**(4), 391–408 (2000)

J. Jacod, A.N. Shiryaev, *Limit Theorems for Stochastic Processes* (Springer, Berlin, 1987)

F. Lindskog, A. McNeil, Common Poisson shock models: applications to insurance and credit risk modelling. ASTIN Bullet. **33**(2), 209–238 (2003)

F. Longstaff, A. Rajan, An empirical analysis of the pricing of collateralized debt obligations. J. Financ. **63**(2), 529–563 (2008)

V. Schmitz, Copulas and stochastic processes, Ph.D. Dissertation, Institute of Statistics, Aachen University, 2003

A. Shiryaev, *Probability*, 2nd edn. (Springer, New York, 1996)

17

Hedging in Incomplete Markets

In this chapter, we present a methodology for hedging basket credit derivatives with single name instruments. Because of the market incompleteness due to the residual correlation risk, perfect replication cannot be achieved. We allow for mean self-financing strategies and use a risk-minimization criterion to find the hedge. Managing credit risk is always a fine balance between hedging the Jump-to-Default exposure (JTD) or the credit spread exposure (CR01). Recently, this credit hedging dilemma (JTD vs CR01) is becoming very topical in the context of managing counterparty credit risk for large derivatives books.

17.1 Introduction

Typical basket products such as first-to-default swaps or CDOs reference a pool of underlying credit entities and their payoff is dependent on the joint default behaviour of the underlying basket. This introduces a default correlation risk, which makes the multi-credit market incomplete: a basket product cannot be completely replicated with single name instruments. Other correlation sensitive instruments are required to offset the residual correlation risk. Furthermore, credit securities have two potential sources of risk: spread risk and default risk. In general, one cannot hedge both at the same time. The hedger would use his judgement and focus mainly on one source of risk depending on the prevailing market conditions. This bi-modal nature of the credit markets introduces another level of complexity to the default correlation incompleteness.

One approach to handle market incompleteness is to use quadratic optimality criteria. Quadratic hedging approaches, such as local risk-minimization

Y. Elouerkhaoui, *Credit Correlation*, Applied Quantitative Finance,
DOI 10.1007/978-3-319-60973-7_17

and mean-variance hedging, have been developed in a series of papers by Fö llmer, Schweizer and Sondermann (see, for example, Föllmer and Sondermann 1986; Föllmer and Schweizer 1991; Schweizer 1993, 1994). Other criteria for determining an optimal hedging strategy have also been developed in the literature. See, for example, El Karoui and Quenez (1995), Davis (1997), Cvitanic (1997). Here, we use the risk-minimization approach of Föllmer and Sondermann to find the optimal hedging strategy for basket products. This analysis is done in the Marshall-Olkin copula framework, where each individual default process is decomposed on a basis of independent Cox processes. The so-called common market factors can trigger joint defaults in the basket, whereas idiosyncratic factors, on the other hand, can only trigger individual defaults. The market incompleteness is the result of the M.O. model, which has simultaneous defaults and hence a particularly large mark space for the point process representing defaults.

17.2 The Model

We work in a financial market represented by a probability space $(\Omega, \mathcal{G}, \mathbb{P})$ and a time horizon $T^* \in (0, \infty)$, on which is given a d-dimensional Brownian motion W and n non-negative random variables $(\tau_1, ..., \tau_n)$ representing the default times of the obligors in the economy.

Assumption 1. We assume that $\mathbb{P}$ is a (risk-neutral) martingale measure. Throughout, we shall work under this martingale measure. We follow the approach of Föllmer and Sondermann (1986) where a "good" martingale measure is chosen, then the minimization of the risk is done with respect to this measure.

Assumption 2. We introduce an $\mathbb{R}^d$-valued Itô process X, describing the state variables in the economy, which solves the following SDE:

$$dX_t = \alpha(X_t)\, dt + \beta(X_t)\, dW_t,$$

for some Lipschitz functions $\alpha_k : \mathbb{R}^d \to \mathbb{R}$ and $\beta_{kj} : \mathbb{R}^d \to \mathbb{R}^d$, $1 \le k \le d$, $1 \le j \le d$.

We denote by $\{\mathcal{F}_t\}$ the filtration generated by X and augmented with the P-null sets of $\mathcal{G}$:

$$\mathcal{F}_t \triangleq \sigma(X_s : 0 \le s \le t) \vee \mathcal{N}.$$

We introduce, for each obligor i, the right-continuous process $D_t^i \triangleq \mathbf{1}_{\{\tau_i \le t\}}$ indicating whether the firm has defaulted or not. We denote by $\{\mathcal{H}_t^i\}$ the filtration generated by this process:

$$\mathcal{H}_t^i \triangleq \mathcal{F}_t^{D^i} \triangleq \sigma\left(D_s^i : 0 \le s \le t\right).$$

The agents' filtration is the one generated by the economic state variables and the default processes

$$\mathcal{G}_t \triangleq \mathcal{F}_t \vee \left[\bigvee_{i=1}^{n} \mathcal{H}_t^i\right]. \tag{17.1}$$

Assumption 3. We assume that the default times are correlated and we allow for multiple instantaneous joint defaults. The multivariate dependence is defined by a (doubly-stochastic) Marshall-Olkin copula.

More precisely, we assume that there exists a set of m independent Cox processes $\left(N_t^{c_j}\right)_{t\ge0}$ with continuous bounded intensities $\lambda^{c_j}(X_t)$, $\lambda^{c_j}(.) : \mathbb{R}^d \to \mathbb{R}_+$, which can trigger simultaneous joint defaults. Each Cox process N^{c_j} can be equivalently represented by the sequence of event trigger times $\left\{\theta_r^{c_j}\right\}_{r\in\{1,2,\ldots\}}$.

For every event type c_j, and for all $t \ge 0$, we define a set of independent Bernoulli variables $\mathbf{A}_t^j = \left(A_t^{1,j}, \ldots, A_t^{n,j}\right)$ with probabilities $\left(p^{1,j}(X_t), \ldots, p^{n,j}(X_t)\right)$, where $p^{i,j}(.) : \mathbb{R}^d \to [0, 1]$ are some continuous functions of the background process X. At the rth occurrence of an event of type c_j, at time $\theta_r^{c_j}$, the variable $A_{\theta_r^{c_j}}^{i,j}$ indicates whether a default of type i has occurred or not. We assume that for $j \ne k$, the vectors $\mathbf{A}_t^j$ and $\mathbf{A}_t^k$ are independent. And we assume that for $t \ne s$, the vectors $\mathbf{A}_t^j$ and $\mathbf{A}_s^j$ are independent.

The process $\left(N_t^i\right)_{t\ge0}$ defined as

$$N_t^i \triangleq \sum_{j=1}^{m} \sum_{\theta_r^{c_j} \le t} A_{\theta_r^{c_j}}^{i,j}, \tag{17.2}$$

is also a Cox process with intensity $\lambda^i(X_t) \triangleq \sum_{j=1}^{m} p^{i,j}(X_t)\lambda^{c_j}(X_t)$. It is obtained by superpositioning m independent (thinned) Cox processes.

The default time τ_i is defined as the first jump time of the Cox process $\left(N_t^i\right)_{t\ge0}$:

$$\tau_i \triangleq \inf\left\{t : N_t^i > 0\right\}. \tag{17.3}$$

This common shock model can also be described formally by the following (more compact) SDE

$$dD_t^i = \left(1 - D_{t-}^i\right) \sum_{j=1}^{m} A_t^{i,j} dN_t^{c_j}. \quad (17.4)$$

Note that the Marshall-Olkin filtration is much larger than the one accessible to the agents in the economy. It contains the information flow of the common trigger events and the "conditional" Bernoulli events:

$$\widetilde{\mathcal{G}}_t = \mathcal{F}_t \vee \left[\bigvee_{j=1}^{m} \mathcal{F}_t^{N^{c_j}} \right] \vee \left[\bigvee_{j=1}^{m} \bigvee_{i=1}^{n} \mathcal{F}_t^{A^{i,j}} \right]. \quad (17.5)$$

This is actually the reason why multi-name credit markets are, in general, incomplete. We have $\{\mathcal{G}_t\} \subset \left\{\widetilde{\mathcal{G}}_t\right\}$; the additional information, which is not captured in the $\{\mathcal{G}_t\}$-filtration is the one associated with the multivariate dependence or the correlation risk. Additional market instruments are required to hedge this type of risk.

17.3 The Problem

In our economy, we assume that we have $(n + 1)$ primary assets available for hedging with price processes $S^i = \left(S_t^i\right)_{0 \leq t \leq T^*}$. The first asset S^0 is the money-market account, i.e., $S_t^0 \triangleq \exp\left(\int_0^t r_s ds\right)$ for some $\mathcal{F}_t$-adapted process r. S^0 will be used as numeraire and all quantities will be expressed in units of S^0. In particular, S^0 will be equal to 1 at all times. We shall consider only zero-coupon credit derivatives or contingent claims of the European type. The hedging asset S^i will represent the zero-coupon defaultable bond maturing at T linked to obligor i; i.e., it pays 1 if obligor i survives until time T, or 0 otherwise. The payoff at maturity is defined as:

$$S_T^i \triangleq 1 - D_T^i.$$

In practice, zero-coupon defaultable bonds are not traded in the market. They can, however, be extracted from the prices of liquid default swap instruments with different maturities. Given some recovery rate assumption and an interpolation method between maturities, a bootstrapping algorithm can be used to extract the value of zero-coupon bonds.

We shall consider here the problem of pricing and hedging zero-coupon contingent claims by dynamically trading the hedging assets S. The contingent claims in this context include credit derivatives of the basket type.

Definition 139 (*Contingent Claim*) A contingent claim is a $\mathcal{G}_T$-measurable random variable $H_T \in L^2(\mathbb{P})$ describing the payoff at maturity T of a financial instrument.

A well-known example is a kth-to-default (zero-coupon note) maturing at T. Its payoff is defined as:

$$H_T^{(k)} \triangleq \mathbf{1}_{\{\sum_{i=1}^{n} D_T^i < k\}},$$

it will pay 1 if there are less than k defaults in the basket or 0 otherwise. The most common structure in this category is a first-to-default, $H_T^{(1)}$, which pays 1 if no obligor in the basket defaults before T. Another popular example of contingent claims is a CDO (zero-coupon note). If we assume that the recovery rate for obligor i is a constant proportion $0 \leq R^i \leq 1$, then the payoff of a CDO tranche covering the portfolio losses, which fall in some range $[K_1, K_2]$, where $0 \leq K_1 < K_2 \leq 1$, is

$$H_T^{(K_1,K_2)} \triangleq \frac{1}{K_2 - K_1} \min\left(\max\left(\frac{1}{n}\sum_{i=1}^{n}\left(1 - R^i\right) D_T^i - K_1, 0\right), K_2 - K_1\right).$$

As usual, the problem of the seller of this contract is twofold: (a) how much should he charge for this security at time 0, (b) how can he cover his position with the available hedging instruments.

This is formalized by considering dynamic strategies $(\alpha, \eta) = (\alpha_t, \eta_t)_{0 \leq t \leq T^*}$, where α is a n-dimensional predictable process and η is adapted. α_t^i represents the number of units of the single-name hedge instrument i held at time t, and η_t is the amount of money invested in the cash account. The value of the portfolio is given by $V_t = (\alpha_t)^{tr} S_t + \eta_t$, and the cumulative gains up to time t are $G_t(\alpha) = \int_{]0,t]} (\alpha_u)^{tr} dS_u$. The cost process is $C_t = V_t - G_t(\alpha)$.

For non-attainable claims, a risk-minimizing strategy is characterized by: (a) its cost process must be a martingale, (b) the cost process is orthogonal to S. As shown in Föllmer and Sondermann (1986), the hedging strategy is obtained by the Kunita-Watanabe decomposition of the $\{\mathcal{G}_t\}$-martingale $H_t = \mathbb{E}\left[H_T \,|\mathcal{G}_t\right]$:

$$H_T = H_0 + \int_{]0,T]} \left(\alpha_t^{H_T}\right)^{tr} dS_t + L_T^{H_T},$$

where L^{H_T} is a martingale orthogonal to S.

Our goal is to find an analytical result for the strategy $\left(\alpha_t^{H_T}\right)$.

17.4 Marked Point Process Representation

In this section, we present the "Equivalent Fatal Shock Model" (see Lindskog and McNeil 2003) used to equivalently describe the Marshall-Olkin set up. This provides an explicit representation of the marked point process, which will be used throughout.

Equivalent Fatal Shock Model. Let $\Pi_{\mathbf{n}}$ be the set of all subsets of $\{1, ..., n\}$. For each $\pi \in \Pi_{\mathbf{n}}$, we introduce the point process N_t^π, which counts the number of shocks in $(0, t]$ resulting in joint defaults of the obligors in π only. For example, if $\pi = \{1, 2\}$, then the process $N_t^{\{1,2\}}$ counts the shocks which trigger simultaneous defaults of obligor 1 and 2 but not the other obligors 3 to n. The process N_t^π is, then, formally defined as

$$N_t^\pi \triangleq \sum_{j=1}^{m} \sum_{r=1}^{N_t^{c_j}} A_{\theta_r^{c_j}}^{\pi,j}, \tag{17.6}$$

where

$$A_t^{\pi,j} \triangleq \prod_{i\in\pi} A_t^{i,j} \prod_{i\notin\pi} \left(1 - A_t^{i,j}\right). \tag{17.7}$$

For $t \geq 0$, $A_t^{\pi,j}$ is a Bernoulli variable, which is equal to 1 if all obligors $i \in \pi$ default and all the others survive. So, at the occurrence of the rth common shock, of type c_j, at time $\theta_r^{c_j}$, the point process N_t^π gets incremented by $\Delta N_{\theta_r^{c_j}}^\pi = A_{\theta_r^{c_j}}^{\pi,j}$.

Note that for $\pi = \{i\}$, $N_t^{\{i\}}$ should not be confused with N_t^i:

$$N_t^i \neq N_t^{\{i\}};$$

N_t^i counts all the shocks that affect obligor i and may affect other obligors as well, on the other hand $N_t^{\{i\}}$ counts the shocks affecting obligor i only and do not affect the others.

Next, we state the key result of the fatal shock representation. We refer to Lindskog and McNeil (2003) for details (see Proposition 4).

Proposition 140 (Fatal shock representation) $(N^{\pi})_{\pi\in\Pi_{\mathbf{n}}}$ *are Cox processes, with intensities*

$$\lambda^{\pi}(X_t) = \sum_{j=1}^{m} p^{\pi,j}(X_t)\,\lambda^{c_j}(X_t),$$

where

$$p^{\pi,j}(X_t) = \prod_{i\in\pi} p^{i,j}(X_t) \prod_{i\notin\pi}\left(1 - p^{i,j}(X_t)\right).$$

Moreover, conditional on $\mathcal{F}_{\infty}$*, the processes* $(N^{\pi})_{\pi\in\Pi_{\mathbf{n}}}$ *are independent.*

We summarize in the following lemma the equivalent description of each obligor i using the fatal shock representation.

Lemma 141 (Obligor description using the fatal shock representation) *The Cox process* N^i *is given by*

$$N_t^i = \sum_{\pi\in\Pi_{\mathbf{n}}} \mathbf{1}_{\{i\in\pi\}} N_t^{\pi}, \tag{17.8}$$

and its intensity can be expressed as the sum[1]

$$\lambda^i(X_t) = \sum_{\pi\in\Pi_{\mathbf{n}}} \mathbf{1}_{\{i\in\pi\}} \lambda^{\pi}(X_t). \tag{17.9}$$

Marked Point Process Representation. The Marshall-Olkin model is defined on the filtration $\{\widetilde{\mathcal{G}}_t\}$, which is larger than the one available to investors, namely $\{\mathcal{G}_t\}$. We shall use the generic tools of the MO model, however, the local characteristics of the MPP representation are derived for $\{\mathcal{G}_t\}$.

---

[1] Here, we have used the fact that if N^i has intensity λ^i, then $\sum a_i N^i$ has intensity $\sum a_i \lambda^i$ when the intensities are calculated in the filtration generated by all N^i.

We define the sequence of ordered default times $(T_0, T_1, ..., T_n) : T_0 = 0 \leq T_1 \leq ... \leq T_n$, and identities of the defaulted obligors as:

$$\begin{aligned} &T_0 = 0, Z_0 = \emptyset; \\ &T_k = \min\{\tau_i : 1 \leq i \leq n, \tau_i > T_{k-1}\}; \\ &Z_k = \pi \text{ if } T_k = \tau_i \text{ for all } i \in \pi, \text{ and } \pi \in \Pi_{\mathbf{n}}; \end{aligned}$$

The mark space of this point process is $E \triangleq \Pi_{\mathbf{n}}$, the set of all subsets of $I_n \triangleq \{1, ..., n\}$.

The double sequence $(T_k, Z_k)_{k \geq 1}$ defines a marked point process with counting measure

$$\mu(\omega, dt \times dz) : (\Omega, \mathcal{G}) \to ((0, \infty) \times E, (0, \infty) \otimes \mathcal{E}),$$

$$\int_0^t \int_E H(\omega, t, z)\, \mu(\omega, dt \times dz) = \sum_{k \geq 1} H(\omega, T_k(\omega), Z_k(\omega))\, \mathbf{1}_{\{T_k(\omega) \leq t\}}.$$

The MPP $(T_k, Z_k)_{k \geq 1}$ can also be described through a family of counting processes $(D_t(\pi))_{\pi \in \Pi_{\mathbf{n}}}$ defined as: $D_t(\pi) \triangleq \sum_{k \geq 1} \mathbf{1}_{\{T_k \leq t, Z_k = \pi\}}$; $D_t(\pi)$ counts the number of events on $(0, t]$ matching a mark equal to the subset π. We can also define $\tau(\pi) \triangleq \sum_{k \geq 1} T_k \mathbf{1}_{\{Z_k = \pi\}}$ as the default time where all obligors in π default simultaneously. If we denote by $\left(\lambda_t^{\pi}\right)_{t \geq 0}$ the $(\mathbb{P}, \{\mathcal{G}_t\})$-intensity of the default time $\tau(\pi)$, then the $(\mathbb{P}, \{\mathcal{G}_t\})$-intensity kernel of the counting measure $\mu(dt \times dz)$ is given by

$$\lambda_t(\omega, dz)\, dt = \lambda_t(\omega)\, \Phi_t(\omega, dz)\, dt,$$

where $(\lambda_t)_{t \geq 0}$ is the non-negative $\{\mathcal{G}_t\}$-predictable process

$$\lambda_t = \sum_{\pi \in \Pi_{\mathbf{n}}} \lambda_t^{\pi}, \tag{17.10}$$

and $\Phi_t(\omega, dz)$ is the probability transition kernel from $(\Omega \times [0, \infty),\ \mathcal{G} \otimes \mathcal{B}_+)$ into $(E, \mathcal{E})$

$$\Phi_t(\omega, \pi) = \frac{\lambda_t^{\pi}}{\lambda_t}, \text{ for } \pi \in \Pi_{\mathbf{n}}, \tag{17.11}$$

with $\Phi_t(.) = 0$ if $\lambda_t = 0$. The pair $(\lambda_t, \Phi_t(dz))$ is the $(\mathbb{P}, \{\mathcal{G}_t\})$-local characteristics of the counting measure $\mu(dt \times dz)$.

For each subset $\pi \in \Pi_{\mathbf{n}}$, the intensity $\left(\lambda_t^{\pi}\right)_{t \geq 0}$ can be computed as

$$\lambda_t^{\pi} = \lim_{h \downarrow 0} \frac{1}{h} \mathbb{P}\left(D_{t+h}(\pi) - D_t(\pi) \,|\mathcal{G}_t\right) = -\frac{\partial}{\partial T} \mathbb{P}\left(\tau(\pi) > T \,|\mathcal{G}_t\right)|_{T=t} \,.$$

Lemma 142 *The process* $\left(\lambda_t^{\pi}\right)_{t \geq 0}$ *is given by*

$$\lambda_t^{\pi} = \left[\prod_{i \in \pi} \left(1 - D_t^i\right)\right] \times \left[\sum_{x \subset (I_n \backslash \pi)} \left[\prod_{i \in x} D_t^i\right] \lambda^{(\pi \cup x)}\left(X_t\right)\right].$$

Proof Recall that the (conditional) multivariate distribution function of default times in the M.O. model is

$$\mathbb{P}\left(\tau_1 > t_1, ..., \tau_n > t_n \,|\mathcal{F}_t\right) = \mathbb{E}\left[\prod_{\pi \in \Pi_{\mathbf{n}}} \exp\left(-\Lambda^{\pi}_{\max\limits_{i \in \pi} t_i}\right) |\mathcal{F}_t\right],$$

where $\Lambda_T^{\pi} \triangleq \int_0^T \lambda^{\pi}\left(X_s\right) ds$, for all $\pi \in \Pi_{\mathbf{n}}$. By defining, for each n-uplet $(t_1, ..., t_n) \in \mathbb{R}_+^n$, the mapping

$$\begin{aligned}
(t_1, ..., t_n) &\rightarrow \left\{\left(\theta_1, \pi_1\right), ..., \left(\theta_k, \pi_k\right), ...\right\}, \\
\theta_0 &= 0, \pi_0 = \emptyset, \\
\theta_k &= \min\left\{t_i : 1 \leq i \leq n, t_i > \theta_{k-1}\right\}, \\
\pi_k &= \pi \text{ if } \theta_k = t_i \text{ for all } i \in \pi, \text{ and } \pi \in \Pi_{\mathbf{n}},
\end{aligned}$$

which sorts and groups the times $(t_1, ..., t_n)$ in a strictly increasing order, we can express the (conditional) density function,

$$f_t\left(t_1, ..., t_n\right) \triangleq \mathbb{P}\left(\tau_1 \in dt_1, ..., \tau_n \in dt_n \,|\mathcal{F}_t\right),$$

as follows

$$\begin{aligned}
f_t\left(t_1, ..., t_n\right) = \mathbb{E}\Bigg[\prod_{k \geq 1} &\left(\sum_{S \subset \{\pi_1 \cup ... \cup \pi_{k-1}\}} \lambda^{\pi_k \cup S}\left(X_{\theta_k}\right)\right) \\
&\exp\left(-\sum_{S \subset \{\pi_1 \cup ... \cup \pi_{k-1}\}} \Lambda_{\theta_k}^{\pi_k \cup S}\right) |\mathcal{F}_t\Bigg],
\end{aligned}$$

where S spans the set of all subsets of $\{\pi_1 \cup \ldots \cup \pi_{k-1}\}$, which contains all the obligors who have defaulted prior to θ_k.

In order to compute the conditional expectation $\mathbb{P}^* (\tau(\pi) > T | \mathcal{G}_t)$, we use the generalized Dellacherie formula

$$\mathbb{P}(\tau(\pi) > T | \mathcal{G}_t) = \sum_{x \in \Pi_\mathbf{n}} D_t^{(x)} \frac{\mathbb{E}\left[\mathbf{1}_{\{\tau(\pi)>T\}} \times \prod_{i \notin x} (1 - D_t^i) \middle| \mathcal{G}_t^{(x)}\right]}{\mathbb{E}\left[\prod_{i \notin x} (1 - D_t^i) \middle| \mathcal{G}_t^{(x)}\right]}. \tag{17.12}$$

The only subsets, $x \in \Pi_\mathbf{n}$, for which the conditional expectation is not equal to zero are the ones of the form $x \subset (I_n \setminus \pi)$, i.e., at least all the obligors in π are alive; here, the notation $(I_n \setminus \pi)$ represents the complement of π in I_n The default state indicator can be expressed as

$$D_t^{(x)} = \left[\prod_{i \in x} D_t^i\right] \left[\prod_{i \in \pi} \left(1 - D_t^i\right)\right] \left[\prod_{i \in (I_n \setminus \pi) \setminus x} \left(1 - D_t^i\right)\right],$$

and Eq. (17.12) becomes

$$\mathbb{P}(\tau(\pi) > T | \mathcal{G}_t) = \left[\prod_{i \in \pi} \left(1 - D_t^i\right)\right] \sum_{x \subset (I_n \setminus \pi)} \left[\prod_{i \in x} D_t^i \prod_{i \in (I_n \setminus \pi) \setminus x} \left(1 - D_t^i\right)\right] H_t^{\pi,(x)}(T),$$

$$H_t^{\pi,(x)}(T) \triangleq \frac{\mathbb{E}\left[\mathbf{1}_{\{\tau(\pi)>T\}} \times \prod_{i \notin x} (1 - D_t^i) \middle| \mathcal{G}_t^{(x)}\right]}{\mathbb{E}\left[\prod_{i \notin x} (1 - D_t^i) \middle| \mathcal{G}_t^{(x)}\right]}.$$

We shall compute each term separately.

On the set $\left[\prod_{i \in x} \mathbf{1}_{\{\tau_i \in ds_i\}}\right] \left[\prod_{i \in \pi} \mathbf{1}_{\{\tau_i > t\}}\right] \left[\prod_{i \in (I_n \setminus \pi) \setminus x} \mathbf{1}_{\{\tau_i > t\}}\right]$, the conditional probability is equal to

$$H_t^{\pi,(x)}(T) = \frac{\mathbb{E}\left[\mathbf{1}_{\{\tau(\pi)>T\}} \times \left[\prod_{i \in \pi} (1 - D_t^i)\right] \times \left[\prod_{i \in (I_n \setminus \pi) \setminus x} (1 - D_t^i)\right] \times \left[\prod_{i \in x} \mathbf{1}_{\{\tau_i \in ds_i\}}\right] | \mathcal{F}_t\right]}{\mathbb{E}\left[\left[\prod_{i \in \pi} (1 - D_t^i)\right] \times \left[\prod_{i \in (I_n \setminus \pi) \setminus x} (1 - D_t^i)\right] \times \left[\prod_{i \in x} \mathbf{1}_{\{\tau_i \in ds_i\}}\right] | \mathcal{F}_t\right]}.$$

Using the expression of the (conditional) density function, we find that the numerator is given by

$$\mathbb{E}\left[\mathbf{1}_{\{\tau(\pi)>T\}} \times \left[\prod_{i\in\pi}\left(1-D_t^i\right)\right] \times \left[\prod_{i\in(I_n\backslash\pi)\backslash x}\left(1-D_t^i\right)\right] \times \left[\prod_{i\in x}\mathbf{1}_{\{\tau_i\in ds_i\}}\right] |\mathcal{F}_t\right]$$

$$= \mathbb{E}\left[\exp\left(-\sum_{S\subset x}\Lambda_T^{\pi\cup S}\right) \times \left[\prod_{q\subset(I_n\backslash\pi)\backslash x}\exp\left(-\sum_{S\subset x}\Lambda_t^{q\cup S}\right)\right] \times \left[f^{(x)}\left((s_i)_{i\in x}\right)\right] |\mathcal{F}_t\right],$$

where $f^{(x)}\left((s_i)_{i\in x}\right) \triangleq \mathbb{P}\left(\bigcap_{i\in x}\{\tau_i \in ds_i\} |\mathcal{F}_\infty\right)$ is the density function of the first $d = |x|$ defaulted obligors before time t .

Similarly, we have for the denominator

$$\mathbb{E}\left[\left[\prod_{i\in\pi}\left(1-D_t^j\right)\right] \times \left[\prod_{i\in(I_n\backslash\pi)\backslash x}\left(1-D_t^j\right)\right] \times \left[\prod_{i\in x}\mathbf{1}_{\{\tau_i\in ds_i\}}\right] |\mathcal{F}_t\right]$$

$$= \mathbb{E}\left[\exp\left(-\sum_{S\subset x}\Lambda_t^{\pi\cup S}\right) \times \left[\prod_{q\subset(I_n\backslash\pi)\backslash x}\exp\left(-\sum_{S\subset x}\Lambda_t^{q\cup S}\right)\right] \times \left[f^{(x)}\left((s_i)_{i\in x}\right)\right] |\mathcal{F}_t\right].$$

Using the fact that $\left[\prod_{q\subset(I_n\backslash\pi)\backslash x}\exp\left(-\sum_{S\subset x}\Lambda_t^{q\cup S}\right)\right]\times\left[f^{(x)}\left((s_i)_{i\in x}\right)\right]$, for all $s_i \leq t$, $i \in x$, is $\mathcal{F}_t$-measurable, we obtain

$$H_t^{\pi,(x)}(T) = \mathbb{E}\left[\exp\left(-\sum_{S\subset x}\left(\Lambda_T^{\pi\cup S} - \Lambda_t^{\pi\cup S}\right)\right) |\mathcal{F}_t\right].$$

Differentiating with respect to T, and evaluating at $T = t$, we get

$$\lambda_t^\pi = \left[\prod_{i\in\pi}\left(1-D_t^i\right)\right] \times \left[\sum_{x\subset(I_n\backslash\pi)}\left[\prod_{i\in x}D_t^i\prod_{i\in(I_n\backslash\pi)\backslash x}\left(1-D_t^i\right)\right]\sum_{S\subset x}\lambda^{(\pi\cup S)}(X_t)\right]. \tag{17.13}$$

After some basic algebra, we find that Eq. (17.13) can be expressed as

$$\lambda_t^\pi = \left[\prod_{i\in\pi}\left(1-D_t^i\right)\right] \times \left[\sum_{x\subset(I_n\backslash\pi)}\left[\prod_{i\in x}D_t^i\right]\lambda^{(\pi\cup x)}(X_t)\right].$$

■

For $1 \leq i \leq n$, the compensated point process $M^i : M_t^i \triangleq D_t^i - \int_0^{t\wedge\tau_i} \lambda^i (X_s)\, ds$, is given by

$$M_t^i = \int_0^t \int_E \mathbf{1}_{\{i\in z\}} \left(\mu \left(ds \times dz\right) - \lambda_s \left(dz\right) ds\right), \tag{17.14}$$

which can also be written as

$$M_t^i = \sum_{\pi\in\Pi_{\mathbf{n}}} \mathbf{1}_{\{i\in z\}} M_t^{\pi}, \tag{17.15}$$

where, for each $\pi \in \Pi_{\mathbf{n}}$, M^{π} is the compensated point process:

$$M_t^{\pi} \triangleq D_t\left(\pi\right) - \int_0^t \lambda_s^{\pi}\, ds. \tag{17.16}$$

This MPP representation makes formal the idea that the mark space of the default times $(\tau_1, ..., \tau_n)$ is $\Pi_{\mathbf{n}}$ since joint defaults are allowed. Here, we have fixed the mark space, but as default events occur, we put zero probability mass for the states of $\Pi_{\mathbf{n}}$, which cannot occur anymore.

17.5 Dynamics of the Zero-Coupon Defaultable Bonds

In this section, we derive the dynamics of the zero-coupon defautable bonds under the martingale measure $\mathbb{P}$. First, we give an explicit formula of the zero-coupon defaultable bond price. Then, we apply Itô's lemma to compute the martingale representation of the price process.

Lemma 143 (Explicit formula of the zero-coupon defaultable bond)

$$S_t^i = \left(1 - D_t^i\right) \mathbb{E}\left[\exp\left(-\int_t^T \lambda^i \left(X_s\right) ds\right) |\mathcal{F}_t\right].$$

Proof The value of the (discounted) zero-coupon single-name bond is

$$S_t^i = \mathbb{E}\left[1 - D_T^i\, |\mathcal{G}_t\right].$$

Since τ^i is the first jump time of a Cox process,

$$\left\{\tau^i > T\right\} \Longleftrightarrow \left\{N_T^i = 0\right\},$$

we can write, for $t = 0$,

$$S_0^i = \mathbb{P}\left(N_T^i = 0\right) = \mathbb{E}\left[\exp\left(-\int_0^T \lambda^i\left(X_s\right) ds\right)\right].$$

In the Marshall-Olkin copula framework, the conditional expectation formula holds as well, i.e.,

$$S_t^i = \mathbb{E}\left[1 - D_T^i \,|\mathcal{G}_t\right] = \left(1 - D_t^i\right) \mathbb{E}\left[\exp\left(-\int_t^T \lambda^i\left(X_s\right) ds\right) |\mathcal{F}_t\right].$$

To verify this property, we need to check that the survival probability does not jump upon the default of the other obligors. For clarity, we shall do the calculations for $n = 2$, the general case is a straightforward extension.

Using the generalized Dellacherie formula, we have

$$\begin{aligned}\mathbb{P}\left(\tau_1 > T \,|\mathcal{G}_t\right) = \mathbf{1}_{\{\tau_1 > t\}} \mathbf{1}_{\{\tau_2 > t\}} \frac{\mathbb{P}\left(\tau_1 > T, \tau_2 > t \,|\mathcal{F}_t\right)}{\mathbb{P}\left(\tau_1 > t, \tau_2 > t \,|\mathcal{F}_t\right)} \\ + \mathbf{1}_{\{\tau_1 > t\}} \mathbf{1}_{\{\tau_2 \leq t\}} \frac{\mathbb{P}\left(\tau_1 > T \,|\mathcal{F}_t \vee \sigma\left(\tau_2\right)\right)}{\mathbb{P}\left(\tau_1 > t \,|\mathcal{F}_t \vee \sigma\left(\tau_2\right)\right)}.\end{aligned}$$

Recall that the M.O. (conditional) multivariate probability function is given by

$$\mathbb{P}\left(\tau_1 > T_1, \tau_2 > T_2 \,|\mathcal{F}_t\right) = \mathbb{E}\left[\exp\left(-\Lambda_{T_1}^{\{1\}} - \Lambda_{T_2}^{\{2\}} - \Lambda_{\max(T_1, T_2)}^{\{1,2\}}\right) |\mathcal{F}_t\right],$$
$$\text{for all } T_1, T_2 \geq 0,$$

where $\Lambda_T^{\pi} \triangleq \int_0^T \lambda^{\pi}(X_t)\,dt$, for $\pi \in \Pi_{\mathbf{n}}$. We shall compute each term in turn. We start with the survival case.

$$\begin{aligned}\frac{\mathbb{P}(\tau_1 > T, \tau_2 > t \,|\mathcal{F}_t)}{\mathbb{P}(\tau_1 > t, \tau_2 > t \,|\mathcal{F}_t)} &= \frac{\mathbb{E}\left[\exp\left(-\Lambda_T^{\{1\}} - \Lambda_t^{\{2\}} - \Lambda_T^{\{1,2\}}\right)|\mathcal{F}_t\right]}{\mathbb{E}\left[\exp\left(-\Lambda_t^{\{1\}} - \Lambda_t^{\{2\}} - \Lambda_t^{\{1,2\}}\right)|\mathcal{F}_t\right]} \\ &= \frac{\mathbb{E}\left[\exp\left(-\Lambda_T^{\{1\}} - \Lambda_t^{\{2\}} - \Lambda_T^{\{1,2\}}\right)|\mathcal{F}_t\right]}{\exp\left(-\Lambda_t^{\{1\}} - \Lambda_t^{\{2\}} - \Lambda_t^{\{1,2\}}\right)} \\ &= \mathbb{E}\left[\frac{\exp\left(-\Lambda_T^{\{1\}} - \Lambda_t^{\{2\}} - \Lambda_T^{\{1,2\}}\right)}{\exp\left(-\Lambda_t^{\{1\}} - \Lambda_t^{\{2\}} - \Lambda_t^{\{1,2\}}\right)}|\mathcal{F}_t\right] \\ &= \mathbb{E}\left[\exp\left(-\left(\Lambda_T^{\{1\}} - \Lambda_t^{\{1\}}\right) - \left(\Lambda_T^{\{1,2\}} - \Lambda_t^{\{1,2\}}\right)\right)|\mathcal{F}_t\right] \\ &= \mathbb{E}\left[\exp\left(-\left(\Lambda_T^1 - \Lambda_t^1\right)\right)|\mathcal{F}_t\right],\end{aligned}$$

the first equality follows from the multivariate probability function; the second and third equalities are due to the fact that $\Lambda_t^{\{1\}}$, $\Lambda_t^{\{2\}}$, and $\Lambda_t^{\{1,2\}}$ are $\mathcal{F}_t$-measurable; the last equality is due to the fatal shock representation: $\lambda^1(X_t) = \lambda^{\{1\}}(X_t) + \lambda^{\{1,2\}}(X_t)$.

For the default case, let us compute the conditional probability on the set $\mathbf{1}_{\{\tau_1 > t\}}\mathbf{1}_{\{\tau_2 \in ds\}}$, for $s \le t$,

$$\frac{\mathbb{P}(\tau_1 > T, \tau_2 \in ds \,|\mathcal{F}_t)}{\mathbb{P}(\tau_1 > t, \tau_2 \in ds \,|\mathcal{F}_t)}$$

The numerator is computed as follows: for $\epsilon > 0$,

$$\begin{aligned}&\mathbb{P}(\tau_1 > T, \tau_2 \in (s-\epsilon, s]\,|\mathcal{F}_t) \\ &= \mathbb{P}(\tau_1 > T, \tau_2 > s\,|\mathcal{F}_t) - \mathbb{P}^*(\tau_1 > T, \tau_2 > s-\epsilon\,|\mathcal{F}_t) \\ &= \mathbb{E}\left[\exp\left(-\Lambda_T^{\{1\}} - \Lambda_s^{\{2\}} - \Lambda_T^{\{1,2\}}\right)|\mathcal{F}_t\right] - \mathbb{E}\left[\exp\left(-\Lambda_T^{\{1\}} - \Lambda_{s-\epsilon}^{\{2\}} - \Lambda_T^{\{1,2\}}\right)|\mathcal{F}_t\right] \\ &= \mathbb{E}\left[\exp\left(-\Lambda_T^{\{1\}} - \Lambda_s^{\{2\}} - \Lambda_T^{\{1,2\}}\right)\left(1 - \exp\left(\Lambda_s^{\{2\}} - \Lambda_{s-\epsilon}^{\{2\}}\right)\right)|\mathcal{F}_t\right] \\ &= \mathbb{E}\left[\left[-\lambda^{\{2\}}(X_s)\,\epsilon\right]\exp\left(-\Lambda_T^{\{1\}} - \Lambda_s^{\{2\}} - \Lambda_T^{\{1,2\}}\right)|\mathcal{F}_t\right] + o(\epsilon).\end{aligned}$$

Similarly, we have for the denominator

$$\begin{aligned}\mathbb{P}(\tau_1 > t, \tau_2 \in (s-\epsilon, s]\,|\mathcal{F}_t) &= \mathbb{E}\Big[\Big[-\lambda^{\{2\}}(X_s)\,\epsilon\Big] \\ &\quad \exp\Big(-\Lambda_t^{\{1\}} - \Lambda_s^{\{2\}} - \Lambda_t^{\{1,2\}}\Big)\,|\mathcal{F}_t\Big] + o(\epsilon). \\ &= \Big[-\lambda^{\{2\}}(X_s)\,\epsilon\Big] \\ &\quad \exp\Big(-\Lambda_t^{\{1\}} - \Lambda_s^{\{2\}} - \Lambda_t^{\{1,2\}}\Big) + o(\epsilon).\end{aligned}$$

Hence,

$$\begin{aligned}\frac{\mathbb{P}(\tau_1 > T, \tau_2 \in ds\,|\mathcal{F}_t)}{\mathbb{P}(\tau_1 > t, \tau_2 \in ds\,|\mathcal{F}_t)} &= \frac{\mathbb{E}\Big[-\lambda^{\{2\}}(X_s)\exp\Big(-\Lambda_T^{\{1\}} - \Lambda_s^{\{2\}} - \Lambda_T^{\{1,2\}}\Big)\,|\mathcal{F}_t\Big]}{\big[-\lambda^{\{2\}}(X_s)\big]\exp\Big(-\Lambda_t^{\{1\}} - \Lambda_s^{\{2\}} - \Lambda_t^{\{1,2\}}\Big)} \\ &= \mathbb{E}\left[\frac{\exp\Big(-\Lambda_T^{\{1\}} - \Lambda_T^{\{1,2\}}\Big)}{\exp\Big(-\Lambda_t^{\{1\}} - \Lambda_t^{\{1,2\}}\Big)}\,|\mathcal{F}_t\right] \\ &= \mathbb{E}\left[\exp\left(-\left(\Lambda_T^1 - \Lambda_t^1\right)\right)|\mathcal{F}_t\right].\end{aligned}$$

Therefore, the conditional survival probability does not jump upon the default of other obligors:

$$S_t^i = \left(1 - D_t^i\right)\mathbb{E}\left[\exp\left(-\int_t^T \lambda^i(X_s)\,ds\right)|\mathcal{F}_t\right].$$

■

Applying Itô's lemma and using the Markovian property of X, we find an explicit expression of the martingale representation of the price process S^i.

Proposition 144 (Single-name price process representation) *We have*

$$\begin{aligned}S_t^i = S_0^i &- \int_0^t \int_E s^i(s, X_s)\,\mathbf{1}_{\{i\in z\}}\left(\mu(ds\times dz) - \lambda_s(dz)\,ds\right) \\ &+ \int_0^t \left(1 - D_s^i\right)\sum_{j=1}^d \sum_{k=1}^d \frac{\partial s^i(s, X_s)}{\partial x_j}\beta_{jk}(X_s)\,dW_s^k,\end{aligned}$$

where $s^i(t,x) : [0,T] \times \mathbb{R}^d \to \mathbb{R}$ *is defined as*

$$s^i(t,x) \triangleq \mathbb{E}_{(t,x)}\left[\exp\left(-\int_t^T \lambda^i(X_s)\,ds\right)\right].$$

Proof The value of the (discounted) zero-coupon single-name bond is

$$S_t^i = \left(1 - D_t^i\right)\mathbb{E}\left[\exp\left(-\int_t^T \lambda^i(X_s)\,ds\right)|\mathcal{F}_t\right].$$

Given the Markovian property of the background process X, we have

$$S_t^i = \left(1 - D_t^i\right)s^i(t, X_t),$$

where $s^i(t,x) : [0,T] \times \mathbb{R}^d \to \mathbb{R}$ is

$$s^i(t,x) \triangleq \mathbb{E}_{(t,x)}\left[\exp\left(-\int_t^T \lambda^i(X_s)\,ds\right)\right].$$

We assume that the function s^i is sufficiently smooth, in particular, that it is continuous, C^1 in the first argument and C^2 in the second argument. X is an $\mathbb{R}^d$-valued diffusion process with drift vector $\alpha(x)$, diffusion matrix $\beta(x)$, and infinitesimal generator

$$\mathcal{A}_t F(x) \triangleq \frac{1}{2}\sum_{j=1}^d \sum_{k=1}^d \sum_{l=1}^d \beta_{jl}(x)\,\beta_{kl}(x)\,\frac{\partial^2 F(x)}{\partial x_j \partial x_k} + \sum_{j=1}^d \alpha_j(x)\,\frac{\partial F(x)}{\partial x_j}.$$

Using the fact that $L_t \triangleq \mathbb{E}\left[\exp\left(-\int_0^T \lambda^i(X_s)\,ds\right)|\mathcal{F}_t\right]$ is an $\{\mathcal{F}_t\}$-martingale, we find that the function s^i is solution of the Feyman-Kac equation

$$-\lambda^i(x)\,s^i(t,x) + \frac{\partial s^i(t,x)}{\partial t} + \mathcal{A}_t s^i(t,x) = 0, \text{ for } (t,x) \in [0,T] \times \mathbb{R}^d,$$

$$s^i(T,x) = 1, \text{ for } x \in \mathbb{R}^d.$$

Applying Itô's lemma to S_t^i, we get

$$\begin{aligned} dS_t^i &= -dD_t^i s^i(t, X_t) + \left(1 - D_t^i\right) ds^i(t, X_t) \\ &= -dD_t^i s^i(t, X_t) + \left(1 - D_t^i\right)\left[\left(\frac{\partial s^i(t, X_t)}{\partial t} + \mathcal{A}_t s^i(t, X_t)\right) dt \right. \\ &\quad \left. + \sum_{j=1}^{d}\sum_{k=1}^{d} \frac{\partial s^i(t, X_t)}{\partial x_j} \beta_{jk}(X_t)\, dW_t^k \right]. \end{aligned}$$

Replacing the term in dt with the Feyman-Kac equation, we get

$$dS_t^i = -dM_t^i s^i(t, X_t) + \left(1 - D_t^i\right)\left[\sum_{j=1}^{d}\sum_{k=1}^{d} \frac{\partial s^i(t, X_t)}{\partial x_j} \beta_{jk}(X_t)\, dW_t^k\right],$$

where M^i is the compensated martingale $M_t^i \triangleq D_t^i - \int_0^{t \wedge \tau_i} \lambda^i(X_s)\, ds$. Using the marked point process representation,

$$M_t^i = \int_0^t \int_E \mathbf{1}_{\{i \in z\}} \left(\mu(ds \times dz) - \lambda_s(dz)\, ds\right),$$

we arrive at the result. ■

Corollary 145 *The dynamics of the zero-coupon defaultable bonds are given by*

$$dS_t^i = S_{t-}^i \left(\mu_t^i dt + \left(\sigma_t^i\right)^{tr} dW_t - dM_t^i\right),$$

where the drift and volatility processes μ_t^i and σ_t^i are

$$\begin{aligned} \mu_t^i &= 0, \\ \left(\sigma_t^i\right)^k &= \left[\sum_{j=1}^{d} \frac{\partial \log s^i(t, X_t)}{\partial x_j} \beta_{jk}(X_t)\right], \textit{for } k = 1, \ldots, d. \end{aligned}$$

17.6 Martingale Representation

In this section, we derive a martingale representation of the $\{\mathcal{G}_t\}$-martingale $H_t = \mathbb{E}\left[H_T \,|\mathcal{G}_t\right]$.

To this end, we use the fatal shock representation of our model in conjunction with the martingale representation result for marked point processes.

The agents' information structure is modelled by the filtered probability space $(\Omega, \mathcal{G}, \{\mathcal{G}_t\}, P^*)$, where $\{\mathcal{G}_t\}$ is the natural filtration generated by the d-dimensional Brownian motion W and the marked point process $\mu(dt \times dz)$ with the $(P^*, \{\mathcal{G}_t\})$-intensity kernel $\lambda_t(dz)$. The Martingale Representation Theorem (see Jacod and Shiryaev 1987, Chap III, Corollary 4.31) shows that the martingale generator in this economy is $\left(W, (\mu(dt \times \{z\}) - \lambda_t(\{z\}))_{z \in \Pi_{\mathbf{n}}}\right)$.

Proposition 146 (Martingale representation of H_t) *The $\{\mathcal{G}_t\}$-martingale $H_t = \mathbb{E}\left[H_T \,| \mathcal{G}_t\right]$, $t \in \left[0, T^*\right]$, where H_T is a $\mathcal{G}_T$-measurable random variable, integrable with respect to $\mathbb{P}$, admits the following integral representation*

$$H_t = H_0 + \int_0^t (\xi_s)^{tr} \, dW_s - \int_0^t \int_E \zeta(s, z) \left(\mu(ds \times dz) - \lambda_s(dz) \, ds\right), \tag{17.17}$$

where ξ is a d-dimensional $\{\mathcal{G}_t\}$-predictable process and $\zeta(s, z)$ is an E-indexed $\{\mathcal{G}_t\}$-predictable process $\zeta(s, z)$ such that

$$\int_0^t \|\xi_s\|^2 \, ds < \infty, \quad \int_0^t \int_E \zeta(s, z) \, \lambda_s(dz) \, ds < \infty,$$

almost surely.

This can be written as

$$H_t = H_0 + \int_0^t (\xi_s)^{tr} \, dW_s - \sum_{\pi \in \Pi_{\mathbf{n}}} \int_{]0,t]} \zeta(s, \pi) \, dM_s^{\pi}. \tag{17.18}$$

In order to replicate the claim H_T, one needs to match the diffusion terms ξ_s^i, $1 \leq i \leq d$, and the jump-to-default terms $[-\zeta(s, \pi)]$ for each possible default state $\pi \in \Pi_{\mathbf{n}}$.

17.7 Computing the Hedging Strategy: The Main Result

In this section, we use the martingale representation of Proposition 146 to derive the risk-minimizing hedging strategy. This is equivalent to finding the

Kunita-Watanabe decomposition of H_t:

$$H_T = H_0 + \int_{]0,T]} (\alpha_t)^{tr} \, dS_t + L_T. \qquad (17.19)$$

Our goal is to establish an analytical result, which derives single-name hedges $(\alpha^i)_{1 \le i \le n}$ in terms of the martingale representation predictable processes ξ and $\zeta(., \pi)$, $\pi \in \Pi_{\mathbf{n}}$.

As shown in Föllmer and Schweizer (1991), the strategy $(\alpha_t)_{t \ge 0}$ can be computed as

$$\alpha_t = d\langle S \rangle_t^{-1} d\langle S, V(\alpha) \rangle_t, \qquad (17.20)$$

where the value process is given by

$$V_t(\alpha) = H_t = \mathbb{E}\left[H_T \,|\mathcal{G}_t \right], \text{ for } t \in [0, T]. \qquad (17.21)$$

This follows from the Kunita-Watanabe decomposition of H and the projection of $V_t(\alpha)$ on the martingale $\int_{]0,t]} (\alpha_s)^{tr} \, dS_s$.

Theorem 147 (Risk-minimizing hedging strategy) *The risk-minimizing hedging strategy of a general (basket) contingent claim with single name instruments is given by the solution of the following linear system,*

for $1 \le k \le n$,

$$\sum_{i=1}^{n} \alpha_t^i S_{t-}^i \left[\left(\sigma_t^i\right)^{tr} \sigma_t^k + \int_E \mathbf{1}_{\{i \in z\}} \mathbf{1}_{\{k \in z\}} \lambda_t(dz) \right]$$
$$= \left(\sigma_t^k\right)^{tr} \xi_t + \int_E \zeta(t, z) \, \mathbf{1}_{\{k \in z\}} \lambda_t(dz).$$

Proof Using the single-name instrument representation of Corollary 145, we have

$$\left(S_t^i\right) = \int_0^t S_{s-}^i \left(\left(\sigma_s^i\right)^{tr} dW_s - \int_E \mathbf{1}_{\{i \in z\}} \left(\mu(ds \times dz) - \lambda_s(dz) \, ds\right) \right),$$

and the predictable covariance is

$$d\langle S \rangle_t^{i,j} = d\left\langle S^i, S^j \right\rangle_t = S_{t-}^i S_{t-}^j \left(\left(\sigma_t^i\right)^{tr} \sigma_t^j + \int_E \mathbf{1}_{\{i \in z\}} \mathbf{1}_{\{j \in z\}} \lambda_t(dz) \right) dt.$$

The value process $V_t(\alpha) = \mathbb{E}\left[H_T \,|\mathcal{G}_t\right]$ is given by the martingale representation

$$V_t(\alpha) = \mathbb{E}\left[H_T \,|\mathcal{G}_t\right] = H_0 + \int_0^t (\xi_s)^{tr} \, dW_s - \int_0^t \int_E \zeta(s,z) \left(\mu(ds \times dz) - \lambda_s(dz)\, ds\right).$$

Hence, we have

$$d\left\langle S, V(\alpha)\right\rangle_t^i = S_{t-}^i \left(\left(\sigma_t^i\right)^{tr} \xi_t + \int_E \mathbf{1}_{\{i \in z\}} \zeta(t,z)\, \lambda_t(dz)\right) dt,$$

and the strategy $(\alpha_t)_{t \geq 0}$ is given by the solution of the following system

$$\sum_{i=1}^n \alpha_t^i \left[S_{t-}^i S_{t-}^k \left[\left(\sigma_t^i\right)^{tr} \sigma_t^k + \int_E \mathbf{1}_{\{i \in z\}} \mathbf{1}_{\{k \in z\}} \lambda_t(dz)\right]\right]$$
$$= S_{t-}^k \left(\sigma_t^k\right)^{tr} \xi_t + \int_E \zeta(t,z)\, S_{t-}^k \mathbf{1}_{\{k \in z\}} \lambda_t(dz).$$

■

Proposition 144 establishes the martingale representation for the single-name securities whose payoff is $H_T = 1 - D_T^i$:

$$\zeta^{1-D_T^i}(t,z) = \mathbf{1}_{\{i \in z\}} s^i(t, X_t), \text{ for } z \in \Pi_{\mathbf{n}},$$
$$\left(\xi_t^{1-D_T^i}\right)^k = \left(1 - D_t^i\right) \sum_{j=1}^d \frac{\partial s^i(t, X_t)}{\partial x_j} \beta_{jk}(X_t).$$

The hedging strategy is solution of
for $1 \leq k \leq n$

$$\sum_{i=1}^n \alpha_t^i \left[\int_E \zeta^{1-D_T^k}(t,z)\, \zeta^{1-D_T^i}(t,z)\, \lambda_t(dz) + \left(\xi_t^{1-D_T^k}\right)^{tr} \xi_t^{1-D_T^i}\right]$$
$$= \int_E \zeta(t,z)\, \zeta^{1-D_T^k}(t,z)\, \lambda_t(dz) + \left(\xi_t^{1-D_T^k}\right)^{tr} \xi_t.$$

Note that this problem combines both default risk and spread risk.

Application. We consider a first-to-default (basket) contingent claim whose payoff is

$$H_T^{(1)} \triangleq \prod_{i=1}^n \left(1 - D_T^i\right).$$

The price of this claim at time t is

$$H_t^{(1)} = \mathbb{E}\left[\prod_{i=1}^{n}\left(1 - D_T^i\right) | \mathcal{G}_t\right].$$

We can show that it can be expressed as

$$H_t^{(1)} = \left[\prod_{i=1}^{n}\left(1 - D_t^i\right)\right] h^{(1)}\left(t, X_t\right),$$

where the function $h^{(1)}\left(t, x\right) : [0, T] \times \mathbb{R}^d \to \mathbb{R}$ is defined as

$$h^{(1)}\left(t, X_t\right) = \mathbb{E}_{(t,x)}\left[\exp\left(-\int_t^T \lambda^{(1)}\left(X_s\right) ds\right)\right],$$

$$\lambda^{(1)}\left(X_t\right) = \sum_{j=1}^{m}\left[1 - \prod_{i=1}^{n}\left(1 - p^{i,j}\left(X_t\right)\right)\right] \lambda^{c_j}\left(X_t\right).$$

Using Itô's lemma and some algebra, we find

$$\begin{aligned} dH_t^{(1)} &= -\int_E h^{(1)}\left(t, X_t\right)\left(\mu\left(dt \times dz\right) - \lambda_t\left(dz\right) dt\right) \\ &+ \left[\prod_{i=1}^{n}\left(1 - D_t^i\right)\right] \sum_{j=1}^{d}\sum_{k=1}^{d} \frac{\partial h^{(1)}\left(t, X_t\right)}{\partial x_j} \beta_{jk}\left(X_t\right) dW_t^k. \end{aligned}$$

This gives the processes of the martingale representation

$$\zeta^{H_T^{(1)}}\left(t, z\right) = h^{(1)}\left(t, X_t\right), \text{ for all } z \in \Pi_{\mathbf{n}},$$

$$\left(\xi_s^{H_T^{(1)}}\right)^k = \left[\prod_{i=1}^{n}\left(1 - D_t^i\right)\right] \sum_{j=1}^{d} \frac{\partial h^{(1)}\left(t, X_t\right)}{\partial x_j} \beta_{jk}\left(X_t\right),$$

which can be plugged into the linear system of Theorem 147. Inverting this latter gives the single-name hedge ratios of the first-to-default basket claim.

17.8 Conclusion

The problem of hedging basket credit derivatives with single name instruments is a very interesting challenge both for academics and practitioners. In this chapter, we have presented a solution based on a risk-minimization approach. We have seen that, in credit markets, we have two sources of uncertainty: spread risk and default risk. We have addressed both types of risk and we have shown how to derive the single name hedge ratios by solving a quadratic minimization problem.

The explosive nature of the default space representation will probably be one of the limiting factors that need to be addressed. In the pricing problem, questions of numerical efficiency were handled by Fourier transform techniques and recursion methods borrowed from actuarial mathematics. These techniques might also prove to be very useful for the hedging problem. One could also consider other hedging approaches, such as quantile hedging (see Föllmer and Leukert 1999), and do a numerical comparison of the effectiveness of each strategy in different market environments.

References

J. Cvitanic, Nonlinear financial markets: hedging and portfolio optimization, in *Mathematics of Derivative Securities*, ed. by M.A.H Dempster, S.R. Pliska (Cambridge University Press, Cambridge), pp. 227–254

M.H.A. Davis, Option pricing in incomplete markets, in Mathematics of Derivative Securities, ed. by M.A.H Dempster, S.R. Pliska (Cambridge University Press, Cambridge, 1997), pp. 216–226

N. El Karoui, M.C. Quenez, Dynamic programming and pricing of contingent claims in an incomplete market. SIAM J. Control Optim. **33**, 29–66 (1995)

H. Föllmer, P. Leukert, Quantile hedging. Financ. Stochast. **3**, 251–274 (1999)

H. Föllmer, D. Sondermann, Hedging of Non-redundant Contingent Claims, in *Contributions to Mathematical Economics*, ed. by W. Hildenbrand, A. Mas-Colell (North-Holland, Amsterdam, 1986), pp. 205–223

H. Föllmer, M. Schweizer, Hedging of contingent claims under incomplete information, in *Applied Stochastic Analysis*, ed. by M.H.A. Davis, R.J. Elliott (Gordon and Breach, London, 1991), pp. 389–414

J. Jacod, A. Shiryaev, *Limit Theorems for Stochastic Processes* (Springer, New York, 1987)

F. Lindskog, A. McNeil, Common Poisson Shock models: applications to insurance and credit risk modelling. ASTIN Bullet. **33**(2), 209–238 (2003)

M. Schweizer, Variance-optimal hedging in discrete time. Math. Oper. Res. **20**, 1–32 (1993)

M. Schweizer, Approximating random variables by stochastic integrals. Ann. Probab. **22**(3), 1536–1575 (1994)

18

Min-Variance Hedging with Carry

In this chapter, we present the construction of the Min-Variance Hedging Delta operator used for basket products. Because of the market incompleteness—i.e., we cannot replicate a basket product with its underlying default swaps—min-variance hedging is the best thing that we can hope for. There will always be a residual correlation risk orthogonal to the sub-space of hedging instruments. We also present an extension of the standard MVH optimization to take into account the drift mismatch between the basket and the hedging portfolio. This defines the "Min-Variance Hedging Deltas with Carry".

18.1 Introduction

We consider the pricing of basket FTD (or NTD) products in a (Marshall-Olkin) common Poisson shock framework. In this model, the joint defaults are defined through the specification of a set of common market factor events, which can trigger the default of specific individual obligors. The probability of default of an issuer conditional on the occurrence of the common factor event is referred to as the loading on the market factor. It is a measure of the leverage of a particular credit with respect to this common market factor. The market factors, in this construction, are assumed to be independent Poisson shocks. The independence here is default independence: i.e., two market factors cannot default at the same time. This decomposition in terms of independent events translates directly into a canonical decomposition of the single-name intensities.

Y. Elouerkhaoui, *Credit Correlation*, Applied Quantitative Finance,
DOI 10.1007/978-3-319-60973-7_18

The common Poisson shock canonical decomposition, for each single-name i, is given by

$$\lambda_i (t) = \sum_{j=1}^{m} p_{ij} \lambda_j^C (t) + \lambda_i^0 (t),$$

where $(\lambda_i)_{1 \leq i \leq n}$, $\left(\lambda_j^C\right)_{1 \leq j \leq m}$ and $\left(\lambda_i^0\right)_{1 \leq i \leq n}$ are the intensities of the single-names, the market factors and the idiosyncratic events respectively. The matrix $\left[p_{ij}\right]_{\substack{1 \leq i \leq n \\ 1 \leq j \leq m}}$ is the matrix of single-name loadings on each market factor, i.e., conditional on a market shock of type j, the single-name issuer i would default with probability p_{ij}.

We can re-write this decomposition in a compact form where we do not distinguish between the common factors and the idiosyncratic events

$$\lambda_i (t) = \sum_{j=1}^{m+n} p_{ij} \Lambda_j (t),$$

where we set the first m factors to the common drivers, $\Lambda_j (t) = \lambda_j^C (t)$, for $1 \leq j \leq m$, and the next n factors to the idiosyncratic ones, $\Lambda_{m+i} (t) = \lambda_i^0 (t)$, for $1 \leq i \leq n$; the corresponding factor loadings are set accordingly as $p_{i,m+i} = \delta_{ij}$ ($\delta_{ij} = 1$ if $i = j$ and $\delta_{ij} = 0$ otherwise).

The value of a basket default swap is a function of the common market factors intensities $\left(\lambda_j^C\right)_{1 \leq j \leq m}$ and the idiosyncratic intensities $\left(\lambda_i^0\right)_{1 \leq i \leq n}$. Then, the natural deltas that one can define are the partial derivatives with respect to the market factors and the idiosyncratic terms. The market factor delta would, in some sense, quantify the correlation risk; and the idiosyncratic deltas would quantify the issuer specific risk.

First, we define the survival probability Q-factors associated with the Poisson drivers' intensities $\Lambda_j (t)$ as follows

$$Q_{t,T}^{\Lambda_j} = \mathbb{E}\left[\exp\left(-\int_t^T \Lambda_j (s)\, ds\right) |\mathcal{F}_t\right],$$

where the conditioning is taking with respect to the background state variables' filtration $\mathcal{F}_t$.

The associated zero-coupon intensity can then be defined in the usual way as

$$Z_{t,T}^{\Lambda_j} = -\frac{\log\left(Q_{t,T}^{\Lambda_j}\right)}{T-t}.$$

Hence, the value of the basket default swap will be a function of the risk-free discount factors $\left(p_{0,t}\right)_{0\le t\le T}$ and the credit drivers' Q-factors :

$$B = B\left(\left(p_{t,T}\right)_{t\le T\le T^*}, \left(Q_{t,T}^{\Lambda_1}\right)_{t\le T\le T^*}, \ldots, \left(Q_{t,T}^{\Lambda_{m+n}}\right)_{t\le T\le T^*}\right).$$

We can, therefore, define the market factor credit deltas as the standard partial derivatives of this function with respect to the Q-factor zero intensity.

Definition 148 (*Market Factor Credit Deltas*) We define the market factor credit deltas as the partial derivatives of the basket default swap value with respect to the market factor zero-coupon intensities

$$\begin{aligned}\Delta_T^{\Lambda_k} &= \frac{\partial}{\partial Z_{t,T}^{Q^{\Lambda_k}}} B\left(\left(p_{t,T}\right)_{t\le T\le T^*}, \left(Q_{t,T}^{\Lambda_1}\right)_{t\le T\le T^*}, \ldots, \left(Q_{t,T}^{\Lambda_{m+n}}\right)_{t\le T\le T^*}\right)\\ &= -\left(T-t\right) Q_{t,T}^{\Lambda_k} \frac{\partial}{\partial Q_{t,T}^{\Lambda_k}} B\left(\left(p_{t,T}\right)_{t\le T\le T^*}, \left(Q_{t,T}^{\Lambda_1}\right)_{t\le T\le T^*}, \ldots, \left(Q_{t,T}^{\Lambda_{m+n}}\right)_{t\le T\le T^*}\right).\end{aligned}$$

Now, the question is: how can we aggregate these market factor credit deltas to produce a delta with respect to the original issuer credit curve intensities. Clearly, a naive derivation with respect to the original $\lambda_i\left(T\right)$ is not possible. A change in the common market factor of $\lambda_j^C \to \lambda_j^C + \delta\lambda$, $1 \le j \le m$, will imply a change in the original intensity of $\lambda_i \to \lambda_i + p_{ij}\delta\lambda$; and similarly, a change in the idiosyncratic term of $\lambda_i^0 \to \lambda_i^0 + \delta\lambda$ will imply a change in the original intensity of $\lambda_i \to \lambda_i + \delta\lambda$. We do not have a one to one mapping between the change in the original spread and the change in value of the basket default swap. In this chapter, we propose a way to solve this problem through the formalism of "Min-Variance Hedging".

18.2 Min-Variance Hedging

The meaning of a "Delta", in the original Black-Scholes world, is, fundamentally, a hedge ratio measure: i.e., the amount of underlying asset that one needs to have in the dynamic hedging portfolio in order to replicate the payoff of the option. By no-arbitrage arguments, the value of the option is then equal to the value of the replicating portfolio. Incidentally, in this arbitrage-free complete market, it turns out that the delta, defined as a hedge ratio, happens to be equal to the partial derivative of the option value with respect to the underlying. By extension, deltas were generally defined as partial derivatives with respect to an underlying variable.

In the basket default swap problem, the situation is very similar. We sell a basket derivative and we hedge it with the underlying single-name credit default swaps. The main difference here is that the market is incomplete. We cannot replicate the payoff of the basket with its underlying default swaps. The correlation risk is completely orthogonal to the hyper-plane generated by the default swaps. The hedging problem becomes, actually, a projection of the basket on the default swaps' hyper-plane. That is the best that we can do, and that is the most that we can hope for.

The projection on the space of contingent claims endowed with the square-root-of-variance norm is known as "Min-Variance Hedging" (Fig. 18.1).

Theorem 149 (The Min-Variance Hedging Delta Operator) *The min-variance hedging delta operator, defined as the single-name credit delta which minimizes the variance of basket hedged portfolio, is given by*

$$\Delta_T^{\lambda_i} = \frac{\partial^{MVH}}{\partial Z_{t,T}^{Q^{\lambda_i}}} B\left((p_{t,T})_{t \le T \le T^*}, \left(Q_{t,T}^{\Lambda_j} \right)_{t \le T \le T^*}^{1 \le j \le m+n} \right) = \alpha_i^*,$$

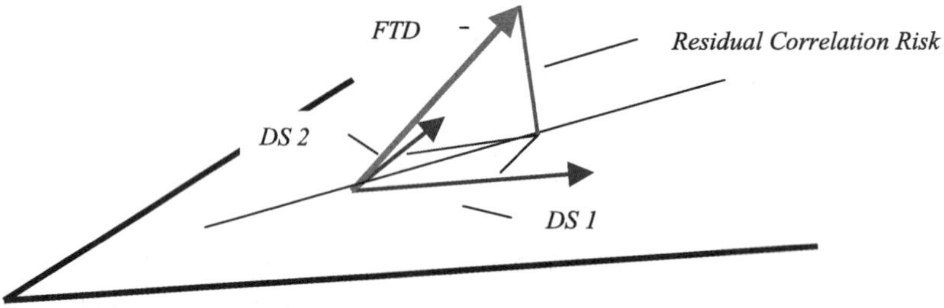

Fig. 18.1 Projection of a two-name FTD on the hyper-plane generated by its underlying default swaps DS1 and DS2. The residual correlation risk is completely orthogonal to the single-name default swaps' hyper-plane

where $\left(\alpha_i^*\right)_{1\le i\le n}$ *is the solution of the following linear system:*

$$\sum_{i=1}^{n}\alpha_i^*\left[\sum_{j=1}^{m+n}p_{ij}p_{kj}Var\left[dZ_{t,T}^{Q^{\Lambda_j}}\right]\right]=\sum_{j=1}^{m+n}\Delta_T^{\Lambda_j}p_{kj}Var\left[dZ_{t,T}^{Q^{\Lambda_j}}\right].$$

Proof Let B, D_1, D_2,..., D_n be the values of the basket default swap and its underlying single-name default swaps respectively. We express these values as a function of $\left(Z_{t,T}^{Q^{\Lambda_j}}\right)_{1\le j\le m+n}$ for a fixed maturity T. Then, we differentiate:

$$dB=\sum_{j=1}^{m+n}\frac{\partial B}{\partial Z_{t,T}^{Q^{\Lambda_j}}}dZ_{t,T}^{Q^{\Lambda_j}}=\sum_{j=1}^{m+n}\Delta_T^{\Lambda_j}dZ_{t,T}^{Q^{\Lambda_j}}.$$

Similarly, for the single-names, we have

$$dD_i=\sum_{j=1}^{m+n}\frac{\partial D_i}{\partial Z_{t,T}^{Q^{\Lambda_j}}}dZ_{t,T}^{Q^{\Lambda_j}}.$$

But a single-name default swap (or any single-name credit product) is, in general, invariant with respect to its market factor decomposition and we have the following relationship between the single-name and market factor shocks:

$$\frac{\partial D_i}{\partial Z_{t,T}^{Q^{\Lambda_j}}}=p_{ij}\frac{\partial D_i}{\partial Z_{t,T}^{Q^{\lambda_i}}}.$$

Hence, the single-name differential is given by

$$dD_i=\frac{\partial D_i}{\partial Z_{t,T}^{Q^{\lambda_i}}}\sum_{j=1}^{m+n}p_{ij}dZ_{t,T}^{Q^{\Lambda_j}}.$$

We construct a portfolio of the basket and its hedge

$$B-\sum_{i=1}^{n}\alpha_iD_i,$$

where $(\alpha_i)_{1\le i\le n}$ are the hedge ratios. And we want to minimize the variance of the change in value of this portfolio

$$\min_{\alpha_i}\left[Var\left[dB-\sum_{i=1}^{n}\alpha_i dD_i\right]\right].$$

We can now define the min-variance hedging deltas as the deltas of the min-variance hedging replicating portfolio on each single-name curve:

$$\alpha_i^* = \alpha_i \frac{\partial D_i}{\partial Z_{t,T}^{Q^{\lambda_i}}}.$$

Replacing the differentials in the expression of the variance and using α_i^*, the problem becomes:

$$\min_{\alpha_i^*}\left[Var\left[\sum_{j=1}^{m+n}\Delta_T^{\Lambda_j} dZ_{t,T}^{Q^{\Lambda_j}} - \sum_{i=1}^{n}\alpha_i^*\sum_{j=1}^{m+n}p_{ij} dZ_{t,T}^{Q^{\Lambda_j}}\right]\right].$$

Assuming independence between the market factors: $Cov\left[dZ_{t,T}^{Q^{\Lambda_j}}, dZ_{t,T}^{Q^{\Lambda_k}}\right] = 0$, for $j \neq k$, we get

$$\min_{\alpha_i^*}\left[\sum_{j=1}^{m+n}\left(\Delta_T^{\Lambda_j} - \sum_{i=1}^{n}\alpha_i^* p_{ij}\right)^2 Var\left[dZ_{t,T}^{Q^{\Lambda_j}}\right]\right].$$

Then, setting the partial derivatives with respect to α_i^* to zero yields: for $1 \leq k \leq n$,

$$0 = \frac{\partial}{\partial \alpha_k^*} = \left[-2\sum_{j=1}^{m+n}p_{kj}\left(\Delta_T^{\Lambda_j} - \sum_{i=1}^{n}\alpha_i^* p_{ij}\right) Var\left[dZ_{t,T}^{Q^{\Lambda_j}}\right]\right].$$

Finally, separating the terms gives the solution:

$$\sum_{i=1}^{n}\alpha_i^*\left[\sum_{j=1}^{m+n}p_{ij}p_{kj} Var\left[dZ_{t,T}^{Q^{\Lambda_j}}\right]\right] = \sum_{j=1}^{m+n}\Delta_t^{\Lambda_j} p_{kj} Var\left[dZ_{t,T}^{Q^{\Lambda_j}}\right].$$

■

Remark 150 Assuming that the market factor intensities follow a log-normal diffusion

$$dZ_{t,T}^{Q^{\Lambda j}} = (\ldots)\,dt + \sigma_{t,T}^{j} Z_{t,T}^{Q^{\Lambda j}}\, dW_t.$$

The Min-Variance Hedging Delta Operator equation then becomes

$$\sum_{i=1}^{n} \alpha_i^* \left[\sum_{j=1}^{m+n} p_{ij} p_{kj} \left(\sigma_{t,T}^{j} Z_{t,T}^{Q^{\Lambda j}} \right)^2 \right] = \sum_{j=1}^{m+n} \Delta_T^{\Lambda j} p_{kj} \left(\sigma_{t,T}^{j} Z_{t,T}^{Q^{\Lambda j}} \right)^2.$$

Assuming that all the intensities have the same volatility, $\sigma_{t,T}^{j} = \sigma_{t,T}^{k}$ for $j \neq k$, we can simplify by the volatility term and the MVH operator is simply defined by:

$$\sum_{i=1}^{n} \alpha_i^* \left[\sum_{j=1}^{m+n} p_{ij} p_{kj} \left(Z_{t,T}^{Q^{\Lambda j}} \right)^2 \right] = \sum_{j=1}^{m+n} \Delta_T^{\Lambda j} p_{kj} \left(Z_{t,T}^{Q^{\Lambda j}} \right)^2.$$

Remark 151 Note that the Min-Variance Hedging Delta Operator is not linear in T. To linearize the operator for all time buckets, we need fix a time horizon T^*, and then define the operator as follows:

$$\sum_{i=1}^{n} \alpha_i^* \left[\sum_{j=1}^{m+n} p_{ij} p_{kj} \left(Z_{t,T*}^{Q^{\Lambda j}} \right)^2 \right] = \sum_{j=1}^{m+n} \Delta_T^{\Lambda j} p_{kj} \left(Z_{t,T*}^{Q^{\Lambda j}} \right)^2.$$

18.3 Hedge and Carry

When we trade first-to-default basket product we need to do two things: first, hedge the credit spread exposure by trading the underlying single-name default swaps, but we also need to finance the basket default swap position. Financing the basket position means that the carry of the hedge portfolio should be, at least, equal to the carry of the basket default swap. In the previous construction of the MVH deltas, we focused our attention on the first part the problem and completely neglected the second one.

To start with, in this section, we present the origin of the carry term, before showing how to account for it in the hedging problem. We present the carry term from two different angles.

18.3.1 The Carry Term

If we step back and reconsider the first line in the proof of the previous section, we will see that we have omitted the time dependence. In fact, the prices of the basket and the hedge instruments $B, D_1, D_2, \ldots, D_n$ are functions of time t and all the credit Q-factors at different maturities. Then, their differentials are:

$$dB = \sum_T \sum_{j=1}^{m+n} \frac{\partial B}{\partial Z_{t,T}^{Q^{\Lambda_j}}} dZ_{t,T}^{Q^{\Lambda_j}} + \frac{\partial B}{\partial t} dt,$$

and

$$\begin{aligned} dD_i &= \sum_T \sum_{j=1}^{m+n} \frac{\partial D_i}{\partial Z_{t,T}^{Q^{\Lambda_j}}} dZ_{t,T}^{Q^{\Lambda_j}} + \frac{\partial D_i}{\partial t} dt \\ &= \sum_T \sum_{j=1}^{m+n} p_{ij} \frac{\partial D_i}{\partial Z_{t,T}^{Q^{\lambda_i}}} dZ_{t,T}^{Q^{\Lambda_j}} + \frac{\partial D_i}{\partial t} dt. \end{aligned}$$

Substituting in the differential of the hedged portfolio, we obtain

$$\begin{aligned} & dB - \sum_{i=1}^{n} \alpha_i dD_i \\ &= \underbrace{\left[\left[\frac{\partial B}{\partial t} - \sum_{i=1}^{n} \alpha_i \frac{\partial D_i}{\partial t} \right] dt \right]}_{CARRY\ TERM} \\ &+ \underbrace{\left[\sum_T \sum_{j=1}^{m+n} \left[\frac{\partial B}{\partial Z_{t,T}^{Q^{\Lambda_j}}} - \sum_{i=1}^{n} \alpha_i p_{ij} \frac{\partial D_i}{\partial Z_{t,T}^{Q^{\lambda_i}}} \right] dZ_{t,T}^{Q^{\Lambda_j}} \right]}_{SPREAD\ TERM}. \end{aligned}$$

Therefore, what we want to do, actually, is to minimize the variance of the spread term under the constraint that the carry term is zero. This will ensure that we are carry flat at all times. In the traditional fixed income world, this carry term is always equal to zero. Since we work under the risk-neutral measure, the lognormal drift for all instruments is equal to the short rate. And this is true for the hedged portfolio as well, that is the reason why, in fixed income, we are flat-carry by construction.

18.3.2 Drift Mismatch

We can also illustrate the carry term from a different angle. Instead of looking at the "functional" differentials, we can write the dynamics of the instruments under the risk-neutral measure and compare the drift terms. For a fixed income exotic product V_t, and its hedge instruments $(H_i)_{1\leq i\leq n}$, we have dynamics of the form

$$
\begin{aligned}
dV_t &= r_t V_t dt + \Sigma_t^V \cdot dW_t, \\
dH_t^i &= r_t H_t^i dt + \Sigma_t^{H^i} \cdot dW_t,
\end{aligned}
$$

with some volatility processes Σ_t^V and $\Sigma_t^{H^i}$. Then, the hedged portfolio will also diffuse with the same drift

$$
\begin{aligned}
dP_t &= d\left(V_t - \sum_{i=1}^{n} \alpha_i H_t^i\right) = r_t \left(V_t - \sum_{i=1}^{n} \alpha_i H_t^i\right) dt + \Sigma_t^P \cdot dW_t \\
&= r_t P_t dt + \Sigma_t^P \cdot dW_t.
\end{aligned}
$$

For a first-to-default swap, the situation is different. We would have following dynamics for the FTD:

$$
dV_t^{ftd} = \left(r_t + \lambda_t^{ftd}\right) V_t^{ftd} dt + \Sigma_t^{V^{ftd}} \cdot dW_t, \text{ for all } t > \tau^{[1]} = \min_{1\leq i\leq n} (\tau_i);
$$

whereas, the single-name hedge instruments would have different (single-name) drifts

$$
dH_t^i = \left(r_t + \lambda_t^i\right) H_t^i dt + \Sigma_t^{H^i} \cdot dW_t, \text{ for all } t > \tau_i.
$$

Thus, the dynamics of the hedged portfolio are given by

$$dP_t = d\left(V_t^{ftd} - \sum_{i=1}^{n} \alpha_i H_t^i\right)$$

$$= \left[\underbrace{r_t P_t}_{funding} + \underbrace{\left(\lambda_t^{ftd} V_t^{ftd} - \sum_{i=1}^{n} \alpha_i \lambda_t^i H_t^i\right)}_{carry} \right] dt + \Sigma_t^P \cdot dW_t,$$

where the drift term will contain the traditional (risk-free) funding term and the additional carry from the FTD hedged position.

Next, we show how to account for this drift mismatch in the min-variance hedging algorithm.

18.4 Min-Variance Hedging with Carry

Because of the drift term, the min-variance hedging procedure needs to be extended to a minimization of the variance while we are flat carry. The minimization under constraint should be performed globally for the sum of all the time buckets. Recall that the hedged portfolio differentials depend on the sum of all the buckets:

$$dB - \sum_{i=1}^{n} \alpha_i dD_i$$

$$= \underbrace{\left[\left[\frac{\partial B}{\partial t} - \sum_{i=1}^{n} \alpha_i \frac{\partial D_i}{\partial t}\right] dt\right]}_{CARRY\ TERM}$$

$$+ \underbrace{\left[\sum_{j=1}^{m+n} \sum_{T \geq t}^{T^*} \left[\left(\frac{\partial B}{\partial Z_{t,T}^{Q^{\Lambda_j}}}\right) - \sum_{i=1}^{n} \alpha_i p_{ij} \left(\frac{\partial D_i}{\partial Z_{t,T}^{Q^{\lambda_i}}}\right)\right] dZ_{t,T}^{Q^{\Lambda_j}}\right]}_{SPREAD\ TERM}.$$

To generate the bucketed deltas, we will resort to the same trick, as before, that we have used to linearize the MVH operator. We fix a time horizon T^*, and we use the variance of the zero-coupon rate, at time T^*, to build the operator.

The minimization problem becomes:

$$\min_{\alpha_i} \left[\sum_{j=1}^{m+n} \left[\left(\sum_{T \geq t}^{T^*} \frac{\partial B}{\partial Z_{t,T}^{Q^{\Lambda_j}}} \right) - \sum_{i=1}^{n} \alpha_i p_{ij} \left(\sum_{T \geq t}^{T^*} \frac{\partial D_i}{\partial Z_{t,T}^{Q^{\lambda_i}}} \right) \right]^2 Var \left[dZ_{t,T^*}^{Q^{\Lambda_j}} \right] \right].$$

In fact, this corresponds to a parallel shift of the credit curves: for all maturities $t \leq T \leq T^*, dZ_{t,T}^{Q^{\Lambda_j}} = dZ_{t,T^*}^{Q^{\Lambda_j}}$.

We shall see later that the transformation, which maps the deltas of the drivers to the single-name MVHC deltas is linear; and we can use the same mapping to obtain bucketed MVHC deltas from bucketed driver deltas. But, at this point, it suffices to write the constraint in a generic way:

$$\frac{\partial B}{\partial t} - \sum_{i=1}^{n} \alpha_i \frac{\partial D_i}{\partial t} = 0.$$

Heuristically, we can think of the carry constraint as:

$$s^{ftd}(t) \simeq \sum_{i=1}^{n} \alpha_i s_i(t),$$

where $s^{ftd}(t)$ and $s_i(t)$ are the break-even spreads of the first-to-default swaps and the underlying single-name default swaps respectively.

Theorem 152 (The Min-Variance Hedging with Carry Delta Operator) *The min-variance hedging with carry delta operator, defined as the single-name credit delta which minimizes the variance of basket hedged portfolio subject to the flat-carry constraint, is given by*

$$\Delta_T^{\lambda_i} = \frac{\partial^{MVHC}}{\partial Z_{t,T}^{Q^{\lambda_i}}} B\left(\left(p_{t,T}\right)_{t \leq T \leq T^*}, \left(Q_{t,T}^{\Lambda_j}\right)_{t \leq T \leq T^*}^{1 \leq j \leq m+n} \right) = \alpha_i^*(T),$$

where

$$\alpha_i^*(T) = \alpha_i^* \frac{\partial D_i}{\partial Z_{t,T}^{Q^{\lambda_i}}} \left(\sum_{T \geq t}^{T^*} \frac{\partial D_i}{\partial Z_{t,T}^{Q^{\lambda_i}}} \right)^{-1},$$

and $\left(\alpha_i^*\right)_{1\le i\le n}$ *is the solution of the following linear system: for* $1 \le k \le n$,

$$\sum_{i=1}^{n}\alpha_i^*\left[\sum_{j=1}^{m+n} p_{ij}p_{kj}Var\left[dZ_{t,T}^{Q^{\Lambda_j}}\right]\right]+\delta\left[\frac{\beta_k}{2}\right]=\sum_{j=1}^{m+n}\left(\sum_{T\ge t}^{T^*}\frac{\partial B}{\partial Z_{t,T}^{Q^{\Lambda_j}}}\right)p_{kj}Var\left[dZ_{t,T}^{Q^{\Lambda_j}}\right],$$

$$\sum_{i=1}^{n}\alpha_i^*\beta_i=\frac{\partial B}{\partial t},$$

where δ *is the Lagrange multiplier and* β_i *is given by*

$$\beta_i=\frac{\partial D_i}{\partial t}\left(\sum_{T\ge t}^{T^*}\frac{\partial D_i}{\partial Z_{t,T}^{Q^{\lambda_i}}}\right)^{-1}.$$

Proof We start by solving the global minimization problem, then we perform the bucketing by linearizing this operator.

The minimization problem for the sum of deltas across all buckets is:

$$\min_{\alpha_i} Var\left[\sum_{j=1}^{m+n}\left[\left(\sum_{T\ge t}^{T^*}\frac{\partial B}{\partial Z_{t,T}^{Q^{\Lambda_j}}}\right)-\sum_{i=1}^{n}\alpha_i p_{ij}\left(\sum_{T\ge t}^{T^*}\frac{\partial D_i}{\partial Z_{t,T}^{Q^{\lambda_i}}}\right)\right]dZ_{t,T^*}^{Q^{\Lambda_j}}\right],$$

subject to the constraint

$$\frac{\partial B}{\partial t}-\sum_{i=1}^{n}\alpha_i\frac{\partial D_i}{\partial t}=0.$$

We define the MVHC deltas as the deltas of the flat carry mean-variance hedging replicating portfolio on each curve,

$$\alpha_i^*=\alpha_i\left(\sum_{T\ge t}^{T^*}\frac{\partial D_i}{\partial Z_{t,T}^{Q^{\lambda_i}}}\right),$$

and the minimization problem becomes

$$\min_{\alpha_i} Var\left[\sum_{j=1}^{m+n}\left[\left(\sum_{T\ge t}^{T^*}\frac{\partial B}{\partial Z_{t,T}^{Q^{\Lambda_j}}}\right)-\sum_{i=1}^{n}\alpha_i^* p_{ij}\right]dZ_{t,T^*}^{Q^{\Lambda_j}}\right],$$

subject to

$$\frac{\partial B}{\partial t} - \sum_{i=1}^{n} \alpha_i^* \beta_i = 0,$$

where

$$\beta_i = \frac{\partial D_i}{\partial t} \left(\sum_{T \geq t}^{T^*} \frac{\partial D_i}{\partial Z_{t,T}^{Q^{\lambda_i}}} \right)^{-1}.$$

we can solve this minimization under constraint problem using the method of Lagrange multipliers.

Let δ be the Lagrange multiplier; then, we can set all the partial derivatives of the following expression to zero:

$$\left[\sum_{j=1}^{m+n} \left[\left(\sum_{T \geq t}^{T^*} \frac{\partial B}{\partial Z_{t,T}^{Q^{\Lambda_j}}} \right) - \sum_{i=1}^{n} \alpha_i^* p_{ij} \right]^2 Var \left[dZ_{t,T^*}^{Q^{\Lambda_j}} \right] \right] - \delta \left[\frac{\partial B}{\partial t} - \sum_{i=1}^{n} \alpha_i^* \beta_i \right].$$

We obtain: for $1 \leq k \leq n$,

$$0 = \frac{\partial}{\partial \alpha_k^*} = \left[\sum_{j=1}^{m+n} \left[-2 p_{kj} \left[\left(\sum_{T \geq t}^{T^*} \frac{\partial B}{\partial Z_{t,T}^{Q^{\Lambda_j}}} \right) - \sum_{i=1}^{n} \alpha_i^* p_{ij} \right] \right] Var \left[dZ_{t,T^*}^{Q^{\Lambda_j}} \right] \right] - \delta \left[-\beta_k \right].$$

Re-arranging the terms gives the system of linear equations to solve: for $1 \leq k \leq n$,

$$\sum_{i=1}^{n} \alpha_i^* \left[\sum_{j=1}^{m+n} p_{ij} p_{kj} Var \left[dZ_{t,T^*}^{Q^{\Lambda_j}} \right] \right] + \delta \left[\frac{\beta_k}{2} \right] = \sum_{j=1}^{m+n} \left(\sum_{T \geq t}^{T^*} \frac{\partial B}{\partial Z_{t,T}^{Q^{\Lambda_j}}} \right) p_{kj} Var \left[dZ_{t,T^*}^{Q^{\Lambda_j}} \right],$$

and

$$\sum_{i=1}^{n} \alpha_i^* \beta_i = \frac{\partial B}{\partial t}.$$

This defines the solution for the parallel delta, i.e., the sum of all bucketed deltas. To do the allocation to individual buckets, we need to linearize the constraint so that the sum of the bucketed MVHC deltas is equal to the solution of the previous system.

Recall that the definition of the bucketed MVHC delta is the bucketed delta of the MVHC replicating portfolio,

$$\alpha_i^*(T) = \alpha_i \frac{\partial D_i}{\partial Z_{t,T}^{Q^{\lambda_i}}},$$

which can be expressed in terms of the sum of the deltas α_i^*,

$$\alpha_i^*(T) = \alpha_i^* \frac{\partial D_i}{\partial Z_{t,T}^{Q^{\lambda_i}}} \left(\sum_{T \geq t}^{T^*} \frac{\partial D_i}{\partial Z_{t,T}^{Q^{\lambda_i}}} \right)^{-1}.$$

This ensures that the sum of the bucketed deltas is equal to the sum of the deltas by construction. ■

Remark 153 In the construction of the MVH operator, we were able to do an optimization for each bucket separately. This was made possible by looking at the credit curve as a string where each bucket can have an independent string shock from the rest of the curve. The shape of the deformation that we apply comes afterwards in the aggregation of the buckets. Summing all the buckets corresponds to a parallel shift of the curve. A different aggregation corresponds to a different curve movement.

Remark 154 To build the MVHC deltas we have a global optimization constraint that we need to satisfy. That is the reason why we cannot look at each bucket independently from the others any longer. We have to assume the string shock that we want to apply then optimize. In the construction described in this chapter, we have assumed a parallel shock. The generalization to a different curve deformation is straightforward.

18.5 Numerical Examples

In this section, we present a few numerical examples to illustrate the intuitive behaviour of the MVHC procedure.

In the first example depicted in Fig. 18.2, we look at the hedge ratios of a first-to-default swap with three names trading at 100 bps and a default correlation of 12%. Then, we increase the spread of the first name and we keep the other spreads unchanged. In the base case scenario where the three spreads are the same the hedge ratios are similar as well; there is no reason to distinguish one from the others. Obviously, in the limit case where the default correlation is equal to 0% the hedge ratios would be (100%, 100%, 100%). At 12% default correlation, the hedge ratios are (85%, 85%, 85%). As the spread of the first

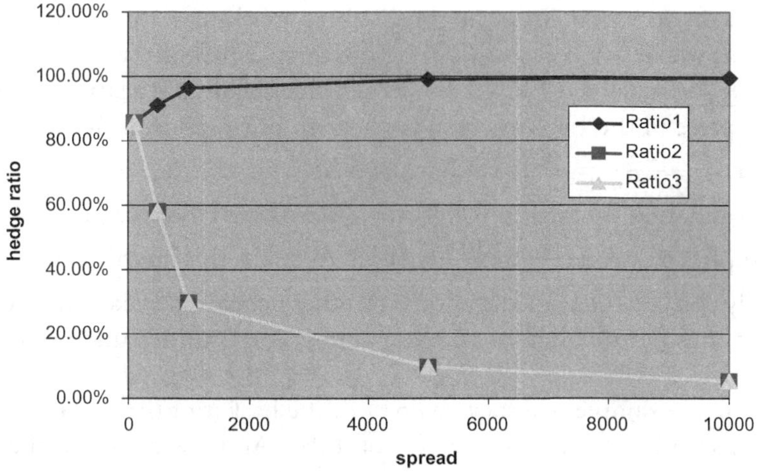

Fig. 18.2 Hedge ratios for the three names in the basket as we increase the spread of the first name

name blows out, the likelihood of default of the first name goes up; this means that we need to hedge more on that name and less on the others. In the extreme case where $s_1 \to +\infty$ the hedge ratios will converge asymptotically to $(100\%, 0\%, 0\%)$.

Figure 18.3 shows how the hedge ratios change with the default correlation. We are using the base case FTD with three names at 100 bps. We change the default inter-dependence between the names and we look at the impact on the

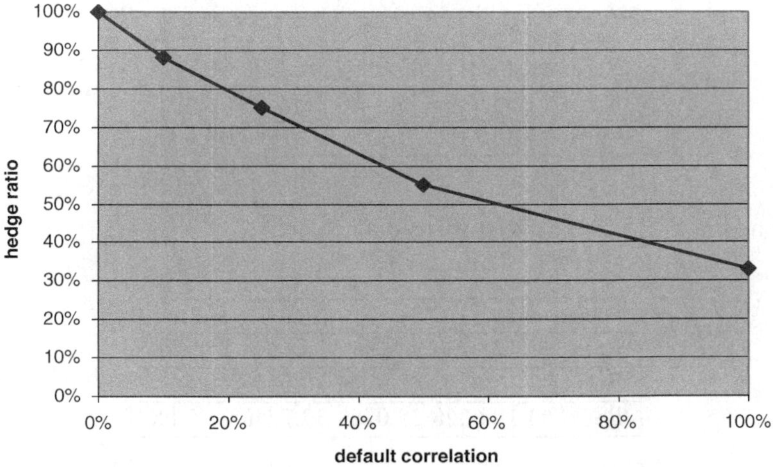

Fig. 18.3 Hedge ratio as a function of default correlation

hedge ratios. Note that since the three names are similar, the hedge ratios are identical as well $\alpha_1 = \alpha_2 = \alpha_3$. Under the carry constraint, we have, roughly, $s^{ftd} \simeq \sum_{i=1}^{n} \alpha_i s_i = \alpha \left[\sum_{i=1}^{n} s_i\right]$. When the default correlation is equal to zero, the price of the FTD is equal to the sum of the spreads 300 bps, which implies $\alpha \simeq \frac{s^{ftd}}{\sum_{i=1}^{n} s_i} = \frac{300}{300} = 1$. At the other end of the spectrum, when the default correlation is equal to 100%, the price of the FTD is equal to 100 bps, which implies $\alpha \simeq \frac{s^{ftd}}{\sum_{i=1}^{n} s_i} = \frac{100}{300} = 0.33$. Actually, this heuristic discussion is not totally correct; there is a duration effect that needs to be taken into account. However, these approximations are useful to get an intuitive understanding of the hedge.

In the next example, we start again with the basket of three names trading at 100 bps and a default correlation of 12%. And we compute the hedge ratios in terms of the underlying default swaps. We look at the dynamics of the hedge ratios for the different basket default swap slices—first-to-default (FTD), second-to-default (STD), third-to-default (TTD)—as one or a few spreads change.

Scenario	1	2	3	4
Curve #1	100	300	300	300
Curve #2	100	100	300	300
Curve #3	100	100	100	300

For the first-to-default, we get the following hedge ratios:

Scenario	1	2	3	4
Ratio #1	85.71%	87.89%	78.28%	68.40%
Ratio #2	85.71%	73.97%	78.28%	68.40%
Ratio #3	85.71%	73.97%	57.19%	68.40%

The hedge ratios for the second-to-default are:

Scenario	1	2	3	4
Ratio #1	12.61%	9.61%	18.64%	25.54%
Ratio #2	12.61%	25.02%	18.64%	25.54%
Ratio #3	12.61%	25.02%	39.77%	25.54%

The hedge ratios for the third-to-default are:

Scenario	1	2	3	4
Ratio #1	1.65%	0.94%	2.21%	6.03%
Ratio #2	1.65%	3.19%	2.21%	6.03%
Ratio #3	1.65%	3.19%	7.72%	6.03%

For the FTD, when one of the spreads goes up, we need to hedge more on that name and less on the others since this name becomes the most likely to default first. For the higher slices, STD and TTD, we need to do exactly the reverse. When a name blows out, we need to hedge less on that name and more on the others. Obviously, if we trade all the three slices FTD, STD and TTD, the perfect hedge would be to do 100% of each single-name credit default swap, which would result in a completely flat position. If we sum up the MVHC hedge ratios on the slices, they do not actually add up to 100%. It is very close but not exactly equal to 100%. The reason is very simple: we are applying the MVHC procedure on a trade by trade basis, which leads to a slightly different result if we were to optimize over the whole portfolio.

18.6 Conclusion

Because of the market incompleteness induced by the correlation risk, the projection of the basket default swap on the sub-space generated by the underlying hedge instruments is the best we can hope for. This chapter presents a useful application of the min-variance hedging technique to basket credit derivatives. We also give an extension of the standard min-variance hedging optimization to account for the drift mismatch between the basket trade and its hedges.

19

Correlation Calibration with Stochastic Recovery

In this chapter, we expand the base correlation framework by enriching it with Stochastic Recovery modelling as a way to address the model limitations observed in a distressed credit environment. We introduce the general class of Conditional-Functional recovery models, which specify the recovery rate as a function of the common conditioning factor of the Gaussian copula. Then, we review some of the most popular ones, such as: the Conditional Discrete model of Krekel (2008), the Conditional Gaussian of Andersen and Sidenius (2005) and the Conditional Mark-Down of Amraoui and Hitier (2008). We also look at stochastic recovery from an aggregate portfolio perspective and present a top-down specification of the problem. By establishing the equivalence between these two approaches, we show that the latter can provide a useful tool for analyzing the structure of various stochastic recovery model assumptions.

19.1 Motivation

During the credit crisis in 2007, we have seen unprecedented levels of volatility and extreme widening of the index level and senior tranches across a number of investment grade indices.

This is due to a technical dislocation in the market that followed a massive sell-off of ABS structured credit assets. In particular, we have seen large write-downs for some AAA assets, the loss of faith in the AAA rating, and a shortage of traditional buyers of AAA risk due to lack of funding and pressures on the balance sheet. This mismatch between supply and demand for AAA risk has pushed spreads wider across the board.

This had a number of consequences for the correlation market:

Y. Elouerkhaoui, *Credit Correlation*, Applied Quantitative Finance,
DOI 10.1007/978-3-319-60973-7_19

- the Gaussian copula model started to fail calibrating at these stressed levels: the base correlation, at the 30% strike, drifted wider gradually until it reached the 100% correlation cap;
- the emergence of a market for the (60–100%)-super-senior tranche, where most dealers were showing quotes in the 10–20 bps range for a tranche that was deemed to be riskless up until that point;
- given the extreme steepness of the base correlation curve, we started to experience a "flipping" in the sign of the single-name deltas for the tightest names in the senior tranche 15–30%.

The reason for this behaviour of the Gaussian copula model is not surprising. Fundamentally, with a typical 40% recovery rate level, the support of the portfolio loss distribution is confined to the (0–60%) interval, by construction, and would therefore put zero-value in the (60–100%) region and not enough probability weight in the (30–60%)-bucket. Dealers have used various ad-hoc adjustments to address the issue, including marking down the recovery rate value to be able to produce fatter tails of the loss distribution. However, this is clearly just a temporary "fudge": the expected value of the single-name recoveries is well-known and provided by the single-name market (from recovery swaps for example); but their distributions is not known. This has motivated the need for introducing stochastic recovery modelling to complement the market standard base correlation framework.

The idea of all Stochastic Recovery models is simple: basically, while the expected value of the recovery rate is fixed, the realized recovery itself will depend on the general state of the economy. In general, in periods of low defaults and a benign credit environment, we will have high levels of recovery, and in a period of high defaults and a stressed credit environment, the realized recoveries will tend to be lower. This view is also supported by empirical evidence (see, for example, Altman et al. (2003)).

There are many Stochastic Recovery model specifications that have been published in the academic and practitioners' literature. This includes:

- Conditional Discrete (Krekel 2008),
- Conditional Gaussian (Andersen and Sidenius 2005),
- Conditional Mark-Down (Amraoui and Hitier 2008).

In the first part of this chapter, we study some popular (conditional) stochastic recovery model choices. Then, in the second part, instead of using the standard Bottom-Up approach, we approach the problem differently and build a Top-Down Stochastic Recovery model instead, where we can specify the portfolio recovery distribution directly. Ultimately, the objective is to build a model that has fatter tails but does not distort the lower part of the capital structure.

By working directly with the loss distribution, we have better control of the stochastic recovery specification, which can also be used in conjunction with a pure defaults base correlation curve. We shall see that both the Conditional bottom-up stochastic recovery approach and the Top-down stochastic recovery specification are equivalent, which provides a useful tool for understanding the behaviour of each model.

To start with, we review the standard stochastic recovery models in the literature in Sect. 19.2. Section 19.3 introduces the concept of average portfolio recovery and the equivalent Marked Point Process (MPP) representation of the conditional stochastic recovery models. And finally, in Sect. 19.4, we study the calibration, pricing and risk properties of the various models.

19.2 Review of Stochastic Recovery Models

We have a set of non-negative random variables $(\tau_1, \ldots, \tau_N)$ representing the default times of a portfolio of N obligors, and a set of random variables in $[0, 1]$, $(R_1, \ldots, R_N)$ representing the recovery value of each obligor.

The most general way of specifying a bottom-up stochastic recovery model is to assume a set of marginal distributions for recoveries and a general copula function of both default times and recoveries:

$$\begin{aligned}&\mathbb{P}\left(\tau_1 \le t_1, \ldots, \tau_N \le t_N, R_1 \le x_1, \ldots, R_N \le x_N\right)\\&= \mathbb{C}\left(\mathbb{P}\left(\tau_1 \le t_1\right), \ldots, \mathbb{P}\left(\tau_N \le t_N\right), \mathbb{P}\left(R_1 \le x_1\right), \ldots \mathbb{P}\left(R_N \le x_N\right)\right).\end{aligned}$$

In a Gaussian copula model, for example,

$$\begin{aligned}&X_i = \sqrt{\rho} Z + \sqrt{1-\rho}\varepsilon_i, \text{ for } i = 1, \ldots, N,\\&\tau_i \le T \Longleftrightarrow X_i \le \Phi^{-1}\left(p_T^i\right),\end{aligned}$$

we could define the dependence structure for recoveries as:

$$\begin{aligned}&Y_i = a_i Z + b_i Y + \sqrt{1 - a_i^2 - b_i^2}\xi_i, \text{ for } i = 1, \ldots, N,\\&R_i \le x \Longleftrightarrow Y_i \le \Phi^{-1}\left(F^{R_i}(x)\right),\end{aligned}$$

where $Y, \varepsilon_1, ..., \varepsilon_N, Z, \xi_1, ..., \xi_N$ are independent standard Gaussian variables, p_T^i is the default probability at time T, $p_T^i = \mathbb{P}(\tau_i \leq T)$, and $F^{R_i}(x)$ is the recovery rate probability distribution function, $F^{R_i}(x) = \mathbb{P}(R_i \leq x)$.

The parameters $(a_i)_{1 \leq i \leq N}$ would control the correlation between recoveries and defaults, and the $(b_i)_{1 \leq i \leq N}$ would add a correlation overlay between recoveries.

19.2.1 Independent Betas

The first naive approach is to assume that the recovery rates are random variables (for example Beta-distributed), which are independent from each other and independent from the default times as well, $a_i = 0$, $b_i = 0$, for all $i = 1, ..., N$, i.e.,

$$\begin{aligned} Cov\left(R_i, R_j\right) &= 0, \text{ for all } i \neq j, \\ Cov\left(R_i, \tau_i\right) &= 0, \text{ for } i = 1, ..., N. \end{aligned}$$

This is the approach used in CreditMetrics for example.

However, for a large portfolio, it turns out that this has no impact on the portfolio loss distribution. Because of the law of large numbers, the effect of independent random recoveries averages out and converges to the mean.

To illustrate this point, we derive the loss distribution for a large homogeneous portfolio,

$$\mathbb{P}(L_T \leq x) = \int_{-\infty}^{+\infty} \mathbb{P}(L_T \leq x \,|Z = z)\, \phi(z)\, dz.$$

The conditional loss variable L_T^Z is the sum of N independent variables, which converges to the mean when N is large,

$$\begin{aligned} L_{T|Z=z} &= \frac{1}{N} \sum_{i=1}^{N} (1 - R_i)\, \mathbf{1}_{\{\tau_i \leq T\}|Z=z} \\ &= \frac{1}{N} \sum_{i=1}^{N} L_{T|Z=z}^i \underset{n \to +\infty}{\to} \mathbb{E}\left[L_{T|Z=z}\right] = (1 - \mathbb{E}[R])\, p_T(z), \end{aligned}$$

where $p_T(z)$ is the conditional default probability

$$p_T(z) = \Phi\left(\frac{\Phi^{-1}(p_T) - \sqrt{\rho} z}{\sqrt{1-\rho}}\right).$$

19.2.2 Recovery Mixture

In the recovery mixture model, we assume that the single name recoveries are correlated between them, but they are uncorrelated with the default times, $a_i = 0, b_i \neq 0$, for all $i = 1, ..., N$, i.e.,

$$\begin{aligned} Cov\left(R_i, R_j\right) &\neq 0, \text{ for all } i \neq j, \\ Cov\left(R_i, \tau_i\right) &= 0, \text{ for all } i = 1, ..., N. \end{aligned}$$

In particular, we assume that recoveries can be in one of two states: either they all take low values, or they all take high values, i.e., we have a Bernoulli state variable S with probability $p_S = \mathbb{P}(S = 1)$, and each recovery takes two values $\left(r_u^i, r_d^i\right)$ for the up and down states:

$$R_i = r_u^i S + r_d^i (1 - S),$$

where the up and down realizations $\left(r_u^i, r_d^i\right)$ are given by solving for the recovery rate mean and variance:

$$\begin{aligned} \overline{R_i} &= r_u^i p_S + r_d^i (1 - p_S), \\ \sigma_{R_i}^2 &= \left(\left(r_u^i\right)^2 - \overline{R_i}^2\right) p_S + \left(\left(r_d^i\right)^2 - \overline{R_i}^2\right)(1 - p_S). \end{aligned}$$

19.2.3 Conditional-Functional Models

In this class of models, we assume that the single names are correlated between themselves and correlated with defaults.

Now that we have assumed correlation between recoveries and defaults, the key object that we need to work with is the recovery conditional on default before a given time horizon T:

$$R_T^i = R_{i|\tau_i \leq T}.$$

The main constraint that we have is the expectation of the recovery conditional on default, which is known from the single name CDS and should be preserved:

$$\mathbb{E}\left[R_T^i\right] = \overline{R_i}.$$

Indeed, we can see this easily by matching the single-name expected loss at time T:

$$\left(1 - \overline{R_i}\right) p_T^i = \mathbb{E}\left[(1 - R_i)\, \mathbf{1}_{\{\tau_i \leq T\}}\right] = \mathbb{E}\left[1 - R_i \,|\tau_i \leq T\,\right] p_T^i = \left(1 - \mathbb{E}\left[R_T^i\right]\right) p_T^i.$$

The random variable R_T^i can have any exogenously specified distribution with mean $\overline{R_i}$.

Assumption. In the conditional-functional models, we assume that the recovery rate is perfectly correlated with the conditioning factor Z and is given by a functional form $f_T^i\,(.)$:

$$R_T^i = f_T^i\,(Z)\,,$$

where $f_T^i\,(.) : \mathbb{R} \to [0, 1]$ is a monotonically increasing function of the conditional factor Z.

Proposition 155 *If $F_T^i\,(x)$ is the probability distribution function of the random variable R_T^i,*

$$F_T^i\,(x) = \mathbb{P}\left(R_T^i \leq x\right) = \mathbb{P}\left(R_i \leq x \,|\tau_i \leq T\,\right),$$

then, the functional form $f_T^i\,(x)$ is given by

$$f_T^i\,(x) = \left(F_T^i\right)^{-1}\left(F_T^Z\,(x)\right),$$

where $F_T^Z\,(x)$ is the probability distribution function of the factor Z conditional on default before time T:

$$F_T^Z\,(x) = \mathbb{P}\left(Z \leq x \,|\tau_i \leq T\,\right).$$

Proof It suffices to write

$$
\begin{aligned}
F_T^i(x) &= \mathbb{P}\left(R_T^i \leq x\right) = \mathbb{P}\left(R_i \leq x \,|\tau_i \leq T\right) = \mathbb{P}\left(f_T^i(Z) \leq x \,|\tau_i \leq T\right) \\
&= \mathbb{P}\left(Z \leq \left(f_T^i\right)^{-1}(x) \,|\tau_i \leq T\right) \\
&= F_T^Z\left(\left(f_t^i\right)^{-1}(x)\right).
\end{aligned}
$$

By setting the variable $z = \left(f_T^i\right)^{-1}(x)$, we obtain

$$
F_T^i\left(f_T^i(z)\right) = F_T^Z(z),
$$

which gives the formula of the functional mapping $f_T^i(.)$:

$$
f_T^i(z) = \left(F_T^i\right)^{-1}\left(F_T^Z(z)\right).
$$

■

The following lemma gives the formula of the distribution function of the factor Z conditional on default.

Lemma 156 *The probability distribution function of the factor Z conditional on default is given by:*

$$
F_T^Z(x) = \frac{\Phi_2\left(\Phi^{-1}\left(p_T^i\right), x; \sqrt{\rho}\right)}{p_T^i}.
$$

Proof We can compute the conditional distribution function as

$$
F_T^Z(x) = \mathbb{P}\left(Z \leq x \,|\tau_i \leq T\right) = \frac{\mathbb{P}\left(Z \leq x, \tau_i \leq T\right)}{\mathbb{P}\left(\tau_i \leq T\right)} = \frac{\mathbb{P}\left(Z \leq x, X_i \leq \Phi^{-1}\left(p_T^i\right)\right)}{p_T^i}.
$$

The joint probability in the denominator is given by

$$
\begin{aligned}
\mathbb{P}\left(Z \leq x, X_i \leq \Phi^{-1}\left(p_T^i\right)\right) &= \mathbb{P}\left(Z \leq x, \sqrt{\rho}Z + \sqrt{1-\rho}\varepsilon_i \leq \Phi^{-1}\left(p_T^i\right)\right) \\
&= \mathbb{P}\left(Z \leq x, \varepsilon_i \leq \frac{\Phi^{-1}\left(p_T^i\right) - \sqrt{\rho}Z}{\sqrt{1-\rho}}\right) \\
&= \int_{-\infty}^{x} \Phi\left(\frac{\Phi^{-1}\left(p_T^i\right) - \sqrt{\rho}z}{\sqrt{1-\rho}}\right)\phi(z)\,dz.
\end{aligned}
$$

If we denote by $\Phi_2(.,.;\rho)$ the bivariate Gaussian distribution function with correlation ρ, the Gaussian integral can be computed as (see, for example, Andersen and Sidenius (2005))

$$\int_{-\infty}^{x} \Phi(\alpha z+\beta)\,\phi(z)\,dz = \Phi_2\left(\frac{\beta}{\sqrt{1+\alpha^2}}, x; \frac{-\alpha}{\sqrt{1+\alpha^2}}\right),$$

which gives the result

$$\mathbb{P}\left(Z \leq x, X_i \leq \Phi^{-1}\left(p_T^i\right)\right) = \Phi_2\left(\Phi^{-1}\left(p_T^i\right), x; \sqrt{\rho}\right).$$

■

19.2.4 Conditional Discrete

In the Conditional Discrete model, the distribution of R_T^i is assumed to be a discrete probability distribution:

$$\mathbb{P}\left(R_T^i = r_j^i\right) = p_j^i, \text{ for } 1 \leq j \leq m.$$

In this case, the cumulative probability function is given by

$$F_T^i(x) = \mathbb{P}\left(R_T^i \leq x\right) = \sum_{j=1}^{m} p_j^i H\left(x - r_j^i\right) = \sum_{k=0}^{m}\left[\sum_{j=1}^{k} p_j^i\right] \mathbf{1}\left\{r_k^i \leq x < r_{k+1}^i\right\},$$

where $H(.)$ is the heavy-side function, and $r_0^i = -\infty$, $r_{m+1}^i = +\infty$. The inverse distribution function is

$$\left(F_T^i\right)^{-1}(x) = \sum_{k=1}^{m} r_k^i \mathbf{1}\left\{\sum_{j=1}^{k-1} p_j^i < x \leq \sum_{j=1}^{k} p_j^i\right\},$$

which can be plugged into the formula of Proposition 155 to get the functional mapping :

$$f_T^i(x) = \left(F_T^i\right)^{-1}\left(F_T^Z(x)\right) = \sum_{k=1}^{m} r_k^i \mathbf{1}\left\{\sum_{j=1}^{k-1} p_j^i < F_T^Z(x) \leq \sum_{j=1}^{k} p_j^i\right\}.$$

19.2.5 Link with Krekel (2008)

In Krekel (2008), instead of making the conditional recovery a function of the common factor, he assumes that it is a function of the single-name Gaussian variates:

$$R_T^i = g_T^i\left(X_i\right),$$

where $g_T^i\left(.\right): \mathbb{R} \to [0, 1]$ is a monotonically decreasing function.

Proposition 157 *If $F_T^i\left(x\right)$ is the probability function of the random variable R_T^i,*

$$F_T^i\left(x\right) = \mathbb{P}\left(R_T^i \leq x\right),$$

then, the functional form $g_T^i\left(x\right)$ is given by

$$g_T^i\left(x\right) = \left(F_T^i\right)^{-1}\left(1 - F_T^{X_i}\left(x\right)\right),$$

where $F_T^{X_i}\left(x\right)$ is the probability distribution function of X_i conditional on default:

$$F_T^{X_i}\left(x\right) = \mathbb{P}\left(X_i \leq x \left| \tau_i \leq T\right.\right) = \frac{1}{p_T^i}\Phi\left(\min\left(x, \Phi^{-1}\left(p_T^i\right)\right)\right).$$

Proof It suffices to write

$$\begin{aligned}
F_T^i\left(x\right) &= \mathbb{P}\left(R_T^i \leq x\right) = \mathbb{P}\left(R_i \leq x \left| \tau_i \leq t\right.\right) = \mathbb{P}\left(g_T^i\left(X_i\right) \leq x \left| \tau_i \leq t\right.\right) \\
&= \mathbb{P}\left(X_i \geq \left(g_T^i\right)^{-1}\left(x\right) \left| \tau_i \leq t\right.\right) \\
&= 1 - F_T^{X_i}\left(\left(g_T^i\right)^{-1}\left(x\right)\right).
\end{aligned}$$

The distribution of X_i conditional on default, $F_T^{X_i}\left(x\right)$, is given by

$$F_T^{X_i}\left(x\right) = \frac{\mathbb{P}\left(X_i \leq x, X_i \leq \Phi^{-1}\left(p_T^i\right)\right)}{p_T^i} = \frac{\Phi\left(\min\left(x, \Phi^{-1}\left(p_T^i\right)\right)\right)}{p_T^i}.$$

■

Proposition 158 *Conditional on Z, the variables $\left(R_T^i\right)_{1\leq i\leq N}$ are independent and their distributions are given by*

$$\mathbb{P}\left(R_T^i \leq x\,|Z=z\right) = \frac{\Phi\left(\frac{\Phi^{-1}(p_T^i)-\sqrt{\rho}z}{\sqrt{1-\rho}}\right) - \Phi\left(\frac{\Phi^{-1}((1-F_T^i(x))p_T^i)-\sqrt{\rho}z}{\sqrt{1-\rho}}\right)}{p_T^i\,(z)}.$$

Proof We can compute the distribution of R_T^i conditional on the common factor Z as:

$$\begin{aligned}\mathbb{P}\left(R_T^i \leq x\,|Z=z\right) &= \mathbb{P}\left(g_T^i\,(X_i) \leq x\,|\tau_i \leq T, Z=z\right)\\ &= \frac{\mathbb{P}\left(g_T^i\,(X_i) \leq x, \tau_i \leq T\,|Z=z\right)}{\mathbb{P}\left(\tau_i \leq T\,|Z=z\right)}\\ &= \frac{\mathbb{P}\left(g_T^i\,(X_i) \leq x, X_i \leq \Phi^{-1}\left(p_T^i\right)|Z=z\right)}{p_T^i\,(z)}.\end{aligned}$$

Replacing the functional mapping with its formula, we have

$$\begin{aligned}\left\{g_T^i\,(X_i) \leq x\right\} &\Longleftrightarrow \left\{\left(F_T^i\right)^{-1}\left(1-F_T^{X_i}\,(X_i)\right) \leq x\right\}\\ &\Longleftrightarrow \left\{\left(1-F_T^i\,(x)\right)p_T^i \leq \Phi\left(\min\left(X_i, \Phi^{-1}\left(p_T^i\right)\right)\right)\right\},\end{aligned}$$

hence,

$$\begin{aligned}&\mathbb{P}\left(g_T^i\,(X_i) \leq x, X_i \leq \Phi^{-1}\left(p_T^i\right)|Z=z\right)\\ &= \mathbb{P}\left(\left(1-F_T^i\,(x)\right)p_T^i \leq \Phi\,(X_i)\,, X_i \leq \Phi^{-1}\left(p_T^i\right)|Z=z\right)\\ &= \mathbb{P}\left(\Phi^{-1}\left(\left(1-F_T^i\,(x)\right)p_T^i\right) \leq X_i, X_i \leq \Phi^{-1}\left(p_T^i\right)|Z=z\right)\\ &= \Phi\left(\frac{\Phi^{-1}\left(p_T^i\right)-\sqrt{\rho}z}{\sqrt{1-\rho}}\right) - \Phi\left(\frac{\Phi^{-1}\left(\left(1-F_T^i\,(x)\right)p_T^i\right)-\sqrt{\rho}z}{\sqrt{1-\rho}}\right).\end{aligned}$$

The last step is to observe that $R_T^i\,|Z=z$ are independent since they depend only on the idiosyncratic variate. ■

In Fig. 19.1, we compare the calibrated base correlation curves for the various models presented in this section. As shown before, the Indepedent Beta model has no impact on the calibration: the Deterministic and Independent Beta

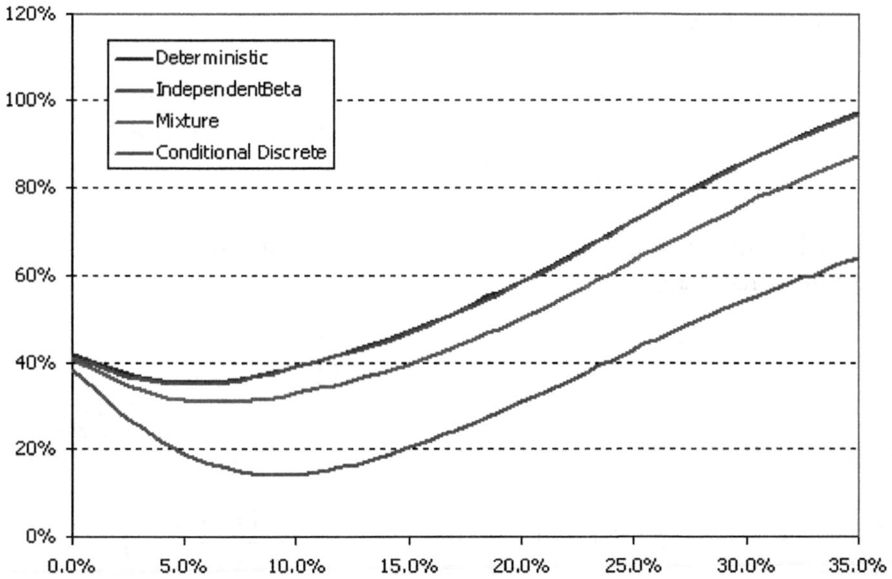

Fig. 19.1 Calibrated base correlation curve for the Independent Beta, Mixture and Conditional Discrete models

curves are on top of each other. The Mixture model lowers the base correlation curve a little bit but its impact is very limited. The Conditional Discrete has the desired impact as it lowers the calibrated base correlation significantly: the correlation at the 35% strike, for example, moves from 98 to 62%.

19.2.6 Conditional Gaussian

In Andersen and Sidenius (2005), the dependence of the recovery rate as a function of the conditioning factor is assumed to be given by the cumulative Gaussian distribution

$$R_T^i = \Phi\left(\mu_T^i + \sigma_T^i Z\right),$$

where the parameter μ_T^i is chosen to match the expected recovery value.

Proposition 159 *The distribution of the recovery conditional on default is given by*

$$\mathbb{P}\left(R_T^i \leq x\right) = \frac{1}{p_T^i}\Phi_2\left(\Phi^{-1}\left(p_T^i\right), \frac{\Phi^{-1}(x) - \mu_T^i}{\sigma_T^i}; \sqrt{\rho}\right).$$

And its expected value is given by

$$\mathbb{E}\left[R_T^i\right] = \frac{1}{p_T^i}\Phi_2\left(\frac{\mu_T^i}{\sqrt{1+\left(\sigma_T^i\right)^2}}, \Phi^{-1}\left(p_T^i\right); \frac{-\sigma_T^i\sqrt{\rho}}{\sqrt{1+\left(\sigma_T^i\right)^2}}\right).$$

Proof Using the result of Proposition 155, we can link the marginal distribution to its functional mapping

$$f_T^i(x) = \left(F_T^i\right)^{-1}\left(F_T^Z(x)\right),$$

hence

$$\begin{aligned}
F_T^i(x) = F_T^Z\left(\left(f_T^i\right)^{-1}(x)\right) &= F_T^Z\left(\frac{\Phi^{-1}(x) - \mu_T^i}{\sigma_T^i}\right) \\
&= \frac{1}{p_T^i}\Phi_2\left(\Phi^{-1}\left(p_T^i\right), \frac{\Phi^{-1}(x) - \mu_T^i}{\sigma_T^i}; \sqrt{\rho}\right).
\end{aligned}$$

The expectation of the conditional recovery is given by

$$\begin{aligned}
\overline{R_i} = \mathbb{E}\left[R_T^i\right] &= \mathbb{E}\left[R_i \,|\tau_i \leq T\right] = \mathbb{E}\left[\Phi\left(\mu_T^i + \sigma_T^i Z\right)|\tau_i \leq T\right] \\
&= \frac{1}{p_T^i}\mathbb{E}\left[\Phi\left(\mu_T^i + \sigma_T^i Z\right)\Phi\left(\frac{\Phi^{-1}\left(p_T^i\right) - \sqrt{\rho}Z}{\sqrt{1-\rho}}\right)\right] \\
&= \frac{1}{p_T^i}\mathbb{E}\left[\Phi\left(\mu_T^i + \sigma_T^i Z\right)p_T^i(Z)\right].
\end{aligned}$$

This Gaussian integral can be computed as (see, for example, Andersen and Sidenius (2005))

$$\int_{-\infty}^{+\infty}\Phi(az+b)\,\Phi(cz+d)\,\phi(z)\,dz = \Phi_2\left(\frac{b}{\sqrt{1+a^2}}, \frac{d}{\sqrt{1+c^2}}; \frac{ac}{\sqrt{1+a^2}\sqrt{1+c^2}}\right),$$

which gives the result

$$\mathbb{E}\left[\Phi\left(\mu_T^i + \sigma_T^i Z\right)p_T^i(Z)\right] = \Phi_2\left(\frac{\mu_T^i}{\sqrt{1+\left(\sigma_T^i\right)^2}}, \Phi^{-1}\left(p_T^i\right); \frac{-\sigma_T^i\sqrt{\rho}}{\sqrt{1+\left(\sigma_T^i\right)^2}}\right).$$

■

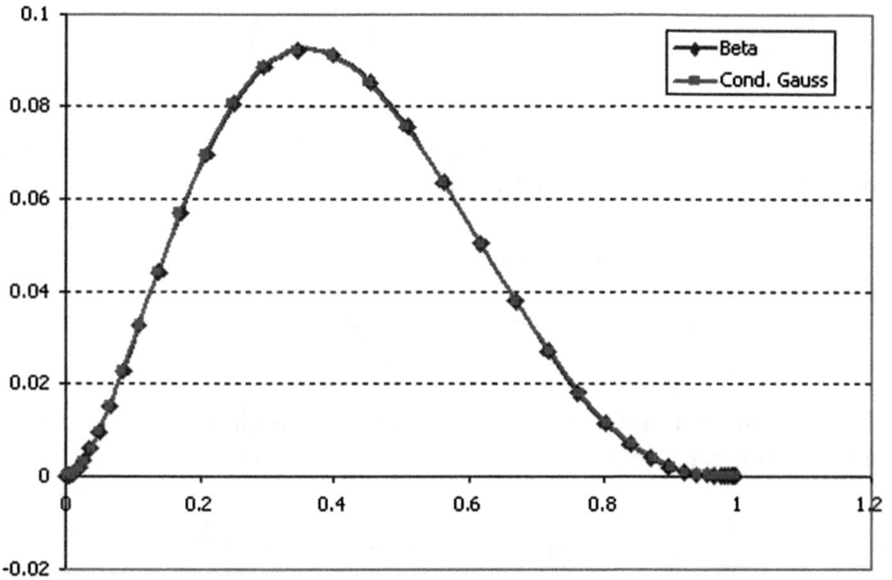

Fig. 19.2 Comparison of the distribution of R_T^i in the Conditional Gaussian model with a Beta distribution

The main advantage of this model is its analytical tractability: by choosing the cumulative Gaussian functional form, there are many well-known analytical "Gaussian integrals" results that can be used. It turns out, in practice, that even with this simple form, the calibrated recovery distributions tend to be very close to a Beta distribution, which is traditionally used to model [0, 1]-bounded recovery distributions. Figure 19.2 shows a comparison of the two distributions.

19.2.7 Conditional Mark-Down

The idea of the conditional Mark-Down model, introduced in Amraoui and Hitier (2008), is to build a self-consistent model that behaves similarly to a deterministic model with a marked-down recovery. This is achieved by choosing a functional form of the stochastic recovery mapping that mimics the loss variable of the deterministic marked-down model.

Let us denote by $\left(\widetilde{R}_i, \widetilde{p}_T^i, \widetilde{p}_T^i(z)\right)$ the value of the mark-down recovery, the mark-down default probability and the conditional mark-down default probability respectively:

$$\widetilde{p}_T^i(z) = \Phi\left(\frac{\Phi^{-1}\left(\widetilde{p}_T^i\right) - \sqrt{\rho}z}{\sqrt{1-\rho}}\right).$$

The functional form of the conditional recovery is chosen such that the single-name conditional expected loss is preserved in the two models:

$$R_T^i = f_T^i(Z),$$
$$f_T^i(z) = 1 - \left(1 - \widetilde{R}_i\right)\frac{\widetilde{p}_T^i(z)}{p_T^i(z)}, \text{ for } z \in \mathbb{R}.$$

Indeed, by using this functional form, we can easily check that we get the same conditional expected loss

$$\mathbb{E}\left[(1 - R_i)\,\mathbf{1}_{\{\tau_i \le T\}} \,|Z = z\right] = \left(1 - f_T^i(z)\right) p_T^i(z) = \left(1 - \widetilde{R}_i\right) \widetilde{p}_T^i(z).$$

Consistency with the pricing of the single-name CDS is ensured by preserving the single-name expected loss

$$\begin{aligned}
\left(1 - \overline{R}_i\right) p_T^i &= \mathbb{E}\left[\left(1 - R^i\right)\mathbf{1}_{\{\tau_i \le T\}}\right] = \mathbb{E}\left[\mathbb{E}\left[\left(1 - R^i\right)\mathbf{1}_{\{\tau_i \le T\}} \,|Z\right]\right] \\
&= \mathbb{E}\left[\left(1 - f_T^i(z)\right) p_T^i(z)\right] = \mathbb{E}\left[\left(1 - \widetilde{R}_i\right) \widetilde{p}_T^i(z)\right] \\
&= \left(1 - \widetilde{R}_i\right) \mathbb{E}\left[\widetilde{p}_T^i(z)\right] = \left(1 - \widetilde{R}_i\right) \widetilde{p}_T^i.
\end{aligned}$$

Hence, to calibrate the single-name CDS mark-down probabilities, we need to set

$$\widetilde{p}_T^i = \left(\frac{1 - \overline{R}_i}{1 - \widetilde{R}_i}\right) p_T^i.$$

Figure 19.3 shows an example of the conditional recovery mapping $f_T^i(z)$ and the conditional probability $p_T^i(z)$ for different values of the conditioning factor z. The two shapes offset each other since the single-name expected loss is always preserved.

Now that we have chosen the functional form $f_T^i(z)$ to mimic the single-name conditional expected loss and we have ensured consistency with the single-name CDS pricing, we need to check that the portfolio loss variable in this model will also mimic the behaviour of the deterministic mark-down model.

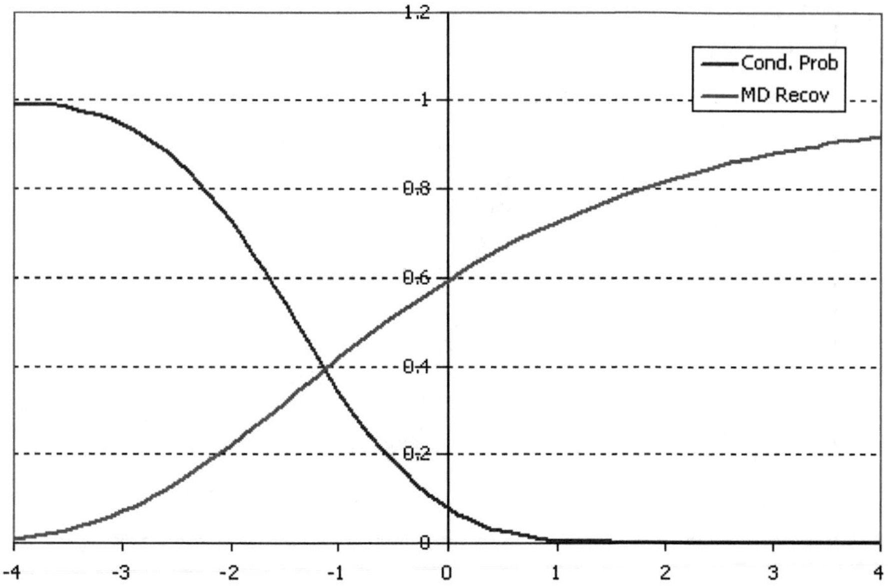

Fig. 19.3 Functional form of the conditional recovery mapping in the Conditional Mark-Down model

It turns out that this is indeed the case when the number of names in the portfolio is large.

We can see this easily for a large homogeneous portfolio:

$$\begin{aligned} L_{T|Z=z} &= \frac{1}{N}\sum_{i=1}^{N}\left(1 - f_T^i(z)\right)\mathbf{1}_{\{\tau_i \le T\}|Z=z} \\ &= \frac{1}{N}\sum_{i=1}^{N} L^i_{T|Z=z} \underset{N\to+\infty}{\to} \mathbb{E}\left[L_{T|Z=z}\right] = (1 - f_T(z))\, p_T(z) \\ &= \left(1 - \widetilde{R}\right)\widetilde{p}_T(z) = \widetilde{L}_{T|Z=z}. \end{aligned}$$

This result holds in the general case as well since the loss distribution becomes a Gaussian mixture where we match the conditional expectations. And although the variances are different, they do average out. However, the convergence is slower for portfolios with the higher-dispersion.

In Fig. 19.4, we benchmark the implied recovery distribution in the Conditional Mark-Down model with a Beta distribution that has the same mean and variance. They both have, broadly, the same shape with slight differences in the tail.

Figure 19.5 compares the calibrated base correlation curves for the Conditional Gaussian and Conditional Mark-Down models. Overall, they move the

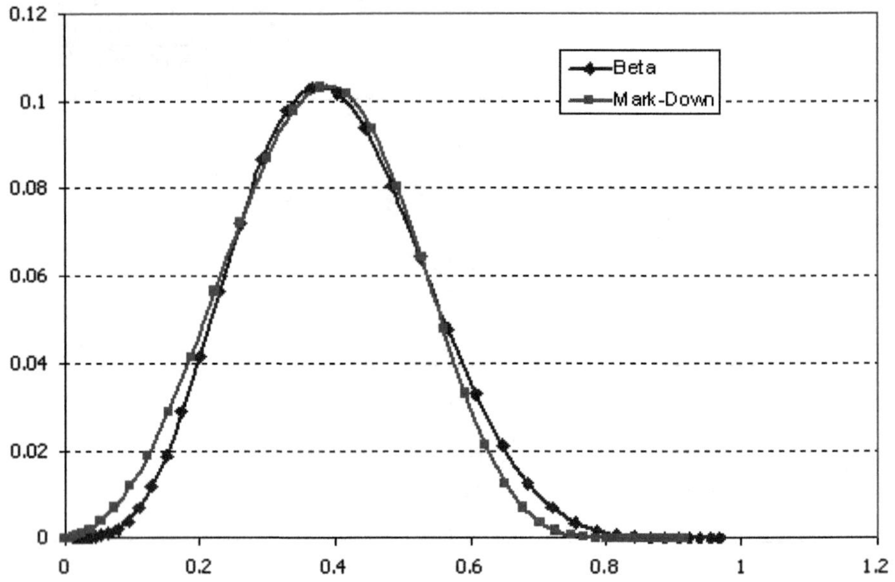

Fig. 19.4 Comparison of in the Conditional Mark-Down model R_T^i and a Beta distribution

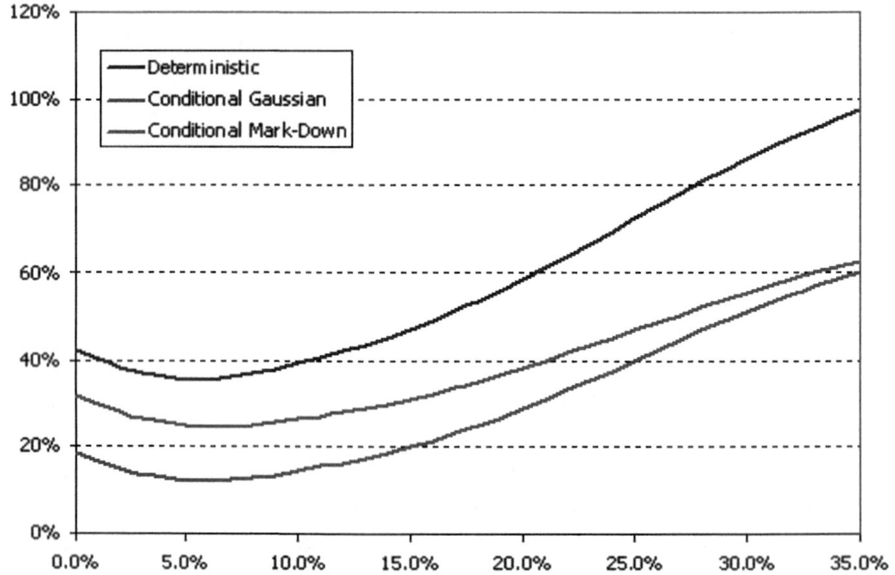

Fig. 19.5 Calibated base correlation curves for the Conditional Gaussian and Conditional Mark-Down models

base correlation down as expected; they both lower the super-senior correlations from 98% to similar levels around 60%, but they also have a big impact on the shape of the base correlation curve in the equity part as well, which is not a desirable feature. Obviously, we need to calibrate the super-senior tranches but we also want to have better control of the shape of the distribution in the equity part. This is one of the main motivations for the top-down stochastic recovery approach introduced next.

19.3 Towards a General Theory of Stochastic Recovery

The inspiration for the approach presented in this section comes from the graphical relationship between the historical average portfolio recovery and the annual default rate (as shown, for example, in Altman et al. (2003)).

In principle, if we have a top-down model, we could use this historically estimated l_n curve—which is defined as the average portfolio loss conditional on n defaults—in the base correlation model and calibrate it. In the Conditional-Functional models presented earlier, we would need to calibrate the single-name functional forms to match the observed l_n curve.

In the top-down approach, we link the portfolio losses directly to the realized default states (as opposed to indirectly via the conditional factor).

In a state of low defaults, we will have small realized portfolio losses; in a state of high defaults, we will have large realized losses, which impact directly the "fatness" of the tail of the distribution.

The main advantage of using a top-down approach is the fact that we can control the shape of the recovery distribution: we can produce fat tails and not distort (or, at least, minimize the impact on) the rest of the loss distribution. In other words, it provides the flexibility needed for pricing super-senior risk while preserving consistency with the junior part of the capital structure.

19.3.1 A Marked Point Process Approach

We define the sequence of strictly ordered default times $(T_0, T_1, ..., T_N)$: $T_0 < T_1 < ... < T_N$ and the identity of the defaulted obligors $(I_0, I_1, ..., I_N)$:

$$\begin{aligned}
&T_0 = 0, \ \ I_0 = 0; \\
&T_k = \min\{\tau_i : 1 \le i \le N, \tau_i > T_{k-1}\}; \\
&I_k = i \text{ if } T_k = \tau_i .
\end{aligned}$$

Let $(N_t)_{t\geq 0}$ be the default counting process

$$N_t = \sum_{i=1}^{N} \mathbf{1}_{\{\tau_i \leq t\}} = \sum_{k=1}^{N} \mathbf{1}_{\{T_k \leq t\}},$$

and let $(L_t)_{t\geq 0}$ be the portfolio loss process

$$L_t = \sum_{i=1}^{N} (1 - R_i)\, \mathbf{1}_{\{\tau_i \leq t\}} = \sum_{k=1}^{N} \left(1 - R_{I_k}\right) \mathbf{1}_{\{T_k \leq t\}} = \sum_{k=1}^{N} l_k \mathbf{1}_{\{T_k \leq t\}} = \sum_{k=1}^{N_t} l_k.$$

We define the "average loss process"

$$1 - r_t = l_t = \frac{L_t}{N_t} \mathbf{1}_{\{N_t > 0\}},$$

and the average loss conditional on the number of defaults l_t^n,

$$l_t^n = l_{t|N_t = n},$$

so that

$$L_t = l_t N_t = \sum_{n\geq 1} l_t^n \mathbf{1}_{\{N_t = n\}}.$$

19.3.2 From Bottom-Up to Top-Down

First, we assume a bottom-up model and we want derive the distributions of the random variables l_T^n.

Starting with a Conditional-Functional stochastic recovery model, the distributions of l_T^n. are given by the following result.

Proposition 160 *If we have a Conditional-Functional model where the recoveries are defined through a mapping of the common factor, then for a large portfolio, the distributions of the random variables* $\left(l_T^n\right)_{n\geq 1}$ *degenerate to deterministic variables. In the homogeneous case, we have*

$$\mathbb{P}\left(l_T^n \leq x\right) = H\left(x - \alpha_n\right),$$

where $H(.)$ is the Heaviside function, and the values (α_n) are given by

$$\alpha_n = 1 - f_T\left(p_T^{-1}\left(\frac{n}{N}\right)\right) = 1 - f_T\left(\frac{\Phi^{-1}(p_T) - \Phi^{-1}\left(\frac{n}{N}\right)\sqrt{1-\rho}}{\sqrt{\rho}}\right).$$

Proof We give a sketch of the proof for a large homogeneous portfolio.

$$\begin{aligned}\mathbb{P}\left(l_T^n \le x\right) &= \mathbb{P}\left(l_T \le x \,|N_T = n\right) = \mathbb{P}\left(\frac{L_T}{N_T}\mathbf{1}_{\{N_T>0\}} \le x \,|N_T = n\right)\\ &= \frac{\mathbb{P}\left((1-R) \le x, N_T = n\right)}{\mathbb{P}\left(N_T = n\right)}\\ &= \frac{\int_{-\infty}^{+\infty} \mathbf{1}_{\{(1-f_T(z))\le x\}}\mathbf{1}_{\{p_T(z)=\frac{n}{N}\}}\phi(z)\,dz}{\int_{-\infty}^{+\infty} \mathbf{1}_{\{p_T(z)=\frac{n}{N}\}}\phi(z)\,dz};\end{aligned}$$

it is convenient to introduce a small ϵ that goes to zero, and re-write as the following limit

$$\begin{aligned}\ldots &= \lim_{\epsilon\to 0}\frac{\int_{-\infty}^{+\infty} \mathbf{1}_{\{(1-f_T(z))\le x\}}\mathbf{1}_{\{\frac{n}{N}\le p_T(z)<\frac{n}{N}+\epsilon\}}\phi(z)\,dz}{\int_{-\infty}^{+\infty} \mathbf{1}_{\{\frac{n}{N}\le p_T(z)<\frac{n}{N}+\epsilon\}}\phi(z)\,dz}\\ &= \lim_{\epsilon\to 0}\frac{\int_{-\infty}^{+\infty} \mathbf{1}_{\left\{z\ge f_T^{-1}(1-x)\right\}}\mathbf{1}_{\left\{p_T^{-1}\left(\frac{n}{N}\right)\ge z> p_T^{-1}\left(\frac{n}{N}+\epsilon\right)\right\}}\phi(z)\,dz}{\int_{-\infty}^{+\infty} \mathbf{1}_{\left\{p_T^{-1}\left(\frac{n}{N}\right)\ge z> p_T^{-1}\left(\frac{n}{N}+\epsilon\right)\right\}}\phi(z)\,dz}\\ &= \lim_{\epsilon\to 0}\frac{\Phi\left(p_T^{-1}\left(\frac{n}{N}\right)\right) - \Phi\left(\min\left(\max\left(f_T^{-1}(1-x), p_T^{-1}\left(\frac{n}{N}+\epsilon\right)\right), p_T^{-1}\left(\frac{n}{N}\right)\right)\right)}{\Phi\left(p_T^{-1}\left(\frac{n}{N}\right)\right) - \Phi\left(p_T^{-1}\left(\frac{n}{N}+\epsilon\right)\right)}\\ &= H(x-\alpha_n),\end{aligned}$$

where

$$\alpha_n = 1 - f_T\left(p_T^{-1}\left(\frac{n}{N}\right)\right) = 1 - f_T\left(\frac{\Phi^{-1}(p_T) - \Phi^{-1}\left(\frac{n}{N}\right)\sqrt{1-\rho}}{\sqrt{\rho}}\right).$$

■

Remark 161 In general, the variables l_T^n are always deterministic, but we have slower convergence when the portfolio is very dispersed.

Figure 19.6 shows the average loss conditional on the number of defaults l_T^n, implied by the Conditional Mark-Down model, for different levels of mark-down recoveries and flat base correlation curve. For a base recovery of 40%,

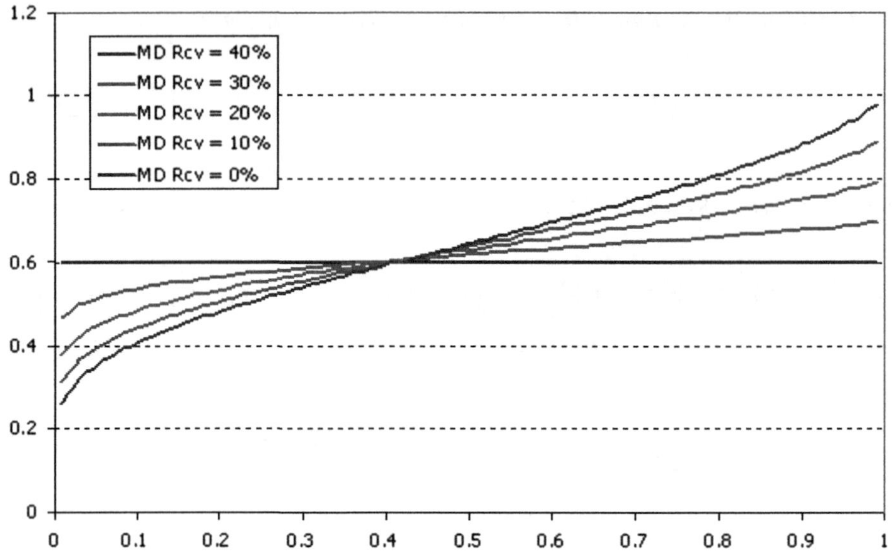

Fig. 19.6 l_T^n curve for the Conditional Mark-Down model with different mark-down recoveries

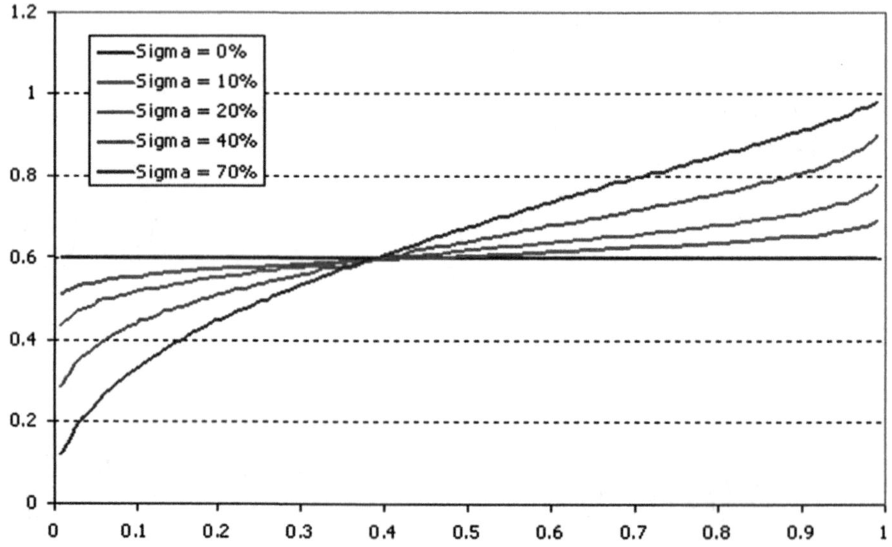

Fig. 19.7 l_T^n curve for the Conditional Gaussian model with different σ_T parameters

we are in the deterministic case and the loss curve l_T^n is flat at 60%. As we mark-down the recovery rate from 40 to 0%, the loss curve l_T^n steepens and covers the interval from 20 to 100%. Figure 19.7 shows the l_T^n-curve for the Conditional Gaussian model. With zero-volatility, we are in the deterministic

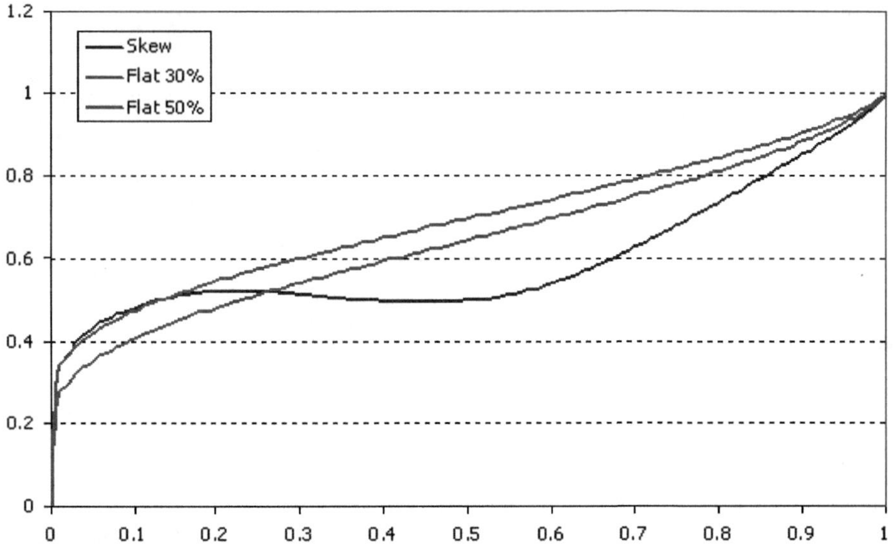

Fig. 19.8 l_T^n curve for the Conditional Mark-Down model with different correlations

recovery set-up and the loss curve is flat. As we increase the volatility parameter, the loss curve steepens and covers a large interval of realized recoveries upon default.

Figures 19.8 and 19.9 show the l_T^n -curve for the Conditional Mark-Down and Conditional Gaussian models with different base correlation curves. As we move from flat correlation to a correlation skew, the shape of the average loss l_T^n becomes much more interesting: there are, typically, three main regions in the curve impacting the various parts of the capital structure: Equity, Mezzanine and Super Senior. In this case, the average recovery rate progresses gradually as we move up the capital structure: for equity, it is around 20%, for Mezzanine, it is 50% and for Super-Senior, it goes closer to 100%.

19.3.3 From Distribution to Functional Form

In the previous section, we started with a stochastic recovery model and we derived the implied average portfolio loss variable l_T^n. Now, we want to do the reverse: i.e., assume a distribution for the average portfolio loss variable l_T, then derive the functional form of the mapping between recovery and defaults.

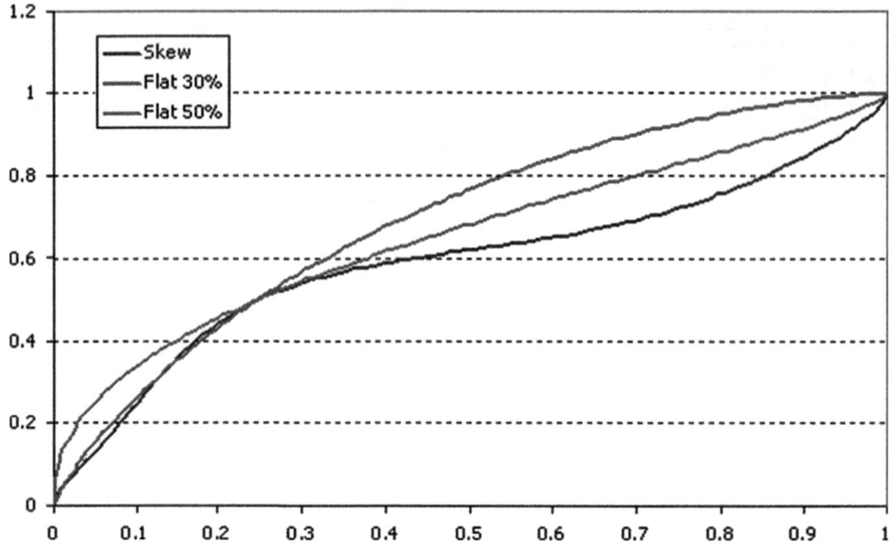

Fig. 19.9 l_T^n curve for the Conditional Gaussian model with different correlations

Here, we express all quantities in terms of the percentage of portfolio defaults

$$X_T = \frac{N_T}{N},$$
$$l_T^n = l_T\left(\frac{n}{N}\right).$$

where $l_T(.) : [0, 1] \to [0, 1]$ is a monotonically increasing function of the percentage of portfolio defaults. Using this notation, for a large portfolio, it is more convenient to re-write the discrete sums as Riemann integrals. Starting with the distribution of l_T, the functional form of $l_T^n = l_T\left(\frac{n}{N}\right)$ is given by the following result.

Proposition 162 *If we assume that the average portfolio loss variable l_T follows a given distribution $F^{l_T}(.)$, then we have*

$$l_T^n = l_T\left(\frac{n}{N}\right) = \left(F^{l_T}\right)^{-1}\left(F^{X_T}\left(\frac{n}{N}\right)\right).$$

Proof We write the probability distribution function of l_T, and we replace with the definition of the functional mapping:

$$\begin{aligned} F^{l_T}(x) &= \mathbb{P}(l_T \le x) = \sum_{n\ge 1} \mathbb{P}(l_T \le x \,|N_T = n)\,\mathbb{P}(N_T = n) + \mathbb{P}(N_T = 0) \\ &= \sum_{n\ge 1} \mathbf{1}\left\{l_T^n \le x\right\} \mathbb{P}(N_T = n) + \mathbb{P}(N_T = 0)\,\mathbf{1}\left\{l_T^0 \le x\right\} \\ &= \sum_{n\ge 0} \mathbf{1}\left\{l_T\left(\frac{n}{N}\right) \le x\right\} \mathbb{P}\left(X_T = \frac{n}{N}\right) \\ &= \int_0^1 \mathbf{1}\{l_T(u) \le x\}\, dF^{X_T}(u) = F^{X_T}\left(l_T^{-1}(x)\right), \end{aligned}$$

hence,

$$l_T(x) = \left(F^{l_T}\right)^{-1}\left(F^{X_T}(x)\right).$$

■

Note that we need to make sure that the expected portfolio loss is preserved. The portfolio loss is given by

$$\mathbb{E}[L_T] = \int_0^1 l_T(x)\, x dF^{X_T}(x) = \int_0^1 \left(F^{l_T}\right)^{-1}\left(F^{X_T}(x)\right) x dF^{X_T}(x),$$

thus, the mean of the distribution l_T is chosen so that we match the portfolio expected loss.

For example, we check the functional form of $l_T(x)$ for a given Beta and a Beta-Kernel distribution. The Beta-Kernel estimator for densities defined on the interval [0, 1] has been developed by Chen (1999). This method was also used, in Renault and Scaillet (2003), to estimate the recovery rate densities.

Figure 19.10 shows examples of recovery rate densities fitted with a Beta and a Beta-Kernel distribution.

In this example, the functional form of $l_T(x)$ is given in Fig. 19.11 for the Beta and Beta-Kernel densities.

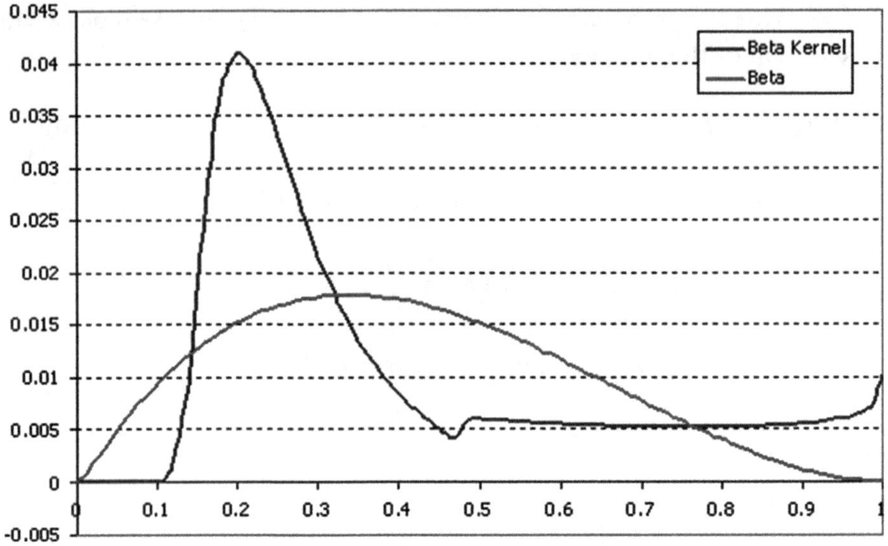

Fig. 19.10 Recovery rate densities with Beta and Beta-Kernel distributions

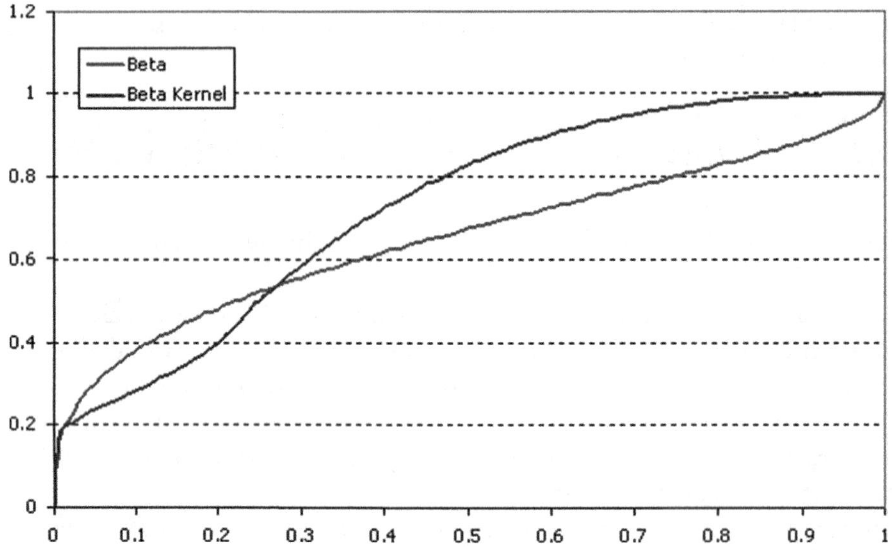

Fig. 19.11 Functional form $l_T(x)$ of for Beta and Beta Kernel densities

19.4 Applications

We calibrate the various models presented on CDX tranches. The 5Y index level is 242 bps, and the 5Y tranche quotes are given below.

0–3%	83.875
3–7%	54.25
7–10%	23.5
10–15%	562.5
15–30%	109.5

In Fig. 19.12, we show the calibrated base correlation curves for the Conditional Gaussian, Conditional Mark-Down and Marked Point Process models. They all do a good job at lowering the base correlation curve, at the 35% strike, to 60%. The Conditional Gaussian distorts the equity correlation substantially: it almost does a parallel translation of the base correlation curve. The Conditional Mark-Down also distorts the pricing in the junior part but

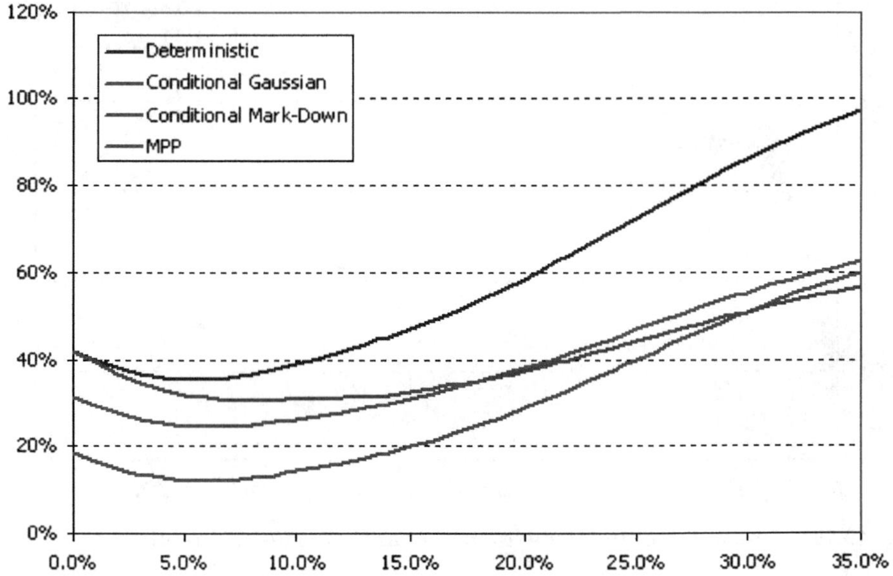

Fig. 19.12 Calibrated base correlation curves for the Conditional Gaussian and Conditional Mark-Down and MPP models

it is not as extreme as the Conditional Gaussian. The MPP model performs better for junior tranches as we have more flexibility in controlling the shape of the curve.

Index Deltas. The index deltas of the various models are given below.

Tranche (%)	Determ.	Cond. Gauss.	Cond. MD	MPP
0–3	1.25	1.45	1.44	1.27
3–7	3.67	3.81	3.93	3.84
7–10	4.10	4.15	4.27	4.32
10–15	3.14	3.20	3.23	3.30
15–30	0.69	0.95	0.91	0.91

Single-Name Deltas. Figures 19.13, 19.14, 19.15, 19.16, 19.17 show a comparison of the single-name deltas for each tranche in the capital structure for the different stochastic recovery models.

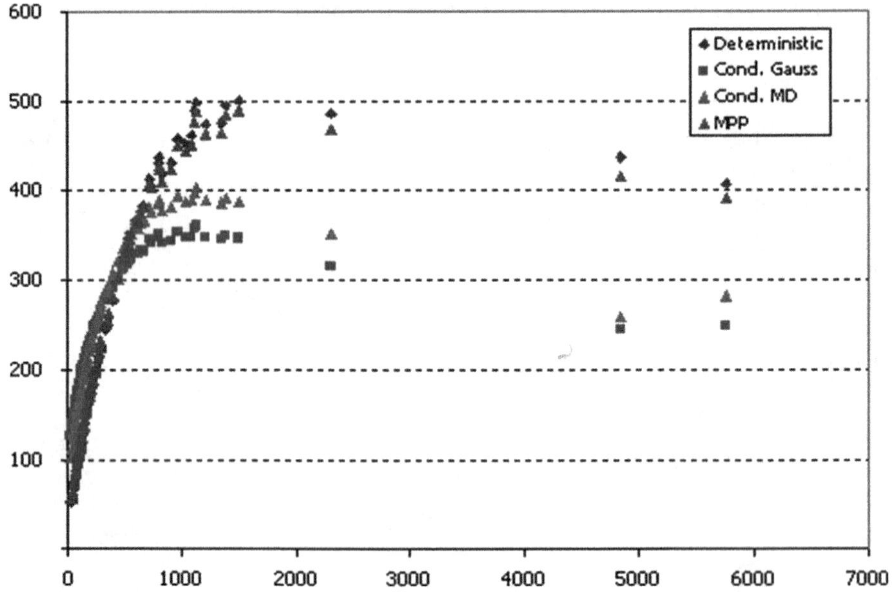

Fig. 19.13 Comparison of the Single-Name Deltas for the (0–3%) tranche

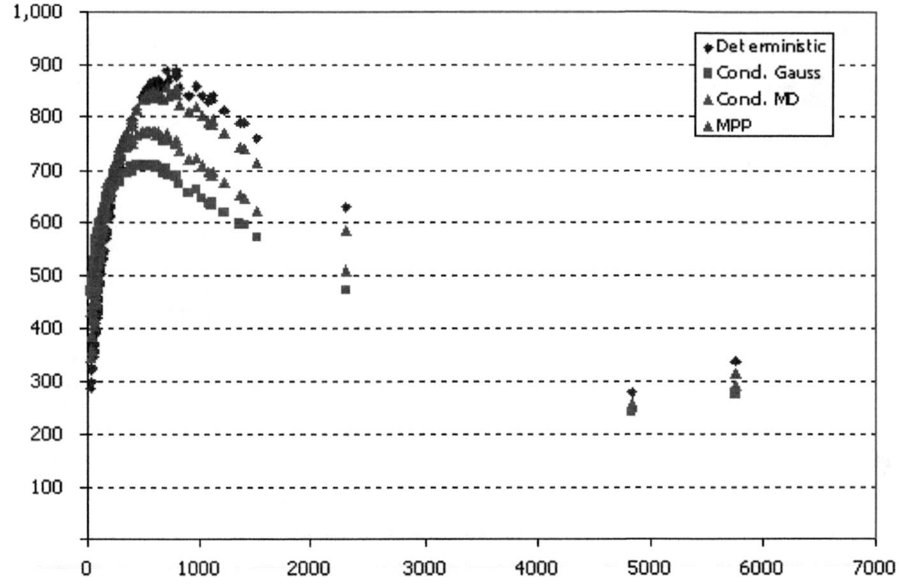

Fig. 19.14 Comparison of the Single-Name Deltas for the (3–7%) tranche

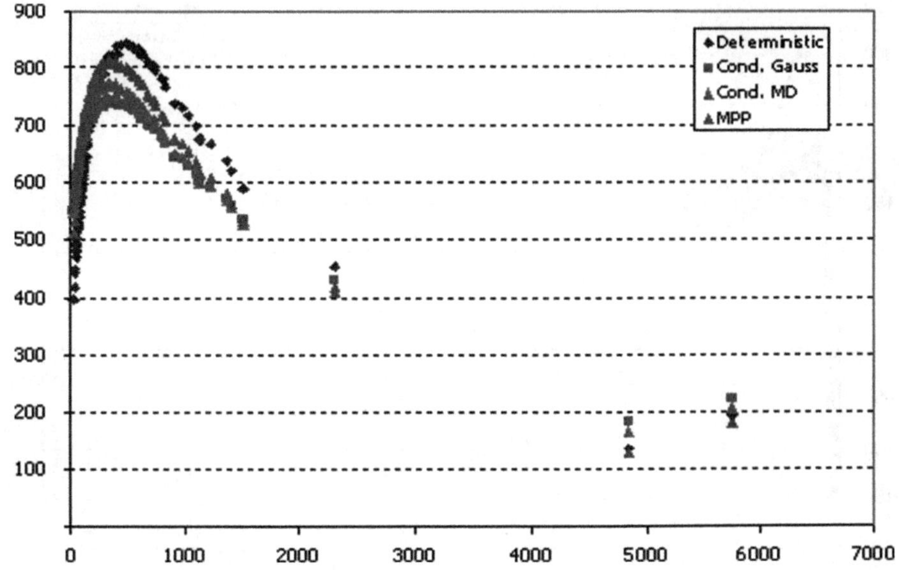

Fig. 19.15 Comparison of the Single-Name Deltas for the (7–10%) tranche

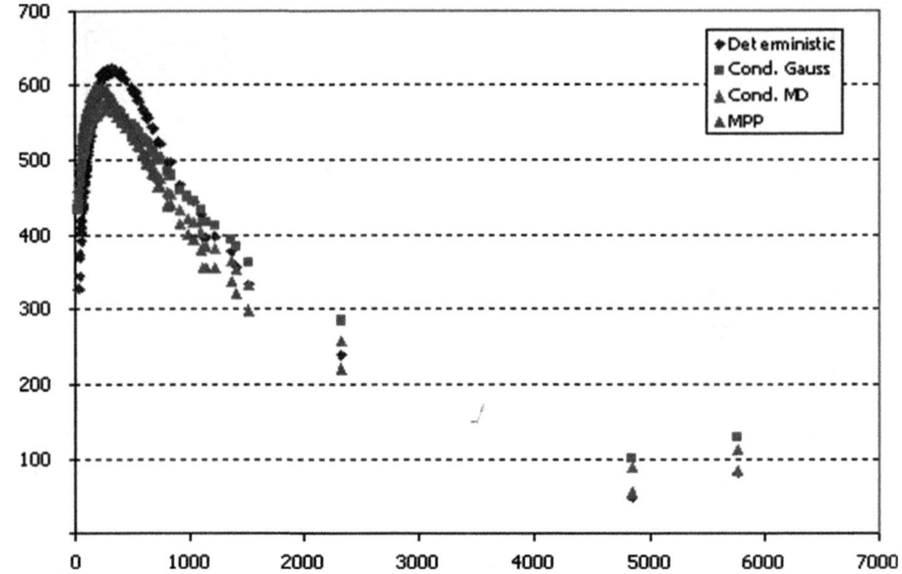

Fig. 19.16 Comparison of the Single-Name Deltas for the (10–15%) tranche

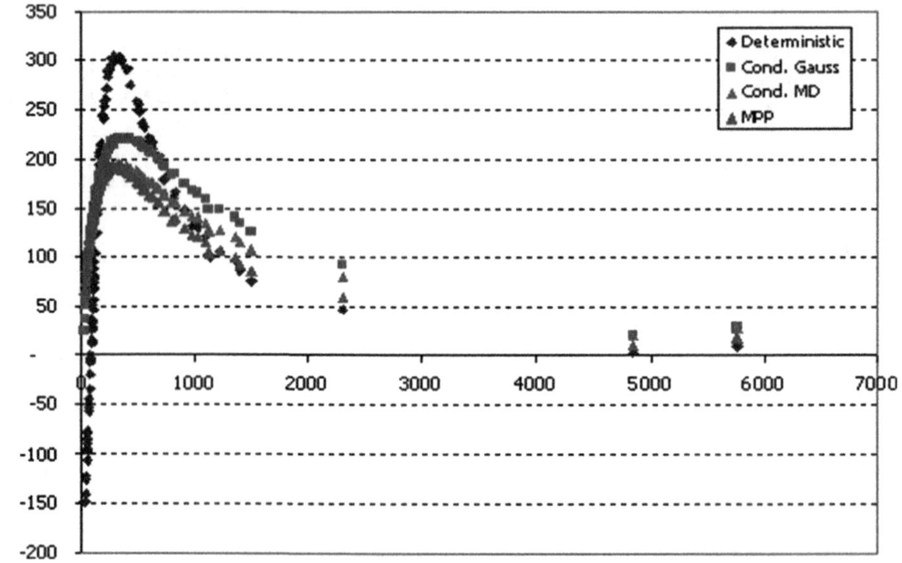

Fig. 19.17 Comparison of the Single-Name Deltas for the (15–30%) tranche

19.5 Summary and Conclusion

In this chapter, we have introduced stochastic recovery as a way to overcome some of the limitations of the base correlation framework and reviewed the various stochastic recovery models in the literature. First, we described a general class of Conditional-Functional Recovery models and studied some of the most commonly used specifications. Then, we took the converse view and constructed a set of stochastic recovery (modelling) building blocks based on a top-down paradigm. In particular, we have shown that the top-down approach provides a very useful tool for analyzing the structure of the standard bottom-up stochastic recovery models. And finally, we have analyzed the calibration behaviour in each model and the implications for risk managing a correlation book.

References

E.I. Altman, B. Brady, A. Resti, A. Sironi, The link between defaults and recovery rates: theory, empirical evidence, and implications (Working Paper, Stern School of Business, 2003)

S. Amraoui, S. Hitier, Optimal stochastic recovery for base correlation (Working Paper, BNP Paribas, 2008)

L. Andersen, J. Sidenius, Extensions to the Gaussian Copula: random recovery and random factor loadings. J. Credit Risk **1**(1), Winter 2004/05, 29–70 (2005)

S. Chen, Beta Kernel estimators for density functions. Comput. Stat. Data Anal. **31**, 131–145 (1999)

Y. Hu, W. Perraudin, The dependence of recovery rates and defaults (Working Paper, Birkbeck College, 2002)

M. Krekel, Pricing distressed CDOs with base correlation and stochastic recovery (Working Paper, Unicredit, 2008)

O. Renault, O. Scaillet, On the way to recovery: a nonparametric bias free estimation of recovery rate densities (FAME Research Paper, No. 83, May 2003, University of Geneva, 2003)

Part IV

The Next Challenge

20

New Frontiers in Credit Modelling: The CVA Challenge

In this chapter, we present a general framework for evaluating the (counterparty) Credit Valuation Adjustment for CDO tranches. We shall see that given the "exotic" nature of the CVA derivative payoff, we will have to leverage a variety of modelling techniques that have been developed over years for the correlation book; this includes: default correlation modelling, credit index options' pricing, dynamic credit modelling and CDO-squared pricing.

20.1 Introduction

Since the beginning of the credit crisis, modelling counterparty credit risk and the correct pricing and hedging of Credit Valuation Adjustments (CVA) has become a critical issue for financial institutions. This is due to a number of reasons:

- given the unprecedented level of volatility in credit markets, the CVA contribution to the quarterly earnings of financial firms is not negligible anymore (both on the asset side and the liability side);
- the CVA numbers are very volatile and have a large impact on reported earnings[1];
- banks have large concentrated exposures to various companies in specific sectors,[2] which, in times of crisis, can experience high default rates and high credit spreads.

This has led to a renewed focus on trading and hedging the CVA Profit and Loss (P&L) reported by the various businesses within the bank. Now, CVA

Y. Elouerkhaoui, *Credit Correlation*, Applied Quantitative Finance,
DOI 10.1007/978-3-319-60973-7_20

is treated like any other derivatives book with Gamma and Cross-Gamma exposures. For the credit correlation business, in particular, this is even more relevant since default correlation is the primary risk traded by the desk. Indeed, the payoff of a CDO tranche with counterparty default risk is similar, in many ways, to the payoff of a CDO-Squared structure, where one of the underlying instruments is the CDO tranche, and the other one is the counterparty reference CDS. Viewed from this angle, CVA is just another "Correlation" credit derivative payoff, which can be traded and risk-managed similarly to the rest of the structured credit book. Therefore, the same mathematical machinery developed for pricing, modelling and hedging structured credit risk can be used for the CVA book as well.

Most of the correlation desk counterparties (or collaterals) have liquidly traded CDS and can be hedged dynamically. This is one of the main differences between credit and the other asset classes—such as Rates, FX, Munis or Commodities—where typical credit exposures are contracted through unmargined swaps to corporates. The key issue for credit is the modelling of the correlation between the CDO portfolio and the counterparty reference entity. In some instances, it can be even trickier when the counterparty reference belongs to the CDO underlying portfolio as well; this raises similar issues encountered when modelling portfolio overlaps in CDO-Squared structures.

Another difference with the other businesses is that many CDO tranches are usually sold to investors in a funded format where a note is issued by a Special Purpose Vehicle (SPV), which holds some collateral bought at inception by the proceeds of the note issuance, and then faces the bank by exchanging the cash flows of the swap. The counterparty risk to the bank is when there is a default of the collateral. Typical questions around Collateral Support Annexes (CSAs), netting agreements, margin posting, and collateral management are not relevant in this case. Usually, there is only one swap in the SPV transaction; the payment at default and the SPV unwind process are documented in the SPV termsheet.

There are a number of good publications on CVA modelling for Credit Default Swaps, but very little on CVA for CDOs. This includes: Brigo and Capponi (2009)—"Unilateral CVA with Correlation and Volatility", Brigo and Chourdakis (2009)—"Bilateral CVA with Stochastic Dynamic Model", Crepey et al. (2010)—"CDS with Counterparty Risk in a Markov Chain Copula Model with Joint Defaults".

The aim of this chapter is to present a general methodology for pricing CVA on CDO tranche securities. Given the exotic nature of the CVA payoff, we will use a combination of default correlation modelling techniques that have been developed, originally, for pricing correlation products; this includes:

- default correlation modelling;
- pricing of credit (index) options;
- dynamic credit modelling;
- CDO-squared pricing.

This chapter is structured as follows. In Sect. 20.2, we review some of the main CVA concepts and definitions. Then, we gradually construct the building blocks needed for evaluating CVA on CDO tranches. In Sect. 20.3, we start with the one-dimensional case and introduce the Conditional Forward Annuity Measure, which is then used to derive the unfunded and funded CVA for CDSs. Then, in Sect. 20.4, we address the same problem for CDOs. To this end, we combine three components: (1) a CDO-Squared model, (2) a Tranche Option model, and (3) a Markovian dynamic model for Forward tranches to derive the CVA for CDOs.

20.2 Review of CVA Concepts and Definitions

In this section, we give some general definitions and derive generic CVA results that do not depend on a particular choice of dynamics or payoff specification.

20.2.1 Set-up

As usual, we work on a probability space $(\Omega, \mathcal{G}, \mathbb{P})$ where we have a set of default times $(\tau_1, ..., \tau_n)$, representing the defaults of a reference portfolio of n obligors. We denote by τ_c, the default time of the counterparty, and we denote by τ_b, the default time of the bank. Their recovery rates are $(R_1, .., R_n)$, R_c and R_b respectively; and their default indicators are denoted by $D_t^i = \mathbf{1}_{\{\tau_i \leq t\}}$, $D_t^c = \mathbf{1}_{\{\tau_c \leq t\}}$, $D_t^b = \mathbf{1}_{\{\tau_b \leq t\}}$ respectively.

The enlarged filtration $\{\mathcal{G}_t\}$ that we work with contains both the defaults filtration $\{\mathcal{H}_t\} = \left(\bigvee_{i=1}^n \mathcal{H}_t^i\right) \vee \mathcal{H}_t^b \vee \mathcal{H}_t^c$ and the background filtration $\{\mathcal{F}_t\}$.

We define a generic derivative contract by its cumulative dividend process C_t. In general, the dividend process is considered to be of the form $C_t = A_t - B_t$, where A and B are bounded increasing adapted right-continuous left-limit (cadlag) processes.

The value of the derivative security C without counterparty risk is given by:

$$V_t = \mathbb{E}\left[\int_t^T p_{t,s} dC_s \,|\mathcal{G}_t\right],$$

where $p_{t,s}$ is the risk-free discount factor $p_{t,s} = \exp\left(-\int_t^s r_u du\right)$, at time t, maturing at s. For example, for a CDS contract, the dividend process is:

$$C_s = (1 - R)\, D_s - \sum_{T_i} \mathbf{1}_{\{T_i \leq s\}} S^0 \delta_i \left(1 - D_{T_i}\right),$$

where S^0 is the running spread, $\{T_i\}$ are the coupon dates, and δ_i are the daycount fractions. For a CDO tranche, the dividend process is

$$C_s = \widehat{L}_s - \sum_{T_i} \mathbf{1}_{\{T_i \leq s\}} S^0 \delta_i \left(N - \widehat{L}_{T_i}\right),$$

here N is the total notional and $\widehat{L}_T$ is the tranche loss variable at time T.

20.2.2 Terminology

When two counterparties enter into a derivatives transaction, they are implicitly long each other's default risk. The bank is short default protection on the counterparty, and is long protection on its own default risk. The risk of losses (or gains) due the default of the counterparty (or one's own default) is referred to as **Counterparty Credit Risk**. The PV adjustment to be added to the value of a default-free derivative is referred to as **Counterparty Valuation Adjustment**.

Unilateral CVA is the PV adjustment where we only consider the default of the counterparty and we assume that the bank is default-free.

Bilateral CVA considers the symmetric effect of both counterparty default and the bank's own default as well.

Asset CVA is the credit charge due to the counterparty default.

Liability CVA is the credit benefit due to the bank's own default.

In the case of default, the **Close-Out Value**, upon termination of the contract, is denoted by $\chi(\tau_c)$. In general, the Close-Out value is assumed to be the value of the contract without counterparty default risk, at the time of default, i.e., $\chi(\tau_c) = V_{\tau_c}$.

20.2.3 Unilateral CVA

When the counterparty defaults, there are two cases: if the Mark-To-Market (MtM) of the swap is in-the-money for us, we can only recover a fraction of this P&L; if the MtM of the swap is in favour of the counterparty then we have to pay back the entire P&L to the liquidators of the defaulted counterparty:

$$\widehat{V}_{\tau_C}^R = \begin{cases} R_C V_{\tau_C} \text{ if } V_{\tau_C} \geq 0, \\ V_{\tau_C} \text{ if } V_{\tau_C} < 0. \end{cases}$$

Thus, by taking the counterparty default risk into account, the value of the contract becomes:

$$\widehat{V}_t = \mathbb{E}\left[\int_t^T \mathbf{1}_{\{\tau_C > s\}} p_{t,s} dC_s + p_{t,\tau_C}\left(R_C V_{\tau_C}^+ + V_{\tau_C}^-\right)\mathbf{1}_{\{\tau_C \leq T\}} |\mathcal{G}_t\right],$$

which has two legs: the value of the extinguishing contract and the value of the recovery leg in the case of early termination,

$$\widehat{V}_t^E = \mathbb{E}\left[\int_t^T \mathbf{1}_{\{\tau_C > s\}} p_{t,s} dC_s \, |\mathcal{G}_t\right],$$
$$\widehat{V}_t^R = \mathbb{E}\left[p_{t,\tau_C}\left(R_C V_{\tau_C}^+ + V_{\tau_C}^-\right)\mathbf{1}_{\{\tau_C \leq T\}} |\mathcal{G}_t\right],$$

where we use the notations $x^+ = \max(x, 0)$ and $x^- = \min(x, 0)$.

Let $\xi_{(\tau_C)}$ denote the loss incurred by the firm at time τ_C due the counterparty default. In the case of a non-margined swap, we have:

$$\xi_{(\tau_C)} = (1 - R_C)\, V_{\tau_C}^+ .$$

The CVA and the Expected Positive Exposure (EPE) are then defined as follows.

Definition 163 The Credit Valuation Adjustment (CVA) is the $\mathcal{G}$-adapted process defined by: for $t \in [0, T]$,

$$CVA_t = \mathbb{E}\left[p_{t,\tau_C}\xi_{(\tau_C)}\mathbf{1}_{\{\tau_C \leq T\}} |\mathcal{G}_t\right].$$

The Expected Positive Exposure (EPE) is the function of time defined as: for $t \in [0, T]$,

$$EPE(t) = \mathbb{E}\left[\xi_{(\tau_C)} \, |\tau_C = t\right].$$

The CVA process is given by the difference of the risky and riskless PVs.

Proposition 164 *For all times before the default of the counterparty, the CVA can be obtained as*

$$CVA_t = V_t - \widehat{V}_t, \textit{ for all } t < \tau_c.$$

Proof We can write the value of the extinguishing leg in terms of the risk-free PV as:

$$\begin{aligned}
\widehat{V}_t^E &= \mathbb{E}\left[\int_t^T \mathbf{1}_{\{\tau_c > s\}} p_{t,s} dC_s \,|\mathcal{G}_t\right] \\
&= \mathbb{E}\left[\mathbf{1}_{\{\tau_c > T\}}\left[\int_t^T p_{t,s} dC_s\right] + \mathbf{1}_{\{\tau_c \le T\}}\left[\int_t^{\tau_c} p_{t,s} dC_s\right] |\mathcal{G}_t\right] \\
&= \mathbb{E}\left[\mathbf{1}_{\{\tau_c > T\}} V_t + \mathbf{1}_{\{\tau_c \le T\}}\left[V_t - \int_{\tau_c}^T p_{t,s} dC_s\right] |\mathcal{G}_t\right] \\
&= \mathbb{E}\left[\mathbf{1}_{\{\tau_c > T\}} V_t + \mathbf{1}_{\{\tau_c \le T\}}\left[V_t - p_{t,\tau_c}\mathbb{E}\left[\int_{\tau_c}^T p_{\tau_c,s} dC_s \,|\mathcal{G}_{\tau_c}\right]\right] |\mathcal{G}_t\right] \\
&= V_t - \mathbb{E}\left[\mathbf{1}_{\{\tau_c \le T\}} p_{t,\tau_c} V_{\tau_c} \,|\mathcal{G}_t\right].
\end{aligned}$$

Hence,

$$\begin{aligned}
\widehat{V}_t &= V_t - \mathbb{E}\left[\mathbf{1}_{\{\tau_c \le T\}} p_{t,\tau_c} V_{\tau_c} - \left[p_{t,\tau_c}\left(R_c V_{\tau_c}^+ + V_{\tau_c}^-\right)\mathbf{1}_{\{\tau_c \le T\}}\right] |\mathcal{G}_t\right] \\
&= V_t - \mathbb{E}\left[\mathbf{1}_{\{\tau_c \le T\}} p_{t,\tau_c}\left(1 - R_c\right) V_{\tau_c}^+ \,|\mathcal{G}_t\right] = V_t - CVA_t.
\end{aligned}$$

■

If we have a portfolio of transactions with the same counterparty, then when a default occurs, the exposures are netted against each other and the recovery value is applied to the net position.

Definition 165 (*Netting*) If we have a portfolio of trades $\left(V_t^i\right)_{1 \le i \le n}$, with a given counterparty where a netting agreement is in place, then the CVA is given by

$$CVA_t = \mathbb{E}\left[p_{t,\tau_c}\left(1 - R_c\right)\left(\sum_{i=1}^n V_{\tau_c}^i\right)^+ \mathbf{1}_{\{\tau_c \le T\}} \,|\mathcal{G}_t\right].$$

20.2.4 Bilateral CVA

We consider the bank's own default probability and the possibility of the bank defaulting before the counterparty. The loss incurred when the counterparty defaults after the bank is referred to as **Asset CVA**. The benefit gained when the bank defaults before the counterparty is called **Liability CVA**.

Let τ_f denote the first-to-default time of the counterparty and the bank $\tau_f = \min(\tau_b, \tau_c)$.

The value of the risky security then becomes

$$\widehat{V}_t = \mathbb{E}\left[\int_t^T \mathbf{1}_{\{\tau_f > s\}} p_{t,s} dC_s + \mathbf{1}_{\{\tau_f \le T\}} \left[\begin{array}{c} \mathbf{1}_{\{\tau_f = \tau_c\}} p_{t,\tau_c}\left(R_c V_{\tau_f}^+ + V_{\tau_f}^-\right) \\ +\mathbf{1}_{\{\tau_f = \tau_b\}} p_{t,\tau_c}\left(V_{\tau_f}^+ + R_b V_{\tau_f}^-\right) \end{array}\right] |\mathcal{G}_t\right].$$

The loss $\xi_{(\tau_f)}^c$ incurred by the firm, at time τ_f, due the counterparty default (conditional on the firm's survival) is

$$\xi_{(\tau_f)}^c = (1 - R_c)\, V_{\tau_f}^+,$$

and the loss $\xi_{(\tau_f)}^b$ incurred by the counterparty due to the firm's default (conditional on the counterparty surviving) is

$$\xi_{(\tau_f)}^b = (1 - R_b)\, V_{\tau_f}^-.$$

To simplify the formulae, we have assumed that we do not have simultaneous defaults.

Definition 166 The (Asset) Credit Valuation Adjustment (CVA) is the $\mathcal{G}$-adapted process defined by, for $t \in [0, T]$,

$$CVA_t^{Asset} = \mathbb{E}\left[p_{t,\tau_c}\xi_{(\tau_f)}^c \mathbf{1}_{\{\tau_f = \tau_c\}}\mathbf{1}_{\{\tau_f \le T\}} |\mathcal{G}_t\right].$$

The (Liability) CVA, also referred to as Debt Valuation Adjustment (DVA), is the $\mathcal{G}$-adapted process defined by, for $t \in [0, T]$,

$$CVA_t^{Liability} = DVA_t = \mathbb{E}\left[p_{t,\tau_c}\xi_{(\tau_f)}^b \mathbf{1}_{\{\tau_f = \tau_b\}}\mathbf{1}_{\{\tau_f \le T\}} |\mathcal{G}_t\right].$$

The Expected Positive Exposure (EPE) is the function of time defined by, for $t \in [0, T]$,

$$EPE(t) = \mathbb{E}\left[\xi_{(\tau_f)}^c \,|\tau_f = t, \tau_f = \tau_c\right].$$

The Expected Negative Exposure (ENE) is the function of time defined as, for $t \in [0, T]$,

$$ENE(t) = \mathbb{E}\left[\xi_{(\tau_f)}^b \,|\tau_f = t, \tau_f = \tau_b\right].$$

The Bilateral CVA is defined as the sum of the Asset CVA and the Liability CVA

$$CVA_t = CVA_t^{Asset} + CVA_t^{Liability}$$

The Bilateral CVA process will be given by the difference between the risky and the risk-free PVs.

Proposition 167 *For all times before the default of either the counterparty or the firm, the Bilateral CVA can be obtained as*

$$CVA_t = V_t - \widehat{V}_t, \; for \; all \; t < \tau_f.$$

Proof The value of the extinguishing leg can be written in terms of the risk-free PV as:

$$\widehat{V}_t^E = \mathbb{E}\left[\int_t^T \mathbf{1}_{\{\tau_f > s\}} p_{t,s} dC_s \,|\mathcal{G}_t\right] = V_t - \mathbb{E}\left[\mathbf{1}_{\{\tau_f \leq T\}} p_{t,\tau_f} V_{\tau_f} \,|\mathcal{G}_t\right].$$

Hence,

$$\begin{aligned}
\widehat{V}_t &= V_t - \mathbb{E}\left[\mathbf{1}_{\{\tau_f \leq T\}} p_{t,\tau_f} V_{\tau_f} - \mathbf{1}_{\{\tau_f \leq T\}} \left[\begin{array}{c} \mathbf{1}_{\{\tau_f = \tau_c\}} p_{t,\tau_f} \left(R_c V_{\tau_f}^{+} + V_{\tau_f}^{-}\right) \\ + \mathbf{1}_{\{\tau_f = \tau_b\}} p_{t,\tau_f} \left(V_{\tau_f}^{+} + R_b V_{\tau_f}^{-}\right) \end{array}\right] |\mathcal{G}_t\right] \\
&= V_t - \mathbb{E}\left[\mathbf{1}_{\{\tau_f \leq T\}} \mathbf{1}_{\{\tau_f = \tau_c\}} p_{t,\tau_f} \left(1 - R_c\right) V_{\tau_c}^{+} \,|\mathcal{G}_t\right] \\
&\quad - \mathbb{E}\left[\mathbf{1}_{\{\tau_f \leq T\}} \mathbf{1}_{\{\tau_f = \tau_b\}} p_{t,\tau_f} \left(1 - R_b\right) V_{\tau_f}^{-} \,|\mathcal{G}_t\right] \\
&= V_t - CVA_t.
\end{aligned}$$

■

The purpose of this section was to define the bilateral CVA and to illustrate the type of calculations involved for this metric. But in the rest of this chapter, we will focus on the unilateral asset-side CVA and develop the modelling framework for this payoff. Similar methods and techniques are also applied to the bilateral payoffs.

20.2.5 SPVs and Funded CVA

When the counterparty of the swap is a collateralized SPV, the default time can be triggered by two events: either default of the collateral or insufficient collateral to cover the losses incurred on the underlying contract. The credit

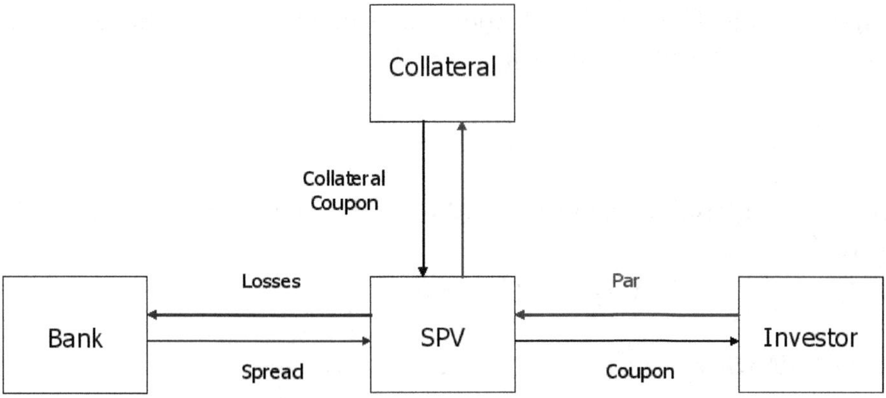

Fig. 20.1 Structure of a note issued by an SPV with collateral

risk in this structure is driven by the SPV collateral; the counterparty, in this case, is the underlying reference of the collateral. When the SPV is unwound because of the complete erosion of collateral this is deemed to be a market risk event and is not included in CVA, but this option is valued in as part of the deal. We can also have a recovery overide on the collateral if additional guarantees are added to the structure (Fig. 20.1).

The default time τ_c in this case refers to the default time of the collateral; and the value of the CVA recovery leg becomes

$$\widehat{V}_t^R = \mathbb{E}\left[p_{t,\tau_f} \min\left(R_c\widehat{N}_{\tau_c}, V_{\tau_c}\right) \mathbf{1}_{\{\tau_c \le T\}} | \mathcal{G}_t\right],$$

where $\widehat{N}_{\tau_c}$ is the remaining notional at the time of default.

The payoff upon default will be different dependent on whether we are long or short the note. Typically, we are issuing the note and we pay the coupons to the client.

If we have a funded note where the counterparty is an SPV with collateral, then the loss incurred by the firm, at time τ_c, due the counterparty default is

$$\xi_{(\tau_c)} = \left(V_{\tau_c} - R_c\widehat{N}_{\tau_c}\right)^+.$$

The payoff upon default can be different between various SPVs depending on the agreed termsheet. For example, we could have: a Par Put feature, or a Deferred Loss Payment feature.

The CVA process in this case is defined as follows.

Definition 168 The Credit Valuation Adjustment (CVA) is the $\mathcal{G}$-adapted process defined by, for $t \in [0, T]$,

$$CVA_t = \mathbb{E}\left[p_{t,\tau_f}\left(V_{\tau_c} - R_c\widehat{N}_{\tau_c}\right)^+ \mathbf{1}_{\{\tau_c \le T\}} |\mathcal{G}_t\right].$$

The Expected Positive Exposure (EPE) is the function of time defined by, for $t \in [0, T]$,

$$EPE(t) = \mathbb{E}\left[\left(V_{\tau_c} - R_c\widehat{N}_{\tau_c}\right)^+ |\tau_c = t\right].$$

The CVA will be equal to the difference between the risky and riskless PVs.

Proposition 169 *For all times before the default of the collateral, the CVA can be obtained as*

$$CVA_t = V_t - \widehat{V}_t, \; for \; all t < \tau_c.$$

Proof Similar to the unfunded case, we have

$$\widehat{V}_t^E = V_t - \mathbb{E}\left[\mathbf{1}_{\{\tau_c \le T\}} p_{t,\tau_c} V_{\tau_c} |\mathcal{G}_t\right],$$

which yields

$$\begin{aligned}
\widehat{V}_t &= V_t - \mathbb{E}\left[\mathbf{1}_{\{\tau_c \le T\}} p_{t,\tau_c} V_{\tau_c} - \left[p_{t,\tau_c} \min\left(R_c\widehat{N}_{\tau_c}, V_{\tau_c}\right) \mathbf{1}_{\{\tau_c \le T\}}\right] |\mathcal{G}_t\right] \\
&= V_t - \mathbb{E}\left[\mathbf{1}_{\{\tau_c \le T\}} p_{t,\tau_c}\left[V_{\tau_c} - \min\left(R_c\widehat{N}_{\tau_c}, V_{\tau_c}\right)\right] |\mathcal{G}_t\right] \\
&= V_t - \mathbb{E}\left[p_{t,\tau_c}\left(V_{\tau_c} - R_c\widehat{N}_{\tau_c}\right)^+ \mathbf{1}_{\{\tau_c \le T\}} |\mathcal{G}_t\right] \\
&= V_t - CVA_t.
\end{aligned}$$

■

20.2.6 CVA Lower Bound

In general, to compute CVA, we need to calculate a sum of swaption prices (i.e. options on the forward value of the trade) weighted by the counterparty default probabilities. Using Jensen's inequality, we can simplify the pricing and get a lower bound on the CVA, which gives a good approximation when the underlying trade in deeply in-the-money.

For a non-margined swap, we have:

$$\begin{aligned}
CVA_t &= \mathbb{E}\left[p_{t,\tau_c}\left(1 - R_c\right) \mathbf{1}_{\{\tau_c \le T\}} \max(V_{\tau_c}, 0) |\mathcal{G}_t\right] \\
&\ge \max\left(\left(1 - R_c\right) \mathbb{E}\left[p_{t,\tau_c} \mathbf{1}_{\{\tau_c \le T\}} V_{\tau_c} |\mathcal{G}_t\right], 0\right).
\end{aligned}$$

For a funded note, we have:

$$
\begin{aligned}
CVA_t &= \mathbb{E}\left[p_{t,\tau_c} \max\left(V_{\tau_c} - R_c\widehat{N}_{\tau_c}, 0\right) \mathbf{1}_{\{\tau_c \le T\}} \,|\mathcal{G}_t\right] \\
&\ge \max\left(\mathbb{E}\left[p_{t,\tau_c}\left(V_{\tau_c} - R_c\widehat{N}_{\tau_c}\right) \mathbf{1}_{\{\tau_c \le T\}} \,|\mathcal{G}_t\right], 0\right) \\
&\ge \max\left(\mathbb{E}\left[p_{t,\tau_c} V_{\tau_c} \mathbf{1}_{\{\tau_c \le T\}} \,|\mathcal{G}_t\right] - R_c\mathbb{E}\left[p_{t,\tau_c}\widehat{N}_{\tau_c} \mathbf{1}_{\{\tau_c \le T\}} \,|\mathcal{G}_t\right], 0\right).
\end{aligned}
$$

The first term can be computed easily as

$$
\begin{aligned}
\mathbb{E}\left[p_{t,\tau_c} V_{\tau_c} \mathbf{1}_{\{\tau_c \le T\}} \,|\mathcal{G}_t\right] = \mathbb{E}\left[p_{t,\tau_c} V_{\tau_c} \,|\mathcal{G}_t\right] &= \mathbb{E}\left[p_{t,\tau_c}\mathbb{E}\left[\int_{\tau_c}^{T} p_{\tau_c,s} dC_s \,|\mathcal{G}_{\tau_c}\right] |\mathcal{G}_t\right] \\
&= \mathbb{E}\left[\int_{\tau_c}^{T} p_{t,s} dC_s \,|\mathcal{G}_t\right] \\
&= \mathbb{E}\left[\int_{t}^{T} p_{t,s} dC_s \,|\mathcal{G}_t\right] - \mathbb{E}\left[\int_{t}^{\tau_c} p_{t,s} dC_s \,|\mathcal{G}_t\right] \\
&= V_t - \widehat{V}_t^E.
\end{aligned}
$$

The second term is given by

$$
\begin{aligned}
\mathbb{E}\left[p_{t,\tau_c}\widehat{N}_{\tau_c} \mathbf{1}_{\{\tau_c \le T\}} \,|\mathcal{G}_t\right] &= \mathbb{E}\left[p_{t,\tau_c} N \mathbf{1}_{\{\tau_c \le T\}} \,|\mathcal{G}_t\right] - \mathbb{E}\left[p_{t,\tau_c}\widehat{L}_{\tau_c} \mathbf{1}_{\{\tau_c \le T\}} \,|\mathcal{G}_t\right] \\
&= N\left[\int_t^T p_{t,s}\mathbb{P}_t\left(\tau_c \in ds\right)\right] - \left[\int_t^T p_{t,s}\mathbb{E}_t\left[\widehat{L}_{s^-} dD_s^c\right]\right].
\end{aligned}
$$

Note that to simplify the formula, we have assumed that the collateral trades at Par during the lifetime of the trade.

So far, we have developed generic CVA formulae that are model-independent. To proceed, we need to make specific choices of model dynamics.

First, to compute the value of the extinguishing leg of the trade, we need a "correlation model" that links the default of the counterparty to the losses on the CDO.

Second, to value the CVA payoff and the recovery leg of the trade with counterparty risk, we need an "option model" since, at the end, CVA pricing boils down to computing an integral of swaption prices weighted by the counterparty default probability.

Next, we show how this is done for CDSs, then for CDOs.

20.3 CVA for CDSs

The filtration in this case contains the information on the CDS reference default and the default of the counterparty $\mathcal{G}_t = \mathcal{F}_t \vee \mathcal{H}_t^i \vee \mathcal{H}_t^c$.

In this section, we omit the index i to lighten up the notations.

We consider a CDS maturing at time T, with payment schedule $(T_0 = 0, T_1, ..., T_n)$; δ_i is the accrual fraction between T_i and T_{i-1}, and S^0 is the CDS running spread.

The value, at time t, of a CDS contract where we buy protection is given by

$$V_{t,t,T}^{CDS} = \mathbb{E}\left[(1-R)\int_t^T p_{t,s} dD_s - \sum_{T_i>t} p_{t,T_i} S^0 \delta_i \left(1 - D_{T_i}\right) |\mathcal{G}_t\right]$$

$$= \mathbf{1}_{\{\tau>t\}}\left[(1-R)\int_t^T p_{t,s} d\mathbb{E}\left[D_s |\mathcal{G}_t\right] - \sum_{T_i>t} p_{t,T_i}\mathbb{E}\left[\left(1 - D_{T_i}\right) |\mathcal{G}_t\right] S^0 \delta_i\right].$$

The unfunded CVA of this contract is

$$CVA_0 = (1-R_c)\int_0^T p_{0.t}\mathbb{E}\left[\left(V_{t,t,T}^{CDS}\right)^+ |\tau_c = t\right]\mathbb{P}\left(\tau_c \in dt\right).$$

To evaluate the CVA for a CDS contract, we need two components: a) a CDS Option model and, b) the correct modelling of the forward value, which captures spread widening effects due the default correlation between the underlying reference credit and the counterparty.

20.3.1 Pricing CDS Options

To start with, we show how the pricing of single-name CDS options is done, before adapting it to CVA pricing by introducing the concept of a Conditional Forward Annuity Measure.

The Filtration in this case is $\mathcal{G}_t = \mathcal{F}_t \vee \mathcal{H}_t$.

We denote by $Q_{t,T}$ the $\{\mathcal{F}_t\}$-adapted process representing the conditional survival probability of the reference entity:

$$Q_{t,T} \triangleq \frac{\mathbb{P}\left(\tau > T |\mathcal{F}_t\right)}{\mathbb{P}\left(\tau > T |\mathcal{F}_t\right)}.$$

We have on the set $\mathbf{1}_{\{\tau>t\}}$:

$$\mathbf{1}_{\{\tau>t\}}\mathbb{P}\left(\tau > T |\mathcal{G}_t\right) = \mathbf{1}_{\{\tau>t\}}\frac{\mathbb{P}\left(\tau > T |\mathcal{F}_t\right)}{\mathbb{P}\left(\tau > T |\mathcal{F}_t\right)} = \mathbf{1}_{\{\tau>t\}} Q_{t,T}.$$

The break-even spread, at time t, of the forward swap starting at time U and maturing at time T is given by

$$S_{t,U,T} = \frac{-(1-R)\int_U^T p_{t,U,s} dQ_{t,U,s}}{\sum_{T_i > U} p_{t,U,T_i} Q_{t,U,T_i} \delta_i},$$

where $p_{t,U,T}$ and $Q_{t,U,T}$ are the forward discount factor and forward survival probabilities,

$$p_{t,U,T} \triangleq \frac{p_{t,T}}{p_{t,U}},$$
$$Q_{t,U,T} \triangleq \frac{Q_{t,T}}{Q_{t,U}}.$$

We also denote by $A_{t,U,T}$ the value of the forward annuity at time t:

$$A_{t,U,T} \triangleq \sum_{T_i > U} p_{t,U,T_i} Q_{t,U,T_i} \delta_i \, .$$

Then, the value of the CDS can be expressed as

$$V_{t,t,T}^{CDS} = \mathbf{1}_{\{\tau > t\}} A_{t,t,T} \left(S_{t,t,T} - S^0 \right);$$

and the value of the option to enter into the CDS swap starting at time t and maturing at T is given by

$$O_{0,t,T} = p_{0,t} \mathbb{E} \left[\left(V_{t,t,T}^{CDS} \right)^+ \right] = p_{0,t} Q_{0,t} \mathbb{E} \left[A_{t,t,T} \left(S_{t,t,T} - S^0 \right) \right].$$

The annuity defined here is strictly positive a.s. (and $\{\mathcal{F}_t\}$-adapted), so we can define a forward annuity measure. The break-even spread is a martingale under this measure, and we can use the standard Black formula to compute this expectation (see, for example, Brigo and Morini (2005) for technical details) :

$$\begin{aligned} O_{0,t,T} &= p_{0,t} Q_{0,t} A_{0,t,T} \mathbb{E}^{Q_A} \left[\left(S_{t,t,T} - S^0 \right) \right] \\ &= p_{0,t} Q_{0,t} A_{0,t,T} \, Black \left(S_{0,t,T}, S^0, \sigma_{BS} \sqrt{T-t} \right) \\ &= \left[\sum_{T_i > U} p_{0,T_i} Q_{0,T_i} \delta_i \right] Black \left(S_{0,t,T}, S^0, \sigma_{BS} \sqrt{T-t} \right). \end{aligned}$$

20.3.2 The Conditional Forward Annuity Measure

We use similar techniques to price the option payoff required for CVA calculations. The filtration in this case is $\mathcal{G}_t = \mathcal{F}_t \vee \mathcal{H}_t \vee \mathcal{H}_t^c$. Here, we will be pricing a set of Conditional CDS Options.

When there is no correlation between the underlying credit and the counterparty, this is simply the integral of the standard CDS option prices given above

$$\begin{aligned} CVA_0 &= (1 - R_c) \int_0^T p_{0,t} \mathbb{E}\left[\left(V_{t,t,T}^{CDS} \right)^+ | \tau_c = t \right] \mathbb{P}(\tau_c \in dt) \\ &= (1 - R_c) \int_0^T p_{0,t} \mathbb{E}\left[\left(V_{t,t,T}^{CDS} \right)^+ \right] \mathbb{P}(\tau_c \in dt) \\ &= (1 - R_c) \int_0^T O_{0,t,T} \mathbb{P}(\tau_c \in dt) . \end{aligned}$$

If there is a correlation, then we need to work under the **Conditional Forward Annuity Measure**.

Using the Generalized Dellacherie Formula (e.g., see Jeanblanc and Rutkowski (2000)), we have, for all $T \geq t$,

$$\begin{aligned} \mathbb{P}(\tau > T | \mathcal{G}_t) = \mathbf{1}_{\{\tau > t\}} \mathbf{1}_{\{\tau_c > t\}} \frac{\mathbb{P}(\tau > T, \tau_c > t | \mathcal{F}_t)}{\mathbb{P}(\tau > t, \tau_c > t | \mathcal{F}_t)} \\ + \mathbf{1}_{\{\tau > t\}} \mathbf{1}_{\{\tau_C \leq t\}} \frac{\mathbb{P}(\tau > T | \mathcal{F}_t \vee \sigma(\tau_c))}{\mathbb{P}(\tau > t | \mathcal{F}_t \vee \sigma(\tau_c))} . \end{aligned}$$

Conditional on the counterparty default, the reference entity survival probability becomes

$$\begin{aligned} &\mathbb{P}(\tau > T | \mathcal{G}_t \vee \{\tau > U\} \vee \{\tau_c = U\}) \\ &= \mathbf{1}_{\{\tau > U\}} \mathbf{1}_{\{\tau_C = U\}} \mathbb{P}(\tau > T | \mathcal{G}_t \vee \{\tau > U\} \vee \{\tau_c = U\}) \\ &= \mathbf{1}_{\{\tau > U\}} \mathbf{1}_{\{\tau_C = U\}} \mathbb{P}(\tau > T | \mathcal{F}_t \vee \{\tau > U\} \vee \{\tau_c = U\}) \\ &= \mathbf{1}_{\{\tau > U\}} \mathbf{1}_{\{\tau_C = U\}} \frac{\mathbb{P}(\tau > T | \mathcal{F}_t \vee \{\tau_c = U\})}{\mathbb{P}(\tau > U | \mathcal{F}_t \vee \{\tau_c = U\})} \\ &= \mathbf{1}_{\{\tau > U\}} \mathbf{1}_{\{\tau_C = U\}} \widetilde{Q}_{t,U,T} , \end{aligned}$$

where we have defined the $\{\mathcal{F}_t\}$-adapted process $\widetilde{Q_{t,U,T}}$ as

$$\widetilde{Q_{t,U,T}} \triangleq \frac{\mathbb{P}(\tau > T \,|\mathcal{F}_t \vee \{\tau_c = U\})}{\mathbb{P}(\tau > U \,|\mathcal{F}_t \vee \{\tau_c = U\})}.$$

To compute the forward survival probability $\widetilde{Q_{t,U,T}}$, we can use a dynamic copula model (see Schönbucher and Schubert (2001)) where the intensities are $\{\mathcal{F}_t\}$-adapted dynamic processes and the default thresholds are linked via a copula function $\mathbb{C}^{\theta}(x_1, x_2)$:

$$\widetilde{Q_{t,U,T}} = \frac{\partial_{x_2}\mathbb{C}^{\theta}\left(\mathbb{P}(\tau > T \,|\mathcal{F}_t), \mathbb{P}(\tau_c > U \,|\mathcal{F}_t)\right)}{\partial_{x_2}\mathbb{C}^{\theta}\left(\mathbb{P}(\tau > U \,|\mathcal{F}_t), \mathbb{P}(\tau_c > U \,|\mathcal{F}_t)\right)}.$$

We consider the value of the forward swap (U, T) conditional on the default of the counterparty at time t:

$$\begin{aligned}\widetilde{V^{CDS}_{t,U,T}} &= \mathbb{E}\left[V^{CDS}_{t,U,T} \,|\tau_c = U\right] = \mathbf{1}_{\{\tau>U\}}\mathbb{E}\left[V^{CDS}_{t,U,T} \,|\tau > U, \tau_c = U\right] \\ &= \mathbf{1}_{\{\tau>U\}}\mathbf{1}_{\{\tau_C=U\}}\left[-(1-R)\int_U^T p_{t,s}d\widetilde{Q_{t,U,s}} - \sum_{T_i>U} p_{t,T_i}S^0\delta_i\widetilde{Q_{t,U,T_i}}\right].\end{aligned}$$

The break-even spread of the forward swap conditional on default is

$$\widetilde{S_{t,U,T}} = \frac{-(1-R)\int_U^T p_{t,U,s}d\widetilde{Q_{t,U,s}}}{\sum_{T_i>U} p_{t,U,T_i}\widetilde{Q_{t,U,T_i}}\delta_i}.$$

The conditional forward annuity $\widetilde{A_{t,U,T}}$ is defined as:

$$\widetilde{A_{t,U,T}} \triangleq \sum_{T_i>U} p_{t,U,T_i}\widetilde{Q_{t,U,T_i}}\delta_i,$$

and the conditional value of the forward CDS is

$$\widetilde{V^{CDS}_{t,t,T}} = \mathbf{1}_{\{\tau>t\}}\mathbf{1}_{\{\tau_C=t\}}\widetilde{A_{t,t,T}}\left(\widetilde{S_{t,t,T}} - S^0\right).$$

The conditional forward CDS option value is then given by

$$\widetilde{O_{0,t,T}} = p_{0,t}\mathbb{E}\left[\left(\widetilde{V^{CDS}_{t,t,T}}\right)^+\right] = p_{0,t}\mathbb{P}(\tau > t, \tau_C = t)\,\mathbb{E}\left[\widetilde{A_{t,t,T}}\left(\widetilde{S_{t,t,T}} - S^0\right)^+\right].$$

The conditional forward annuity is $\{\mathcal{F}_t\}$ -adapted and strictly positive a.s., so we can do a change of measure and use the Black-Scholes derivations:

$$\begin{aligned}\widetilde{O_{0,t,T}} &= p_{0,t}\mathbb{P}\left(\tau > t, \tau_C = t\right)\widetilde{A_{0,t,T}}\, Black\left(\widetilde{S_{0,t,T}}, S^0, \widetilde{\sigma_{BS}}\sqrt{T-t}\right)\\ &= \left[\sum_{T_i > t} p_{0,T_i}\mathbb{P}\left(\tau > t, \tau_C = t\right)\widetilde{Q_{0,t,T_i}}\delta_i\right] Black\left(\widetilde{S_{0,t,T}}, S^0, \widetilde{\sigma_{BS}}\sqrt{T-t}\right).\end{aligned}$$

Proposition 170 *The CVA for a CDS is given by the sum of conditional (Knock-in) CDS options weighted by the default probability of the counterparty,*

$$CVA_0 = (1 - R_c)\int_0^T \widetilde{O_{0,t,T}}\mathbb{P}\left(\tau_c \in dt\right),$$

where $\widetilde{O_{0,t,T}}$ is the value of the conditional (t, T)-forward CDS option

$$\widetilde{O_{0,t,T}} = \left[\sum_{T_i > t} p_{0,T_i}\widetilde{Q^*_{0,t,T_i}}\delta_i\right] Black\left(\widetilde{S_{0,t,T}}, S^0, \widetilde{\sigma_{BS}}\sqrt{T-t}\right),$$

with $\widetilde{Q^*_{0,t,T_i}}$ and $\widetilde{S_{0,t,T}}$ are the conditional forward survival probabilities and conditional forward break-even spread respectively

$$\begin{aligned}\widetilde{Q^*_{0,t,T}} &= \mathbb{P}\left(\tau > T \,|\tau_c = t\right),\\ \widetilde{S_{0,t,T}} &= \frac{-(1-R)\int_t^T p_{0,s}d\widetilde{Q^*_{0,t,s}}}{\sum_{T_i > t} p_{0,T_i}\widetilde{Q^*_{0,t,T_i}}\delta_i}.\end{aligned}$$

The main ingredients that are needed to value the CVA are the conditional forward curves $\widetilde{Q^*_{0,t,T}}$ and the conditional forward CDS break-even spreads $\widetilde{S_{0,t,T}}$. The term-structure of BS volatilities can be provided as an input or can be based on a term-structure model for the hazard rates.

In Fig. 20.2, we give an example of Conditional Forward Spreads, where we compute the (average) CDS spread widening for different correlations. The counterparty CDS is 1000 bps. The base CDS is: low = 100 bps, medium = 1000 bps, high = 2000 bps.

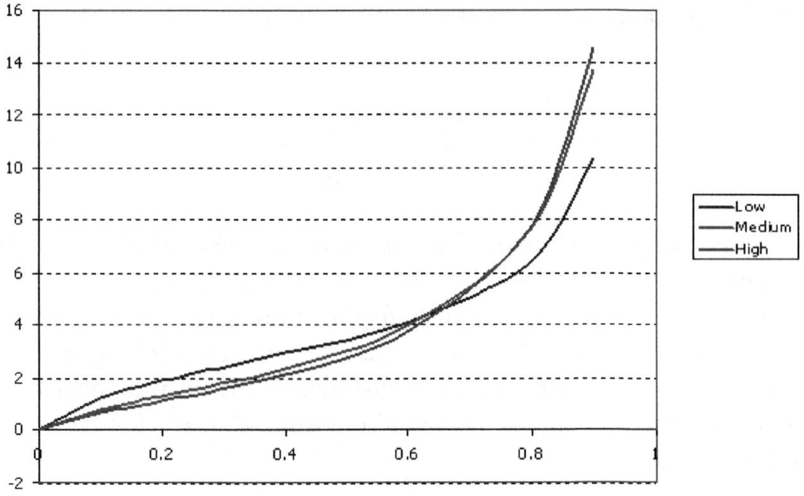

Fig. 20.2 Spread widening for different values of correlation

20.3.3 Funded CVA Formula

When we have a CDS facing a collateralized SPV, the CVA is given by

$$CVA_0 = \mathbb{E}\left[p_{t,\tau_c}\left(V_{\tau_c} - R_c N\right)^+ \mathbf{1}_{\{\tau_c \le T\}}\right].$$

For a single-name CDS, there is no erosion of collateral due to defaults; if the reference entity defaults the losses on the CDS will be covered by the collateral, but this is considered to be market risk and not credit risk included in CVA.

We need to price the following option payoff

$$\widetilde{O_{0,t,T}} = p_{0,t}\mathbb{E}\left[\mathbf{1}_{\{\tau_c = t\}}\left(V_{t,t,T}^{CDS} - R_c N\right)^+ | \tau_c = t\right].$$

The strike of the option will change because we have the additional cash buffer provided by the collateral at the time of default. The recovery value of the collateral is a cash-dollar amount that needs to be converted into a running spread strike.

We can re-write the option payoff as

$$\begin{aligned}
\mathbf{1}_{\{\tau_c = t\}}\left(V_{t,t,T}^{CDS} - R_c N\right)^+ &= \mathbf{1}_{\{\tau > t\}}\mathbf{1}_{\{\tau_c = t\}}\left(\widetilde{A_{t,t,T}}\left(S_{t,t,T} - S^0\right) - R_c N\right)^+ \\
&= \mathbf{1}_{\{\tau > t\}}\mathbf{1}_{\{\tau_c = t\}}\left(\widetilde{A_{t,t,T}}\left(S_{t,t,T} - S^*\right)\right)^+,
\end{aligned}$$

where S^* is the option strike adjusted with the cash-settled component

$$S^* = S^0 + S^c,$$
$$S^c \triangleq \frac{R_c N}{\widetilde{A_{t,t,T}}} \simeq \frac{R_c N}{\widetilde{A_{0,t,T}}},$$

which can be simplified further by approximating the equivalent running spread with its forward value. This is a reasonable approximation, to first-order, since typically the volatility of the annuity process is much lower than the volatility of the break-even spread. The correct modelling of this term would add a second-order convexity adjustment, which can be neglected for our purposes. Note that this definition also ensures that the forward value of the trade is preserved.

Proposition 171 *The CVA for a CDS facing a collateralized SPV is given by the sum of conditional CDS options, with collateral-adjusted strikes, weighted by the default probability of the collateral,*

$$CVA_0 = (1 - R_c) \int_0^T \widetilde{O_{0,t,T}} \mathbb{P}(\tau_c \in dt),$$

where $\widetilde{O_{0,t,T}}$ *is the value of the conditional* (t, T)*-forward CDS option*

$$\widetilde{O_{0,t,T}} = \left[\sum_{T_i > t} p_{0,T_i} \widetilde{Q^*_{0,t,T_i}} \delta_i \right] Black\left(\widetilde{S_{0,t,T}}, S^*, \widetilde{\sigma_{BS}} \sqrt{T - t} \right),$$

with $\widetilde{Q^*_{0,t,T_i}}$, $\widetilde{S_{0,t,T}}$ *and* S^* *are the conditional forward survival probability, the conditional forward break-even spread and the collateral-adjusted strike respectively*

$$\widetilde{Q^*_{0,t,T}} = \mathbb{P}(\tau > T \mid \tau_c = t),$$
$$\widetilde{S_{0,t,T}} = \frac{-(1 - R) \int_t^T p_{0,s} d\widetilde{Q^*_{0,t,s}}}{\sum_{T_i > t} p_{0,T_i} \widetilde{Q^*_{0,t,T_i}} \delta_i},$$
$$S^* = S^0 + \frac{R_c N}{\sum_{T_i > t} \frac{p_{0,T_i}}{p_{0,t}} \frac{\widetilde{Q^*_{0,t,T_i}}}{\widetilde{Q^*_{0,t,t}}} \delta_i}.$$

We compare the funded and unfunded CVA for a 5Y CDS with 100 bps running spread, marked at 0.921 upfront (SNAC level = 132 bps). The counterparty curve is marked at 2557 bps. In Fig. 20.3, we look at the impact of correlation. In Fig. 20.4, we look at the impact of volatility.

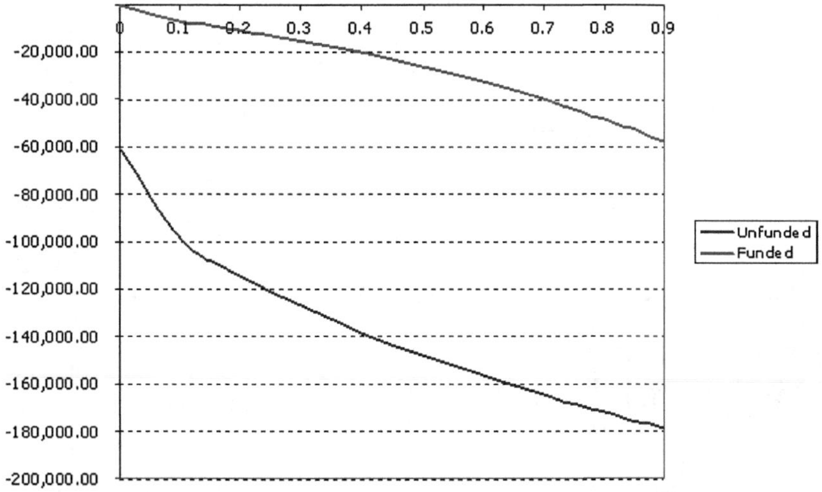

Fig. 20.3 Funded vs Unfunded CVA for a CDS as a function of correlation

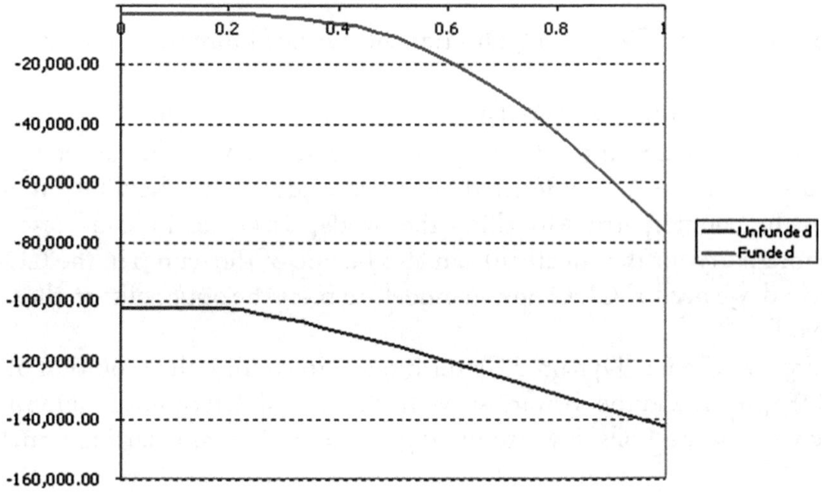

Fig. 20.4 Funded vs Unfunded CVA for a CDS as a function of volatility

20.4 CVA for CDOs

The filtration in this case contains the information on the underlying credits in the portfolio and the default of the counterparty (or the collateral).

We denote by L_t and $\widehat{L}_t$ the (normalized) portfolio loss and tranche loss variables respectively

$$L_t = \frac{1}{n}\sum_{i=1}^{n}(1 - R_i)\, D_t^i,$$
$$\widehat{L}_t = \frac{1}{\beta - \alpha}\left[(L_t - \alpha)^+ - (L_t - \beta)^+\right].$$

The value of a CDO tranche where we are a buyer of protection is given by

$$V_{t,t,T}^{CDO} = \mathbb{E}\left[\int_t^T p_{t,s} d\widehat{L}_s - \sum_{T_i > t} p_{t,T_i} S^0 \delta_i \left(1 - \widehat{L}_{T_i}\right) | \mathcal{G}_t\right].$$

The unfunded CVA is

$$CVA_0 = (1 - R_c)\int_0^T p_{0,t}\mathbb{E}\left[\left(V_{t,t,T}^{CDO}\right)^+ | \tau_c = t\right]\mathbb{P}\left(\tau_c \in dt\right).$$

To evaluate the CVA for a CDO tranche, we need three components.

- First, a "CDO-Squared model" that correlates the counterparty with the underlying names in the CDO portfolio. The CVA Payoff can be viewed as a single-name CDO-Squared with two underlyings: the CDO tranche and the counterparty. Modelling the overlap is key as in some cases the counterparty (or the collateral) can also be one of the names in the CDO.
- Second, we need a "CDO option model" to price the optionality in the CVA payoff.
- Third, we need a "Dynamic Credit model" to capture the evolution of the CDS spreads and the forward skew. In the case of deterministic intensities, the CVA payoff boils down to pricing a series of forward-starting tranches.

20.4.1 Forward Tranches

To compute the unfunded CVA, first, we generate the conditional forward tranche profile

$$
\begin{aligned}
FV_{0,t,T}^{CDO} &\triangleq \mathbb{E}\left[V_{t,t,T}^{CDO}\,|\tau_c = t\right] \\
&= \mathbb{E}\left[\mathbb{E}\left[\int_t^T p_{t,s} d\widehat{L}_s - \sum_{T_i > t} p_{t,T_i} S^0 \delta_i \left(1 - \widehat{L}_{T_i}\right) |\mathcal{G}_t\right] |\tau_c = t\right] \\
&= \mathbb{E}\left[\int_t^T p_{t,s} d\widehat{L}_s - \sum_{T_i > t} p_{t,T_i} S^0 \delta_i \left(1 - \widehat{L}_{T_i}\right) |\tau_c = t\right].
\end{aligned}
$$

Basically, this boils down to computing the following forward tranche loss curve for each default time t,

$$\mathbb{E}\left[\widehat{L}_T\,|\tau_c = t\right], \text{ for all } T \geq t.$$

To this end, we need to use a CDO-Squared model.

20.4.2 CDO-Squared Approach

In a nutshell, the CDO-Squared model defines a family of copula functions of the loss variables, parametrized by ρ, which is, then, combined with the marginal (skewed) distributions of the loss variables to get the multivariate joint loss distribution.

To create the correlation dependence between the underlying CDO loss variables, we use a Gaussian copula with a flat correlation parameter ρ for all the names in the CDO-Squared master portfolio.

We have a portfolio of $(n + 1)$ credits (including the counterparty reference) linked through a Gaussian copula

$$
\begin{aligned}
X_i &= \sqrt{\rho} Y + \sqrt{1-\rho}\varepsilon_i, \text{ for } 1 \leq i \leq n+1, \\
\{\tau_i \leq t\} &\Leftrightarrow \left\{X_i \leq \Phi^{-1}\left(\mathbb{P}\left(\tau_i \leq t\right)\right)\right\}, \\
\tau_{n+1} &= \tau_c.
\end{aligned}
$$

Step 1. Using this default dependence of the single-names, we can generate the (unskewed) univariate loss distributions and the joint loss distribution

$$\begin{aligned}\phi(K) &= \mathbb{P}(L_T \leq K),\\ \phi_c(t) &= \mathbb{P}(\tau_c \leq t),\\ \Phi(K,t) &= \mathbb{P}(L_T \leq K, \tau_c \leq t).\end{aligned}$$

Step 2. Using the base correlation skew, we generate the skewed marginal loss distribution

$$\phi^*(K) = \mathbb{P}\left(L_T^* \leq K\right).$$

Step 3. The skewed and unskewed loss variables are then mapped via this functional-mapping

$$L_T^* = \left(\phi^*\right)^{-1}(\phi(L_T)).$$

The joint loss distribution is given by

$$\mathbb{P}\left(L_T^* \leq K, \tau_c \leq t\right) = \Phi\left(\left(\phi^*\right)^{-1}(\phi(K)), t\right).$$

The skewed and unskewed marginal distributions are pre-computed and stored.

Conditional on Y, we can generate the distribution of L_T using standard methods (such as recursion, FFT, normal proxy,...), and the expected tranche loss is obtained by using the functional-mapping and integrating over the values of L_T:

$$\begin{aligned}\mathbb{E}\left[\widehat{L}_T \,|Y\right] = \mathbb{E}\left[f\left(L_T^*\right)|Y\right] &= \mathbb{E}\left[f\left(\left(\phi^*\right)^{-1}(\phi(L_T))\right)|Y\right]\\ &= \int f\left(\left(\phi^*\right)^{-1}(\phi(x))\right)\mathbb{P}(L_T \in dx\,|Y).\end{aligned}$$

Now, having defined the correlation structure, we can compute the **Conditional Forward Tranche Loss Curve**: for all $T \geq t$,

$$\mathbb{E}\left[\widehat{L}_T \,|\tau_c = t\right] = \frac{\mathbb{E}\left[\widehat{L}_T \mathbf{1}_{\{\tau_c = t\}}\right]}{\mathbb{P}(\tau_c = t)}.$$

We have two cases.

1. If the counterparty reference is not included in the CDO portfolio (i.e., there is no overlap), we can compute this expectation by conditioning on the common factor

$$\mathbb{E}\left[\widehat{L}_T \mathbf{1}_{\{\tau_c=t\}}\right] = \int \mathbb{E}\left[\widehat{L}_T \,|Y\right] \mathbb{P}\left(\tau_c = t\,|Y\right) \phi\left(Y\right) dY.$$

The expected tranche loss $\mathbb{E}\left[\widehat{L}_T \,|Y\right]$ is given by the functional-mapping formula; and the probability $\mathbb{P}\left(\tau_c = t\,|Y\right)$ is given by the Gaussian copula.

2. If there is overlap (i.e., the counterparty belongs also to the underlying portfolio), then this will have an impact the expectation and needs to be included in the conditioning, we get

$$\mathbb{E}\left[\widehat{L}_T \mathbf{1}_{\{\tau_c=t\}}\right] = \int \mathbb{E}\left[\widehat{L}_T \,|Y, \tau_c = t\right] \mathbb{P}\left(\tau_c = t\,|Y\right) \phi\left(Y\right) dY.$$

Note that to compute the expectation, we need to set the counterparty default probability to 1, re-generate the distribution of L_T and use the mapping to get the tranche loss

$$\mathbb{E}\left[\widehat{L}_T \,|Y, \tau_c = t\right] = \int f\left(\left(\phi^*\right)^{-1}\left(\phi\left(x\right)\right)\right) \mathbb{P}\left(L_T \in dx\,|Y, \tau_c = t\right).$$

20.4.3 Extinguishing CDOs

Similar calculations can be used to price an Extinguishing CDOs. This is useful for computing the CVA lower bound or for checking the call/put parity relationship.

The value of the extinguishing CDO is given by

$$V_0^E = \mathbb{E}\left[\int_0^T p_{0,s}\left(1 - D_{s^-}^c\right) d\widehat{L}_s - \sum_{T_i>0} p_{0,T_i} S^0 \delta_i \left(1 - D_{T_i}^c\right)\left(1 - \widehat{L}_{T_i}\right)\right].$$

Integrating by parts, we can re-write the Loss leg as

$$\mathbb{E}\left[\int_0^T p_{0,s}\left(1 - D_{s^-}^c\right) d\widehat{L}_s\right] = \mathbb{E}\left[p_{0,T}\left(1 - D_T^c\right)\widehat{L}_T\right]$$
$$-\mathbb{E}\left[\int_0^T \left(1 - D_s^c\right)\widehat{L}_s dp_{0,s} - \int_0^T p_{0,s}\widehat{L}_{s^-} dD_s^c\right].$$

Similarly, the premium leg is given by

$$\mathbb{E}\left[\sum_{T_i>0} p_{0,T_i} S^0 \delta_i \left(1 - D_{T_i}^c\right)\left(1 - \widehat{L}_{T_i}\right)\right]$$
$$= \sum_{T_i>0} p_{0,T_i} S^0 \delta_i \left[\mathbb{P}\left(\tau_c > T_i\right) - \mathbb{E}\left[\left(1 - D_{T_i}^c\right)\widehat{L}_{T_i}\right]\right].$$

So, we need to compute two terms: the Tranche Loss Conditional on Survival and the Tranche Loss Conditional on Default

$$\mathbb{E}\left[\mathbf{1}_{\{\tau_c>t\}}\widehat{L}_t\right], \text{ for all } t \geq 0,$$
$$\mathbb{E}\left[\widehat{L}_{t^-}\mathbf{1}_{\{\tau_c=t\}}\right], \text{ for all } t \geq 0.$$

1. If the counterparty is not included in the portfolio, we condition on the common factor, and we get

$$\mathbb{E}\left[\mathbf{1}_{\{\tau_c>t\}}\widehat{L}_t\right] = \int \mathbb{E}\left[\widehat{L}_t \,|Y\right] \mathbb{P}\left(\tau_c > t \,|Y\right) \phi\left(Y\right) dY,$$
$$\mathbb{E}\left[\widehat{L}_{t^-}\mathbf{1}_{\{\tau_c=t\}}\right] = \int \mathbb{E}\left[\widehat{L}_{t^-} \,|Y\right] \mathbb{P}\left(\tau_c = t \,|Y\right) \phi\left(Y\right) dY.$$

2. If the counterparty belongs to the underlying portfolio, then we have to account for it in the conditioning

$$\mathbb{E}\left[\mathbf{1}_{\{\tau_c>t\}}\widehat{L}_t\right] = \int \mathbb{E}\left[\widehat{L}_t \,|Y, \tau_c > t\right] \mathbb{P}\left(\tau_c > t \,|Y\right) \phi\left(Y\right) dY,$$
$$\mathbb{E}\left[\widehat{L}_{t^-}\mathbf{1}_{\{\tau_c=t\}}\right] = \int \mathbb{E}\left[\widehat{L}_{t^-} \,|Y, \tau_c = t\right] \mathbb{P}\left(\tau_c = t \,|Y\right) \phi\left(Y\right) dY.$$

Note that, in this case, when computing the expected tranche loss, we need to set the counterparty default probability to 0.

20.4.4 Options on Tranches

To price an option on a tranche, we can use one of the following methods:

- use a simple **Black formula** conditional on survival of the tranche. But in this case, we would typically have an implied volatility skew since there will be jumps induced by the conditioning on the $\{\mathcal{G}_t\}$-filtration and the default of the other names;

- use a **Bottom-up model** where the default of the individual names and their dynamics are modelled. But this would be numerically very intensive;
- use a **Top-down approach** without random-thinning, and assume that $\mathcal{G}_t = \sigma(L_t)$. This leads to a combination (or a mixture) of Black formulas by conditioning on the realizations of the portfolio loss variable L_t.

For the purposes of valuing CVA for a correlation book, the main factor that needs to be modelled properly is the correlation between the portfolio and the collateral. Getting the forward values correctly priced is key, the volatility modelling aspect is a second step. Moreover, the impact of volatility is typically very limited when trades are deeply-in-the-money.

20.4.5 Black Formula

The trick here is to condition on the survival of the tranche (i.e., the tranche has not been wiped out before the option maturity date). This is necessary to be able to define an (a.s.) strictly positive annuity.

The value of a (U, T)-tranche, at time at t, is

$$\begin{aligned} V_{t,U,T}^{CDO} &= \mathbb{E}\left[\int_U^T p_{t,U,s} d\widehat{L}_s - \sum_{T_i>U} p_{t,U,T_i} S^0 \delta_i \left(1-\widehat{L}_{T_i}\right) | \mathcal{G}_t\right] \\ &= \mathbf{1}_{\{\widehat{N}_U>0\}} \mathbb{E}\left[\int_U^T p_{t,U,s} d\widehat{L}_s - \sum_{T_i>U} p_{t,U,T_i} S^0 \delta_i \left(1-\widehat{L}_{T_i}\right) | \mathcal{G}_t \vee \{\widehat{N}_U>0\}\right]. \end{aligned}$$

The break-even spread is given by

$$S_{t,U,T} = \frac{\mathbb{E}\left[\int_U^T p_{t,U,s} d\widehat{L}_s \, | \mathcal{G}_t \vee \{\widehat{N}_U>0\}\right]}{\mathbb{E}\left[\sum_{T_i>U} p_{t,U,T_i} \delta_i \left(1-\widehat{L}_{T_i}\right) | \mathcal{G}_t \vee \{\widehat{N}_U>0\}\right]}.$$

And the Forward Annuity can be defined as

$$A_{t,U,T} = \mathbb{E}\left[\sum_{T_i>U} p_{t,U,T_i} \delta_i \left(1-\widehat{L}_{T_i}\right) | \mathcal{G}_t \vee \{\widehat{N}_U>0\}\right].$$

The value of the tranche can be written in terms of the break-even spread and the annuity,

$$V_{t,U,T}^{CDO} = \mathbf{1}_{\{\widehat{N}_t > 0\}} A_{t,U,T} \left(S_{t,U,T} - S^0 \right),$$

and we can derive the value of the CDO option as

$$\begin{aligned} O_0 &= p_{0,t} \mathbb{E}\left[\left(V_{t,t,T}^{CDO} \right)^+ \right] = p_{0,t} \mathbb{P}\left(\widehat{N}_t > 0 \right) \mathbb{E}\left[A_{t,t,T} \left(S_{t,t,T} - S^0 \right)^+ \right] \\ &= p_{0,t} \mathbb{P}\left(\widehat{N}_t > 0 \right) \mathbb{E}\left[\sum_{T_i > t} p_{0,t,T_i} \delta_i \left(1 - \widehat{L}_{T_i} \right) | \{ \widehat{N}_t > 0 \} \right] \mathbb{E}^{Q_A}\left[\left(S_{t,t,T} - S^0 \right)^+ \right] \\ &= \mathbb{E}\left[\sum_{T_i > t} p_{0,T_i} \delta_i \left(1 - \widehat{L}_{T_i} \right) \mathbf{1}_{\{\widehat{N}_t > 0\}} \right] Black\left(S_{0,t,T}, S^0, \sigma_{BS} \sqrt{T - t} \right) \\ &= \mathbb{E}\left[\sum_{T_i > t} p_{0,T_i} \delta_i \left(1 - \widehat{L}_{T_i} \right) \right] Black\left(S_{0,t,T}, S^0, \sigma_{BS} \sqrt{T - t} \right); \end{aligned}$$

the last step is due to the fact that $\left\{ \widehat{N}_t = 0 \right\} \Longrightarrow \left\{ \widehat{N}_T = 1 - \widehat{L}_T = 0 \right\}$, for all $T \geq t$.

20.4.6 Black CVA

For CVA calculations, we need to compute the value of the forward tranche on the set $\mathbf{1}_{\{\widehat{N}_t > 0\}} \mathbf{1}_{\{\tau_c = t\}}$:

$$\begin{aligned} \widetilde{V_{t,U,T}^{CDO}} &= \mathbf{1}_{\{\tau_c = U\}} \mathbb{E}\left[\int_U^T p_{t,U,s} d\widehat{L}_s - \sum_{T_i > U} p_{t,U,T_i} S^0 \delta_i \left(1 - \widehat{L}_{T_i} \right) | \mathcal{G}_t \right] \\ &= \mathbf{1}_{\{\widehat{N}_U > 0\}} \mathbf{1}_{\{\tau_c = U\}} \mathbb{E}\left[\int_U^T p_{t,U,s} d\widehat{L}_s - \ldots | \mathcal{G}_t \vee \left\{ \widehat{N}_U > 0 \right\} \vee \{ \tau_c = U \} \right]. \end{aligned}$$

The break-even spread is

$$\widetilde{S_{t,U,T}} = \frac{\mathbb{E}\left[\int_U^T p_{t,U,s} d\widehat{L}_s \, | \mathcal{G}_t \vee \left\{ \widehat{N}_U > 0 \right\} \vee \{ \tau_c = U \} \right]}{\mathbb{E}\left[\sum_{T_i > U} p_{t,U,T_i} \delta_i \left(1 - \widehat{L}_{T_i} \right) | \mathcal{G}_t \vee \left\{ \widehat{N}_U > 0 \right\} \vee \{ \tau_c = U \} \right]},$$

and the Forward Annuity is given by

$$\widetilde{A_{t,U,T}} = \mathbb{E}\left[\sum_{T_i > U} p_{t,U,T_i}\delta_i\left(1 - \widehat{L}_{T_i}\right)|\mathcal{G}_t \vee \left\{\widehat{N}_U > 0\right\} \vee \{\tau_c = U\}\right],$$

which yields the price of the CVA knock-in option

$$\begin{aligned}
\widetilde{O}_0 &= p_{0,t}\mathbb{P}(\tau_c = t)\,\mathbb{E}\left[\left(V_{t,t,T}^{CDO}\right)^+ |\{\tau_c = t\}\right] = p_{0,t}\mathbb{E}\left[\mathbf{1}_{\{\tau_c = t\}}\left(V_{t,t,T}^{CDO}\right)^+\right] \\
&= p_{0,t}\mathbb{P}\left(\widehat{N}_t > 0, \tau_c = t\right)\mathbb{E}\left[\widetilde{A_{t,t,T}}\left(\widetilde{S_{t,t,T}} - S^0\right)^+\right] \\
&= p_{0,t}\mathbb{P}\left(\widehat{N}_t > 0, \tau_c = t\right)\widetilde{A_{0,t,T}}\mathbb{E}^{Q_{\tilde{A}}}\left[\left(\widetilde{S_{t,t,T}} - S^0\right)^+\right] \\
&= \mathbb{E}\left[\sum_{T_i > t} p_{0,T_i}\delta_i\left(1 - \widehat{L}_{T_i}\right)\mathbf{1}_{\{\widehat{N}_t > 0\}}\mathbf{1}_{\{\tau_c = t\}}\right] Black\left(\widetilde{S_{0,t,T}}, S^0, \widetilde{\sigma_{BS}}\sqrt{T-t}\right) \\
&= \mathbb{E}\left[\sum_{T_i > t} p_{0,T_i}\delta_i\left(1 - \widehat{L}_{T_i}\right)\mathbf{1}_{\{\tau_c = t\}}\right] Black\left(\widetilde{S_{0,t,T}}, S^*, \widetilde{\sigma_{BS}}\sqrt{T-t}\right).
\end{aligned}$$

20.4.7 Dynamic Loss Model

Instead of pricing credit (index) options in a simple Black model with appropriate conditioning, we can also use a well-posed Dynamic Loss model to work out the option prices, which we can then plug into our CVA calculations

We use a Top-Down approach (without random-thinning) and assume that $\mathcal{G}_t = \mathcal{F}_t \vee \sigma(L_t)$, so that we can condition on realizations of the portfolio loss variable L_t thereby obtaining a mixture of Black formulas.

The value of the forward tranche (U, T) is

$$\begin{aligned}
V_{t,U,T}^{CDO} &= \mathbb{E}\left[\int_U^T p_{t,U,s}d\widehat{L}_s - \sum_{T_i > U} p_{t,U,T_i}S^0\delta_i\left(1 - \widehat{L}_{T_i}\right)|\mathcal{F}_t \vee \sigma(L_U)\right] \\
&= \mathbf{1}_{\{\widehat{N}_U > 0\}}\mathbb{E}\left[\int_U^T p_{t,U,s}d\widehat{L}_s - \ldots|\mathcal{F}_t \vee \sigma(L_U) \vee \left\{\widehat{N}_U > 0\right\}\right].
\end{aligned}$$

We define the break-even spread conditional on a given realization of the loss variable $\{L_U = x\}$:

$$S^x_{t,U,T} = \frac{\mathbb{E}\left[\int_U^T p_{t,U,s} d\widehat{L}_s \,|\mathcal{F}_t \vee \{L_U = x\} \vee \{\widehat{N}_U > 0\}\right]}{\mathbb{E}\left[\sum_{T_i > U} p_{t,U,T_i} \delta_i \left(1 - \widehat{L}_{T_i}\right) |\mathcal{F}_t \vee \{L_U = x\} \vee \{\widehat{N}_U > 0\}\right]}.$$

The Forward Annuity conditional on $\{L_U = x\}$ can be defined as

$$A^x_{t,U,T} = \mathbb{E}\left[\sum_{T_i > U} p_{t,U,T_i} \delta_i \left(1 - \widehat{L}_{T_i}\right) |\mathcal{F}_t \vee \{L_U = x\} \vee \{\widehat{N}_U > 0\}\right].$$

We re-write the value of the tranche as

$$V^{CDO}_{t,t,T} = \sum_x \mathbf{1}_{\{\widehat{N}_t > 0\}} \mathbf{1}_{\{L_t = x\}} A^x_{t,t,T} \left(S^x_{t,t,T} - S^0\right).$$

And the option value becomes

$$\begin{aligned}
O_0 &= p_{0,t} \mathbb{E}\left[\left(V^{CDO}_{t,t,T}\right)^+\right] = p_{0,t} \mathbb{E}\left[\int_x \mathbf{1}_{\{\widehat{N}_t > 0\}} \mathbf{1}_{\{L_t \in dx\}} A^x_{t,t,T} \left(S^x_{t,t,T} - S^0\right)^+\right] \\
&= p_{0,t} \left[\int_x \mathbb{P}\left(\widehat{N}_t > 0, L_t \in dx\right) A^x_{0,t,T} \mathbb{E}^{Q_{A^x}} \left[\left(S^x_{t,t,T} - S^0\right)^+\right]\right] \\
&= \int_x \mathbb{E}\left[\sum_{T_i > t} p_{0,T_i} \delta_i \left(1 - \widehat{L}_{T_i}\right) \mathbf{1}_{\{\widehat{N}_t > 0\}} \mathbf{1}_{\{L_t \in dx\}}\right] Black\left(S^x_{0,t,T}, S^0, \sigma^x_{BS} \sqrt{T - t}\right) \\
&= \int_x \mathbb{E}\left[\sum_{T_i > t} p_{0,T_i} \delta_i \left(1 - \widehat{L}_{T_i}\right) \mathbf{1}_{\{L_t \in dx\}}\right] Black\left(S^x_{0,t,T}, S^0, \sigma^x_{BS} \sqrt{T - t}\right).
\end{aligned}$$

To compute the terms used in the Annuity and the Break-even spread, we need a dynamic loss model (or at least a time-copula) to generate the conditional forward tranche loss curve:

$$\mathbb{E}\left[\widehat{L}_T \,|L_t = x\right], \text{ for all } T \geq t.$$

This will be a double-integral on (L_t, L_T), which can be computed numerically in a dynamic loss model.

To construct a dynamic loss model, we typically have a low-dimensional Markov process z_t and a mapping function to convert it into the loss variable at each time step by matching their marginals

$$\begin{aligned} dz_t &= ..., \\ L_t &= \phi_{L_t}^{-1}\left(\phi_{z_t}(z_t)\right). \end{aligned}$$

To compute the tranche loss $\mathbb{E}\left[\widehat{L}_T \,|L_t = x\right]$, we integrate over the conditional densities of the Markovian process

$$\mathbb{E}\left[\widehat{L}_T \,|L_t = x\right] = \mathbb{E}\left[f\left(L_T\right)|L_t = x\right] = \mathbb{E}\left[f\left(\phi_{L_T}^{-1}\left(\phi_{z_T}\left(z_T\right)\right)\right)|z_t = \phi_{z_t}^{-1}\left(\phi_{L_t}\left(x\right)\right)\right].$$

20.4.8 Dynamic CVA

As we have seen previously, for CVA calculations, we need to compute the value of the tranche on the set $\mathbf{1}_{\{\widehat{N}_t>0\}}\mathbf{1}_{\{L_t=x\}}\mathbf{1}_{\{\tau_c=t\}}$:

$$\begin{aligned} \widetilde{V_{t,U,T}^{CDO}} &= \mathbf{1}_{\{\tau_c=U\}}\mathbb{E}\left[\int_U^T p_{t,U,s}d\widehat{L}_s - ... |\mathcal{F}_t \vee \sigma\left(L_U\right) \vee \{\tau_c = U\}\right] \\ &= \mathbf{1}_{\{\widehat{N}_U>0\}}\mathbf{1}_{\{\tau_c=U\}}\mathbb{E}\left[... |\mathcal{F}_t \vee \sigma\left(L_U\right) \vee \left\{\widehat{N}_U > 0\right\} \vee \{\tau_c = U\}\right]. \end{aligned}$$

The break-even spread conditional on $\{L_U = x\}$ is

$$\widetilde{S_{t,U,T}^x} = \frac{\mathbb{E}\left[\int_U^T p_{t,U,s}d\widehat{L}_s \,|\mathcal{F}_t \vee \{L_U = x\} \vee \left\{\widehat{N}_U > 0\right\} \vee \{\tau_c = U\}\right]}{\mathbb{E}\left[\sum_{T_i>U} p_{t,U,T_i}\delta_i\left(1 - \widehat{L}_{T_i}\right)|\mathcal{F}_t \vee \{L_U = x\} \vee \left\{\widehat{N}_U > 0\right\} \vee \{\tau_c = U\}\right]},$$

and the Forward Annuity conditional on $\{L_U = x\}$ is given by

$$\widetilde{A_{t,U,T}^x} = \mathbb{E}\left[\sum_{T_i>U} p_{t,U,T_i}\delta_i\left(1 - \widehat{L}_{T_i}\right)|\mathcal{F}_t \vee \{L_U = x\} \vee \left\{\widehat{N}_U > 0\right\} \vee \{\tau_c = U\}\right].$$

The CVA (knock-in) option price is

$$\widetilde{O}_0 = p_{0,t}\mathbb{P}(\tau_c = t)\,\mathbb{E}\left[\left(V_{t,t,T}^{CDO}\right)^+ | \{\tau_c = t\}\right] = p_{0,t}\mathbb{E}\left[\mathbf{1}_{\{\tau_c=t\}}\left(V_{t,t,T}^{CDO}\right)^+\right]$$

$$= p_{0,t}\left[\int \mathbb{P}\left(\widehat{N}_t > 0, L_t \in dx, \tau_c = t\right)\mathbb{E}\left[\widetilde{A_{t,t,T}^x}\left(\widetilde{S_{t,t,T}^x} - S^0\right)^+\right]\right]$$

$$= p_{0,t}\left[\int \mathbb{P}\left(\widehat{N}_t > 0, L_t \in dx, \tau_c = t\right)\widetilde{A_{0,t,T}^x}\mathbb{E}^{Q_{\widetilde{A}^x}}\left[\left(\widetilde{S_{t,t,T}^x} - S^0\right)^+\right]\right]$$

$$= \int \mathbb{E}\left[\sum_{T_i>t} p_{0,T_i}\delta_i\left(1-\widehat{L}_{T_i}\right)\mathbf{1}_{\{\widehat{N}_t>0\}}\mathbf{1}_{\{L_t\in dx\}}\mathbf{1}_{\{\tau_c=t\}}\right] Black\left(\widetilde{S_{0,t,T}^x}, S^0, ...\right)$$

$$= \int \mathbb{E}\left[\sum_{T_i>t} p_{0,T_i}\delta_i\left(1-\widehat{L}_{T_i}\right)\mathbf{1}_{\{L_t\in dx\}}\mathbf{1}_{\{\tau_c=t\}}\right] Black\left(\widetilde{S_{0,t,T}^x}, S^0, \widetilde{\sigma_{BS}^x}\sqrt{T-t}\right).$$

Now, to compute the conditional expected tranche loss curve involved here,

$$\mathbb{E}\left[\widehat{L}_T \,|L_t = x, \tau_c = t\right], \text{ for all } T \geq t,$$

we need to combine both the CDO-Squared copula and the dynamic loss model.

We show how this can be done in the next section.

20.4.9 Markovian Dynamics

To compute the tranche loss $\mathbb{E}\left[\widehat{L}_T \,|L_t = x, \tau_c = t\right]$, we need to condition on Y, and compute the loss densities conditional on $\{\tau_c = t\}$, then integrate using the z_t-densities

$$\mathbb{E}\left[\widehat{L}_T \,\middle|L_t^* = x, \tau_c = t\right] = \frac{\mathbb{E}\left[f\left(L_T^*\right)\mathbf{1}_{\{L_t^*=x\}}\mathbf{1}_{\{\tau_c=t\}}\right]}{\mathbb{P}\left(L_t^* = x, \tau_c = t\right)}.$$

Before starting the main calculations, we pre-compute and store the skew mappings $g_T^* = \phi_{L_T^*}^{-1} \circ \phi_{L_T}$.

The denominator is given by

$$\mathbb{P}\left(L_t^* = x, \tau_c = t\right) = \int_Y \mathbb{P}\left(L_t^* = x, \tau_c = t\,|Y\right)\phi(Y)\,dY$$

$$= \int_Y \mathbb{P}\left(L_t = \left(g_t^*\right)^{-1}(x)\,|Y, \tau_c = t\right)\mathbb{P}\left(\tau_c = t\,|Y\right)\phi(Y)\,dY.$$

The numerator is computed as

$$\begin{aligned}
&\mathbb{E}\left[f\left(L_T^*\right)\mathbf{1}_{\{L_t^*=x\}}\mathbf{1}_{\{\tau_c=t\}}\right]\\
&=\int_Y \mathbb{E}\left[f\left(L_T^*\right)\mathbf{1}_{\{L_t^*=x\}}\mathbf{1}_{\{\tau_c=t\}}\,|Y\right]\phi(Y)\,dY\\
&=\int_Y \mathbb{E}\left[f\left(g_T^*\left(L_T\right)\right)\mathbf{1}_{\{g_t^*(L_t)=x\}}\,|Y,\,\tau_c=t\right]\mathbb{P}\left(\tau_c=t\,|Y\right)\phi(Y)\,dY.
\end{aligned}$$

Then, to calculate the conditional expectation in the integrand, we need to use the z_t-time-dependence and the conditional marginals of (L_t, L_T):

$$\begin{aligned}
&\mathbb{E}\left[f\left(L_T^*\right)\mathbf{1}_{\{L_t^*=x\}}\,|Y,\,\tau_c=t\right]\\
&=\mathbb{E}\left[f\left(g_T^*\left(L_T\right)\right)\mathbf{1}_{\{g_t^*(L_t)=x\}}\,|Y,\,\tau_c=t\right]\\
&=\mathbb{E}\left[f\left(g_T^*\left(\phi_{L_T|Y,\tau_c=t}^{-1}\left(\phi_{z_T}\left(z_T\right)\right)\right)\right)\mathbf{1}_{\left\{g_t^*\left(\phi_{L_t|Y,\tau_c=t}^{-1}\left(\phi_{z_t}(z_t)\right)\right)=x\right\}}\right]\\
&=\mathbb{E}\left[f\left(g_T^*\left(\phi_{L_T|Y,\tau_c=t}^{-1}\left(\phi_{z_T}\left(z_T\right)\right)\right)\right)\Big|z_t=\phi_{z_t}^{-1}\left(\phi_{L_t|Y,\tau_c=t}\left(\left(g_t^*\right)^{-1}(x)\right)\right)\right]\\
&\times\mathbb{P}\left(L_t=\left(g_t^*\right)^{-1}(x)\,|Y,\,\tau_c=t\right).
\end{aligned}$$

20.5 Applications

We run the CVA on the tranches of the CDX10 portfolio. The 5Y index level is 132 bps, and the tranche upfronts and break-even spreads are given below.

For the counterparty (or collateral) curve, we have: the 5Y upfront is 38.71 and the corresponding SNAC level is 2557 bps; the recovery rate is 32%.

In Figs. 20.5 and 20.6, we show the impact of correlation and volatility on the unfunded CVA for different tranches in the capital structure. As expected, the credit CVA is very sensitive to the correlation assumptions and less so to the choice of volatility. This is not surprising, as we increase default correlation, the CDO portfolio spreads widen after the counterparty default, which creates a so-called wrong-way risk effect. The volatility has also an impact, but it is not as pronounced as the impact of correlation, especially when the trades are Deep-In-The-Money. In our example, the [7%-10%] tranche is closer to the At-The-Money strike, and therefore exhibits the most sensitivity to volatility. We also have a similar pattern for the funded CVA shown in Figs. 20.7 and 20.8.

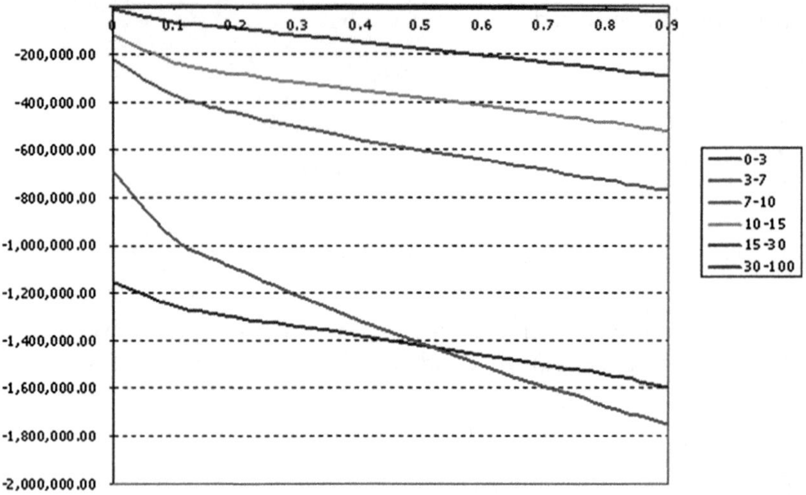

Fig. 20.5 Unfunded CVA for a non-margined swap as function of default correlation, for a volatility of 50%

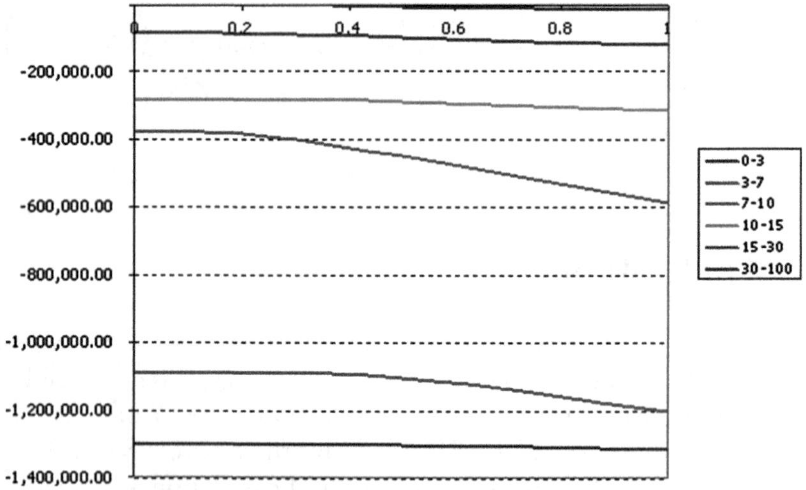

Fig. 20.6 Unfunded CVA for a non-margined swap as function of volatility, for a default correlation of 20%

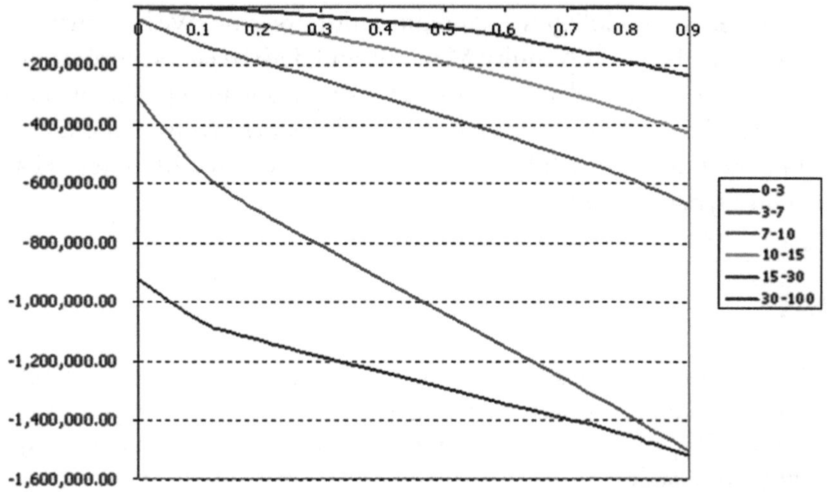

Fig. 20.7 Funded CVA for a swap facing an SPV as function of default correlation, for a volatility of 50%

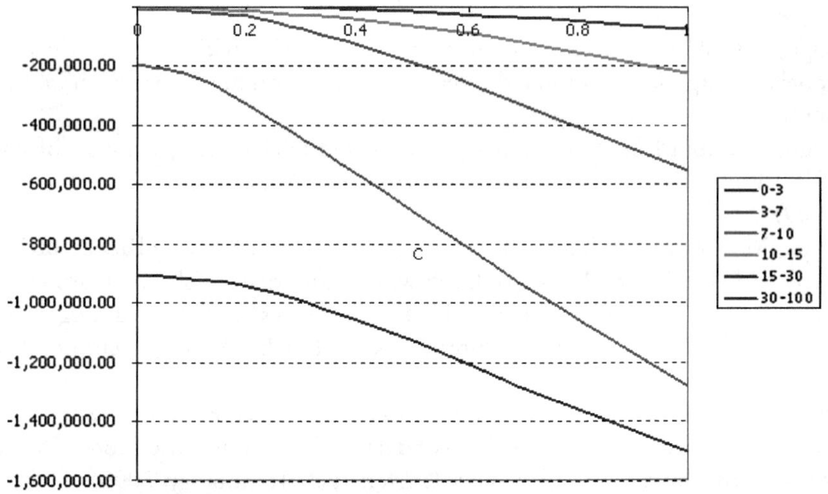

Fig. 20.8 Funded CVA for a swap facing an SPV as function of volatility, for a default correlation of 20%

20.6 Summary and Conclusion

We have reviewed the most commonly used CVA concepts and definitions, and we have described a general methodology for pricing CVA on CDO Tranches. To do so, we have used a variety of modelling techniques developed over the years for credit correlation books; this includes: dynamic loss models,

CDO-Squared methodologies and credit options pricing. We have introduced the Conditional Forward Annuity Measure, and derived closed-form solutions for funded and unfunded CVA on CDS using standard change of measure methods. Finally, we have addressed some of the issues involved in CDO CVA modelling, and derived useful pricing formulas based on a Black model and a Markovian Dynamic model.

Notes

1. This is one of the reasons why many institutions use both pre-CVA and post-CVA numbers when comparing quarter-on-quarter earnings.
2. During the credit crisis in 2007, it was mono-lines and re-insurers; today, it is energy companies.

References

D. Brigo, A. Capponi, Bilateral counterparty risk valuation with stochastic dynamical models and applications to credit default swaps (Working Paper, FitchSolutions, 2009)

D. Brigo, K. Chourdakis, Counterparty risk for credit default swaps: impact of spread volatility and default correlation. Int. J. Theor. Appl. Financ. **12**, 1007–1026 (2009)

D. Brigo, M. Morini, CDS market formulas and models (Working Paper, 2005)

S. Crepey, M. Jeanblanc, B. Zagari, CDS with counterparty risk in a markov chain copula model with joint defaults, Recent Advances in Financial Engineering, ed. by M.Kijima, C. Hara, Y. Muromachi, K. Tanaka (World Scientific, 2010), pp. 91–126

C. Dellacherie, *Capacites et Processus Stochastiques* (Springer, Berlin, 1972)

M. Jeanblanc, M. Rutkowski, Modelling of default risk: mathematical tools (Working Paper, Université d'Evry and Warsaw University of Technology, 2000)

P.J. Schönbucher, D Schubert, Copula-dependent default risk in intensity models (Working Paper, Department of Statistics, Bonn University, 2001)

Bibliography

References

N. Agmon, Y. Alhassid, R. Levine, An algorithm for finding the distribution of maximal entropy. J. Comput. Phys. **30**, 250–259 (1979)

M. Akahira, K. Takahashi, A higher order large deviation approximation for the discrete distributions. J. Japan Statist. Soc. **31**(2), 257–267 (2001)

E.I. Altman, B. Brady, A. Resti, A. Sironi, The link between defaults and recovery rates: theory, empirical evidence, and implications (Working Paper, Stern School of Business, 2003)

S. Amraoui, S. Hitier, Optimal stochastic recovery for base correlation (Working Paper, BNP Paribas, 2008)

L. Andersen, J. Sidenius, Extensions to the Gaussian Copula: random recovery and random factor loadings. J. Credit Risk **1**(1, Winter 2004/05), 29–70

L. Andersen, J. Sidenius, S. Basu, All your hedges in one basket. Risk (November 2003), 67–70

A. Antonov, S. Mechkov, T. Misirpashaev, Analytical techniques for synthetic CDOs and credit default risk measures (Working Paper, Numerix, 2005)

S. Assefa, T. Bielecki, S. Crepey, M. Jeanblanc, CVA computation for counterparty risk assessment in credit portfolios, in Credit Risk Frontiers, ed. by T. Bielecki, D. Brigo, F. Patras (Wiley), pp. 397–436

M. Avellaneda, Minimum-entropy calibration of asset pricing models. Int. J. Theor. Appl. Financ. **1**(4), 447 (1998)

M. Avellaneda, C. Friedman, R. Holmes, D. Samperi, Calibrating volatility surfaces via relative entropy minimization. Appl. Math. Financ. **4**(1), 37–64 (1997)

T. Aven, A theorem for determining the compensator of a counting process. Scand. J. Stat. **12**(1), 69–72 (1985)

S. Babbs, The term structure of interest rates: stochastic processes and contingent claims, Ph.D. Thesis, University of London, 1990

Y. Elouerkhaoui, *Credit Correlation*, Applied Quantitative Finance,
DOI 10.1007/978-3-319-60973-7

K. Back, Asset pricing for general processes. J. Math. Econ. **20**, 371–395 (1991)
R. Barlow, F. Proschan, *Statistical Theory of Reliability and Life Testing* (Silver Spring, Maryland, 1981)
M. Baxter, Levy process dynamic modelling of single-name credits and CDO tranches (Working Paper, Nomura International plc, 2006)
D. Becherer, M. Schweizer, Classical solutions to reaction diffusion systems for hedging problems with interacting itô and point processes. Ann. Appl. Probab. **15**(2), 1111–1144 (2005)
A. Bélanger, S. Shreve, D. Wong, A general framework for pricing credit risk. Math. Financ. **14**, 317–350 (2004)
N. Bennani, The forward loss model: a dynamic term structure approach for the pricing of portfolio credit derivatives (Working Paper, 2005)
J. Beumee, D. Brigo, D. Schiemert, G. Stoyle, *Charting a Course Through the CDS Big Bang* (Fitch Solutions, Quantitative Research, 2009)
T. Bielecki, M. Rutkowski, *Credit Risk: Modeling, Valuation and Hedging* (Springer, New York, 2002a)
T. Bielecki, M. Rutkowski, Intensity-based valuation of basket credit derivatives, in *by J*, ed. by Mathematical Finance (Singapore, Yong (World Scientfic, 2002), pp. 12–27
T. Björk, Y. Kabanov, W. Runggaldier, Bond market structure in the presence of marked point processes. Math. Financ. **7**, 211–239 (1997)
T. Björk, G.D. Masi, Y. Kabanov, W. Runggaldier, Towards a general theory of bond markets. Financ. Stoch. **1**, 141–174 (1997)
N. Bouleau, D. Lamberton, Residual risks and hedging strategies in markovian markets. Stoch. Process. Appl. **33**, 131–150 (1989)
E. Bouyé, V. Durrleman, A. Nikeghbali, G. Riboulet, T. Roncalli, Copulas for finance: a reading guide and some applications (Working Paper, Groupe de Recherche Operationnelle, Credit Lyonnais, 2000)
J.P. Boyd, *Chebyshev and Fourier Spectral Methods* (Dover Publications, New York, 2000)
D.T. Breeden, R.H. Litzenberger, Prices of state-contingent claims implicit in options prices. J. Bus. **51**(4) (1978)
P. Brémaud, *Point Processes and Queues: Martingale Dynamics* (Springer, New York, 1980)
D. Brigo, A. Capponi, Bilateral counterparty risk valuation with stochastic dynamical models and applications to credit default swaps (Working Paper, FitchSolutions, 2009)
D. Brigo, K. Chourdakis, Counterparty risk for credit default swaps: impact of spread volatility and default correlation. Int. J. Theor. Appl. Financ. **12**, 1007–1026 (2009)
D. Brigo, M. Morini, CDS market formulas and models (Working Paper, 2005)
D. Brigo, A. Pallavicini, R. Torresetti, Calibration of CDO tranches with the dynamical generalized-poisson loss model. Risk (May 2007)

D. Brigo, A. Pallavicini, R. Torresetti, Credit models and the crisis: default cluster dynamics and the generalized Poisson loss model. J. Credit Risk **6**(4), 39–81 (2010)

D. Brigo, A. Pallavicini, R. Torresetti, *Credit Models and the Crisis: A Journey into CDOs, Copulas, Correlations and Dynamic Models* (Wiley, Chichester, 2010)

P.W. Buchen, M. Kelly, The maximum entropy distribution of an asset inferred from option prices. J. Financ. Quant. Anal. **31**(1), 143–159 (1996)

H. Buhlmann, H.U. Gerber, Discussion of "The Aggregate Claims Distribution and Stop-Loss Reinsurance" by H.H. Panjer. Trans. Soc. Actuaries **32**, 537–538 (1980)

X. Burtschell, J. Gregory, J.P. Laurent, Beyond the Gaussian Copula: stochastic and local correlation. J. Credit Risk **3**(1), 31–62 (2007)

X. Burtschell, J. Gregory, J.P. Laurent, A comparative analysis of CDO pricing models. J. Deriv. **16**(4), 9–37 (2009)

L.H.Y. Chen, Poisson approximations for dependent trials. Ann. Probab. **3**, 534–545 (1975)

S. Chen, Beta Kernel estimators for density functions. Comput. Stat. Data Anal. **31**, 131–145 (1999)

C.S. Chou, P.A. Meyer, Sur la représentation des martingales comme intégrales stochastiques dans les processus ponctuels, in Séminaire de Probabilités IX, Lecture Notes in Math. 124 (Springer, Berlin, 1975), pp. 226–236

P. Collin-Dufresne, R. Goldstein, J. Hugonnier, A general formula for valuing defaultable securities. Econometrica **72**(5), 1377–1407 (2004)

S. Crepey, M. Jeanblanc, B. Zagari, CDS with counterparty risk in a markov chain copula model with joint defaults, in *Recent Advances in Financial Engineering*, ed. by M.Kijima, C. Hara, Y. Muromachi, K. Tanaka (World Scientific, 2010), pp. 91–126

J. Cvitanic, Nonlinear financial markets: hedging and portfolio optimization, in *Mathematics of Derivative Securities*, ed. by M.A.H. Dempster, S.R. Pliska (Cambridge University Press, Cambridge, 1997), pp. 227–254

K. Dambis, On the decomposition of continuous sub-martingales. Theor. Probab. Appl. **10**, 401–410 (1965)

W. Darsow, B. Nguyen, E. Olsen, Copulas and Markov processes. Ill. J. Math. **36**(4), 600–642 (1992)

R. Davies, Numerical inversion of a characteristic function. Biometrika **60**, 415–417 (1973)

M.H.A. Davis, Option pricing, in *incomplete markets, in Mathematics of Derivative Securities*, ed. by M.A.H. Dempster, S.R. Pliska (Cambridge University Press, Cambridge, 1997), pp. 216–226

M. Davis, V. Lo, Infectious defaults. Quant. Financ. **1**, 305–308 (2001a)

M. Davis, V. Lo, Modelling default correlation in bond portfolios, in *Mastering Risk, Volume 2: Applications*, ed. by Carol Alexander (Financial Times Prentice Hall, 2001b), pp. 141–151

A. Decarreau, D. Hilhorst, C. Lemarechal, J. Navaza, Dual methods in entropy maximization: application to some problems in crystallography. SIAM J. Optim. **2**(2), 173–197 (1992)

F. Delbaen, W. Schachermayer, The existence of absolute continuous local martingales measures. Ann. Appl. Probab. **5**(4), 926–945 (1995)

C. Dellacherie, Un exemple de la théorie géné rale des processus, in *Séminaires de Probabilités IV*. Lecture Notes in Math. 124 (Springer, Berlin, 1970), pp. 60–70

C. Dellacherie, *Capacités et processus stochastiques* (Springer, Berlin, 1972)

C. Dellacherie, P.A. Meyer, *Probabilités et potentiel* (Hermann, Paris, 1975)

C. Dellacherie, P.A. Meyer, A propos du travail de Yor sur les grossissements des tribus, in *Séminaire de Probabilités XII*. Lecture Notes in Math. 649 (Springer, Berlin, 1978), pp. 69–78

A. Dembo, J.D. Deuschel, D. Duffie, Large portfolio losses. Financ. Stoch. **8**, 3–16 (2004)

M.A.H. Dempster, E.A. Medova, S.W. Yang, Empirical Copulas for CDO tranche pricing using relative entropy. Int. J. Theor. Appl. Financ. **10**(4), 679–701 (2007)

N. De Pril, On the exact computation of the aggregate claims distribution in the individual life model. ASTIN Bullet. **16**, 109–112 (1986)

L. Dubins, G. Schwarz, On continuous Martingales. Proc. Nat. Acad. Sci. USA **53**, 913–916 (1965)

D. Duffie, First-to-default valuation (Working Paper, Graduate School of Business, Stanford University, 1998)

D. Duffie, Credit swap valuation. Financ. Anal. J. (Jan-Feb, 1999), 73–87

D. Duffie, N. Garleanu, Risk and valuation of collateralized debt obligations. Financ. Anal. J. **57**(1), 41–59 (2001)

D. Duffie, J. Pan, Analytical Value-At-Risk with jumps and credit risk. Financ. Stoch. **5**, 155–180 (2001)

D. Duffie, J. Pan, K. Singleton, Transform analysis and asset pricing for affine jump-diffusions. Econometrica **68**(6), 1343–1376 (2000)

D. Duffie, H.R. Richardson, Mean-variance hedging in continuous time. Ann. Appl. Probab. **1**, 1–15 (1991)

D. Duffie, K. Singleton, Modeling term structures of defaultable bonds. Rev. Financ. Stud. **12**(4), 687–720 (1999a)

D. Duffie, K. Singleton, Simulating correlated defaults (Working Paper, Graduate School of Business, Stanford University, 1999b)

F. Dufresne, Between the individual and collective models revisited (Working Paper, Ecole des HEC, University of Lausanne, 2002)

B. Dupire, Pricing with a smile. Risk **7**, 18–20 (1994)

N. El Karoui, Y. Jiao, D. Kurtz, Gauss and Poisson approximation: applications to CDOs tranche pricing. J. Comput. Financ. **12**(2), 31–58 (2008)

N. El Karoui, M.C. Quenez, Dynamic programming and pricing of contingent claims in an incomplete market. SIAM J. Control Optim. **33**, 29–66 (1995)

R.J. Elliott, M. Jeanblanc, M. Yor, On models of default risk. Math. Financ. **10**(2), 179–195 (2000)

P. Embrechts, F. Lindskog, A. McNeil, Modelling dependence with copulas and applications to risk management, in *Handbook of Heavy Tailed Distributions in Finance*, ed. by S.T. Rachev (Elsevier, Amsterdam, North-Holland, 2003)

P. Embrechts, A. McNeil, D. Straumann, Correlation: pitfalls and alternatives. Risk (May 1999), 69–71

Y. Elouerkhaoui, Etude des problèmes de corré lation and d'incomplétude dans les marchés de crédit. Ph.D. Thesis, Université Paris-Dauphine (2006)

Y. Elouerkhaoui, Pricing and hedging in a dynamic credit model. Int. J. Theor. Appl. Financ. **10**(4), 703–731 (2007)

Y. Elouerkhaoui, Marshall-Olkin Copula-based models, in *The Oxford Handbook of Credit Derivatives*, ed. by A. Lipton, A. Rennie (2013), pp. 257–284

E. Errais, K. Giesecke, L. Goldberg, Pricing credit from the top down with affine point processes (Working Paper, 2006)

H. Föllmer, P. Leukert, Quantile hedging. Financ. Stoch. **3**(3), 251–273 (1999)

H. Föllmer, D. Sondermann, Hedging of non-redundant contingent claims, in *Contributions to Mathematical Economics*, ed. by W. Hildenbrand, A. Mas-Colell (North-Holland, Amsterdam, 1986), pp. 205–223

H. Föllmer, M. Schweizer, Hedging by sequential regression: an introduction to the mathematics of options trading. ASTIN Bullet. **18**, 147–160 (1989)

H. Föllmer, M. Schweizer, Hedging of contingent claims under incomplete information, in *Applied Stochastic Analysis*, ed. by M.H.A. Davis, R.J. Elliott (Gordon and Breach, London, 1991), pp. 389–414

M. Fréchet, Les tableaux de corrélation dont les marges sont données. Annales de l'Université de Lyon, Sciences Math ématiques et Astronomie **20**, 13–31 (1957)

R. Frey, J. Backhaus, Portfolio credit risk models with interacting default intensities: a Markovian approach (Working Paper, Department of Mathematics, University of Leipzig, 2004)

R. Frey, A. McNeil, M. Nyfeler, Copulas and credit models. Risk (October 2001), 111–114

R. Frey, A. McNeil, VaR and expected shortfall in portfolios of dependent credit risks: conceptual and practical insights. J. Bank. Financ. **26**, 1317–1334 (2002)

R. Frey, A. McNeil, Dependent defaults in models of portfolio credit risk. J. Risk **6**(1), 59–92 (2003)

D. Gatarek, J. Jablecki, Modeling joint defaults in correlation-sensitive instruments. J. Credit Risk **12**(3), 2016 (2016)

K. Giesecke, A simple exponential model for dependent defaults. J. Fixed Income **13**(3), 74–83 (2003)

K. Giesecke, Correlated default with incomplete information. J. Bank. Financ. **28**(7), 1521–1545 (2004)

K. Giesecke, L. Goldberg, A top down approach to multi-name credit (Working Paper, 2005)

K. Giesecke, S. Weber, Cyclical correlations, credit contagion, and portfolio losses. J. Bank. Financ. **28**(12), 3009–3036 (2004)

J. Gil-Pelaez, Note on the inversion theorem. Biometrika **38**, 481–482 (1951)

P. Glasserman, *Monte Carlo methods in financial engineering* (Springer, New York, 2004)

P. Glasserman, J. Li, Importance sampling for a mixed poisson model of portfolio credit risk, in Proceedings of the 2003 Winter Simulation Conference, IEEE Press, Piscataway, NJ (2003a)

P. Glasserman, J. Li, Importance sampling for portfolio credit risk (Working Paper, Columbia Business School, 2003b)

G.M. Gupton, C.C. Finger, M. Bhatia, *CreditMetrics-Technical Document* (Morgan Guaranty Trust Co, New York, 1997)

P. Hagan, D. Kumar, A. Lesniewski, D. Woodward, Managing smile risk (Wilmott Magazine, September, 2002), pp. 84–108

P. Hagan, D. Kumar, A. Lesniewski, D. Woodward, Arbitrage free SABR (Wilmott Magazine, January, 2014), pp. 60–75

P. Hagan, D. Woodward, Equivalent black volatilities. Appl. Math. Financ. **6**, 147–157 (1999)

A. Hawkes, Spectra of some self-exciting and mutually exciting point processes. Biometrika **58**(1), 83–90 (1971)

A. Hawkes, D. Oakes, A cluster process representation of a self-exciting process. J. Appl. Probab. **11**, 493–503 (1974)

D. Heath, R. Jarrow, A. Morton, Bond pricing and the term structure of interest rates: a new methodology for contingent claims valuation. Econometrica **60**, 77–106 (1992)

D. Heath, E. Platen, M. Schweizer, A comparaison of two quadratic approaches to hedging in incomplete markets. Math. Financ. **11**(4), 385–413 (2001a)

D. Heath, E. Platen, M. Schweizer, Numerical comparison of local risk-minimisation and mean-variance hedging, in *Options Pricing, Interest Rates and Risk Management*, ed. by E. Jouini, J. Cvitanic (Cambridge University Press, Musiela M., 2001b), pp. 509–537

C. Hipp, Improved approximations for the aggregate claims distribution in the individual model. ASTIN Bullet. **16**, 89–100 (1986)

Y. Hu, W. Perraudin, The dependence of recovery rates and defaults (Working Paper, Birkbeck College, 2002)

J. Hull, A. White, Valuation of a CDO and n^{th} -to-default cds without monte carlo simulation. J. Deriv. **2**, 8–23 (2004)

J. Hull, A. White, Valuing credit derivatives using an implied copula approach. J. Deriv. **14**(2), 8–28 (2006)

P. Hunt, J. Kennedy, A. Pelsser, Markov-functional interest rate models. Financ. Stoch. **4**(4), 391–408 (2000)

J. Jacod, A. Shiryaev, *Limit Theorems for Stochastic Processes* (Springer, New York, 1987)

F. Jamshidian, Bond and option evaluation in the gaussian interest rate model. Res. Financ. **9**, 131–710 (1991)

R. Jarrow, D. Lando, S. Turnbull, A Markov model for the term structure of credit risk spreads. Rev. Financ. Stud. **10**(2), 481–523 (1997)

R. Jarrow, S. Turnbull, Pricing derivatives on financial securities subject to credit risk. J. Financ. **50**(1), 53–85 (1995)

R. Jarrow, F. Yu, Counterparty risk and the pricing of defaultable securities. J. Financ. LV **I**(5), 1765–1800 (2001)

E.T. Jaynes, Information theory and statistical mechanics. Phys. Rev. **106**, 620–630 (1957)

M. Jeanblanc, M. Rutkowski, Modelling of default risk: an overview, in *Mathematical Finance: Theory and Practice*, ed. by J. Yong, R. Cont (Higher Education Press, Beijing, 2000a), pp. 171–269

M. Jeanblanc, M. Rutkowski, Modelling of default risk: mathematical tools (Working Paper, Université d'Evry and Warsaw University of Technology, 2000b)

J.L. Jenson, *Saddlepoint Approximations* (Oxford University Press, Oxford, UK, 1995)

H. Joe, *Multivariate Models and Dependence Concepts* (Chapman & Hall, London, 1997)

N. Johnson, S. Kotz, *Distributions in Statistics: Continuous Multivariate Distributions* (Wiley, New York, 1972)

M. Joshi, Applying importance sampling to pricing single tranches of CDOs in a one-factor Li model (Working Paper, QUARC, Royal Bank of Scotland, 2004)

M. Joshi, D. Kainth, Rapid computation of prices and deltas of nth to default swaps in the Li model. Quant. Financ. **4**(3), 266–275 (2004)

J.F. Jouanin, G. Rapuch, G. Riboulet, T. Roncalli, Modelling dependence for credit derivatives with Copulas (Working Paper, Groupe de Recherche Operationnelle, Credit Lyonnais, 2001)

J. Kevorkian, J.D. Cole, *Multiple Scale and Singular Perturbation Methods* (Springer, New York, 1996)

C. Kimberling, A probabilistic interpretation of complete monotonicity. Aequationes Mathematicae **10**, 152–164 (1974)

D.E. Knuth, The art of computer programming, vol. 2, Semi-numerical Algorithms Addison-Wesley Series in Computer Science and Information Processing (1969)

P.S. Kornya, Distribution of the aggregate claims in the individual risk theory model, Trans. Soc. Actuaries **35**, 823–836. Discussion pp. 837–858

M. Krekel, Pricing distressed CDOs with base correlation and stochastic recovery (Working Paper, Unicredit, 2008)

S. Kuon, A. Reich, L. Reimers, Panjer vs Kornya vs De Pril comparison from a practical point of view. ASTIN Bullet. **17**, 183–191 (1987)

S. Kusuoka, A remark on default risk models. Adv. Math. Econ. **1**, 69–82 (1999)

D. Lando, Three essays on contingent claims pricing. Ph.D. thesis, Cornell University, 1994

D. Lando, On Cox processes and credit risky securities. Rev. Deriv. Res. **2**(2/3), 99–120 (1998)

J.P. Laurent, J. Gregory, I will survive. Risk (June 2003), pp. 103–107

J.P. Laurent, J. Gregory, Basket default swaps. CDOs and factor Copulas. J. Risk **7**(4), 103–122 (2005)

D.X. Li, On default correlation: a Copula function approach. J. Fixed Income **9**, 43–54 (2000)

D.X. Li, M. Liang, CDO squared pricing using a gaussian mixture model with transformation of loss distribution (Working Paper, Barclays Capital, 2005)

Y. Li, Z. Zheng, A top-down model for cash CLO (Working Paper, 2009)

F. Lindskog, A. McNeil, Common Poisson Shock models: applications to insurance and credit risk modelling. ASTIN Bullet. **33**(2), 209–238 (2003)

A. Lipton, A. Rennie, *Credit Correlation: Life After Copulas* (World Scientific, Singapore, 2008)

A. Lipton, A. Rennie, *The Oxford Handbook of Credit Derivatives* (Oxford University Press, Oxford, 2013)

A. Lipton, I. Savescu, CDSs, CVA and DVA—a structural approach. Risk (April 2013), 60–65

A. Lipton, A. Sepp, Credit value adjustment for credit default swaps via the structural default model. J. Credit Risk **5**, 123–146 (2009)

F. Longstaff, A. Rajan, An empirical analysis of the pricing of collateralized debt obligations. J. Financ. **63**(2), 529–563 (2008)

A.W. Marshall, I. Olkin, *A multivariate exponential distribution* (J. Am. Stat, Assoc, 1967)

R. Martin, K. Thompson, C. Browne, Taking to the saddle. Risk (June 2001), 91–94

R. Mashal, M. Naldi, Pricing multiname credit derivatives: heavy tailed hybrid approach (Working Paper, Lehman Brothers, 2002a)

R. Mashal, M. Naldi, Extreme events and default baskets. Risk (June 2002), 119–122

L. McGinty, E. Beinstein, R. Ahluwalia, M. Watts, *Introducing Base Correlations* (Credit Derivatives Strategy, JP Morgan, 2004)

S. Merino, M. Nyfeler, Calculating portfolio loss. Risk (August 2002), 82–86

K.R. Miltersen, An arbitrage theory of the term structure of interest rates. Ann. Appl. Probab. **4**(4), 953–967 (1994)

T. Møller, Risk-minimizing hedging strategies for insurance payment processes. Financ. Stochast. **5**, 419–446 (2001)

M. Morini, *Understanding and Managing Model Risk: A Practical Guide for Quants, Traders and Validators* (Wiley, Chichester, 2011)

K. Nagpal, R. Bahar, Measuring default correlation. Risk (March 2001), 129–132

R. Nelsen, *An Introduction to Copulas* (Springer, New York, 1999)

H. Niederreiter, *Random Number Generation and Quasi-Monte Carlo Methods* (Society for Industrial and Applied Mathematics, Philadelphia, 1992)

I. Norros, A compensator representation of multivariate life length distributions with applications. Scand. J. Stat. **13**, 99–112 (1986)

D. O'Kane, *Modelling Single-Name and Multi-Name Credit Derivatives* (Wiley, Chichester, 2008)

H. Panjer, Recursive evaluation of a family of compound distributions. ASTIN Bullet. **12**, 22–26 (1981)

A. Patton, Modelling time-varying exchange rate dependence using the conditional Copula (Working Paper 2001–09, University of California, San Diego, 2001)

W.H. Press, S.A. Teukolsky, W.T. Vetterling, B.P. Flannery, *Numerical Recipes in C* (Cambridge University Press, Cambridge, 1992)

P. Protter, Stochastic integration and differential equations, 2nd edn., Version 2.1 (Springer, New York, 2005)
O. Renault, O. Scaillet, On the way to recovery: a nonparametric bias free estimation of recovery rate densities (FAME Research Paper, No. 83, May 2003, University of Geneva, 2003)
A. Reyfman, K. Ushakova, W. Kong, *How to Value Bespoke Tranches Consistently with Standard Ones* (Credit Derivatives Research, Bear Stearns, 2004)
E. Rogge, P.J. Schönbucher, Modelling dynamic portfolio credit risk (Working Paper, 2003)
T. Rolski, H. Schmidli, V. Schmidt, J. Teugels, *Stochastic Processes for Insurance and Finance* (Wiley, New York, 1999)
S.M. Ross, *Simulation* (Academic Press, Boston, 1997)
P. Santa-Clara, D. Sornette, The dynamics of the forward interest rate curve with stochastic string shocks. Rev. Financ. Stud. **14**(1), 149–85 (2001)
V. Schmitz, Copulas and stochastic processes, Ph.D. Dissertation, Institute of Statistics, Aachen University, 2003
P.J. Schönbucher, The term structure of defaultable bond prices. Rev. Deriv. Res. **2**(2/3), 161–192 (1998)
P.J. Schönbucher, Factor models for portfolio credit risk. J. Risk Financ. **3**(1), 45–56 (2001)
P.J. Schönbucher, Taken to the limit: simple and not-so-simple loan loss distributions (Working Paper, Department of Statistics, Bonn University, 2002)
P.J. Schönbucher, *Credit Derivatives Pricing Models: Models, Pricing and Implementation* (Wiley, Chichester, 2003)
P.J. Schönbucher, Portfolio losses and the term structure of loss transition rates: a new methodology for the pricing of portfolio credit derivatives (Working Paper, ETH Zurich, 2005)
P.J. Schönbucher, D. Schubert, Copula-dependent default risk in intensity models (Working Paper, Department of Statistics, Bonn University, 2001)
M. Schweizer, Hedging of options in a general semimartingale model (Dissertation, no 8615, ETH Zurich, 1988)
M. Schweizer, Option hedging for semimartingales. Stoch. Process. Appl. **37**, 339–363 (1991)
M. Schweizer, Martingales densities for general asset prices. J. Math. Econ. **21**(4), 363–378 (1992)
M. Schweizer, Variance-optimal hedging in discrete time. Math. Oper. Res. **20**, 1–32 (1993)
M. Schweizer, Approximating random variables by stochastic integrals. Ann. Probab. **22**(3), 1536–1575 (1994)
M. Schweizer, A guided tour through quadratic hedging approaches, in *Option Pricing, Interest Rates and Risk Management*, ed. by E. Jouini, J. Cvitanic, M. Musiela (Cambridge University Press, Cambridge, 2001)
A. Servigny, O. Renault, Default correlation: empirical evidence (Working Paper, Standard and Poors, 2002)

M. Shaked, G. Shanthikumar, The multivariate hazard construction. Stoch. Process. Appl. **24**, 241–258 (1987)

D. Shelton, *Back To Normal* (Global Structured Credit Research, Citigroup, 2004)

A. Shiryaev, *Probability*, 2nd edn. (Springer, New York, 1996)

S. Shreve, *Stochastic Calculus for Finance*, vol. 2 (Springer, New York, 2004)

J. Sidenius, V. Piterbarg, L. Andersen, A new framework for dynamic credit portfolio loss modeling. Int. J. Theor. Appl. Financ. **11**(2), 163–197 (2008)

A. Sklar, Fonctions de répartition à *n* dimensions et leurs marges. Publications de l'Institut de Statistique de l'Université de Paris **8**, 229–231 (1959)

I.M. Sobol, On the distribution of points in a cube and the approximate evaluation of integrals. USSR. Comput. Math. Math. Phys. **7**(4), 86–112 (1967)

C. Stein, A bound for the error in the normal approximation to the distribution of a sum of dependent random variables, in *Proceedings of Sixty Berkely Symp. Math. Statis. Probab* (University of California Press, Berkeley, 1972), pp. 583–602

R. Torresetti, D. Brigo, A. Pallavicini, Implied expected tranche loss surface from CDO data (Working Paper, 2007)

R. Torresetti, D. Brigo, A. Pallavicini, Risk-neutral versus objective loss distribution and CDO tranche valuation. J. Risk Manag. Financ. Inst. **2**(2), 175–192 (2009)

J. Turc, D. Benhamou, B. Hertzog, M. Teyssier, *Pricing Bespoke CDOs: Latest Developments* (Quantitative Strategy, Credit Research, Societe Generale, 2006)

J. Turc, P. Very, D. Benhamou, *Pricing CDOs with a Smile* (Quantitative Strategy, Credit Research, Societe Generale, 2005)

L. Vacca, Unbiased risk-neutral loss distributions. Risk (November 2005), 97–101

O. Vasicek, The loan loss distribution (Working Paper, KMV Corporation, 1997)

M. Walker, CDO models—towards the next generation: incomplete markets and term structure (Working Paper, 2006)

S. Watanabe, Additive functionals of Markov processes and levy systems. Jpn. J. Math. **34**, 53–79 (1964)

R.P. White, T.N.E. Greville, On computing the probability that exactly *k* of *n* independent events will occur. Trans. Soc. Actuaries **11**, 88–95. Discussion pp. 96–99 (1959)

D. Wong, Copula from the limit of a multivariate binary model (Working Paper, Bank of America Corporation, 2000)

F. Yu, Correlated defaults and the valuation of defaultable securities (Working Paper, University of California, Irvine, 2004)

Applied Quantitative Finance

"This book covers several topics, which are important to know in order to stay *au courant* with current developments in credit modelling. It is written by Dr. Youssef Elouerkhaoui—a prominent thought-leader in the field. The book strikes a fine balance between theoretical and practical considerations. I recommend it wholeheartedly to anyone who wishes to master the complex art and science of credit derivatives."

—Prof. Alexander Lipton, *Founder and CEO, Stronghold Labs Fellow Connection Science and Engineering, Massachusetts Institute of Technology*

"Over the last two decades credit models have undergone a steady advancement, and today serve a foundational role in a variety of important financial calculations, including credit derivatives pricing, counterparty exposure metrics, funding value adjustments, and regulatory credit risk capital. In this exciting new contribution to the field, Youssef Elouerkhaoui condenses lessons learned from his many years of work as a leading practitioner into a holistic exposition of multi-name credit modeling theory. I highly recommend this book to academics and practitioners looking for a deeper and more advanced treatment of default co-dependence than what can be found in standard textbook material."

—Leif Andersen, *Global Co-Head of the Quantitative Strategies Group at Bank of America Merrill Lynch*

"This book provides exceptionally broad coverage of the critical topic of credit correlation modeling by one of the field's leading experts. It combines mathematical rigor with industry insight and should be a valuable resource to researchers and practitioners."

—Paul Glasserman, *Jack R. Anderson Professor of Business, Research Director Program for Financial Studies, Columbia Business School*

"This is probably the most complete book on credit risk modeling from the market perspective. Covering the large range of topics from credit default swaps thru collateralized debt obligations to credit value adjustment gives a comprehensive insight of the subject. Mathematically deep and rigorous but still compatible with the market practice is an absolute must for both university researchers and industry quants."

—Dariusz Gatarek, *Professor at Polish Academy of Sciences*

"I know Youssef as an honest and precise researcher, and I think that in this book he has taken up a long overdue task. Credit correlation is the spectre that has been haunting mathematical finance for the last 15 years. At the centre of research until the crisis, it was then neglected after the fall of the credit derivatives market. Yet, it is still central to much of present modelling, in particular XVAs. Youssef does a great job in helping us understanding the subtleties and the advances in this field."

—Massimo Morini, *Head of Interest Rate and Credit Models at Banca IMI Professor at Bocconi University, MSc Director at Milan Polytechnic, Research Fellow at Cass Business School*

"This book provides a comprehensive analysis of the credit modeling issues that practitioners have faced over the last 15 years. Covering in detail credit correlation from simple basket products to counterparty risk, through Index Tranches and CDO^2, this volume fills a conspicuous gap in the credit modeling literature and is likely to become a standard reference for both practitioners and academics."

—Luca Capriotti, *Global Head Quantitative Strategies Credit and Financing at Credit Suisse*

Applied Quantitative Finance is a new series developed to bring to readers the very latest market tested tools, techniques and developments in quantitative finance. Written for practitioners who need to understand how things work 'on the floor', the series will deliver the most cutting-edge applications in quantitative finance in areas such as asset pricing, risk management and financial derivatives. Written for quantitative-minded practitioners, this series will also appeal to researchers and students who want to see how quantitative finance is applied in practice.

More information about this series
at http://www.springer.com/series/14774